Interdisciplinary Biotechnological Advances

This series is an authoritative resource that explains recent advances and emerging directions in biotechnology, reflecting the forefront of research clearly and reliably, without excessive hype. Each volume is written by authors with excellent reputations and acknowledged expertise in the topic under discussion. The volumes span the entire field from an interdisciplinary perspective, covering everything from biotechnology principles and methods to applications in areas including genetic engineering, transgenic plants and animals, environmental problems, genomics, proteomics, diagnosis of disease, gene therapy, and biomedicine. The significance of these applications for the achievement of UN Sustainable Development Goals is highlighted. The series will be highly relevant for Master's and PhD students in Biotechnology, Nanochemistry, Biochemical Engineering, and Microbiology, medical students, academic and industrial researchers, agricultural scientists, farmers, clinicians, industry personnel, and entrepreneurs.

Parul Chaudhary • Anuj Chaudhary
Editors

Microbial Inoculants

Applications for Sustainable Agriculture

Springer

Editors
Parul Chaudhary
School of Agriculture
Graphic Era Hill University
Dehradun, Uttarakhand, India

Anuj Chaudhary
School of Agriculture & Environmental Sciences
Shobhit University
Gangoh, Uttar Pradesh, India

ISSN 2730-7069 ISSN 2730-7077 (electronic)
Interdisciplinary Biotechnological Advances
ISBN 978-981-97-0632-7 ISBN 978-981-97-0633-4 (eBook)
https://doi.org/10.1007/978-981-97-0633-4

This Springer imprint is published by the registered company Springer Nature Singapore Pte Ltd.
The registered company address is: 152 Beach Road, #21-01/04 Gateway East, Singapore 189721, Singapore

Paper in this product is recyclable.

Preface

This book discusses the role of microbes in agriculture for plant attributes, soil fertility, and bioremediation, which aids in sustainable agriculture. Nowadays, due to an increase in the human population, it is essential to enhance food development in the near future, but exhaustive nonsustainable agricultural practices such as the use of agrochemicals threaten food security, the economy, and the environment globally. Soil deterioration poses the most serious environmental threat to food production, resulting in poverty and hunger in poor countries. As a result, the global community faces challenges regarding the development of ecologically sound, efficient, and long-term alternative options to meet rising food requirement. Therefore, contributing to food security and advancing sustainable and innovative modern agricultural aimed at addressing environmental, economic, and social challenges connected with present intense nonsustainable agriculture practices is essential. As a result, beneficial microbial inoculants will be widely used in developing new strategies to increase sustainable food production. The application of bioinoculants helps to provide nutrients that directly support soil health and sustainable food production. Hence, this book explores the role of microbial inoculants for better agronomical performance, promoting sustainable advancement in agriculture and emphasizing soil health improvement for long-term benefits. The book is highly recommended for agriculture microbiologists, agronomists, plant pathologists, and related areas.

We appreciate all of the contributing author's hard work in providing insightful contributions to the book. We would especially want to express our gratitude to the university administration and our respected colleagues for their unwavering support and encouragement during the assignment. Lastly, we would especially like to express our gratitude to our family and close companions for their assistance and collaboration in coordinating this.

Dehradun, India — Parul Chaudhary
Gangoh, India — Anuj Chaudhary

About the Book

The book *Microbial Inoculants: Applications for Sustainable Agriculture* is an edited collection of 17 chapters on current developments and the application of microbial inoculants for sustainable agriculture, written by renowned specialists from around the globe. In order to improve the quantity and quality of agricultural produce, while reducing the negative effects on soil health and the environment. This book presents the most effective applications and practices of microbial inoculants. Professionals, academicians, and researchers in the area will find the book useful to improve crop productivity and soil fertility.

Contents

Editors and Contributors

About the Editors

Parul Chaudhary is working as an Assistant Professor in the School of Agriculture, Graphic Era Hill University, Dehradun, Uttarakhand, India and also associated as an Adjunct Professor with the Department of Microbiology, Graphic Era (Deemed to be) University, Dehradun, India. She holds an extensive 'Research and Teaching' experience of more than 6 years in the field of Agricultural Microbiology. Dr. Chaudhary received her PhD in Microbiology at the Govind Ballabh Pant University of Agriculture & Technology in Pantnagar, U.S. Nagar, India. Her areas of interests in the agricultural field include microbiology, nanotechnology, bioremediation, soil metagenomics and biofertilizers. She has published more than 50 research and review articles in international journals. She co-authored 19 book chapters with internationally well-known publishers. Dr. Chaudhary had worked at ICAR- National Dairy Research Institute, Karnal, Haryana, India. She has reviewed publications for various National and International Journals. She is also a member of the editorial board in BMC Microbiology and Associate Editor in Frontiers in Microbiology. Email: parulchaudhary1423@gmail.com

Anuj Chaudhary has worked as Assistant Professor at Shobhit University, Gangoh, Saharanpur, Uttar Pradesh, India. He has completed his Bachelor of Technology in Biotechnology and Master of Technology in Agricultural Processing & Food Engineering from Sardar Vallabhbhai Patel University of Agriculture & Technology, Modipuram, Meerut, India. His area of interests is in agriculture microbiology, biotechnology, agricultural processing and food technology. He has published 12 research articles in reputed International Journals and authored 13 book chapters. He has edited a book entitled "Advances in Nanotechnology for Smart Agriculture: Techniques & Applications" published in CRC Press, Taylor & Francis. Email: anujchaudharysvp@gmail.com

Contributors

Mosimininuoluwa T. Adebajo Department of Criminology and Security Studies, Kings University, Ode-Omu, Nigeria

School of Political Science, University of Haifa, Haifa, Israel

Bartholomew Saanu Adeleke Department of Biological Sciences, Microbiology Unit, Faculty of Science, Olusegun Agagu University of Science and Technology, Okitipupa, Nigeria

Food Security and Safety Focus Area, Faculty of Natural and Agricultural Sciences, North-West University, Mmabatho, South Africa

Uswat T. Adeyemi Department of Agricultural Economics and Farm Management, Faculty of Agriculture, University of Ilorin, Ilorin, Nigeria

Lukman Ahamad Section of Plant Pathology and Nematology, Department of Botany, Aligarh Muslim University, Aligarh, Uttar Pradesh, India

Saheed A. Akinola Food Security and Safety Focus Area, Faculty of Natural and Agricultural Sciences, North-West University, Mmabatho, South Africa

Department of Microbiology and Parasitology, School of Medicine and Pharmacy, College of Medicine and Health Sciences, University of Rwanda, Butare, Rwanda

S. T. M. Aravindharajan Division of Microbiology, ICAR-Indian Agricultural Research Institute, New Delhi, India

Modupe Stella Ayilara Food Security and Safety Focus Area, Faculty of Natural and Agricultural Sciences, North-West University, Mmabatho, South Africa

Department of Biological Sciences, Kings University, Ode-Omu, Nigeria

Mayank Bahuguna Department of Biosciences, Himalayan School of Biosciences, Swami Rama Himalayan University, Dehradun, Uttarakhand, India

Geeta Bhandari Department of Biosciences, Himalayan School of Biosciences, Swami Rama Himalayan University, Dehradun, Uttarakhand, India

Shruti Bhasin Department of Biotechnology, Banasthali Vidyapith, Jaipur, Rajasthan, India

Shivani Bhutani Department of Biotechnology, Invertis Village, Bareilly, Uttar Pradesh, India

Neha Bisht Department of Microbiology, Govind Ballabh Pant University of Agriculture and Technology, Pantnagar, Uttarakhand, India

Anuj Chaudhary School of Agriculture & Environmental Sciences, Shobhit University, Gangoh, Uttar Pradesh, India

Parul Chaudhary School of Agriculture, Graphic Era Hill University, Dehradun, Uttarakhand, India

Shalu Chaudhary Department of Biosciences, Himalayan School of Biosciences, Swami Rama Himalayan University, Dehradun, Uttarakhand, India

Manoj Kumar Chitara Department of Plant Pathology, College of Agriculture, Chandra Shekhar Azad University of Agriculture and Technology, Kanpur, Uttar Pradesh, India

Hemant Dasila Department of Microbiology, Akal College of Basic Sciences, Sirmaur, Himachal Pradesh, India

Sergio de los Santos Villalobos Laboratorio de Biotecnología del Recurso Microbiano, Instituto Tecnológico de Sonora, Ciudad Obregón, Sonora, Mexico

Vaibhav Dhaigude Animal Nutrition Division, ICAR-National Dairy Research Institute, Karnal, Haryana, India

Nisha Dinkar Department of Biotechnology, Khandelwal College of Management Science and Technology, Bareilly, Uttar Pradesh, India

Sami Abou Fayssal Department of Agronomy, University of Forestry, Sofia, Bulgaria

Department of Plant Production, Lebanese University, Beirut, Lebanon

Saurabh Gangola School of Agriculture, Graphic Era Hill University, Bhimtal Campus, Bhimtal, Uttarakhand, India

Aditi Ghosh Animal Biochemistry Division, ICAR-National Dairy Research Institute, Karnal, Haryana, India

Gopinath V Department of Plant Pathology and Nematology, PGCA, Dr. Rajendra Prasad Central Agricultural University, Pusa, Samastipur, Bihar, India

Pratap Gore Animal Physiology Division, ICAR-National Dairy Research Institute, Karnal, Haryana, India

Rekha Goswami Department of Environmental Sciences, Graphic Era Hill University, Dehradun, Uttarakhand, India

Sanjay Gupta Department of Biosciences, Himalayan School of Biosciences, Swami Rama Himalayan University, Dehradun, Uttarakhand, India

Vandana Jaggi Department of Microbiology, CBSH, Govind Ballabh Pant University of Agriculture and Technology, Pantnagar, Uttarakhand, India

Department of Pediatrics and Human Development, College of Human Medicine, Michigan State University, Grand Rapids, Michigan, United States

Durgesh Kumar Jaiswal Department of Biotechnology, Graphic Era Deemed to be University, Dehradun, Uttarakhand, India

Divya Joshi Patanjali Herbal Research Division, Patanjali Research Institute, Haridwar, Uttarakhand, India

Department of Microbiology, Govind Ballabh Pant University of Agriculture and Technology, Pantnagar, Uttarakhand, India

Samiksha Joshi School of Agriculture, Graphic Era Hill University, Bhimtal Campus, Bhimtal, Uttarakhand, India

Deeksha Kashyap Section of Plant Pathology and Nematology, Department of Botany, Aligarh Muslim University, Aligarh, Uttar Pradesh, India

Jaskiran Kaur Animal Physiology Division, ICAR-National Dairy Research Institute, Karnal, Haryana, India

Arshad Khan Section of Plant Pathology and Nematology, Department of Botany, Aligarh Muslim University, Aligarh, Uttar Pradesh, India

Hina Khatoon Department of Environmental Science, Dev Bhoomi Uttarakhand University, Dehradun, Uttarakhand, India

Bharti Kukreti Department of Microbiology, CBSH, Govind Ballabh Pant University of Agriculture and Technology, Pantnagar, Uttarakhand, India

Ashish Kumar Department of Microbiology, Govind Ballabh Pant University of Agriculture and Technology, Pantnagar, Uttarakhand, India

Govind Kumar Crop Production Division, ICAR-Central Institute of Subtropical Horticulture, Lucknow, Uttar Pradesh, India

J. Kumar School of Agriculture, Graphic Era Hill University, Dehradun, Uttarakhand, India

Damini Maithani School of Biotechnology, IFTM University, Moradabad, Uttar Pradesh, India

Anshi Mehra Department of Chemical Engineering, B.R Ambedkar National Institute of Technology, Jalandhar, Punjab, India

Anita Mishra Vidyadayini Institute of Science, Management and Technology, Bhopal, Madhya Pradesh, India

Yukti Nautiyal Department of Botany and Microbiology, H. N. B. Garhwal University, Srinagar, Uttarakhand, India

Manoj Kumar Prajapati Department of Plant Pathology and Nematology, PGCA, Dr. Rajendra Prasad Central Agricultural University, Pusa, Samastipur, Bihar, India

Preeti Department of Veterinary Public Health, IVRI, Izzatnagar, Bareilly, Uttar Pradesh, India

Jyoti Rajwar Department of Microbiology, Govind Ballabh Pant University of Agriculture and Technology, Pantnagar, Uttarakhand, India

Himanshu Rawal Department of Botany and Microbiology, H. N. B. Garhwal University, Srinagar, Uttarakhand, India

Neha Saini Department of Microbiology, Govind Ballabh Pant University of Agriculture and Technology, Pantnagar, Uttarakhand, India

Anamita Sen Department of Plant Pathology and Nematology, PGCA, Dr. Rajendra Prasad Central Agricultural University, Pusa, Samastipur, Bihar, India

Adita Sharma College of Fisheries, Dholi, Dr. Rajendra Prasad Central Agricultural University, Pusa, Muzaffarpur, Bihar, India

Anita Sharma Department of Microbiology, Govind Ballabh Pant University of Agriculture and Technology, Pantnagar, Uttarakhand, India

Barkha Sharma Department of Microbiology, Govind Ballabh Pant University of Agriculture and Technology, Pantnagar, Uttarakhand, India

Minerva Sharma Department of Biochemistry, School of Life Sciences, Dr. Bhimrao Ambedkar University, Agra, Uttar Pradesh, India

Portia Sharma Department of Botany, St. John's College, Agra, Uttar Pradesh, India

Ajay Veer Singh Department of Microbiology, CBSH, Govind Ballabh Pant University of Agriculture and Technology, Pantnagar, Uttarakhand, India

Dipti Singh Department of Biochemistry, Govind Ballabh Pant University of Agriculture and Technology, Pantnagar, Uttarakhand, India

Jyoti Singh Department of Microbiology, Govind Ballabh Pant University of Agriculture and Technology, Pantnagar, Uttarakhand, India

Manali Singh Department of Life Sciences, Parul Institute of Applied Sciences, Parul University, Waghodia, Vadodara, Gujarat, India

Shivani Singh Department of Microbiology, Govind Ballabh Pant University of Agriculture and Technology, Pantnagar, Uttarakhand, India

Nishtha Srivastava Department of Applied Science and Humanities, Invertis Village, Bareilly, Uttar Pradesh, India

Pragati Srivastava Department of Microbiology, Govind Ballabh Pant University of Agriculture and Technology, Pantnagar, Uttarakhand, India

Deep Chandra Suyal Vidyadayini Institute of Science, Management and Technology, Bhopal, Madhya Pradesh, India

Shalini Tiwari Department of Microbiology, Govind Ballabh Pant University of Agriculture and Technology, Pantnagar, Uttarakhand, India

Vishal Tripathi Department of Biotechnology, Graphic Era Deemed to be University, Dehradun, Uttarakhand, India

Bhavya Trivedi Department of Microbiology, Maya Group of Colleges, Dehradun, Uttarakhand, India

Viabhav Kumar Upadhayay Department of Microbiology, Dr. Rajendra Prasad Central Agricultural University, Pusa, Samastipur, Bihar, India

Qudus O. Uthman Soil, Water and Ecosystem Sciences, University of Florida, Gainesville, FL, USA

Exploring Beneficial Microbes and Their Multifaceted Applications: An Overview

Anuj Chaudhary, Parul Chaudhary, Sami Abou Fayssal, Shivani Singh, Durgesh Kumar Jaiswal, Vishal Tripathi, and J. Kumar

Abstract Modern agricultural practices have long relied on chemical inputs, including fertilizers, pesticides, and herbicides. However, these inputs come at a cost to crop nutritional value and soil health. Overuse of such chemicals has led to environmental degradation, the emergence of weed and disease resistance, and the contamination of food supplies. Furthermore, it has contributed to the accumulation of hazardous substances in soil, exacerbating soil health degradation. In response to these challenges, there is a pressing need for eco-friendly approaches to address the growing concern. Microbial inoculants, functioning as biofertilizers, bioherbicides, biopesticides, and biocontrol agents, offer a promising alternative to chemical inputs. By introducing beneficial microorganisms into soil or plants, the overall health of both soil and crops can be significantly improved. Microbial inoculants play a pivotal role in pest management, consequently contributing to improved human health. Comprising diverse microorganisms, such as bacteria, fungi, and algae, microbial inoculants interact with soil through various mechanisms, effectively reducing the dependency on chemical fertilizers and mitigating their adverse

A. Chaudhary (✉)
School of Agriculture and Environmental Sciences, Shobhit University, Gangoh, Uttar Pradesh, India

P. Chaudhary · J. Kumar
School of Agriculture, Graphic Era Hill University, Dehradun, Uttarakhand, India

S. Abou Fayssal
Department of Agronomy, Faculty of Agronomy, University of Forestry, Sofia, Bulgaria

Department of Plant Production, Faculty of Agriculture, Lebanese University, Beirut, Lebanon

S. Singh
Department of Microbiology, Govind Ballabh Pant University of Agriculture and Technology, Pantnagar, Uttarakhand, India

D. K. Jaiswal · V. Tripathi
Department of Biotechnology, Graphic Era Deemed to be University, Dehradun, Uttarakhand, India

P. Chaudhary, A. Chaudhary (eds.), *Microbial Inoculants*, Interdisciplinary Biotechnological Advances, https://doi.org/10.1007/978-981-97-0633-4_1

effects. This chapter comprehensively explores the vital role of microbial inoculants in sustaining long-term soil microbial populations and elucidates their distinctive modes of action.

Keywords Bioremediation · Composting · Microbes · Plant productivity · Soil health

1 Introduction

As the global population continues to rise, the need for sustainable agricultural systems to meet increased food demands is more pressing than ever. The imperative for environmentally friendly farming practices and sustainable methods of crop enhancement is undeniable. A key aspect of achieving sustainable agriculture involves reducing reliance on and replacing chemical fertilizers and pesticides, which pose significant risks to plant, animal, and human health (Chodak et al. 2015). The utilization of agricultural by-products resulting from the incorporation of beneficial bacteria is highly recommended, as these microbes colonize various plant components, including roots, stems, leaves, fruits, and nodules (Liu et al. 2017; Agri et al. 2021). The intricate interactions established between plants and microorganisms underpin mutual benefits (Nadarajah and Abdul Rahman 2021; Manzar et al. 2022). In the pursuit of enhancing plant growth and development by harnessing soil microorganisms, researchers and botanists have pioneered genetic modifications that confer resilience to a range of challenging environmental conditions, from extreme temperatures to sunlight exposure and salt or drought stresses (Kumar and Verma 2018).

Moreover, repeated cultivation results in the depletion of essential nutrients from the soil, causing a decline in crop yield and quality. While chemical fertilizers and pesticides have been commonly employed to mitigate these adverse effects, their prolonged use has led to significant health and environmental issues, disrupting ecosystems and fostering resistance among insects (Chaudhary et al. 2022a). A more ecologically friendly approach lies in the adoption of biopesticides and biofertilizers, which promote sustainable crop production while reducing chemical inputs. Biofertilizers, in addition to stimulating plant growth and increasing yield, offer a cost-effective and environmentally sound means of restoring soil fertility (Khati et al. 2018; Kumari et al. 2020).

The deterioration of soil health is a paramount challenge in achieving agricultural sustainability, often stemming from improper farming practices and excessive chemical usage, including pesticides and fertilizers (Kumar et al. 2017a, b, 2018). The integration of microbial inoculants into soil or plants holds promise as a strategy to curb soil quality degradation (Banik et al. 2019). Realizing the full potential of microorganisms necessitates a deep understanding of their roles in processes such as pathogen suppression, systemic induction of resistance, antioxidant activities, toxin reduction, nutrient cycling, and biogeochemical cycling within the plant-soil system. Further research is required to unlock the synergistic potential of these microorganisms and their underlying mechanisms (Chaudhary et al. 2023a).

Soil fertility decline, and the subsequent reduction in crop yield, is primarily attributed to the diminished activity of soil microorganisms (Gouda et al. 2018). The soil serves as a dynamic repository of microorganisms, animals, and plants, collaborating to support biodiversity and ecosystem stability. Soil microbial diversity bolsters ecosystem function in the face of changing and challenging environmental conditions, enhancing its resilience (Aguilar-Paredes et al. 2020). These microorganisms play vital roles in organic matter availability, nutrient cycling, soil structure, control of pathogens, pollutant degradation, cold stress tolerance, and reduction of greenhouse gas emissions (Chaudhary et al. 2022b). Prioritizing the care and cultivation of soil microbes is paramount in shaping fertilizer strategies, as understanding these communities informs sustainable and ecologically friendly agricultural practices (Yang et al. 2017; Ma et al. 2020; Dinca et al. 2022).

The central role of soil microbes in recycling various nutrients, particularly carbon, nitrogen, phosphorus, and sulfur, cannot be understated. Among these microbes, bacteria play vital role in maintaining soil health and productivity (Kukreti et al. 2020; Agri et al. 2022). Introducing soil and plant microbial inoculants is a modern farming practice aimed at promoting soil health and boosting crop productivity. These microbes not only enhance plant growth and resistance to phytopathogens but also influence overall agro-production efficiency, as these factors are intertwined. The correlation between increased agro-productivity and environmental sustainability was previously incongruent, but current agricultural trends emphasize leveraging soil microflora for enhanced growth, yield, and quality (Khati et al. 2017; Chaudhary et al. 2021a, b, 2023b, c).

The incorporation of compost serves as a catalyst for soil biodiversity and restoration (Scotti et al. 2016), contributing to enhanced soil structure and acting as a vital nutrient source (Toundou et al. 2021). Compost plays a pivotal role in maintaining and improving soil fertility, consequently elevating agricultural productivity (Luo et al. 2017). Moreover, it facilitates the proliferation of beneficial microorganisms while curbing soil pathogens (Siedt et al. 2021). The potential applications of compost are directly linked to the quality of the manufacturing process and the resulting product (Gao et al. 2019). Beneficial microorganisms, such as biofertilizers, biopesticides, bioherbicides, bioinsecticides, compost, and manure, stand as vital contributors to sustainable agriculture (Kumari et al. 2021; Kumar et al. 2021a, b). These microorganisms find applications not only in the agriculture sector but also in the food and pharmaceutical industries, contributing to processes such as biological transformation, biological nutrient fixation, and bioremediation of soil and water pollutants (Kumar et al. 2021a, b, 2022; Singh et al. 2023). The core focus of this chapter revolves around exploring the multifaceted roles of microbes in agriculture (Table 1).

Table 1 Applications of microbes in different fields

S. no.	Microorganisms	Benefits	References
1.	*Azospirillum*, *Rhizobium*	Removing organic pollutants, like imidacloprid from wastewater	Kulkarni et al. (2022)
2.	*Azotobacter chroococcum*	Increasing soil fertility and crop production achieved through nitrogen fixation	Shrestha et al. (2022)
3.	*Trichoderma* spp.	Increases resistance against biotic and abiotic stress	Hermosa et al. (2012)
4.	*Chaetomium globosum*	Plant growth stimulant	Jehangir et al. (2017)
5.	*Penicillium* spp.	Phosphate solubilization	El-Ghwas et al. (2023)
6.	*Aspergillus niger*	Root growth promoter, zinc solubilizer	Sharma et al. (2013)
7.	*Mesorhizobium* sp.	Plant growth-promoting bacteria from *Cicer arietinum*	Chandra and Chakraborty (2023)
8.	*Baculoviruses*	Bioinsecticides for lepidopteran caterpillars	Thakur et al. (2020)
9.	*Trichoderma harzianum*, *T. viridae*	Biopesticides	Damalas and Koutroubas 2018
10.	*Lactobacillus delbrueckii*, *Streptococcus thermophilus*, *Lactobacillus rhamnosus GG*	Freeze-dried herbal bio-yogurt	Heema et al. (2022)
11.	*Lactobacillus paracasei*, *Lactobacillus crispatus*, *Leuconostoc citreum*	Give aroma to Hurood cheese (namely, Hurood)	Wang et al. (2023)
12.	*Saccharomyces cerevisiae* and *Lactobacillus lactis*	Explore the diverse range of chocolate flavors that arise from the fermentation	Ho et al. (2015)
13.	Yeast, lactic acid bacteria, acetic acid bacteria	Sourdough production	Haifeng et al. (2021)
14.	*Aspergillus*, *Corynebacterium*, *Brevibacterium*	Dietary feed supplements provide sources of amino acids	Karimi et al. (2021)
15.	*Propionibacterium*, *Pseudomonas*	Riboflavin production	Noor-Hassim et al. (2023)
16.	*Aphanizomenon*, *Chlamydomonas scenedesmus*, *Chlorella*, *Spirulina*	Nutrient source for livestock	Wang et al. (2021)
17.	*Streptomyces davaonensis*, *Streptomyces cinnabarinus*	Broad-spectrum antibiotic roseoflavin	Dmytruk et al. (2023)

2 Soil Microbiomes

Microbial life thrives across the Earth's terrestrial expanse, significantly influencing the ecosystems they inhabit. In just 1 g of soil, an astonishing 10 billion microbes representing a multitude of distinct species can be found. Each ecosystem boasts unique soil attributes that harbor diverse microbial communities, predominantly composed of bacteria but also encompassing archaea, protists, fungi, viruses, and other microscopic entities in varying proportions. The disruption of soil microbiome equilibrium holds profound consequences, as the well-being and composition of microorganisms are interlinked with soilborne microbial partners (Omotayo and Babalola 2021).

In the quest for environmental and agricultural sustainability, preserving ecosystem equilibrium, biodiversity, and the health of plants, animals, and humans is paramount. A paradigm shift toward greater environmental awareness is essential, which can mitigate the escalating anthropogenic impact on climate and environment, accommodating the expanding human population. As soil forms the foundation of agricultural practices and subsequently food production, deeper insights into diverse soil microbiomes are imperative for future applications and methodologies (Tosi et al. 2020). The utilization of symbiotic nitrogen-fixing bacteria as plant nitrogen sources instead of relying solely on synthetic nitrogen fertilizers stands as a viable approach.

3 Microbial Application for Sustainable Agriculture

Modern agriculture's industrialization has driven intensive crop practices heavily reliant on synthetic pesticides and fertilizers. To meet global food demands without overburdening water and fertilizer consumption, adopting agricultural techniques that prioritize sustainability is vital. The microbiome, consisting of myriad bacteria, inhabits both soil and plants and plays a pivotal role in plant growth (Ray et al. 2020). The ability of plants to sustainably feed an expanding global population hinges on soil quality. In the past five decades, cultivated land area has surged by over 500%, accompanied by a 700% upsurge in fertilizer use and a substantial rise in pesticide usage. Root exudation and rhizo-deposition contribute to aboveground and belowground biomass diversity (Banerjee et al. 2019; Morella et al. 2020), yielding essential carbohydrates that nourish enriched soil microbes. In agriculture, specific microbial strains are employed to balance nitrogen and phosphorus levels. Furthermore, the desorption and adhesion of inorganic nutrients from surfaces and the breakdown of organic waste for subsequent soil incorporation are facilitated by native microbial populations (Finkel et al. 2017; Kumar and Dubey 2020; Chaudhary et al. 2021a, b).

3.1 Application of Beneficial Microbes for Soil Sustainability

Rhizobia, specialized bacteria forming nodules on plant roots, have recently received extensive attention due to their potential symbiotic benefits. Nodule-infecting rhizobia possess the capacity to fix biological nitrogen for plant utilization, collaborating with host plants to exchange nitrogenous compounds and carbon from photosynthesis (Mngongo et al. 2023). Despite progress, there is much to uncover about the functionality, taxonomic diversity of these symbiotic microbes, and their impact on nitrogen cycling and plant yield enhancement. Areas degraded by agricultural practices necessitate regeneration for improved soil health and productivity. Any initiative aiming for soil health restoration must involve thorough taxonomic and functional soil microbiome analyses. Current endeavors are exploring synthetic biology techniques to establish symbiotic partnerships in staple crops like wheat, maize, and others (Ryu et al. 2020).

Beyond growth promotion, plant growth-promoting (PGP) microorganisms confer resilience against various biotic and abiotic stresses (Olenska et al. 2020; Tyagi et al. 2022a, b). While many studies focus on PGP attributes such as stress tolerance, ACC deaminase synthesis, biofilm formation, nitrogen fixation, phosphate solubility, plant hormone production, and siderophore synthesis (Sarkar et al. 2018), metagenomic investigations reveal previously unknown microorganisms harboring novel or improved traits relevant to bioremediation and agriculture (Parveen et al. 2022). Integrating next-generation sequencing and advanced culture techniques has unveiled the functions and compositions of microbial populations (Mueller and Sachs 2015). For example, by assessing the metagenome of tomato plants resistant to a pathogen, researchers identified a flavobacteria genome that was more prevalent in the rhizosphere microbiome of resistant plants, offering insights into plant disease resistance (Kwak et al. 2018). Other studies have identified bacteria aiding plant growth under low-temperature conditions through root microbiome analysis using 16S rRNA gene sequencing (Beirinckx et al. 2020). The creation of intelligent consortia is also being explored, capitalizing on microbial community design (Paredes et al. 2018).

3.2 Rhizosphere Microbiomes for Sustainable Agriculture

Significant research has been dedicated to deciphering the interactions between plants and rhizosphere microbes, spanning fungi, bacteria, archaea, actinomycetes, protozoa, and viruses that bolster plant growth. While greenhouse studies offer controlled environments for comprehensive investigations, their findings may not entirely translate to field conditions. To harness the full potential of these strategies, field assessments of rhizosphere competence are necessary (Lekberg and Helgason 2018). Although successful in promoting plant growth under controlled conditions, the interactions between introduced microorganisms and native mycorrhizal populations remain understudied (Turrini et al. 2018). Soil microbial populations greatly influence the breakdown of resistant soil organic matter and the weathering of minerals. Soil bacteria significantly impact the decomposition and transformation of organic matter, with specific groups feeding on different forms of organic carbon and nitrogen (Tian et al. 2018). Organic soil fertilization, compared to mineral fertilization, fosters microbial diversity and biomass (Schmid et al. 2018; Banerjee et al. 2019). Nevertheless, long-term agroecosystem studies comparing organic and traditional farming practices are limited in assessing their effects on soil health and restoration (Khatoon et al. 2020). Organic farming has been associated with increased soil richness, reduced evenness, and improved microbial population density compared to traditionally managed soils with mineral fertilization (Hartman et al. 2018).

3.3 *Role of Rhizosphere in the Promotion of Soil Microbes*

The rhizosphere, defined as the zone within a few millimeters of the roots, teems with biological activity fueled by root exudates that nourish soil microorganisms. Nutrient accessibility to plants hinges on microbial activity, influencing root exudate characteristics (Berendsen et al. 2012). The rhizosphere houses an array of microbes, including fungi, bacteria, archaea, actinomycetes, protozoa, and nematodes. Fungi can be saprophytic, mutualistic, or pathogenic, while bacteria include decomposers and lithotrophs. *Archaea* encompass *Euryarchaeota* and *Crenarchaeota*, and actinomycetes are dominated by *Nocardia*. A diverse microbiome forms the rhizosphere's vibrant ecosystem (Mendes et al. 2013). The role of the rhizosphere in promoting soil microbes is depicted in Fig. 1.

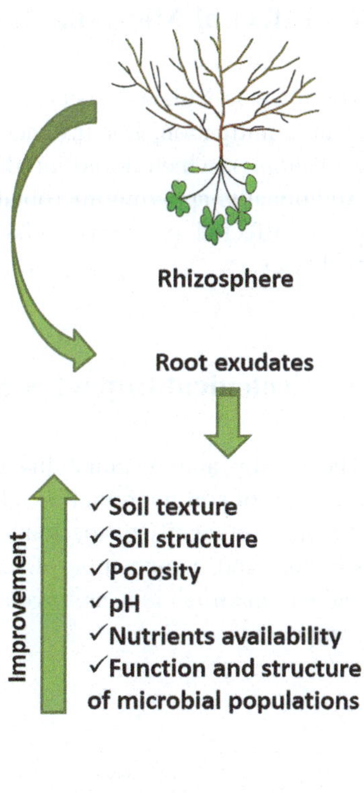

Fig. 1 Enhanced soil microbial populations in the rhizosphere

Table 2 Mitigation of climate change by soil- and plant-amended microbial inoculants

Microbial inoculant type	Location of inoculation	Role/effect	Reference
CO_2-fixing bacteria	Soil	Reduce soil CO_2 emissions and promote microbial CO_2 sequestration	Li et al. (2022)
N_2O-reducing bacteria	Soil	Reduce soil N_2O emissions	Usyskin-Tonne et al. (2019)
EPS-producing microorganisms	Soil	Improve the ability to endure drought and salt stresses	Dar et al. (2021)
Bacillus and *Arthrobacter* spp.	Soil	Reduce soil acidification	Orozco-Mosqueda et al. (2021)
Pseudomonas spp.	Soil	Increase the production of siderophores in soil	Orozco-Mosqueda et al. (2021)
PGPR	Plant	Improve the ability of roots to absorb water and enhance their ability to adapt to drought and salinity stress	Zhang et al. (2020)
IAA-producing bacteria	Plant	Improve gene expression and tolerance to salt and drought stresses	Barnawal et al. (2017)
PGPR-producing volatile organic composites	Plant	Improve antioxidant defense response against salt and drought stresses	Yasmin et al. (2021)
Plant-nodulating rhizobia	Plant	Reduce the impact of salt stress on shoot weight of plants	Benidire et al. (2020)

4 Effect of Microbial Inoculants on Climate Change

Microbial inoculants' specific functions directly determine their role in climate change mitigation. For instance, soil-based N_2O-reducing bacteria (*Dyadobacter fermentans*), which do not produce N_2O, could potentially reduce N_2O emissions. Additionally, applying microbial plant growth-promoting rhizobacteria (PGPR) to plants affected by climate change can stimulate root growth and development (Table 2).

5 Ecological Impact of Soil Microbes

The vitality and sustainability of global ecosystems are intrinsically tied to the activities of soil microbial populations (Wagg et al. 2011; Bowles et al. 2014). The functional diversity of these microbial populations, in turn, profoundly shapes the structure and composition of plants. Molecular transformations, nutrient cycling, and soil modifications are pivotal for the development and productivity of all plant components.

5.1 Molecular Transformation and Nitrogen Availability

Plant growth heavily relies on inorganic nitrogen (Hobara et al. 2014), despite 95% of required nitrogen (N) already existing in the soil as amines and amides (Pellitier and Zak 2018). Soil microbes play a pivotal role in enzymatically liberating soil-bound nitrogen, rendering it accessible for plant use (Nicolas et al. 2019). Mycorrhizal fungi, a significant group of soil microbes, excel in nitrogen absorption from the soil. In particular, ectomycorrhizal fungi (ECM) and arbuscular mycorrhizal fungi (AMF) profoundly influence plant-mycorrhizal interactions (Hannula and Träger 2020; Zhang et al. 2019). These fungi assimilate sugars and amino acids, releasing embodied nitrogen for the host plant's benefit. For example, ECM fungi employ lignocellulolytic enzymes to liberate inorganic nitrogen from organic sources, a capability not shared by saprophytic fungi and soil bacteria (Pellitier and Zak 2018). Consequently, mycorrhizal fungi alter plant uptake of inorganic soil nitrogen by expanding the accessible soil area for fine roots to absorb. Comprehending soil biogeochemical processes and ecosystem sustainability hinges on grasping this type of microbial-mediated plant organic nitrogen uptake.

5.2 Microbes' Role in Nutrient Processes and Soil Formation

Plant community productivity, ecosystem stability, and nutrient dynamics—both above and below the ground—are profoundly influenced by nutrient availability (Bever et al. 2012). To address deficiencies in nutrients like nitrogen and phosphorus and mitigate their scarcity, microbial interactions are of paramount importance (Mitter et al. 2016; Liang et al. 2022). One noteworthy plant-soil microbial interaction is the leguminous-rhizobia symbiosis, which fixes nitrogen biologically in the soil (Bano and Iqbal 2016).

Soil development and fertility result from intricate interactions between physical, chemical, and biological processes (Semenov and Đukić 2020). Elements like topography, climate, parent material, and soil microbes collectively shape soil quality (Schulz et al. 2013). However, the type, function, and quantity of microbial biomass significantly contribute to the availability of nutrients for plant growth. Thus, microbial biomass and functional diversity substantially affect nutrient cycling, storage, and transformation.

6 Microbes as Biofertilizers

Strides have been made in advancing sustainable agriculture through environmentally conscious and cleaner approaches. Under optimal conditions, biofertilizers can be directly applied to soil, inoculated onto seeds, or introduced to the roots of various crops. Biofertilizers harbor live microbes that enhance nutrient availability and stimulate plant growth through biological activities when applied to soil, seeds, or plant surfaces (Ortiz and Sansinenea 2021). Processes like atmospheric nitrogen fixation, phosphorus solubilization, and the synthesis of plant growth-promoting chemicals are natural ways through which nutrients are supplied to the soil. Biofertilizers, therefore, stand as organic and economical alternatives to chemical fertilizers, bolstering nutrient accessibility to agricultural crops (Husson et al. 2021).

A diverse array of microbial taxa, predominantly bacteria and fungi, inhabits the plant rhizosphere (De la Fuente Cantó et al. 2020). Rhizobacteria, free-living soil microorganisms colonizing plant roots, play a pivotal role in plant growth. These microbes produce and release various regulatory compounds in proximity to plant roots, impacting plant development (Basu et al. 2021; Khoshru et al. 2020). Bacteria and fungi present in the rhizosphere can act as biofertilizers, bolstering plant growth by enhancing resistance against both abiotic and biotic stresses and assisting plant nutrition (Chaudhary et al. 2022c, d). They also function as biopesticides against various crop-damaging pests. Moreover, microbes possess the capability to decompose and detoxify potentially toxic organic and inorganic compounds (contaminants) resulting from diverse human activities, including agriculture, contributing to bioremediation and improved soil health (Tarekegn et al. 2020).

Bacterial biofertilizers contribute to supplementing various essential nutrients for plants (Fasusi et al. 2021). They can fix nitrogen, solubilize micronutrients such as potassium and phosphorus, and release chemicals promoting plant growth or organic compounds suppressing plant diseases. *Rhizobium*, *Azospirillum*, *Azotobacter*, and *Bacillus* are widely used bacteria as biofertilizers (Bhardwaj et al. 2014; Gupta et al. 2015). *Azotobacter* and *Azospirillum* are employed with nonleguminous crops, while *Rhizobium* is utilized with leguminous crops. Utilizing these microbes as biofertilizers fosters soil fertility, boosts plant growth and crop yield, and provides a natural substitute for synthetic agrochemicals like chemical fertilizers and pesticides, ultimately promoting sustainable agriculture. Fungi also play a role as biofertilizers, particularly in the form of mycorrhizal associations that enhance nutrient release and utilization, especially phosphorus (Liu et al. 2021). These fungi extend their hyphae into nutrient-depleted zones, increasing the plant's absorption surface and nutrient availability (Diagne et al. 2020). The symbiotic bond between arbuscular mycorrhizal fungi and plants improves the rhizosphere's microenvironment, mineral absorption capacity, stress and disease tolerance, and overall plant growth (Mitra et al. 2021).

While microbial biofertilizers offer numerous advantages, such as affordability, ease of use, and positive impacts on soil health and plant productivity, they face challenges. Effective biofertilizer strains necessitate thorough initial laboratory screening to identify potent candidates. Manufacturing and quality assurance require advanced technology and skilled personnel. Challenges also include limited financial resources for transportation and storage, potential issues like strain quality and shelf-life, inadequate awareness among farmers, and environmental constraints (Itelima et al. 2018). Successful bioinoculants must endure soil conditions

Table 3 Potentiality of microbial inoculants as biofertilizers

Microbial inoculant type	Role	Reference
Cyanobacteria	Release of plant growth-promoting hormone protection against pathogens and pests through chemical synthesis	Kollmen and Strieth (2022); Christi (2023)
Arbuscular mycorrhizae	Improvement of plants' nitrogen fixation and phosphate solubilization	Kuila and Ghosh (2022)
PGPRs	An increase in amino acids and protein synthesis as a result of enhanced nitrogen absorption	Mokabel et al. (2022)
Potassium-solubilizing bacteria	Increased seedling vigor	Shahwar et al. (2023)
Siderophore-producing *Bacillus*	Increased plant growth	Sarwar et al. (2020)
Pseudomonas taiwanensis spp.	Increased phosphate solubilization	Chaudhary and Sharma (2019)
PGPRs	Increased IAA synthesis by plants, hastening of flower induction and fruit setting	Numan et al. (2018)
Pseudomonas and *Xanthomonas* spp.	Increased gibberellin production and improved plant stem elongation	AlAli et al. (2021)
Bradyrhizobium and *Azospirillum* spp.	Improved cytokinin production by plants and reduced root to shoot ratio	AlAli et al. (2021)
Azospirillum spp.	Increased abscisic acid synthesis by plants	Shahwar et al. (2023)

compatible with crops and navigate abiotic and biotic variables, which necessitates well-rounded efficacy. Table 3 illustrates the improvements observed as a result of using microbial inoculants as biofertilizers.

7 Microbial Inoculants as Biocontrol Agents

Microbial inoculants, with their inherent antibacterial and antifungal properties, play a vital role in biocontrol strategies within agricultural ecosystems (Rani et al. 2018). The biocontrol mechanisms of most microbial inoculants are closely tied to their capacity to produce extracellular hydrolytic enzymes. These enzymes facilitate competition for resources and release secondary metabolites that are detrimental to plant pathogens even at minuscule concentrations (Rani et al. 2018). Moreover, certain microbial inoculants induce defense responses in host plants, resulting in reduced damage from pathogenic agents and modulation of plant hormone levels, such as ethylene and auxin. Notably, some microbial inoculants even exhibit herbicidal activity, exemplified by *Colletotrichum coccodes* functioning as a mycoherbicide against velvet leaf. Refer to Fig. 2 for an overview of the key facets of microbial inoculants as biocontrol agents.

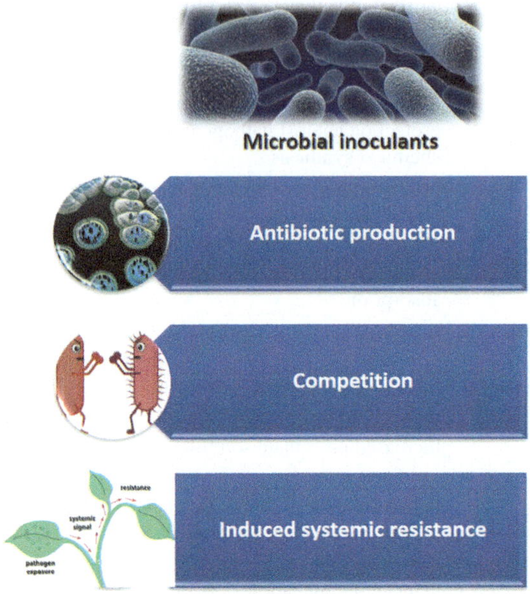

Fig. 2 Key aspects of microbial inoculants as biocontrol agents

8 Role of Microbes in the Composting Process for Soil Sustainability

Composting, a biological process, involves the aerobic decomposition of complex biodegradable organic matter like household waste, animal waste, and suitable industrial waste by bacteria, fungi, and worms. This process results in the production of valuable and stabilized end products known as compost, composed mainly of fulvic acid, humin, and humic acid, along with various saccharides, aliphatic sugars, methylated sugars, amino sugars, aromatic substances, uronic acids, purines, pyrimidines, and amino acids. These components enhance compost's agricultural utility, contributing to improved soil health, enhanced biochemical cycling capacity, increased soil diversity, and mineralization (Pane et al. 2014; Luo et al. 2017). Compost also finds use in erosion control, drainage improvement, weed suppression, plant disease management, bioremediation, and pollution control, serving as a substitute for chemical fertilizers (Pane et al. 2019; Uyizeye et al. 2019). Composting not only reduces the weight, volume, and color of raw materials but also stands distinct from native soil, peats, and coals (Yu et al. 2019).

Two primary types of composting, aerobic and anaerobic, depend on oxygen requirements and microbial nature. Aerobic composting necessitates oxygen for the decomposition of organic matter into carbon dioxide, ammonia, water, heat, and humus. This heat production creates an inhospitable environment for pathogens and weed seeds, leading to their elimination. In anaerobic composting, microorganisms break down organic matter into products like ammonia, methane, organic acids, and hydrogen sulfides without requiring oxygen (Jiang et al. 2011). Aerobic composting is preferred due to its shorter duration, pathogen and weed seed elimination through increased temperature, and faster decomposition rate. Other crucial parameters in the composting process include temperature, pH, moisture content, composting method, carbon-to-nitrogen ratio, and organic waste (Pezzolla et al. 2021).

Aerobic composting is preferred over anaerobic composting for the production of compost or manure. Aerobic composting follows four distinct phases, each with its predominant microbial community and worms:

1. **Decomposition Phase/Active Phase/Initial Mesophilic Phase:** In this stage, a microbial community breaks down readily available substrate, generating heat through microbial respiration. This heat raises the temperature from the mesophilic stage (20–40 °C) to the thermophilic stage (50–70 °C).
2. **Thermophilic Phase:** During this phase, elevated temperatures accelerate decomposition rates and eliminate mesophilic microorganisms. The heat produced is sufficient to eliminate pathogens, weed seeds, and larvae.
3. **Second Mesophilic Phase/Cooling Phase:** Depletion of high-energy compounds initiates this stage, leading to a decrease in thermophilic microbe populations and temperature. Mesophilic microorganisms reestablish themselves during this phase.
4. **Maturing/Curing Phase/Finishing Phase:** In this stage, stable and biologically active nontoxic humus forms as toxic and resistant compounds stabilize.

Major bacterial phyla involved in composting include *Firmicutes*, *Actinobacteria*, *Proteobacteria*, *Bacteroidetes*, and *Chloroflexi*. Bacteroidetes are vital for breaking down complex polysaccharides, while *Proteobacteria* are directly associated with nitrogenous organic substrate mineralization (Kabisch et al. 2014). *Actinobacteria* produce hydrolytic enzymes essential for breaking down lignocellulose and refractory cellulose, alongside secreting antibiotics that suppress pathogenic microorganisms. Chloroflexi are pivotal in xylan degradation under thermophilic conditions, as well as the biogeochemical cycling of chlorine and the decomposition of soluble microbial products (Aguilar-Paredes et al. 2023).

Fungi are also significant contributors to composting, aiding in the breakdown of organic materials. *Ascomycota* and *Basidiomycota* are the dominant fungal phyla, with various genera playing roles in different composting stages (Li et al. 2020a, b). *Candida*, *Pichia*, and *Trichosporon* are prominent in early stages, while *Candida*, *Cercophora*, *Conocybe*, *Cryptococcus*, *Leucoagaricus*, *Malassezia*, *Mortierella*, *Phialopora*, and *Thermomyces* appear in later stages. For a visual summary of microbial roles in composting, refer to Figs. 3 and 4.

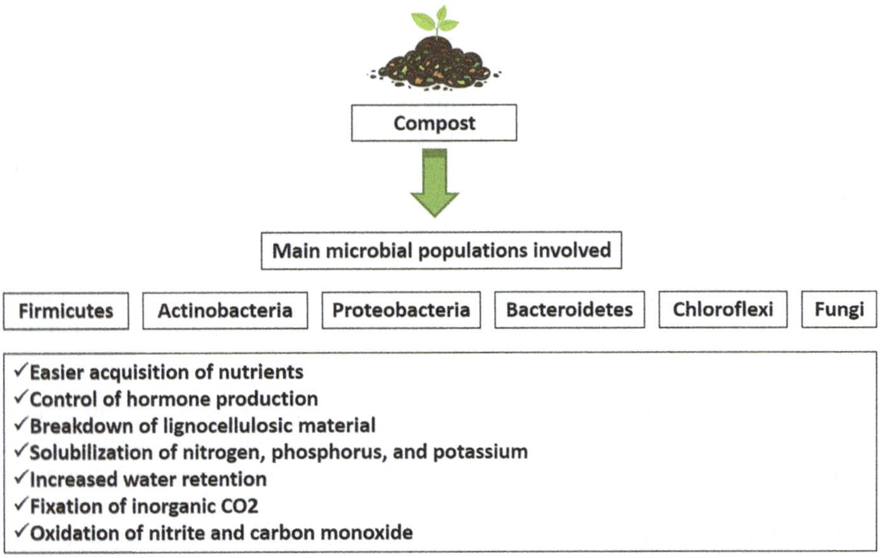

Fig. 3 Role of microbes during composting

8.1 Firmicutes

Firmicutes, gram-positive bacteria found abundantly in plant rhizospheres, have emerged as a focal point of extensive research due to their potential to foster sustainable agricultural systems (Moubareck et al. 2023). These microorganisms play a pivotal role in stimulating plant development by facilitating the acquisition of essential resources such as nitrogen, phosphorus, and iron. Moreover, *Firmicutes* exert influence over plant hormone production and, in some instances, produce direct analogs. Their significance extends beyond plant growth, encompassing their role as biological agents for managing plant pathogens. Additionally, *Firmicutes* offer solutions to various soil-related challenges, including soil desertification, salinity reduction, and the degradation of recalcitrant organic and inorganic pollutants (Moubareck et al. 2023).

During the mesophilic stage of the composting process, *Firmicutes* dominate the microbial community. However, as the composting process progresses, their prevalence gradually diminishes (Yang et al. 2020). Notably, many *Firmicutes* species thrive in high-temperature environments, existing as either thermophiles (above 50 °C) or thermotolerant (30–50 °C) organisms (Biyada et al. 2021). Taxa such as *Brevibacillus*, *Geobacillus*, *Bacillus*, and *Aeribacillus* are frequently encountered and are known to withstand a broad temperature range spanning from 20 to 60 °C (Awasthi et al. 2020; Lopez et al. 2021; Moreno et al. 2021).

A distinguishing feature of *Firmicutes* is their ability to degrade nondigestible carbohydrates like cellulose, accomplished by the action of proteases and pectinases (Xu et al. 2017; Yin et al. 2017). Noteworthy genera, including *Caldibacillus*,

Fig. 4 Enhancing soil fertility through the composting process and microbial contributions

Pseudoclostridium, *Thermoanaerobacterium*, and *Thermohydrogenium*, are recognized for their capacity to produce endoglucanases and exoglucanases, belonging to the GH1, GH5, and GH10 enzyme families. This functional diversity is instrumental in the breakdown of complex substrates, a process vital in composting. Additionally, some of these species were formerly classified under the name *Clostridium thermosaccharolyticum* (Gavande et al. 2021).

8.2 Actinobacteria

Actinobacteria constitute a substantial family of gram-positive bacteria, widely distributed across the globe. Their distinctive features include an appearance reminiscent of fungi due to elongated cell branching, a high DNA content rich in cytosine and guanine, and the emission of a distinctive earthy aroma when present in moist soil (Barka et al. 2015). Consequently, they display a pleomorphic shape, and certain variants even manifest coccoid components, resulting in extensive filaments that infiltrate the soil and disperse detachable spores.

Actinobacteria species hold pivotal roles as essential microorganisms crucial for enhancing plant growth and yield. They offer a multitude of advantages for sustainable long-term agricultural production (Boubekri et al. 2022). Notably, actinobacterial species are recognized for their ability to produce free-living symbiotic or diazotrophic bacteria, which contribute to the solubilization of essential nutrients. Additionally, they serve as effective biocontrol agents, displaying antiviral and antifungal properties. *Actinobacteria* also function as plant probiotics, fostering beneficial relationships with plants, and offer solutions to mitigate abiotic stress factors commonly faced in agriculture.

Among *Actinobacteria*, species within the genus *Streptomyces* stand out as key players in enhancing crop health and promoting robust growth (Boubekri et al. 2022). They are widely regarded as plant growth-promoting bacteria (PGPB) due to their capability to produce phytohormones, facilitate phosphate production, and promote nitrogen fixation, all of which contribute significantly to the overall well-being and productivity of plants (Ngalimat et al. 2021).

8.3 Proteobacteria

Proteobacteria represent the most diverse and widely distributed bacterial phylum, encompassing a spectrum of phylogenetic, ecological, and pathogenic diversity (Zhou et al. 2020). These bacteria are characterized by their gram-negative nature and the presence of an outer membrane containing lipopolysaccharides. Proteobacteria stand out for their remarkable resilience in the face of various agricultural practices, making them integral to agroecosystems. Moreover, they possess the unique ability to retain moisture, a trait with significant implications for nutrient cycling within ecosystems.

One distinctive feature of *Proteobacteria* is their pivotal role in quorum sensing, facilitated by the production of a chemical known as n-acyl-homoserine lactone (AHL). This quorum-sensing mechanism serves a dual purpose in agricultural contexts. On one hand, it can be beneficial by contributing to the intricate dynamics of plant diseases, aiding in the understanding and management of these agricultural challenges (Ishaq 2017). During the composting process, *Proteobacteria* reign supreme as the most prevalent and abundant phylum, particularly during the mesophilic stage. However, as the composting journey progresses, their relative abundance gradually diminishes, paving the way for other microbial players to take center stage (Tortosa et al. 2021).

8.4 *Bacteroidetes*

Bacteroidetes constitute a highly diverse group of bacteria, exhibiting a wide spectrum of characteristics. They are universally characterized as gram-negative bacteria, displaying a remarkable array of physiological traits. Found ubiquitously across various environments, with a particular stronghold in soil ecosystems, these microorganisms exhibit varied mobility patterns, including non-motility, flagellar propulsion, and gliding motility (Larsbrink and McKee 2020).

One of the key ecological roles played by *Bacteroidetes* is their significant contribution to the degradation of complex, polymeric organic matter. Their activities in soil and environmental contexts are pivotal in breaking down recalcitrant substances. During the composting process, *Bacteroidetes* take center stage in the breakdown of lignocellulosic biomass, particularly under anaerobic conditions, where they excel at degrading complex organic compounds (Awasthi et al. 2017). As composting progresses through its phases, especially during the cooling and maturation stages, *Bacteroidetes* shine in utilizing cellulose and xylan as their primary carbon sources. This preference for these abundant substrates significantly contributes to the transformation of organic matter during the composting process, ultimately yielding valuable, nutrient-rich compost (Lopez-González et al. 2015).

8.5 *Chloroflexi*

Chloroflexi members are a fascinating group of bacteria known for their versatility and unique characteristics. These microorganisms encompass both non-motile and motile filamentous species, displaying gram-negative attributes (Islam et al. 2019). A striking feature of *Chloroflexi* is their adaptability to various ecological niches, with many of them thriving as mild thermophiles. Despite their intriguing nature, *Chloroflexi* have presented challenges in terms of isolation in pure cultures, primarily due to their distinct habitat preferences, often remote from conventional laboratory settings. In terms of metabolic prowess, *Chloroflexi* are truly exceptional. They stand out for their ability to harness inorganic CO2, demonstrating autotrophic capabilities. Additionally, they possess the capacity to oxidize carbon monoxide and nitrite while reducing nitrate and ferric ions under anaerobic conditions, making them integral players in various biogeochemical cycles (Islam et al. 2019).

During the composting process, *Chloroflexi* play a prominent role, especially during the maturation stage. Here, they are closely linked to the biogeochemical cycling of chlorine, participating in the biological removal of nutrients and the breakdown of soluble microbial products (Li et al. 2019). However, their primary function in the thermophilic phase of composting is noteworthy. *Chloroflexi* excel in the hydrolysis of internal glycosidic bonds, a crucial step in the degradation of xylan, a complex carbohydrate found in plant cell walls (Wang et al. 2016). This explains the substantial and gradual increase in *Chloroflexi* abundance throughout the composting period, underscoring their vital role in this intricate process (Wan et al. 2021; Werner et al. 2022).

8.6 Fungi

Fungi play a pivotal role in the intricate process of organic material decomposition during composting (Wang et al. 2020). Within the fungal kingdom, the *Ascomycota* phylum, followed closely by the subphylum *Basidiomycota*, reign as the most prevalent participants in this essential ecological process. The breakdown of diverse compostable materials, including food waste, sewage sludge, garden waste, and maize straw, has consistently demonstrated the involvement of *Mucoromycotina* fungi (Meng et al. (2019). As composting journeys through its early stages, a distinct roster of fungi takes the forefront. *Candida*, *Pichia*, and *Trichosporon*, in tandem with yeasts belonging to *Saccharomycetales* and *Tremellales*, prominently occupy these initial phases, actively initiating the degradation process. As compost matures and undergoes transformations, a shift in fungal genera becomes evident. *Candida*, *Cercophora*, *Conocybe*, *Cryptococcus*, *Leucoagaricus*, *Malassezia*, *Mortierella*, *Phialopora*, and *Thermomyces*, among others, emerge as noteworthy contributors during the later stages of composting, demonstrating the dynamic nature of fungal communities throughout this ecological journey (Wang et al. 2018a, b).

9 Conclusion

The integration of microbial inoculants into food production holds promising potential as an eco-friendly alternative to conventional agrochemicals. Embracing microbial inoculants necessitates a comprehensive understanding of their underlying mechanisms of action. This shift away from chemical interventions in agriculture can lead to a reduction in their presence within the human food supply. Microbial inoculants offer dual benefits as both biocontrol agents and bioherbicides, effectively managing pests and weeds. Leveraging natural resources such as beneficial microorganisms present a sustainable avenue to enhance agricultural yield and product quality. In the long run, the advancement of microbial inoculant technology holds the key to ensuring food security for future generations. Nevertheless, the intricate mechanisms governing the interactions of microbial inoculants within soil environments remain insufficiently elucidated, demanding greater focus and inquiry from researchers.

References

Agri U, Chaudhary P, Sharma A (2021) In vitro compatibility evaluation of agriusable nanochitosan on beneficial plant growth-promoting rhizobacteria and maize plant. Natl Acad Sci Lett 44:555–559. https://doi.org/10.1007/s40009-021-01047-w

Agri U, Chaudhary P, Sharma A, Kukreti B (2022) Physiological response of maize plants and its rhizospheric microbiome under the influence of potential bioinoculants and nanochitosan. Plant Soil 474:451–468. https://doi.org/10.1007/s11104-022-05351-2

Aguilar-Paredes A, Valdés G, Nuti M (2020) Ecosystem functions of microbial consortia in sustainable agriculture. Agronomy 10:1902. https://doi.org/10.3390/agronomy10121902

Aguilar-Paredes A, Valdés G, Araneda N, Valdebenito E, Hansen F, Nuti M (2023) Microbial community in the composting process and its positive impact on the soil biota in sustainable agriculture. Agronomy 13(2):542. https://doi.org/10.3390/agronomy13020542

AlAli HA, Khalifa A, Al-Malki M (2021) Plant growth promoting rhizobacteria from *Ocimum basilicum* improve growth of *Phaseolus vulgaris* and *Abelmoschus esculentus*. S Afr J Bot 139:200–209. https://doi.org/10.1016/j.sajb.2021.02.019

Awasthi MK, Zhang Z, Wang Q, Shen F, Li R, Li D-S, Ren X, Wang M, Chen H, Zhao J (2017) New insight with the effects of biochar amendment on bacterial diversity as indicators of biomarkers support the thermophilic phase during sewage sludge composting. Bioresour Technol 238:589–601. https://doi.org/10.1016/j.biortech.2017.04.100

Awasthi MK, Duan Y, Awasthi SK, Liu T, Zhang Z (2020) Effect of biochar and bacterial inoculum additions on cow dung composting. Bioresour Technol 297:122407. https://doi.org/10.1016/j.biortech.2019.122407

Banerjee S, Walder F, Büchi L, Meyer M, Held AY, Gattinger A, Keller T, Charles R, van der Heijden MGA (2019) Agricultural intensification reduces microbial network complexity and the abundance of keystone taxa in roots. ISME J 13:1722–1736. https://doi.org/10.1038/s41396-019-0383-2

Banik A, Dash GK, Swain P, Kumar U, Mukhopadhyay SK, Dangar TK (2019) Application of rice (Oryza sativa l.) root endophytic diazotrophic azotobacter sp. strain Avi2 (MCC 3432) can increase rice yield under green house and field condition. Microbiol Res 219:56–65. https://doi.org/10.1016/j.micres.2018.11.004

Bano S, Iqbal SM (2016) Biological nitrogen fixation to improve plant growth and productivity. Int J Agric Innov Res 4:597–599

Barka EA, Vatsa P, Sanchez L, Gaveau-Vaillant N, Jacquard C, Meier-Kolthoff JP, Klenk HP, Clément C, Ouhdouch Y, van Wezel GP (2015) Taxonomy, physiology, and natural products of actinobacteria. Microbiol Mol Biol Rev 80(1):1–43. https://doi.org/10.1128/MMBR.00019-15

Barnawal D, Bharti N, Pandey SS, Pandey A, Chanotiya CS, Kalra A (2017) Plant growth-promoting rhizobacteria enhance wheat salt and drought stress tolerance by altering endogenous phytohormone levels and TaCTR1/TaDREB2 expression. Physiol Plant 161(4):502–514. https://doi.org/10.1111/ppl.12614

Basu A, Prasad P, Das SN, Kalam S, Sayyed RZ, Reddy MS, El Enshasy H (2021) Plant growth promoting rhizobacteria (PGPR) as green bioinoculants: recent developments, constraints, and prospects. Sustainability 13:1140. https://doi.org/10.3390/su13031140

Beirinckx S, Viaene T, Haegeman A, Debode J, Amery F, Vandenabeele S (2020) Tapping into the maize root microbiome to identify bacteria that promote growth under chilling conditions. Microbiome 8:54. https://doi.org/10.1186/s40168-020-00833-w

Benidire L, El Khalloufi F, Oufdou K, Barakat M, Tulumello J, Ortet P, Heulin T, Achouak W (2020) Phytobeneficial bacteria improve saline stress tolerance in *Vicia faba* and modulate microbial interaction network. Sci Total Environ 139020:139020. https://doi.org/10.1016/j.scitotenv.2020.139020

Berendsen RL, Pieterse CMJ, Bakker PAHM (2012) The Rhizosphere microbiome and plant health. Trends Plant Sci 17:478–486. https://doi.org/10.1016/j.tplants.2012.04.001

Bever JD, Platt TG, Morton ER (2012) Microbial population and community dynamics on plant roots and their feedbacks on plant communities. Annu Rev Microbiol 66:265–283. https://doi.org/10.1146/annurev-micro-092611-150107

Bhardwaj D, Ansari MW, Sahoo RK, Tuteja N (2014) Biofertilizers function as key player in sustainable agriculture by improving soil fertility, plant tolerance and crop productivity. Microb Cell Factories 13:66. https://doi.org/10.1186/1475-2859-13-66

Biyada S, Merzouki M, Dėmčėnko T, Vasiliauskienė D, Ivanec-Goranina R, Urbonavičius J, Marčiulaitienė E, Vasarevičius S, Benlemlih M (2021) Microbial community dynamics in the mesophilic and thermophilic phases of textile waste composting identified through next-generation sequencing. Sci Rep 11:23624. https://doi.org/10.1038/s41598-021-03191-1

Boubekri K, Soumare A, Mardad I, Lyamlouli K, Ouhdouch Y, Hafidi M, Kouisni L (2022) Multifunctional role of Actinobacteria in agricultural production sustainability: a review. Microbiol Res 261:127059. https://doi.org/10.1016/j.micres.2022.127059

Bowles TM, Acosta-Martínez V, Calderón F, Jackson LE (2014) Soil enzyme activities, microbial communities, and carbon and nitrogen availability in organic agroecosystems across an intensively-managed agricultural landscape. Soil Biol Biochem 68:252–262. https://doi.org/10.1016/j.soilbio.2013.10.004

Chandra S, Chakraborty D (2023) Isolation and characterization of plant growth promoting bacteria from root nodules of *Cicer arientinum* L. Int J Plant Soil Sci 35:404–409. https://doi.org/10.9734/ijpss/2023/v35i143064

Chaudhary P, Sharma A (2019) Response of Nanogypsum on the performance of plant growth promotory bacteria recovered from nanocompound infested agriculture field. Environ Ecol 37:363–372

Chaudhary A, Parveen H, Chaudhary P, Bhatt P, Khatoon H (2021a) Rhizospheric microbes and their mechanism. In: Bhatt P, Gangola S, Udayanga D, Kumar G (eds) Microbial technology for sustainable environment. Springer, Singapore. https://doi.org/10.1007/978-981-16-3840-4_6

Chaudhary P, Parveen H, Gangola S, Kumar G, Bhatt P, Chaudhary A (2021b) Plant Growth Promoting Rhizobacteria and their application in sustainable crop production. In: Bhatt P, Gangola S, Udayanga D, Kumar G (eds) Microbial technology for sustainable environment. Springer, Singapore. https://doi.org/10.1007/978-981-16-3840-4_13

Chaudhary P, Singh S, Chaudhary A, Agri U, Bhandari G (2022a) Chapter 3: Agrochemicals and their effects on soil microbial population. In: Soni R, Suyal D, Goel R (eds) Plant protection: from chemicals to biologicals. De Gruyter, Berlin, Boston, pp 45–58. https://doi.org/10.1515/9783110771558-003

Chaudhary P, Chaudhary A, Agri U, Khatoon H, Singh A (2022b) Recent trends and advancements for agro-environmental sustainability at higher altitudes. In: Goel R, Soni R, Suyal DC, Khan M (eds) Survival strategies in cold-adapted microorganisms. Springer, Singapore. https://doi.org/10.1007/978-981-16-2625-8_19

Chaudhary P, Agri U, Chaudhary A, Kumar A, Kumar G (2022c) Endophytes and their potential in biotic stress management and crop production. Front Microbiol 13:933017. https://doi.org/10.3389/fmicb.2022.933017

Chaudhary P, Singh S, Chaudhary A, Sharma A, Kumar G (2022d) Overview of biofertilizers in crop production and stress management for sustainable agriculture. Front Plant Sci 13:930340. https://doi.org/10.3389/fpls.2022.930340

Chaudhary P, Xu M, Ahamad L, Chaudhary A, Kumar G, Adeleke BS, Verma KK, Hu D-M, Širić I, Kumar P, Popescu SM, Abou Fayssal S (2023a) Application of synthetic consortia for improvement of soil fertility, pollution remediation, and agricultural productivity: a review. Agronomy 13:643. https://doi.org/10.3390/agronomy13030643

Chaudhary P, Chen S, Rajput VD, Jaiswal DK (2023b) Editorial: bioinoculants with nano-compounds to improve soil health: a step toward sustainable agriculture. Front Environ Sci 11:1270002. https://doi.org/10.3389/fenvs.2023.1270002

Chaudhary P, Chen S, Rajput VD, Jaiswal DK (eds) (2023c) Bioinoculants with nano-compounds to improve soil health: a step toward sustainable agriculture. Frontiers Media SA, Lausanne. https://doi.org/10.3389/978-2-8325-3287-4

Chodak M, Gołębiewski M, Morawska-Płoskonka J, Kuduk K, Niklinska M (2015) Soil chemical properties affect the reaction of forest soil bacteria to drought and rewetting stress. Ann Microbiol 65:1627–1637. https://doi.org/10.1007/s13213-014-1002-0

Christi Y (2023) Biotechnology for sustainable production of food. In: Ferranti P (ed) Sustainable food science—a comprehensive approach, pp 1–29. https://doi.org/10.1016/B978-0-12-823960-5.00089-5

Damalas CA, Koutroubas SD (2018) Current status and recent developments in biopesticide use. Agriculture 8(1):13. https://doi.org/10.3390/agriculture8010013

Dar A, Zahir ZA, Iqbal M, Mehmood A, Javed A, Hussain A, Ahmad M (2021) Efficacy of rhizobacterial exopolysaccharides in improving plant growth, physiology, and soil properties. Environ Monit Assess 193:1–15

De la Fuente Cantó C, Simonin M, King E, Moulin L, Bennett MJ, Castrillo G, Laplaze L (2020) An extended root phenotype: the rhizosphere, its formation and impacts on plant fitness. Plant J 103:951–964. https://doi.org/10.1111/tpj.14781

Diagne N, Ngom M, Djighaly PI, Fall D, Hocher V, Svistoonoff S (2020) Roles of arbuscular mycorrhizal fungi on plant growth and performance: importance in biotic and abiotic stressed regulation. Diversity 12:370. https://doi.org/10.3390/d12100370

Dinca LC, Grenni P, Onet C, Onet A (2022) Fertilization and soil microbial community: a review. Appl Sci 12:1198. https://doi.org/10.3390/app12031198

Dmytruk KV, Ruchala J, Fayura LR, Chrzanowski G, Dmytruk OV, Tsyrulnyk AO, Andreieva YA, Fedorovych DV, Motyka OI, Mattanovich D, Marx H, Sibirny AA (2023) Efficient production of bacterial antibiotics aminoriboflavin and roseoflavin in eukaryotic microorganisms, yeasts. Microb Cell Factories 22:1–13. https://doi.org/10.1186/s12934-023-02129-8

El-Ghwas DE, Elkhateeb WA, Daba GM (2023) Fungi: the next generation of biofertilizers. Environ Sci 2:34. https://doi.org/10.5281/zenodo.7500741

Fasusi OA, Cruz C, Babalola OO (2021) Agricultural sustainability: microbial biofertilizers in rhizosphere management. Agriculture 11:163. https://doi.org/10.3390/agriculture11020163

Finkel OM, Castrillo G, Paredes SH, González IS, Dangl JL (2017) Understanding and exploiting plant beneficial microbes. Curr Opin Plant Biol 38:155–163. https://doi.org/10.1016/j.pbi.2017.04.018

Gao X, Tan W, Zhao Y, Wu J, Sun Q, Qi H, Xie XY, Wei Z (2019) Diversity in the mechanisms of humin formation during composting with different materials. Environ Sci Technol 53:3653–3662. https://doi.org/10.1021/acs.est.8b06401

Gavande PV, Basak A, Sen S, Lepcha K, Murmu N, Rai V, Mazumdar D, Saha SP, Das V, Ghosh S (2021) Functional characterization of thermotolerant microbial consortium for lignocellulolytic enzymes with central role of Firmicutes in rice straw depolymerization. Sci Rep 11:3032. https://doi.org/10.1038/s41598-021-82163-x

Gouda S, Kerry RG, Das G, Paramithiotis S, Shin HS, Patra JK (2018) Revitalization of plant growth promoting rhizobacteria for sustainable development in agriculture. Microbiol Res 206:131–140. https://doi.org/10.1016/j.micres.2017.08.016

Gupta G, Parihar SS, Ahirwar NK, Snehi SK, Singh V (2015) Plant growth promoting Rhizobacteria (PGPR): current and future prospects for development of sustainable agriculture. J Microb Biochem Technol 7:96–102. https://doi.org/10.4172/1948-5948.1000188

Haifeng L, Hu S, Fu J (2021) Effects of acetic acid bacteria in starter culture on the properties of sourdough and steamed bread. Grain Oil Sci Technol 5:13–21. https://doi.org/10.1016/j.gaost.2021.11.003

Hannula SE, Träger S (2020) Soil fungal guilds as important drivers of the plant richness–productivity relationship. New Phytol 226:947–949. https://doi.org/10.1111/nph.16523

Hartman K, van der Heijden MG, Wittwer RA, Banerjee S, Walser J-C, Schlaeppi K (2018) Cropping practices manipulate abundance patterns of root and soil microbiome members paving the way to smart farming. Microbiome 6:1–14. https://doi.org/10.5810/kentucky/9780813175843.003.0001

Heema R, Murugan B, Narayanan R, Pugazhenthi TR (2022) Development of herbal bio-yoghurt powder using freeze drying technology. Madras Agric J. https://doi.org/10.29321/MAJ.10.000581

Hobara S, Osono T, Hirose D, Noro K, Hirota M, Benner R (2014) The roles of microorganisms in litter decomposition and soil formation. Biogeochemistry 118:471–486. https://doi.org/10.1007/s10533-014-9956-3

Ho VTT, Zhao J, Fleet G (2015) The effect of lactic acid bacteria on cocoa bean fermentation. Int J Food Microbiol 205:54–67. https://doi.org/10.1016/j.ijfoodmicro.2015.03.031

Husson O, Sarthou JP, Bousset L, Ratnadass A, Schmidt HP, Kempf J, Husson B, Tingry S, Aubertot JN, Deguine JP, Goebel FR, Lamichhane JR (2021) Soil and plant health in relation to dynamic sustainment of Eh and pH homeostasis: a review. Plant Soil 466:391–447. https://doi.org/10.1007/s11104-021-05047-z

Ishaq SL (2017) Plant-microbial interactions in agriculture and the use of farming systems to improve diversity and productivity. AIMS Microbiol 3:335–353. https://doi.org/10.3934/microbiol.2017.2.335

Islam ZF, Cordero PRF, Feng J, Chen Y-J, Bay SK, Jirapanjawat T, Gleadow RM, Carere CR, Stott MB, Chiri E, Greening C (2019) Two Chloroflexi classes independently evolved the ability to persist on atmospheric hydrogen and carbon monoxide. ISME J 13:1801–1813. https://doi.org/10.1038/s41396-019-0393-0

Itelima JU, Bang WJ, Onyimba IA (2018) A review: biofertilizer; a key player in enhancing soil fertility and crop productivity. J Microbiol Biotechnol 2:22–28. https://doi.org/10.26765/DRJAFS.2018.4815

Jehangir IA, Mir MA, Bhat MA (2017) Biofertilizers an approach to sustainability in agriculture: a review. Int J Pure Appl Biosci 5:327–334. https://doi.org/10.18782/2320-7051.5011

Jiang T, Schuchardt F, Li G, Guo R, Zhao Y (2011) Effect of C/N ratio, aeration rate and moisture content on ammonia and greenhouse gas emission during the composting. J Environ Sci 23(10):1754–1760. https://doi.org/10.1016/S1001-0742(10)60591-8

Kabisch A, Otto A, König S, Becher D, Albrecht D, Schüler M, Teeling H, Amann RI, Schweder T (2014) Functional characterization of polysaccharide utilization loci in the marine Bacteroidetes 'Gramella forsetii' KT0803. ISME J 8(7):1492–1502. https://doi.org/10.1038/ismej.2014.4

Karimi S, Mahboobi Soofiani N, Mahboubi A, Ferreira JA, Lundh T, Kiessling A, Taherzadeh MJ (2021) Evaluation of nutritional composition of pure filamentous fungal biomass as a novel ingredient for fish feed. Fermentation 7(3):152. https://doi.org/10.3390/fermentation7030152

Khati P, Chaudhary P, Gangola S, Bhatt P, Sharma A (2017) Nanochitosan supports growth of *Zea mays* and also maintains soil health following growth. 3 Biotech 7:81. https://doi.org/10.1007/s13205-017-0668-y

Khati P, Parul, Bhatt P, Nisha Kumar R, Sharma A (2018) Effect of nanozeolite and plant growth promoting rhizobacteria on maize. 3 Biotech 8:141. https://doi.org/10.1007/s13205-018-1142-1

Khatoon Z, Huang S, Rafique M, Fakhar A, Kamran MA, Santoyo G (2020) Unlocking the potential of plant growth-promoting rhizobacteria on soil health and the sustainability of agricultural systems. J Environ Manag 273:111118. https://doi.org/10.1016/j.jenvman.2020.111118

Khoshru B, Mitra D, Khoshmanzar E, Myo EM, Uniyal N, Mahakur B, Mohapatra PK, Panneerselvam P, Boutaj H, Alizadeh M, Cely MVT, Senapati A, Rani A (2020) Current scenario and future prospects of plant growth-promoting rhizobacteria: an economic valuable resource for the agriculture revival under stressful conditions. J Plant Nutr 43:3062–3092. https://doi.org/10.1080/01904167.2020.1799004

Kollmen J, Strieth D (2022) The beneficial effects of cyanobacterial co-culture on plant growth. Life 12:223. https://doi.org/10.3390/life12020223

Kuila D, Ghosh S (2022) Aspects, problems and utilization of Arbuscular Mycorrhizal (AM) application as bio-fertilizer in sustainable agriculture. Curr Res Microb Sci 3:100107. https://doi.org/10.1016/j.crmicr.2022.100107

Kukreti B, Sharma A, Chaudhary P, Agri U, Maithani D (2020) Influence of nanosilicon dioxide along with bioinoculants on *Zea mays* and its rhizospheric soil. 3 Biotech 10:345. https://doi.org/10.1007/s13205-020-02329-8

Kulkarni K, Chawan A, Kulkarni A, Gharat S (2022) Bioremediation of imidacloprid using Azospirillium biofertilizer and Rhizobium biofertilizer. Korean J Chem Eng 39:2702–2712. https://doi.org/10.1007/s11814-022-1149-5

Kumar A, Dubey A (2020) Rhizosphere microbiome: engineering bacterial competitiveness for enhancing crop production. J Adv Res 24:337–352. https://doi.org/10.1016/j.jare.2020.04.014

Kumar A, Verma JP (2018) Does plant—microbe interaction confer stress tolerance in plants: a review? Microbiol Res 207:41–52. https://doi.org/10.1016/j.micres.2017.11.004

Kumar U, Berliner J, Adak T, Rath PC, Dey A, Pokhare SS (2017a) Non-target effect of continuous application of chlorpyrifos on soil microbes, nematodes and its persistence under subhumid tropical rice-rice cropping system. Ecotoxicol Environ Saf 135:225–235. https://doi.org/10.1016/j.ecoenv.2016.10.003

Kumar U, Panneerselvam P, Govindasamy V, Vithalkumar L, Senthilkumar M, Banik A, Annapurna K (2017b) Long-term aromatic rice cultivation effect on frequency and diversity of diazotrophs in its rhizosphere. Ecol Eng 101:227–236. https://doi.org/10.1016/j.ecoleng.2017.02.010

Kumar U, Nayak AK, Shahid M, Gupta VV, Panneerselvam P, Mohanty S (2018) Continuous application of inorganic and organic fertilizers over 47 years in paddy soil alters the bacterial community structure and its influence on rice production. Agric Ecosyst Environ 262:65–75. https://doi.org/10.1016/j.agee.2018.04.016

Kumar G, Lal S, Maurya SK, Bhattacherjee AK, Chaudhary P, Gangola S, Rajan S (2021a) Exploration of *Klebsiella pneumoniae* M6 for paclobutrazol degradation, plant growth attributes, and biocontrol action under subtropical ecosystem. PLoS One 16:e0261338. https://doi.org/10.1371/journal.pone.0261338

Kumar A, Sharma A, Chaudhary P, Gangola S (2021b) Chlorpyrifos degradation using binary fungal strains isolated from industrial waste soil. Biologia 10:3071–3081. https://doi.org/10.1007/s11756-021-00816-8

Kumar G, Lal S, Soni SK, Maurya SK, Shukla PK, Chaudhary P, Bhattacherjee AK, Garg N (2022) Mechanism and kinetics of chlorpyrifos co-metabolism by using environment restoring microbes isolated from rhizosphere of horticultural crops under subtropics. Front Microbiol 13:891870. https://doi.org/10.3389/fmicb.2022.891870

Kumari S, Sharma A, Chaudhary P, Khati P (2020) Management of plant vigor and soil health using two agriusable nanocompounds and plant growth promotory rhizobacteria in Fenugreek. 3 Biotech 10:1–11. https://doi.org/10.1007/s13205-020-02448-2

Kumari H, Khati P, Gangola S, Chaudhary P, Sharma A (2021) Performance of plant growth promotory rhizobacteria on Maize and soil characteristics under the influence of TiO_2 nanoparticles. Pantnagar J Res 19:28–39

Kwak M-J, Kong HG, Choi K, Kwon S-K, Song JY, Lee J (2018) Rhizosphere microbiome structure alters to enable wilt resistance in tomato. Nat Biotechnol 36:1100–1109. https://doi.org/10.1038/nbt.4232

Larsbrink J, McKee LS (2020) Bacteroidetes bacteria in the soil: glycan acquisition, enzyme secretion, and gliding motility. Adv Appl Microbiol 110:63–98. https://doi.org/10.1016/bs.aambs.2019.11.001

Lekberg Y, Helgason T (2018) In situ mycorrhizal function-knowledge gaps and future directions. New Phytol 220:957–962. https://doi.org/10.1111/nph.15064
Li X, Shi X-S, Lu M-Y, Zhao Y-Z, Li X, Peng H, Guo R-B (2019) Succession of the bacterial community and functional characteristics during continuous thermophilic composting of dairy manure amended with recycled ceramsite. Bioresour Technol 294:122044. https://doi.org/10.1016/j.biortech.2019.122044
Li J, Bao H, Xing W, Yang J, Liu R, Wang X, Lv L, Tong X, Wu F (2020a) Succession of fungal dynamics and their influence on physicochemical parameters during pig manure composting employing with pine leaf biochar. Bioresour Technol 297:122377. https://doi.org/10.1016/j.biortech.2019.122377
Li Y, Tremblay J, Bainard LD, Cade-Menun B, Hamel C (2020b) Long-term effects of nitrogen and phosphorus fertilization on soil microbial community structure and function under continuous wheat production. Environ Microbiol 22:1066–1088. https://doi.org/10.1111/1462-2920.14824
Li C, Li X, Shi Y, Yang Y, Li H (2022) Effects of nitrogen addition on soil carbon-fixing microbial diversity on different slopes in a degraded alpine meadow. Front Plant Sci 13:921278. https://doi.org/10.3389/fpls.2022.921278
Liang Y, Pan F, Jiang Z, Li Q, Pu J, Liu K (2022) Accumulation in nutrient acquisition strategies of arbuscular mycorrhizal fungi and plant roots in poor and heterogeneous soils of karst shrub ecosystems. BMC Plant Biol 22:188. https://doi.org/10.1186/s12870-022-03514-y
Liu Y, Jianwei Guo J, Li L, Asem MD, Zhang Y, Mohamad OA (2017) Endophytic bacteria associated with endangered plant *Ferula sinkiangensis* K. M. Shen in an arid land: diversity and plant growth-promoting traits. J Arid Land 9:432–445. https://doi.org/10.1007/s40333-017-0015-5
Liu RC, Xiao ZY, Hashem A, AbdAllah EF, Wu QS (2021) Mycorrhizal fungal diversity and its relationship with soil properties in *Camellia oleifera*. Agriculture 11:470. https://doi.org/10.3390/agriculture11060470
Lopez MJ, Jurado MM, López-González JA, Estrella-González MJ, Martínez-Gallardo MR, Toribio A, Suárez-Estrella F (2021) Characterization of thermophilic lignocellulolytic microorganisms in composting. Front Microbiol 12:697480. https://doi.org/10.3389/fmicb.2021.697480
Lopez-González J, Suárez-Estrella F, Vargas-García M, López M, Jurado M, Moreno J (2015) Dynamics of bacterial microbiota during lignocellulosic waste composting: studies upon its structure, functionality and biodiversity. Bioresour Technol 175:406–416. https://doi.org/10.1016/j.biortech.2014.10.123
Luo X, Liu G, Xia Y, Chen L, Jiang Z, Zheng H, Wang Z (2017) Use of biochar-compost to improve properties and productivity of the degraded coastal soil in the Yellow River Delta, China. J Soils Sediments 17:780–789. https://doi.org/10.1007/s11368-016-1361-1
Ma S, Xiong J, Cui R, Sun X, Han L, Xu Y, Kan Z, Gong X, Huang G (2020) Effects of intermittent aeration on greenhouse gas emissions and bacterial community succession during large-scale membrane-covered aerobic composting. J Clean Prod 266:121551. https://doi.org/10.1016/j.jclepro.2020.121551
Manzar N, Jayaswal D, Kishan G, Chauhan S, Joshi D, Singh J, Chaudhary P, Jamali H, Kushwaha P, Parveen H, Kashyap AS, Sahu PK, Singh UB (2022) Application of molecular ecology approaches in sustainable agriculture for a better understanding of plant–microbiome interactions. In: Singh UB, Sahu PK, Singh HV, Sharma PK, Sharma SK (eds) Rhizosphere microbes. Microorganisms for sustainability, vol 40. Springer, Singapore. https://doi.org/10.1007/978-981-19-5872-4_3
Mendes R, Garbeva P, Raaijmakers JM (2013) The rhizosphere microbiome: significance of plant beneficial, plant pathogenic, and human pathogenic microorganisms. FEMS Microbiol Rev 37:634–663. https://doi.org/10.1111/1574-6976.12028
Meng Q, Yang W, Men M, Bello A, Xu X, Xu B, Deng L, Jiang X, Sheng S, Wu X, Han Y, Zhu H (2019) Microbial community succession and response to environmental variables during cow manure and corn straw composting. Front Microbiol 10:529. https://doi.org/10.3389/fmicb.2019.00529

Mitra D, Rad KV, Chaudhary P, Ruparelia J, Sagarika MS, Boutaj H et al (2021) Involvement of strigolactone hormone in root development, influence and interaction with mycorrhizal fungi in plant: mini-review. Curr Res Microbial Sci 2:100026. https://doi.org/10.1016/j.crmicr.2021.100026
Mitter B, Pfaffenbichler N, Sessitsch A (2016) Plant–microbe partnerships in 2020. Microb Biotechnol 9:635–640. https://doi.org/10.1111/1751-7915.12382
Mngongo ME, Ojija F, Aloo BN (2023) The role of Rhizobia toward food production, food and soil security through microbial agro-input utilization in developing countries. Case Stud Chem Environ Eng 8:100404. https://doi.org/10.1016/j.cscee.2023.100404
Mokabel S, Olama Z, Ali S, El-Dakak R (2022) The role of plant growth promoting rhizosphere microbiome as alternative biofertilizer in boosting *Solanum melongena* L. adaptation to salinity stress. Plants 11:659. https://doi.org/10.3390/plants11050659
Morella NM, Weng FC-H, Joubert PM, Metcalf CJE, Lindow S, Koskella B (2020) Successive passaging of a plant-associated microbiome reveals robust habitat and host genotype-dependent selection. Proc Natl Acad Sci USA 117:1148–1159. https://doi.org/10.1073/pnas.1908600116
Moreno J, López-González J, Arcos-Nievas M, Suárez-Estrella F, Jurado M, Estrella-González M, López M (2021) Revisiting the succession of microbial populations throughout composting: a matter of thermotolerance. Sci Total Environ 773:145587. https://doi.org/10.1016/j.scitotenv.2021.145587
Moubareck CA, Alawlaqi B, Alhajeri S (2023) Characterization of physicochemical parameters and bacterial diversity of composted organic food wastes in Dubai. Heliyon 9:e16426. https://doi.org/10.1016/j.heliyon.2023.e16426
Mueller UG, Sachs JL (2015) Engineering microbiomes to improve plant and animal health. Trends Microbiol 23:606–617. https://doi.org/10.1016/j.tim.2015.07.009
Nadarajah K, Abdul Rahman NSN (2021) Plant-microbe interaction: aboveground to belowground, from the good to the bad. Int J Mol Sci 22(19):10388. https://doi.org/10.3390/ijms221910388
Ngalimat MS, Mohd Hata E, Zulperi D, Ismail SI, Ismail MR, Mohd Zainudin NAI, Said INB, Yusof MT (2021) Plant growth-promoting bacteria as an emerging tool to manage bacterial rice pathogens. Microorganisms 9:682. https://doi.org/10.3390/microorganisms9040682
Nicolas C, Martin-Bertelsen T, Floudas D, Bentzer J, Smits M, Johansson T, Troein C, Persson P, Tunlid A (2019) The soil organic matter decomposition mechanisms in ectomycorrhizal fungi are tuned for liberating soil organic nitrogen. ISME J 13:977–988. https://doi.org/10.1038/s41396-018-0331-6
Noor-Hassim M, Ng C, Teo H, Azmi W, Muhamad-Zalan N, Binti Karim N, Ahmad A (2023) The utilization of microbes for sustainable food production. BioTechnologia 104(2):209–216. https://doi.org/10.5114/bta.2023.127209
Numan M, Bashir S, Khan Y, Mumtaz R, Shinwari ZK, Khan AL, Kham A, Al-Harrasi A (2018) Plant growth promoting bacteria as an alternative strategy for salt tolerance in plants: a review. Microbiol Res 209:21–32. https://doi.org/10.1016/j.micres.2018.02.003
Olenska E, Małek W, Wójcik M, Swiecicka I, Thijs S, Vangronsveld J (2020) Beneficial features of plant growth-promoting rhizobacteria for improving plant growth and health in challenging conditions: a methodical review. Sci Total Environ 32:429–448
Omotayo OP, Babalola OO (2021) Resident rhizosphere microbiome's ecological dynamics and conservation: towards achieving the envisioned Sustainable Development Goals, a review. Int Soil Water Conserv Res 9(1):127–142. https://doi.org/10.1016/j.iswcr.2020.08.002
Orozco-Mosqueda MC, Flores A, Rojas-Sánchez B, Urtis-Flores CA, Morales-Cedeño LR, Valencia-Marin MF, Chávez-Avila S, Rojas-Solis D, Santoyo G (2021) Plant growth-promoting bacteria as bioinoculants: attributes and challenges for sustainable crop improvement. Agronomy 11:1167. https://doi.org/10.3390/agronomy11061167
Ortiz A, Sansinenea E (2021) Recent advancements for microorganisms and their natural compounds useful in agriculture. Appl Microbiol Biotechnol 105:891–897. https://doi.org/10.1007/s00253-020-11030-y

Pane C, Palese AM, Celano G, Zaccardelli M (2014) Effects of compost tea treatments on productivity of lettuce and kohlrabi systems under organic cropping management. Ital J Agron 9(3):153–156. https://doi.org/10.4081/ija.2014.596

Pane C, Spaccini R, Piccolo A, Celano G, Zaccardelli M (2019) Disease suppressiveness of agricultural greenwaste composts as related to chemical and bio-based properties shaped by different on-farm composting methods. Biol Control 137:104026. https://doi.org/10.1016/j.biocontrol.2019.104026

Paredes SH, Gao T, Law TF, Finkel OM, Mucyn T, Teixeira PJPL et al (2018) Design of synthetic bacterial communities for predictable plant phenotypes. PLoS Biol 16:e2003962. https://doi.org/10.1371/journal.pbio.2003962

Parveen H, Chaudhary A, Chaudhary P, Sultana R, Khati P, Kumar G, Khati P, Rana M, Bhatt P (2022) Omics approaches for the production of the microbial enzymes and applications. In: Bhatt P (ed) Industrial applications of microbial enzymes, 1st edn. CRC Press. https://doi.org/10.1201/9781003202998-16

Pellitier PT, Zak DR (2018) Ectomycorrhizal fungi and the enzymatic liberation of nitrogen from soil organic matter: why evolutionary history matters. New Phytol 217:68–73. https://doi.org/10.1111/nph.14598

Pezzolla D, Cucina M, Proietti P, Calisti R, Regni L, Gigliotti G (2021) The use of new parameters to optimize the composting process of different organic wastes. Agronomy 11(10):2090. https://doi.org/10.3390/agronomy11102090

Rani A, Singh R, Kumar P, Shukla G (2018) Pros and cons of fungicides: an overview. Int J Eng Sci Res Technol 6:112–117. https://doi.org/10.5281/zenodo.233295

Ray P, Lakshmanan V, Labbé JL, Craven KD (2020) Microbe to microbiome: a paradigm shift in the application of microorganisms for sustainable agriculture. Front Microbiol 11:622926. https://www.frontiersin.org/article/10.3389/fmicb.2020.622926

Ryu M-H, Zhang J, Toth T, Khokhani D, Geddes BA, Mus F (2020) Control of nitrogen fixation in bacteria that associate with cereals. Nat Microbiol 5:314–330. https://doi.org/10.1038/s41564-019-0631-2

Sarkar A, Pramanik K, Mitra S, Soren T, Maiti TK (2018) Enhancement of growth and salt tolerance of rice seedlings by ACC deaminase-producing *Burkholderia* sp. MTCC 12259. J Plant Physiol 231:434–442. https://doi.org/10.1016/j.jplph.2018.10.010

Sarwar S, Khaliq A, Yousra M, Sultan T, Ahmad N, Khan MZ (2020) Screening of siderophore-producing PGPRs isolated from groundnut (*Arachis hypogaea* L.) rhizosphere and their influence on iron release in soil. Commun Soil Sci Plant Anal 51:1680–1692. https://doi.org/10.1080/00103624.2020.1791159

Schmid CA, Schröder P, Armbruster M, Schloter M (2018) Organic amendments in a long-term field trial—consequences for the bulk soil bacterial community as revealed by network analysis. Microb Ecol 76:226–239. https://doi.org/10.1007/s00248-017-1110-z

Schulz S, Brankatschk R, Dümig A, Kögel-Knabner I, Schloter M, Zeyer J (2013) The role of microorganisms at different stages of ecosystem development for soil formation. Biogeosciences 10:3983–3996. https://doi.org/10.5194/bg-10-3983-2013

Scotti R, Pane C, Spaccini R, Palese AM, Piccolo A, Celano G, Zaccardelli M (2016) On-farm compost: a useful tool to improve soil quality under intensive farming systems. Appl Soil Ecol 107:13–23. https://doi.org/10.1016/j.apsoil.2016.05.004

Semenov AM, Đukić DA (2020) The role of microbial communities in soil formation and soil ecosystem health. Paleontol J 54:843–852. https://doi.org/10.1134/S0031030120080146

Shahwar D, Mushtaq Z, Mushtaq H, Alqarawi AA, Park Y, Alshahrani TS, Faizan S (2023) Role of microbial inoculants as bio fertilizers for improving crop productivity: a review. Heliyon 9(6):e16134. https://doi.org/10.1016/j.heliyon.2023.e16134

Sharma SB, Sayyed RZ, Trivedi MH, Gobi TA (2013) Phosphate solubilizing microbes: sustainable approach for managing phosphorus deficiency in agricultural soils. Springerplus 2:1–14. https://doi.org/10.1186/2193-1801-2-587

Shrestha S, Devkota D, Paudel B (2022) Effect of biofertilizer (*Azotobacter chroococum*) at different nitrogen level on growth and yield of cauliflower in palung, makwanpur. Plant Physiol Soil Chem 2:46–51. https://doi.org/10.26480/ppsc.01.2022.46.51

Siedt M, Schäffer A, Smith KE, Nabel M, Roß-Nickoll M, van Dongen JT (2021) Comparing straw, compost, and biochar regarding their suitability as agricultural soil amendments to affect soil structure, nutrient leaching, microbial communities, and the fate of pesticides. Sci Total Environ 751:141607. https://doi.org/10.1016/j.scitotenv.2020.141607

Singh S, Chaudhary P, Sharma A (2023) Efficient decolourization of toluidine blue and methylene blue using *Bacillus cereus*. Vegetos. https://doi.org/10.1007/s42535-023-00747-6

Tarekegn MM, Salilih FZ, Ishetu AI (2020) Microbes used as a tool for bioremediation of heavy metal from the environment. Cogent Food Agric 6:1783174. https://doi.org/10.1080/23311932.2020.1783174

Thakur N, Kaur S, Tomar P, Thakur S, Yadav AN (2020) Microbial biopesticides: current status and advancement for sustainable agriculture and environment. Elsevier, In New and future developments in microbial biotechnology and bioengineering, pp 243–282. https://doi.org/10.1016/B978-0-12-820526-6.00016-6

Tian J, He N, Hale L, Niu S, Yu G, Liu Y (2018) Soil organic matter availability and climate drive latitudinal patterns in bacterial diversity from tropical to cold temperate forests. Funct Ecol 32:61–70. https://doi.org/10.1111/1365-2435.12952

Tortosa G, Fernández-González AJ, Lasa AV, Aranda E, Torralbo F, González-Murua C, Fernández-López M, Benítez E, Bedmar EJ (2021) Involvement of the metabolically active bacteria in the organic matter degradation during olive mill waste composting. Sci Total Environ 789:147975. https://doi.org/10.1016/j.scitotenv.2021.147975

Tosi M, Mitter EK, Gaiero J, Dunfield K (2020) It takes three to tango: the importance of microbes, host plant, and soil management to elucidate manipulation strategies for the plant microbiome. Can J Microbiol 66(7):413–433. https://doi.org/10.1139/cjm-2020-0085

Toundou O, Pallier V, Feuillade-Cathalifaud G, Tozo K (2021) Impact of agronomic and organic characteristics of waste composts from Togo on *Zea mays* L. nutrients contents under water stress. J Environ Manag 285:112158. https://doi.org/10.1016/j.jenvman.2021.112158

Turrini A, Avio L, Giovannetti M, Agnolucci M (2018) Functional complementarity of arbuscular mycorrhizal fungi and associated microbiota: the challenge of translational research. Front Plant Sci 9:1407. https://doi.org/10.3389/fpls.2018.01407

Tyagi J, Chaudhary P, Jyotsana, Bhagwati U, Bhandari G, Chaudhary A (2022a) Chapter 17: Impact of endophytic fungi in biotic stress management. In: Soni R, Suyal D, Goel R (eds) Plant protection: from chemicals to biologicals. De Gruyter, Berlin, Boston, pp 447–462. https://doi.org/10.1515/9783110771558-017

Tyagi J, Chaudhary P, Mishra A, Khatwani M, Dey S, Varma A (2022b) Role of endophytes in abiotic stress tolerance: with special emphasis on *Serendipita indica*. Int J Environ Res 16:62. https://doi.org/10.1007/s41742-022-00439-0

Usyskin-Tonne A, Hadar Y, Minz D (2019) Altering N_2O emissions by manipulating wheat root bacterial community. Sci Rep 9:7613. https://doi.org/10.1038/s41598-019-44124-3

Uyizeye OC, Thiet RK, Knorr MA (2019) Effects of community-accessible biochar and compost on diesel-contaminated soil. Biorem J 23(2):107–117. https://doi.org/10.1080/10889868.2019.1603139

Wagg C, Jansa J, Schmid B, van der Heijden MG (2011) Belowground biodiversity effects of plant symbionts support aboveground productivity. Ecol Lett 14:1001–1009. https://doi.org/10.1111/j.1461-0248.2011.01666.x

Wan J, Wang X, Yang T, Wei Z, Banerjee S, Friman V-P, Mei X, Xu Y, Shen Q (2021) Livestock manure type affects microbial community composition and assembly during composting. Front Microbiol 12:621126. https://doi.org/10.3389/fmicb.2021.621126

Wang C, Dong D, Wang H, Müller K, Qin Y, Wang H, Wu W (2016) Metagenomic analysis of microbial consortia enriched from compost: new insights into the role of Actinobacteria in lignocellulose decomposition. Biotechnol Biofuels Bioprod 9:22. https://doi.org/10.1186/s13068-016-0440-2

Wang K, Mao H, Li X (2018a) Functional characteristics and influence factors of microbial community in sewage sludge composting with inorganic bulking agent. Bioresour Technol 249:527–535. https://doi.org/10.1016/j.biortech.2017.10.034

Wang K, Yin X, Mao H, Chu C, Tian Y (2018b) Changes in structure and function of fungal community in cow manure composting. Bioresour Technol 255:123–130. https://doi.org/10.1016/j.biortech.2018.01.064

Wang X, Kong Z, Wang Y, Wang M, Liu D, Shen Q (2020) Insights into the functionality of fungal community during the large scale aerobic co-composting process of swine manure and rice straw. J Environ Manag 270:110958. https://doi.org/10.1016/j.jenvman.2020.110958

Wang Y, Tibbetts SM, McGinn PJ (2021) Microalgae as sources of high-quality protein for human food and protein supplements. Foods 10:3002. https://doi.org/10.3390/foods10123002

Wang Y, Zeng H, Cao Y, Wang S, Wang B (2023) A study of key aroma compounds in Hurood cheese and their potential correlations with lactic acid bacteria. Fermentation 9(7):670. https://doi.org/10.3390/fermentation9070670

Werner KA, Poehlein A, Schneider D, El-Said K, Wöhrmann M, Linkert I, Hübner T, Brüggemann N, Prost K, Daniel R, Grohmann E (2022) Thermophilic composting of human feces: development of bacterial community composition and antimicrobial resistance gene pool. Front Microbiol 13:824834. https://doi.org/10.3389/fmicb.2022.824834

Xu S, Lu W, Liu Y, Ming Z, Liu Y, Meng R, Wang H (2017) Structure and diversity of bacterial communities in two large sanitary landfills in China as revealed by high-throughput sequencing (MiSeq). Waste Manag 63:41–48. https://doi.org/10.1016/j.wasman.2016.07.047

Yang W, Guo Y, Wang X, Chen C, Hu Y, Cheng L, Gu S, Xu X (2017) Temporal variations of soil microbial community under compost addition in black soil of Northeast China. Appl Soil Ecol 121:214–222. https://doi.org/10.1016/j.apsoil.2017.10.005

Yang Y, Awasthi MK, Bao H, Bie J, Lei S, Lv J (2020) Exploring the microbial mechanisms of organic matter transformation during pig manure composting amended with bean dregs and biochar. Bioresour Technol 313:123647. https://doi.org/10.1016/j.biortech.2020.123647

Yasmin H, Rashid U, Hassan MN, Nosheen A, Naz R, Ilyas N, Sajjad M, Azmat A, Alyemeni MN (2021) Volatile organic compounds produced by *Pseudomonas pseudoalcaligenes* alleviated drought stress by modulating defense system in maize (*Zea mays* L.). Physiol Plant 172(2):896–911. https://doi.org/10.1111/ppl.13304

Yin Y, Gu J, Wang X, Song W, Zhang K, Sun W, Zhang X, Zhang Y, Li H (2017) Effects of copper addition on copper resistance, antibiotic resistance genes, and intl1 during swine manure composting. Front Microbiol 8:344. https://doi.org/10.3389/fmicb.2017.00344

Yu H, Xie B, Khan R, Shen G (2019) The changes in carbon, nitrogen components and humic substances during organic-inorganic aerobic co-composting. Bioresour Technol 271:228–235. https://doi.org/10.1016/j.biortech.2018.09.088

Zhang Z, Yuan Y, Liu Q, Yin H (2019) Plant nitrogen acquisition from inorganic and organic sources via root and mycelia pathways in ectomycorrhizal alpine forests. Soil Biol Biochem 136:107517. https://doi.org/10.1016/j.soilbio.2019.06.013

Zhang M, Yang L, Hao R, Bai X, Wang Y, Yu X (2020) Drought-tolerant plant growth-promoting rhizobacteria isolated from jujube (*Ziziphus jujuba*) and their potential to enhance drought tolerance. Plant Soil 452(1–2):423–440. https://doi.org/10.1007/s11104-020-04582-5

Zhou Z, Tran PQ, Kieft K, Anantharaman K (2020) Genome diversification in globally distributed novel marine Proteobacteria is linked to environmental adaptation. ISME J 14(8):2060–2077. https://doi.org/10.1038/s41396-020-0669-4

Influence of Microbial Inoculants on Soil Health

Pragati Srivastava, Parul Chaudhary, Anita Mishra, Hemant Dasila, and Damini Maithani

Abstract The escalating world population is directly affecting food production leading to a rapid fall of agriculture land resources which is posing a major commination to sustainable agriculture. The resident soil bacteria and use of microbial inoculants play a cognitive role in determining soil health. Soil microbial diversity is significantly affected by bioinoculant types and diversity and its mediating disturbance regime, which directly affects plant health. Extensive agriculture practices such as tillage, mono-cropping, and use of chemical fertilizers have adversely influenced resident microbial community and deteriorating soil structure. A proper balance of soil resident microbial communities in particular functional communities and use of bioinoculants should be in pace to maintain soil health so that soil nutrient status and beneficial microbes within soil regime are not disturbed and plant performance is also inflated.

Keywords Bacteria · Soil fertility · Soil health indicators · Beneficial microbes

P. Srivastava (✉)
Department of Microbiology, G. B. Pant University of Agriculture and Technology, Pantnagar, Uttarakhand, India

P. Chaudhary
School of Agriculture, Graphic Era Hill University, Dehradun, Uttarakhand, India

A. Mishra
Vidyadayini Institute of Science, Management and Technology, Bhopal, Madhya Pradesh, India

H. Dasila
Department of Microbiology, Akal College of Basic Sciences, Sirmaur, Himachal Pradesh, India

D. Maithani
School of Biotechnology, IFTM University, Moradabad, Uttar Pradesh, India

P. Chaudhary, A. Chaudhary (eds.), *Microbial Inoculants*, Interdisciplinary Biotechnological Advances, https://doi.org/10.1007/978-981-97-0633-4_2

1 Introduction

Soil health is defined by its indispensable potential to succor all living forms including plants, animals, and humans. Clean air, abundant crops with excellent yields, lush grasslands, diverse fauna, and breathtaking sceneries are all indicators of healthy soil. In essence, it aids in the management of water, which indirectly supports a variety of life forms. Hazardous contaminants are filtered and buffered by soil. Organic and inorganic elements are vigorously detoxified by the soil's microbial community (Srivastava et al. 2020; Javed et al. 2022). The recycling of both micro- and macronutrients is aided by it, also offering structural support and stability to plant roots as well as preservation of archaeological relics (Srivastava et al. 2020; Davis et al. 2023). Physical, chemical, and biological perturbations are detrimental to the wellness of the soil. Overgrazing by wild animals reduces plant's capacity to replace themselves, which results in desertification and a quick drop in net output. Pesticides and fertilizers used in excess impair soil health and harm the native microbial community, which is crucial for the soil food webs (Chaudhary et al. 2022a). Tillage is a major component in the reduction of pore spaces, which limits the process of water and air filtration and destroys the habitat of resident organisms. These factors collectively cause the soil to become deteriorated and infertile (Krasilnikov and Taboada 2022).

Microbial inoculants are consortium of microbes as a sustainable tool for green revolution. Plant associated microorganism forms holobionts which maintain a good pace with plant's growth and development (Khati et al. 2019; Chaudhary et al. 2021a). Rhizo-microbiome essentially participates in nutrient acquisition and enhances soil physicochemical properties (Agri et al. 2022). Moreover, some microorganisms produce antibiotics, hormones, and siderophores that impede the growth of disease-causing organisms and mediate biocontrol activities (Massalha et al. 2017; Chaudhary and Sharma 2019; Agri et al. 2021; Srivastava et al. 2022). For typical growth and development, plants need macromolecules such phosphorus P, potassium K, nitrogen N, calcium Ca, magnesium Mg, and sulfur. Moreover, albeit in smaller amounts, micronutrients including Fe, Mo, Cu, Mo, Cl, and Zn are crucial for the operation of the photosynthetic machinery in plants. The bioavailability of these micro-/macronutrients is increased by microbial inoculants (Rashid et al. 2016; Kumari et al. 2021). Biological nitrogen fixation carried out by certain group of bacteria and methanogenic archaea helps in fixing nitrogen present in the atmosphere into utilizable forms to the plants. Root nodule symbiosis between plants and microbes is a comprehensive mechanism of N_2 fixation.

For centuries, agriculture has been practiced to feed mankind to meet the expanding food demands of the growing population. The main concern is the sustainable use of ever limiting land resources for agriculture practice and its judicious use keeping an eye on the environment and human health. During the 1960s era of green revolution where enhanced crop yield was the major concern, which leads to uncontrollable usage of agrochemicals, resulted in fall of in the soil quality and its pollution. Lack of soil testing facilities led to imprudent application of chemical fertilizers without knowledge of the recommended dose for application required for increased crop production (Li et al. 2022). As a result of continuous application of toxic chemical, pesticides delimited the microbiome of the soil, and its natural microbiota was also disturbed. In order to replenish the soil nutrient status and resident microbial community, a new way forward came into existence: microbial inoculation. The recurrent microbial inoculation of soil is a major breakthrough in enhancing the soil fertility. Microbial inoculant types with different properties such as phosphate solubilizing and nitrogen fixing bacteria as individual or as consortia exhibit succession in the resident microbial population that ultimately raise (Chaudhary et al. 2022b, c, d). Soil ecosystem is mainly influenced by soil microbial population, which mitigates organic matter mineralization, nutrient cycling, and resistance capacity to soilborne pathogens. Exogenous anthropogenic disturbances such as fertilization, pesticide application, irrigation practices, and natural climatic change (temperature and heavy rainfall or drought) contribute in the composition of the resident microbial community. Soil is always prone to foreign invaders whether it is beneficial microorganism or soilborne pathogens, which can alter the succession in resident microbial population (Wang et al. 2021).

Instigating plant growth promoting microbes, into plant soil ecosystem that encourages lucrative effect on plant growth and productivity, is an environmentally sound approach (Khati et al. 2017, 2018). Endless number of research studies has proven beneficial effects of efficient microbial inoculants on various commercial crops (Chaudhary et al. 2021b). However, its quantity and its application interval time are not determined. This information needs to be expanded for its effective results on plant growth promotion. How does the inoculation impact the native microbial assemblage needs to be answer that will help in thorough understanding between inhabitant microbes and externally inoculated intruders.

Soil is not just simply the provenance of micro- and macronutrient to plants, but it is actually a complex ecosystem which consists of bacteria, protest, fungi, and animals. Plants are known to exhibit a diverse interaction with microorganisms which could be either competitive, neutral, commensalism, or amensalism in nature. In recent decades, interaction studies focused on soil ecosystem has been much concentrated on abiotic stress condition with respect to plant response and characterizing the positive interaction that are regulating plant growth and development (Yaish et al. 2016; Meena et al. 2017; Tyagi et al. 2022a, b).

To improve soil nutrient quality, most used traditional method is to use increased input of inorganic fertilizers. However, use of unsustainable fertilization in contemporary agriculture system led to the large-scale alternation in soil biogeochemical cycle emissions (Steffen et al. 2015). Here comes the role of microorganisms that naturally enhance soil macro- and micronutrient availability via mineral solubilization mechanisms (Dasila et al. 2022).

2 Root Exudate Signal and Microbe's Relation

Unique microorganisms connected to particular plants are crucial for successful agriculture. The microbiome of a certain plant host is shaped by a variety of variables, for instance, changing weather patterns, biotic and abiotic elements, and numerous environmental changes brought on by human activity. Plant root exudates are a key factor in the interaction between plants and microbial flora in rhizosphere

(Kalia et al. 2020; Manzar et al. 2022). The rhizosphere soil consists of 10^{10}–10^{12} cells per gram of the soil, whereas bulk soil has less than 10^{8} cell density per gram of soil. *Proteobacteria* and *Actinobacteria* make up the largest group in the rhizosphere. *Pseudomonas* makes the biggest contribution. The most significant genera found in the rhizosphere microbiome include *Bacillus*, *Arthrobacter*, *Rhizobium*, *Mycobacterium*, *Cellulomonas*, *Flavobacterium*, *Alcaligenes*, *Micrococcus*, and *Azotobacter*. *Bacillus*, *Enterobacter*, *Paenibacillus*, *Acinetobacter*, and *Burkholderia* are further groups that are either directly or indirectly engaged in promoting plant growth (Zuluaga et al. 2020).

Microorganism's growth in soil is usually carbon limiting in nature, and thus deposition of amino acids, organic acids, and sugars via plants in the roots provides favorable condition for microbes (Jacoby et al. 2017). Since deposition of rich carbon source material is not restricted to harbors only beneficial microbes as pathogens can also invade under favorable condition hence it has been postulated that plants have evolved certain kind of mechanisms to discriminate between beneficial and pathogenic microorganisms. This discriminatory behavior can be results via release of specific root exudates that are responsible for shaping the microbial community structure in the rhizospheric system (Fig. 1). Plant root exudates such as flavonoids, terpenoids, or strigolactones are known to shape microbial community structure (Massalha et al. 2017; Mondal et al. 2023) (Fig. 2). Root exudates are root derived signals that seem to attract beneficial microbes (Lareen et al. 2016). Most of the research in terms of signals perception came from field of phytopathology where plant receptor like kinase

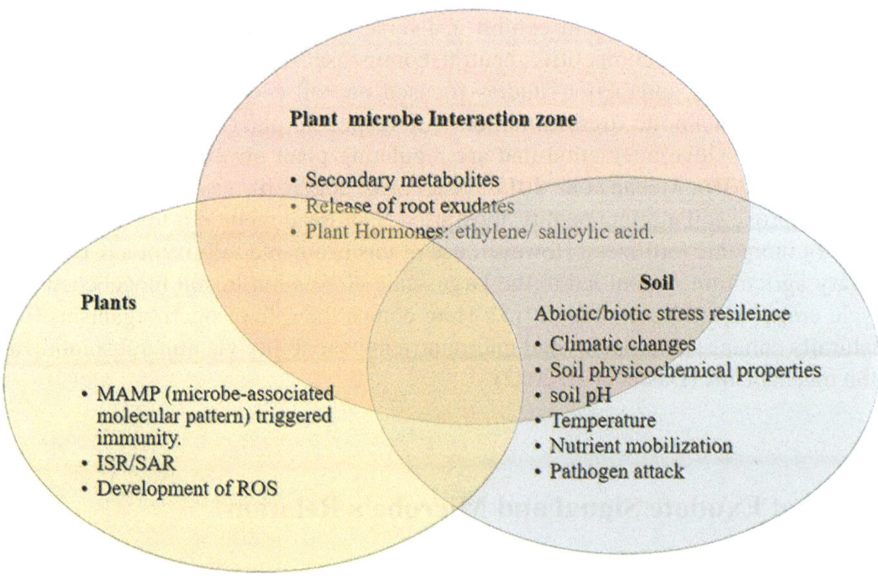

Fig. 1 The processes that contribute to the formation of the plant rhizosphere microbiota. The forces that promote the establishment of the rhizosphere microbiome under the influence of soil, plants, and plant-microbe interactions

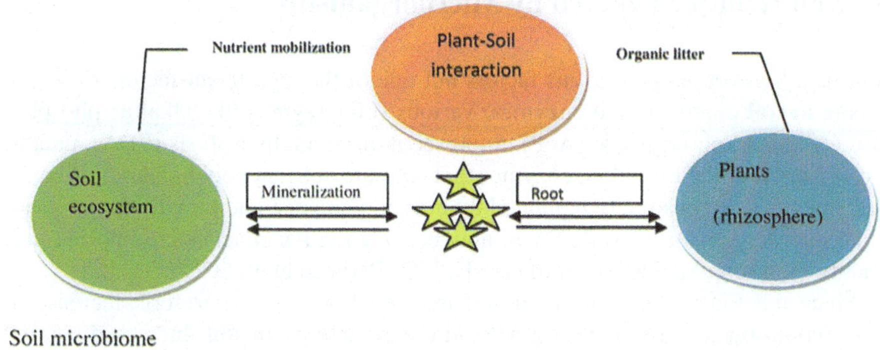

Fig. 2 Plant-soil interaction that mediates nutrient mobilization in soil ecosystem

acts as signal (Antolín-Llovera et al. 2012; Venturi and Fuqua 2013). Mutually beneficial relationship model like in mycorrhizal interactions and nodulation system was used to identify recognition mechanisms associated with plant and microbial interaction (Lagunas et al. 2015; Dolatabadian 2020). Microbial interaction studies revealed that root exudates impart significant effect in the inauguration of root microbiome (Delaux et al. 2015; Massalha et al. 2017; Manzar et al. 2022).

Soil is the utmost appropriate, dynamic, and composite habitation for plant, animal, and microbial growth. Because microbes are metabolically diverse and soil contains a wide range of microhabitats, microbial range in soil is substantially higher than in other environments (Jacoby et al. 2017). Soil quality indicators like microbial population, soil enzymes, soil pH, and nutrients all play an important role in maintaining soil quality and health (Chaudhary et al. 2021c, d). Soil physicochemical parameters vary seasonally and influence the structure of bacterial community. Enzyme activity can be used as a biomarker to recognize fluctuations in soil quality, as well as a measure of microbial diversity (Chaudhary et al. 2021e, f).

3 Physical and Chemical Indicators for Soil Health

It is essential to have healthy soil which depends on the physical, chemical, and biological indicators in order to achieve long-term sustainable agriculture (Tahat et al. 2020). Bulk density, porosity, texture, and aggregate stability of the soil are physical markers that have a role in water retention and erosion. Different compounds are thought to offer nutrients to the soil and play a role in defining the soil health. Soil organic matter and ion-exchange capacity aid in the supply of nutrients. The pH of the soil has an impact on mineral solubility, microbial activity, and nutritional status, all of which affect soil health (Msimbira and Smith 2020). Chemical indicators are useful for preserving organic materials and nutrient cycles.

4 Soil Health Mediated by Microorganisms

Soil health governed by various factors but one of the significant factors that contribute to soil quality is soil enzymes. Various soil enzymes like alkaline phosphatase (AP), acid phosphatase (AcP), fluorescein diacetate hydrolysis (FDA), urease, etc. are some of the important enzymes that regulate nutrient pool in the soil (Fig. 3). Soil enzymes are positively regulated by microorganisms, and inoculation of beneficial microorganisms in soil system now became regular practices to improve soil quality (Kukreti et al. 2020; Dasila et al. 2022; Parveen et al. 2022) (Table 1).

Since the FDA represents total soil microbial activity. Therefore, increase in FDA would further stimulate soil AP and urease activity in soil. Increased soil AP and AcP activity represents the potential of microorganisms to convert organic bound phosphate form to available inorganic form of phosphorous which is later available for plants for its growth and development. Soil urease activity represents the available nitrogen pool mediated by soil microorganisms. Recent reports of improving soil health via inoculation of *Pseudomonas* strains promote soil health and establish the positive correlation among soil enzyme and beneficial bioinoculants (Dasila et al. 2018, 2022). Application of *Bacillus* spp., *Pseudomonas*, and *Pantoea agglomerans* improved soil enzyme activities, soil pH, NPK, and microbial count (Chaudhary et al. 2021e, f).

Microbial species belonging to bacteria, fungi, algae, and actinomycetes that potentially maintain soil nutrient pool include various strains of *Agrobacterium* spp., *Azotobacter*, *Bacillus*, *Burkholderia*, *Bradyrhizobium*, *Enterobacter*, *Kushneria*, *Paenibacillus*, *Pseudomonas* spp., *Rhizobium*, *Serratia*, *Salmonella*, and *Thiobacillus* (Moreno-Espíndola et al. 2018). Although there are a number of genera that contribute to maintain soil nutrient pool, these genera are common in nature, but their impact in improving soil health is significant as compared to others (Table 2).

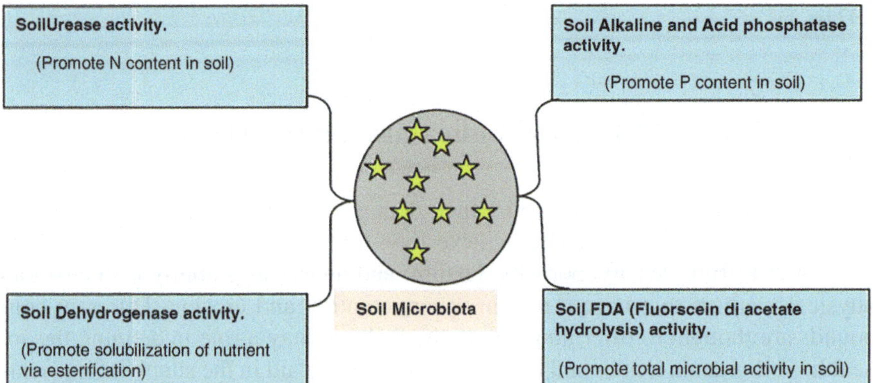

Fig. 3 Soil microorganism mediated promotion of different soil enzyme activity that is essential for maintaining nutrient pool in soil

Table 1 Soil enzymes as soil health indicators

Enzymes	Microbes involved	Functions	References
Fluorescein diacetate	*Bacillus* spp.	Involved in lipase, esterase, and protease hydrolysis in soil	Khati et al. (2017)
Dehydrogenase	*Bacillus* and *Pseudomonas* spp.	Improves soil health and helps in C-cycling and soil respiration	Khati et al. (2018)
Alkaline phosphatase	*Bacillus*, *Pseudomonas taiwanensis*	Involved in solubilization of phosphate and nutrient cycling	Kukreti et al. (2020); Kumari et al. (2020)
Cellulase	*Bacillus subtilis*, *Cellulomonas*	Helped in C-cycling and degradation of cellulose	Kumar et al. (2023)
Amylase	*Bacillus cereus*, *B. amyloliquefaciens*	Involved in starch degradation into simpler forms	Abo-Kamer et al. (2023)
β-Glucosidase	*Pseudomonas*, *Bacillus subtilis*	Involved in cellobiose hydrolysis and C-cycling	Sritongon et al. (2023)

Table 2 Effect of microbes on soil health parameters

Microbes	Soil parameters	References
Consortium of *Trichoderma reesei*, *Bacillus megaterium*, *Acinetobacter calcoaceticus*	Improved soil organic carbon content, microbial diversity, and humic and fulvic acid ratio	Jurys and Feizienė (2021)
Rhizobia, *Lactobacilli*	Variations in soil microbial diversity	Nuzzo et al. (2020)
Bacillus subtilis	Having positive impact on soil health	Mahapatra et al. (2022)
Azotobacter sp., *Chlorella pyrenoidosa*	Improved soil fertility	Zhou et al. (2023)
T. viridae, *Anabaena* sp.	Improved available nitrogen in soil	Triveni et al. (2015)
Bacillus subtilis, *Paecilomyces lilacinus*	Improved *Pseudomonas* and *Flavobacteria* in soil	Chen et al. (2023)
Trichoderma	Altered microbial population and increased *Archaeorhizomyces* sp. abundance	Zhang et al. (2018)
B. subtilis, *Pseudomonas aeruginosa*	Improved soil quality	Sharma et al. (2023)
B. paralicheniformis	Improved soil fertility	Zhao et al. (2023)
Enterobacter, *Kluyvera* sp.	Improved soil enzyme activities	Chauchan and Upadhyay (2023)
Pseudomonas fluorescens, *Bacillus subtilis*	Improved soil microbial biomass and urease activity	NG et al. (2022)

5 Screening of Potential Strains for Developing Effective Bioinoculants

Given that bacteria and plants have been interacting via evolution in nature, using them singly or in combination as consortium increases the probability of enhancing the productivity in many crop system (Alori and Babalola 2018; Chaudhary et al. 2023a, b). Plant growth promoting traits such as P solubilization, N fixation, siderophore production, biocontrol activity, and secretory functions (phytohormones, VOC, etc.) are a promising tool for testing the efficacy or competitive ability of the abovementioned traits on field evaluation (Chaudhary et al. 2021, 2022). Factors unlike root colonization and metabolite synthesis and release are critical step in the application of PGPR on host plant for phytostimulation and biofertilization. The coherent viability of applied bioinoculants is achieved when specific cell densities up to 10^8–10^9 CFU/mL of active cells for root soaking and microbial shots for seed are reached. The competence efficacy of the inoculated bacteria is dependent upon the host plant. Reinoculation of PGPR at regular interval is required for maintaining the viable cell for longer stretch of time (Sultana and Hossain 2022).

Through the use of carrier-based microbial inoculations, the shelf life of growing microorganisms is increased. It is important to examine the cost and accessibility of a carrier-based inoculant before using it. The carrier should have the following qualities: no lumps, good moisture absorption, affordability, processing ease, and buffering capacity. Compost, clay, peat, coal, organic materials, sawdust, wheat bran, and inert substances as silicates, bentonite, and vermiculite materials are a few examples of solid-based dry formulations. Synthesis of Aqueous formulations out of mineral oil, organic oils, oil-in-water suspensions, landfill leachates, humic acid and molasses (Bargaz et al. 2018).

When probiotic is employed as a carrier, the beneficial microorganisms receive additional nutrition. Microbial compositions based on probiotics make storage and application easy and efficient, boosting commercialization (Naik et al. 2020). Solid-based carriers such as talc, charcoal, and inert carrier material CMC can be used. Liquid based formulations utilizing oil and aqueous or liquid polymer base with effective microbes are one of the criteria to enhance plant growth promotion via foliar application or seed dipping. *Bacillus subtilis* and *Pseudomonas* spp. consortia significantly promoted seed germination and increment in plant's height of tomato plants (Kumar et al. 2020). Different beneficial effects of effective microbial inoculants in major cereal, leguminous, and horticultural crops are discussed in Table 3.

Table 3 Different beneficial effects of effective microbial inoculants in major cereal, leguminous, and horticultural crops

Bacterial strains	Crop	Growth condition	Effect on plants	References
Streptomyces albus, *Candida utilis*, and *Lactobacillus plantarum*	Black gram (*Vigna mungo*)	Effective microorganism with 2 kg pure sugarcane and 20 L water; NPK 980, 901, and 3014 mg kg^{-1}	Significant increment in total chlorophyll content, shoot length, and plant height	Raja Namasivayam and Bharani (2012)
Bacillus sp. AJK-3, *Stenotrophomonas* AJK-9, *Acetobacter pasteurianus* AJK-7	Wheat (*Triticum aestivum* L.)	Two seedlings per growth pouch were accommodated in incubation chamber at 20 °C. Seed germination on water agar plates was done and aseptically transferred on growth pouches having one-half strength Hogland solution	Nutrient content and enhanced plant growth	Majeed et al. (2015)
Lactic acid bacteria, actinomycetes, photosynthetic bacteria and yeast	Wheat (*Triticum aestivum* L.)	Effective microorganism application with parthenium weed (*Parthenium hysterophorus* L.) green manure for pot trial. Soil pH, 7.5; organic matter, 0.87%; NPK, 0.045%, 6.50 mg kg^{-1}, 100 mg kg^{-1}, and B 1.08 mg kg^{-1}	Increment in grain per spike, crop yield, shoot length, and spike length	Javaid and Shah (2010)
Candida tropicalis Y6, *Phanerochaete chrysosporium* VV18, *Streptomyces globisporous* C3	Calendula (*Calendula* L.) Marigold (*Tagetes* L.)	pH, 8.26–9.17; available NPK, 220–226 kg hm^{-2}, 27–38 kg hm^{-2}, 85–601 mg kg^{-1}, 0.31%, OC 0.35–0.36%	Flower number per plant and carotenoid pigment content increased exponentially and enhanced soil fertility	Sharma et al. (2017)

(continued)

Table 3 (continued)

Bacterial strains	Crop	Growth condition	Effect on plants	References
Trichoderma viride, *Nocardia coralline*, *Saccharomyces cerevisiae*, *Pseudomonas fluorescens*, *Pseudomonas putida*, *B. licheniformis*, *B. megaterium*, *B. polymyxa*	Blueberry (*Vaccinium corymbosum* L.)	Clay content, 22.7; sand content, 40.5; silt, 36.80; pH, 6.74; temperature, 13 °C; OM, 5.77%; total NPK 11.6 mg kg^{-1}, 406.3 mg kg^{-1}, 77.3 mg kg^{-1}	Promoted in up gradation of shoot and root weight	Schoebitz et al. (2016)
Azospirillum brasilense, *Bradyrhizobium japonicumstrain* SEMIA5079, *B. diazoefficiens* SEMIA 5080, *Bacillus subtilis* QST713	Soyabean	pH, 5.10; clay, 502 gKg^{-1} and 117 gKg^{-1}; silt, 281 gKg^{-1}; NPK 1.00, 57, 3.1 gKg^{-1}	NPK content; nodule number was increased; gram yield also improved up to 516 kg/hectare	Moretti et al. (2020)
Pseudomonas fluorescens + *Azospirillum* AZ204 consortium	Mexican cotton/*Gossypium hirsutum*	pH, 3.76–6.69; sand clay loam Clay, 30.51; silt, 23.17; coarse sand, 29.13	Lowest root rot incidence and highest seed cotton yield 2010, 1894 kg/hectare	Marimuthu et al. (2013)
Bacillus megaterium, *Arthrobacter chlorophenolicus*, *Enterobacter* sp.	Wheat (*Triticum aestivum* L.)	Pot soil: pH, 7.84; EC, 0.164 dsm^{-1}; OC, 0.50%; NPK 0.095 g/kg, 0.019 g/kg, 0.118 g/kg Field soil: pH, 7.8; EC, 161 dsm^{-1}; OC, 0.502; pH, 7.8; NPK, 202.45 kg/hec, 19.28 kg/hec, 250.16 kg/hec	Increased growth parameters, total protein content, higher soil physicochemical parameter and soil enzyme activities	Kumar et al. (2014)
Pseudomonas aeruginosa BHUJY 16, *P. aeruginosa* BHUJY 20, *P. putida* BHUJY 13, *P. putida* BHUVY 23, *P. fluorescence* BHUVY 29	Rice (*Oryzae sativa*)	–	Enhanced P solubilization and IAA production	Yadav et al. (2014)
Azospirillum brasilense sp7, *Pseudomonas putida* KT2440, *Acinetobacter* sp. EMM02, *Sphingomonas* spOF178A	Corn (*Zea mays*)	Max temp, 16 °C; min temp, 2 °C; annual precipitation, 900–1100 mm	Dry weight of aerial parts of plants increased up to 5.43 g; RDW, 5.33 g	Molina-Romero et al. (2021)

6 Host Plant That Affects Its Microbiome Structure

Undoubtedly plant species has a tremendous effect on shaping its microbiome. Bouffaud et al. (2014) have reported different commercial crop plants unlike wheat, sorghum, maize, and rice that have different microbial composition. 16S rRNA sequencing of root and rhizosphere of grape wine elucidated different indigenous microbial species residing in the root and rhizosphere (Samad et al. 2017). Phylogenetically distinct plants exhibit variation in the makeup of their microbial communities, particularly when endophytic populations are taken into account. The bacterial endophytes that were most common in potato plants were uncommon in eucalyptus plants and vice versa (Manter et al. 2010). The microbial structure is influenced by host parameters like species, health state, developmental stage, secretion (root exudates, metabolites), genotypes, and canopy. Moreover, environmental factors in some way influence how the microbiome is created. pH, soil type, water content, soil organic matter, and cultivar practices are also factors.

Plant species with different genotypes have different microbiome assemblage (Lundberg et al. 2012; Bulgarelli et al. 2015). A research study reported that in maize OUT, α-diversity richness and β-diversity are both interlinked with plant genotype (Peiffer et al. 2013). Commercial crops including sweet potatoes, oats, peas, and wheat have structural analyses of their microbiomes that indicate the assembly of the microorganisms is genotype dependent (Marques et al. 2014; Tsoi et al. 2019). In comparison with the one with high starch content, the one with low starch content had a different microbial assemblage. *Pseudomonas*, *Sphingobium*, *Acinetobacter*, *Chryseobacterium*, and other dominating species with low starch content have been identified. The age and development stage of the plant host are other factors that affect microbiome structure. Depending on the plant's developmental stage, such as seedling growth, root growth, shoot development, root architecture, morphology, and root exudates, the distinct microbial community changes as the plant ages (Bednarek 2012; Sugiyama et al. 2014). Different microbial assemblages were revealed by DGGE fingerprinting study of sweet potato cultivars at different plant ages. At the seedling and uprooting stages, the same was also observed in *Arabidopsis thaliana* (Chaparro et al. 2014; Xun et al. 2021).

7 Technical Intervention for Studying Host Microbiome Assemblage

7.1 Plating Method

A quick method for metabolite profiling and species identification that permits a more thorough investigation of colonies. Because of the method's poor resolution capability, colonies cannot be separated morphologically. You can only find culturable microorganisms (Gamalero et al. 2022).

7.2 Microscopy (TEM/SEM)

It's an appropriate and effective method for the identification and examination of surface structures of various microbes (locomotive organs). It also develops an understanding of different ecological niches and discovering bacteria's potential for pathogenesis. This method's primary drawback is that the sample size must be constrained in order to endure the strong vacuum pressure, allow for electron transmission, and fit inside the analyzer. Sample preparation is an expensive and tedious job, and the machinery requires a large setup and space for its installation with constant voltage requirement (Relucenti et al. 2021).

7.3 Molecular Fingerprinting Techniques (DGGE)

DGGE is a very powerful tool for the identification of microbial community directly from its ecological niches, which devoid the step for culturing of microbes. Its three types—PCR DGGE, phylogenetic DGGE, and functional gene DGGE—are the most frequently used forms. Fewer amount of sampling is required for identification and characterization of microbial assemblage in PCR-DGGE (Douterelo et al. 2014). Unique and specific clusters are recognized from various plant zone: roots and shoot, leaves, etc. It's a boon for simultaneous identification and assessment of microbiome assemblage according to ecological niches and ancient differences (Enespa and Chandra 2022).

7.4 Next-Generation Sequencing

Meta-genomics is currently regarded as one of the most significant scientific advancements in the study of microbial diversity. It provides extensive information on genetic diversity, species interactions, and the evolution of species, as well as species richness and interspecies richness. Plants and microorganisms can establish a tight loop to mutually benefit from coevolution. This caused the microorganisms that are linked with plants to adapt, performing new functions such as colonization, defensive mechanisms, and C and N metabolism. With the use of next-generation sequencing, we can learn more about the plant microbiome and gain a clearer understanding of this complex ecosystem (Singh et al. 2020).

8 Technological Intervention for Improving Soil Health

An endeavor to reduce harmful effects on soil health due to extensive farming practices to meet the food requirement of the growing population is done. A lot of effort is put into promoting soil health-based agricultural management techniques in order to decrease erosion, improve soil structure, increase nutrient use efficiency, and maintain or increase yields (Carey et al. 2022). Less commonly, pest and disease management is taken into account as part of a soil health management system. Agronomic, environmental, and economic objectives could all be met by cross-sectoral, integrated farming solutions produced by such a strategy (O'Callaghan et al. 2022). One of these approaches is nanotechnology, a newly developed revolutionary technology that has arguably demonstrated its successful application as nanofertilizers (Parveen et al. 2023; Srivastava et al. 2023; Chaudhary et al. 2023c). Due to its physical, chemical, and biological characteristics, which have dimensions less than 100 nm, it may be applied more effectively in a variety of fields, including agriculture and medicine (Kalia et al. 2020; Babu et al. 2022). Its primary technological interventions target the food processing sector, postharvest monitoring, and the development of nano biosensors to create novel crop protection and production techniques (Dhaigude et al. 2023; Rani et al. 2023).

Nanofertilizers are a boon for increasing agricultural output as their nutrient use efficiency is on par with conventional chemical fertilizers, which exhibit bioaccumulation capacity and cause serious environmental problems including eutrophication and the poisoning of water bodies through runoff. Nanofertilizers are resilient under a variety of abiotic and biotic conditions and exhibit a threefold increase in nutrient usage efficiency (Chaudhary et al. 2023d, e, f). According to recent studies, its application greatly raises crop yield by increasing metabolic pathways linked to N metabolism, antioxidant system, photosynthetic machinery, nutrient intake, and translocation (Kumari et al. 2019).

9 Conclusion

Microbial inoculants have the potential to substitute synthetic agricultural pesticides by helping to increase crop yields. Despite the immense power of effective inoculants, previously conducted studies on the basis of field trial suggest that favorable outcomes are possible in greenhouse settings. Nevertheless, in most cases, these inoculants fall short in field's trial situations. The primary barrier to the widespread use of bioinoculant technology is the variability in field performance that may be brought on by the cumulative effect of dilution of inoculated microbes within the continuum of the soil-microbe-plant interaction with the soil components and existing environmental conditions. As a result, developing prospective PGPR acceptable strains, determining the most effective ways of inoculant application, and understanding the interactions between inoculated and native microbes are all prerequisites for a bioinoculant that functions effectively in the field. A comprehensive strategy should be devised to increase agricultural production in a way that is environmentally friendly and has least side effects.

References

Abo-Kamer AM, Abd-El-salam IS, Mostafa FA, Mustafa ARA, Al-Madboly LA (2023) A promising microbial α-amylase production, and purification from *Bacillus cereus* and its assessment as antibiofilm agent against *Pseudomonas aeruginosa* pathogen. Microb Cell Factories 22:141. https://doi.org/10.1186/s12934-023-02139-6

Agri U, Chaudhary P, Sharma A (2021) In vitro compatibility evaluation of agriusable nanochitosan on beneficial plant growth-promoting rhizobacteria and maize plant. Natl Acad Sci Lett 44:555–559. https://doi.org/10.1007/s40009-021-01047-w

Agri U, Chaudhary P, Sharma A, Kukreti B (2022) Physiological response of maize plants and its rhizospheric microbiome under the influence of potential bioinoculants and nanochitosan. Plant Soil 474:451–468. https://doi.org/10.1007/s11104-022-05351-2

Alori ET, Babalola OO (2018) Microbial inoculants for improving crop quality and human health in Africa. Front Microbiol 19:2213. https://doi.org/10.3389/fmicb.2018.02213

Antolín-Llovera M, Ried MK, Binder A, Parniske M (2012) Receptor kinase signaling pathways in plant-microbe interactions. Annu Rev Phytopathol 8:451–473. https://doi.org/10.1146/annurev-phyto-081211-173002

Babu S, Singh R, Yadav D, Rathore SS, Raj R, Avasthe R, Yadav SK, Das A, Yadav V, Yadav B, Shekhawat K (2022) Nanofertilizers for agricultural and environmental sustainability. Chemosphere 292:133451. https://doi.org/10.1016/j.chemosphere.2021.133451

Bargaz A, Lyamlouli K, Chtouki M, Zeroual Y, Dhiba D (2018) Soil microbial resources for improving fertilizers efficiency in an integrated plant nutrient management system. Front Microbiol 9:1606. https://doi.org/10.3389/fmicb.2018.01606

Bednarek P (2012) Chemical warfare or modulators of defence responses—the function of secondary metabolites in plant immunity. Curr Opin Plant Biol 15(4):407–414. https://doi.org/10.1016/j.pbi.2012.03.002

Bouffaud ML, Poirier MA, Muller D, Moënne-Loccoz Y (2014) Root microbiome relates to plant host evolution in maize and other Poaceae. Environ Microbiol 16(9):2804–2814. https://doi.org/10.1111/1462-2920.12442

Bulgarelli D, Garrido-Oter R, Münch PC, Weiman A, Dröge J, Pan Y, McHardy AC, Schulze-Lefert P (2015) Structure and function of the bacterial root microbiota in wild and domesticated barley. Cell Host Microbe 17(3):392–403. https://doi.org/10.1016/j.chom.2015.01.011

Carey CJ, Strohm H, Smith F, Biaggi M (2022) Foliar application of a microbial inoculant and alkali-extracted humic product boosts forage productivity and quality on a central coast California rangeland. Front Sustain Food Syst 6:106. https://doi.org/10.3389/fsufs.2022.847096

Chaparro JM, Badri DV, Vivanco JM (2014) Rhizosphere microbiome assemblage is affected by plant development. ISME J 8(4):790–803. https://doi.org/10.1038/ismej.2013.196

Chaudhary P, Sharma A (2019) Response of nanogypsum on the performance of plant growth promotory bacteria recovered from nanocompound infested agriculture field. Environ Ecol 37:363–372

Chaudhary A, Parveen H, Chaudhary P, Bhatt P, Khatoon H (2021) Rhizospheric microbes and their mechanism. In: Bhatt P, Gangola S, Udayanga D, Kumar G (eds) Microbial technology for sustainable environment. Springer, Singapore. https://doi.org/10.1007/978-981-16-3840-4_6

Chaudhary P, Parveen H, Gangola S, Kumar G, Bhatt P, Chaudhary A (2021a) Plant Growth Promoting Rhizobacteria and their application in sustainable crop production. In: Bhatt P, Gangola S, Udayanga D, Kumar G (eds) Microbial technology for sustainable environment. Springer, Singapore. https://doi.org/10.1007/978-981-16-3840-4_13

Chaudhary P, Khati P, Chaudhary A, Gangola S, Kumar R, Sharma A (2021b) Bioinoculation using indigenous *Bacillus* spp. improves growth and yield of *Zea mays* under the influence of nanozeolite. 3 Biotech 11:11. https://doi.org/10.1007/s13205-020-02561-2

Chaudhary P, Khati P, Chaudhary A, Maithani D, Kumar G, Sharma A (2021c) Cultivable and metagenomic approach to study the combined impact of nanogypsum and *Pseudomonas taiwanensis* on maize plant health and its rhizospheric microbiome. PLoS One 16:e0250574. https://doi.org/10.1371/journal.pone.0250574

Chaudhary P, Khati P, Gangola S, Kumar A, Kumar R, Sharma A (2021d) Impact of nanochitosan and *Bacillus* spp. on health, productivity and defence response in *Zea mays* under field condition. 3 Biotech 11:237. https://doi.org/10.1007/s13205-021-02790-z

Chaudhary P, Chaudhary A, Parveen H, Rani A, Kumar G, Kumar A, Sharma A (2021e) Impact of nanophos in agriculture to improve functional bacterial community and crop productivity. BMC Plant Biol 21:519. https://doi.org/10.1186/s12870-021-03298-7

Chaudhary P, Sharma A, Chaudhary A, Khati P, Gangola S, Maithani D (2021f) Illumina based high throughput analysis of microbial diversity of maize rhizosphere treated with nanocompounds and *Bacillus* sp. Appl Soil Ecol 159:103836. https://doi.org/10.1016/j.apsoil.2020.103836

Chaudhary A, Chaudhary P, Upadhyay A, Kumar A, Singh A (2022) Effect of gypsum on plant growth promoting rhizobacteria. Environ Ecol 39(4A):1248–1256

Chaudhary P, Singh S, Chaudhary A, Agri U, Bhandari G (2022a) Chapter 3: Agrochemicals and their effects on soil microbial population. In: Soni R, Suyal D, Goel R (eds) Plant protection: from chemicals to biologicals. De Gruyter, Berlin, Boston, pp 45–58. https://doi.org/10.1515/9783110771558-003

Chaudhary P, Agri U, Chaudhary A, Kumar A, Kumar G (2022b) Endophytes and their potential in biotic stress management and crop production. Front Microbiol 13:933017. https://doi.org/10.3389/fmicb.2022.933017

Chaudhary P, Singh S, Chaudhary A, Sharma A, Kumar G (2022c) Overview of biofertilizers in crop production and stress management for sustainable agriculture. Front Plant Sci 13:930340. https://doi.org/10.3389/fpls.2022.930340

Chaudhary P, Chaudhary A, Agri U, Khatoon H, Singh A (2022d) Recent trends and advancements for agro-environmental sustainability at higher altitudes. In: Goel R, Soni R, Suyal DC, Khan M (eds) Survival strategies in cold-adapted microorganisms. Springer, Singapore. https://doi.org/10.1007/978-981-16-2625-8_19

Chaudhary P, Xu M, Ahamad L, Chaudhary A, Kumar G, Adeleke BS, Verma KK, Hu D-M, Širić I, Kumar P, Popescu SM, Abou Fayssal S (2023a) Application of synthetic consortia for improvement of soil fertility, pollution remediation, and agricultural productivity: a review. Agronomy 13:643. https://doi.org/10.3390/agronomy13030643

Chaudhary P, Chen S, Rajput VD, Jaiswal DK (2023b) Editorial: bioinoculants with nanocompounds to improve soil health: a step toward sustainable agriculture. Front Environ Sci 11:1270002. https://doi.org/10.3389/fenvs.2023.1270002

Chaudhary P, Chaudhary A, Nadda AK, Khati P (eds) (2023c) Advances in nanotechnology for smart agriculture: techniques and applications, 1st edn. CRC Press. https://doi.org/10.1201/9781003345565

Chaudhary P, Chaudhary A, Dhaigude V, Fayssal SA, Kumar P, Sharma A, Chen S (2023d) Environmental-friendly nanoparticles in agriculture. In: Chaudhary P, Chaudhary A, Nadda AK, Khati P (eds) Advances in nanotechnology for smart agriculture: techniques and applications, 1st edn. CRC Press. https://doi.org/10.1201/9781003345565-13

Chaudhary P, Chaudhary A, Khati P, Kumar G, Tyagi J, Behera M (2023e) Nanotechnology and omics approach in agrobiotechnology. In: Fernandez-Luqueno F, Patra JK (eds) Agricultural and environmental nanotechnology. Interdisciplinary biotechnological advances. Springer, Singapore. https://doi.org/10.1007/978-981-19-5454-2_12

Chaudhary P, Chen S, Rajput VD, Jaiswal DK (eds) (2023f) Bioinoculants with nano-compounds to improve soil health: a step toward sustainable agriculture. Frontiers Media SA, Lausanne. https://doi.org/10.3389/978-2-8325-3287-4

Chen P, Zhang J, Li M, Fang F, Hu J, Sun Z, Zhang A, Gao X, Li J (2023) Synergistic effect of *Bacillus subtilis* and *Paecilomyces lilacinus* in alleviating soil degradation and improving watermelon yield. Front Microbiol 13:1101975. https://doi.org/10.3389/fmicb.2022.1101975

Dasila H, Anjul R, Damini AR, Manvika S, Salil T (2018) Interaction between *Dalbergia sissoo* Roxb. and *Pseudomonas koreensis* AS15 strain is cultivar specific. Int J Curr Microbiol App Sci 7(10):297–306. https://doi.org/10.20546/ijcmas.2018.710.031

Dasila H, Sah VK, Jaggi V, Sahgal M (2022) Impact of phosphate solubilizing bacteria on soil enzyme activity and plant vigor in four wheat genotypes. Indian. J Ecol 49(4):1341–1350. https://doi.org/10.55362/IJE/2022/3669

Davis AG, Huggins DR, Reganold JP (2023) Linking soil health and ecological resilience to achieve agricultural sustainability. Front Ecol Environ 21:131. https://doi.org/10.1002/fee.2594
Delaux PM, Radhakrishnan G, Oldroyd G (2015) Tracing the evolutionary path to nitrogen-fixing crops. Curr Opin Plant Biol 26:95–99. https://doi.org/10.1016/j.pbi.2015.06.003
Dhaigude V, Naliyapara HB, Chaudhary P, Fayssal SA, Nadda AK, Minkina T, Rajput VD (2023) Role of nanobiosensors in agriculture advancement. In: Chaudhary P, Chaudhary A, Nadda AK, Khati P (eds) Advances in nanotechnology for smart agriculture: techniques and applications, 1st edn. CRC Press. https://doi.org/10.1201/9781003345565-9
Dolatabadian A (2020) Plant-microbe interaction. Biology (Basel) 10(1):15. https://doi.org/10.3390/biology10010015
Douterelo I, Boxall JB, Deines P, Sekar R, Fish KE, Biggs CA (2014) Methodological approaches for studying the microbial ecology of drinking water distribution systems. Water Res 65:134–156. https://doi.org/10.1016/j.watres.2014.07.008
Enespa, Chandra P (2022) Tool and techniques study to plant microbiome current understanding and future needs: an overview. Commun Integr Biol 15(1):209–225. https://doi.org/10.1080/19420889.2022.2082736
Gamalero E, Bona E, Glick BR (2022) Current techniques to study beneficial plant-microbe interactions. Microorganisms 10(7):1380. https://doi.org/10.3390/microorganisms10071380. PMID: 35889099; PMCID: PMC9317800
Jacoby R, Peukert M, Succurro A, Koprivova A, Kopriva S (2017) The role of soil microorganisms in plant mineral nutrition—current knowledge and future directions. Front Plant Sci 8:1617. https://doi.org/10.3389/fpls.2017.01617
Javaid A, Shah MB (2010) Growth and yield response of wheat to EM (effective microorganisms) and parthenium green manure. Afr J Biotechnol 9(23):3373–3381
Javed A, Ali E, Afzal KB, Osman A, Riaz S (2022) Soil fertility: factors affecting soil fertility, and biodiversity responsible for soil fertility. Int J Plant Anim Environ Sci 12(1):21–33. https://doi.org/10.26502/ijpaes.202129
Jurys A, Feizienė D (2021) The effect of specific soil microorganisms on soil quality parameters and organic matter content for cereal production. Plants 10(10):2000. https://doi.org/10.3390/plants10102000
Kalia A, Sharma SP, Kaur S, Kaur H (2020) Bacterial inoculants: how can these microbes sustain soil health and crop productivity? In: Soil health, pp 337–372. https://doi.org/10.1007/978-3-030-44364-1_18
Khati P, Chaudhary P, Gangola S, Bhatt P, Sharma A (2017) Nanochitosan supports growth of *Zea mays* and also maintains soil health following growth. 3 Biotech 7:81. https://doi.org/10.1007/s13205-017-0668-y
Khati P, Parul, Bhatt P, Nisha Kumar R, Sharma A (2018) Effect of nanozeolite and plant growth promoting rhizobacteria on maize. 3 Biotech 8:141. https://doi.org/10.1007/s13205-018-1142-1
Khati P, Chaudhary P, Gangola S, Sharma P (2019) Influence of nanozeolite on plant growth promotory bacterial isolates recovered from nanocompound infested agriculture field. Environ Ecol 37:521–527
Krasilnikov P, Taboada MA (2022) Fertilizer use, soil health and agricultural sustainability. Agriculture 12(4):462. https://doi.org/10.3390/agriculture12040462
Kukreti B, Sharma A, Chaudhary P, Agri U, Maithani D (2020) Influence of nanosilicon dioxide along with bioinoculants on *Zea mays* and its rhizospheric soil. 3 Biotech 10:345. https://doi.org/10.1007/s13205-020-02329-8
Kumar A, Maurya BR, Raghuwanshi R (2014) Isolation and characterization of PGPR and their effect on growth, yield and nutrient content in wheat (*Triticum aestivum* L.). Biocatalysis and agricultural. Biotechnology 3(4):121–128. https://doi.org/10.1016/j.bcab.2014.08.003
Kumar P, Aeron A, Shaw N, Singh A, Bajpai VK, Pant S, Dubey RC (2020) Seed bio-priming with tri-species consortia of phosphate solubilizing rhizobacteria (PSR) and its effect on plant growth promotion. Heliyon 6(12):e05701. https://doi.org/10.1016/j.heliyon.2020.e05701

Kumar HKN, Mohana NC, Rakshith D, Abhilash MR, Satish S (2023) Multicomponent assessment and optimization of the cellulase activity by Serratia marcescens inhabiting decomposed leaf litter soil. Sustain Chem Pharm 31:100951. https://doi.org/10.1016/j.scp.2022.100951

Kumari B, Mallick MA, Solanki MK, Hora A, Mani M (2019) Applying nanotechnology to bacteria: an emerging technology for sustainable agriculture. In: Role of plant growth promoting microorganisms in sustainable agriculture and nanotechnology. Woodhead Publishing, pp 121–143. https://doi.org/10.1016/B978-0-12-817004-5.00008-7

Kumari S, Sharma A, Chaudhary P, Khati P (2020) Management of plant vigor and soil health using two agriusable nanocompounds and plant growth promotory rhizobacteria in Fenugreek. 3 Biotech 10:1–11. https://doi.org/10.1007/s13205-020-02448-2

Kumari H, Khati P, Gangola S, Chaudhary P, Sharma A (2021) Performance of plant growth promotory rhizobacteria on Maize and soil characteristics under the influence of TiO_2 nanoparticles. Pantnagar J Res 19:28–39

Lagunas B, Schäfer P, Gifford M (2015) Housing helpful invaders: the evolutionary and molecular architecture underlying plant root-mutualist microbe interactions. J Exp Bot 66(8):2177–2186. https://doi.org/10.1093/jxb/erv038

Lareen A, Burton F, Schäfer P (2016) Plant root-microbe communication in shaping root microbiomes. Plant Mol Biol 90:575–587. https://doi.org/10.1007/s11103-015-0417-8

Li J, Wang J, Liu H, Macdonald CA, Singh BK (2022) Application of microbial inoculants significantly enhances crop productivity: a meta-analysis of studies from 2010 to 2020. J Sustain Agric Environ 1(3):216–225. https://doi.org/10.1002/sae2.12028

Lundberg DS, Lebeis SL, Paredes SH, Yourstone S, Gehring J, Malfatti S, Tremblay J, Engelbrektson A, Kunin V, Rio TG, Edgar RC (2012) Defining the core Arabidopsis thaliana root microbiome. Nature 488(7409):86–90. https://doi.org/10.1038/nature11237

Mahapatra S, Yadav R, Ramakrishna W (2022) *Bacillus subtilis* impact on plant growth, soil health and environment: Dr. Jekyll and Mr. Hyde. J Appl Microbiol 132:3543. https://doi.org/10.1111/jam.15480

Majeed A, Abbasi MK, Hameed S, Imran A, Rahim N (2015) Isolation and characterization of plant growth-promoting rhizobacteria from wheat rhizosphere and their effect on plant growth promotion. Front Microbiol 17(6):198. https://doi.org/10.3389/fmicb.2015.00198

Manter DK, Delgado JA, Holm DG, Stong RA (2010) Pyrosequencing reveals a highly diverse and cultivar-specific bacterial endophyte community in potato roots. Microb Ecol 60:157–166. https://doi.org/10.1007/s00248-010-9658-x

Manzar N, Jayaswal D, Kishan G, Chauhan S, Joshi D, Singh J, Chaudhary P, Jamali H, Kushwaha P, Parveen H, Kashyap AS, Sahu PK, Singh UB (2022) Application of molecular ecology approaches in sustainable agriculture for a better understanding of plant–microbiome interactions. In: Singh UB, Sahu PK, Singh HV, Sharma PK, Sharma SK (eds) Rhizosphere microbes. Microorganisms for sustainability, vol 40. Springer, Singapore. https://doi.org/10.1007/978-981-19-5872-4_3

Marimuthu S, Ramamoorthy V, Samiyappan R, Subbian P (2013) Intercropping system with combined application of *Azospirillum* and *Pseudomonas fluorescens* reduces root rot incidence caused by *Rhizoctonia bataticola* and increases seed cotton yield. J Phytopathol 161(6):405–411. https://doi.org/10.1111/jph.12084

Marques JM, da Silva TF, Vollu RE, Blank AF, Ding GC, Seldin L, Smalla K (2014) Plant age and genotype affect the bacterial community composition in the tuber rhizosphere of field-grown sweet potato plants. FEMS Microbiol Ecol 88(2):424–435. https://doi.org/10.1111/1574-6941.12313

Massalha H, Korenblum E, Tholl D, Aharoni A (2017) Small molecules below-ground: the role of specialized metabolites in the rhizosphere. Plant J 90(4):788–807. https://doi.org/10.1111/tpj.13543

Meena KK, Sorty AM, Bitla UM, Choudhary K, Gupta P, Pareek A, Singh DP, Prabha R, Sahu PK, Gupta VK, Singh HB (2017) Abiotic stress responses and microbe-mediated mitigation in plants: the omics strategies. Front Plant Sci 9(8):172. https://doi.org/10.3389/fpls.2017.00172

Molina-Romero D, Juárez-Sánchez S, Venegas B, Ortíz-González CS, Baez A, Morales-García YE, Muñoz-Rojas J (2021) A bacterial consortium interacts with different varieties of maize, promotes the plant growth, and reduces the application of chemical fertilizer under field conditions. Front Sustain Food Syst 4:616757. https://doi.org/10.3389/fsufs.2020.616757

Mondal S, Pramanik K, Pal P, Mitra S, Ghosh SK, Mondal T, Soren T, Maiti TK (2023) Multifaceted roles of root exudates in light of plant-microbe interaction. In: Unravelling plant-microbe synergy 1. Academic, pp 49–76. https://doi.org/10.1016/B978-0-323-99896-3.00003-5

Moreno-Espíndola IP, Ferrara-Guerrero MJ, Luna-Guido ML, Ramírez-Villanueva DA, De León-Lorenzana AS, Gómez-Acata S, González-Terreros E, Ramírez-Barajas B, Navarro-Noya YE, Sánchez-Rodríguez LM, Fuentes-Ponce M, Macedas-Jímenez JU, Dendooven L (2018) The bacterial community structure and microbial activity in a traditional organic Milpa farming system under different soil moisture conditions. Front Microbiol 9:2737. https://doi.org/10.3389/fmicb.2018.02737

Moretti LG, Crusciol CA, Bossolani JW, Momesso L, Garcia A, Kuramae EE, Hungria M (2020) Bacterial consortium and microbial metabolites increase grain quality and soybean yield. J Soil Sci Plant Nutr 20(4):1923–1934. https://doi.org/10.1007/s42729-020-00263-5

Msimbira LA, Smith DL (2020) The roles of plant growth promoting microbes in enhancing plant tolerance to acidity and alkalinity stresses. Front Sustain Food Syst 4:106. https://doi.org/10.3389/fsufs.2020.00106

Naik K, Mishra S, Srichandan H, Singh PK, Choudhary A (2020) Microbial formulation and growth of cereals, pulses, oilseeds and vegetable crops. Sustain Environ Res 30:1–8. https://doi.org/10.1186/s42834-020-00051-x

Ng CWW, Yan WH, Tsim KWK, So PS, Xia YT, To TC (2022) Effects of *Bacillus subtilis* and *Pseudomonas fluorescens* as the soil amendment. Heliyon 8:e11674. https://doi.org/10.1016/j.heliyon.2022.e11674

Nuzzo A, Satpute A, Albrecht U, Strauss SL (2020) Impact of soil microbial amendments on tomato rhizosphere microbiome and plant growth in field soil. Microb Ecol 80:398–409. https://doi.org/10.1007/s00248-020-01497-7

O'Callaghan M, Ballard RA, Wright D (2022) Soil microbial inoculants for sustainable agriculture: limitations and opportunities. Soil Use Manag 38(3):1340–1369. https://doi.org/10.1111/sum.12811

Parveen H, Chaudhary A, Chaudhary P, Sultana R, Khati P, Kumar G, Khati P, Rana M, Bhatt P (2022) Omics approaches for the production of the microbial enzymes and applications. In: Bhatt P (ed) Industrial applications of microbial enzymes, 1st edn. CRC Press. https://doi.org/10.1201/9781003202998-16

Parveen H, Chaudhary P, Srivastava P, Priya S, Bhandari G, Chaudhary A (2023) Role of nanoparticles in agriculture. In: Chaudhary P, Chaudhary A, Nadda AK, Khati P (eds) Advances in nanotechnology for smart agriculture: techniques and applications, 1st edn. CRC Press. https://doi.org/10.1201/9781003345565-1

Peiffer JA, Spor A, Koren O, Jin Z, Tringe SG, Dangl JL, Buckler ES, Ley RE (2013) Diversity and heritability of the maize rhizosphere microbiome under field conditions. Proc Natl Acad Sci 110(16):6548–6553. https://doi.org/10.1073/pnas.1302837110

Raja Namasivayam SK, Bharani RS (2012) Effect of compost derived from decomposed fruit wastes by effective microorganism (EM) technology on plant growth parameters of Vigna mungo. J Bioremed Biodegr 3(11):167. https://doi.org/10.4172/2155-6199.1000167

Rani A, Kumar T, Trivedi B, Chaudhary P (2023) Nanotechnology in climate smart agriculture. In: Chaudhary P, Chaudhary A, Nadda AK, Khati P (eds) Advances in nanotechnology for smart agriculture: techniques and applications, 1st edn. CRC Press. https://doi.org/10.1201/9781003345565-14

Rashid MI, Mujawar LH, Shahzad T, Almeelbi T, Ismail IM, Oves M (2016) Bacteria and fungi can contribute to nutrients bioavailability and aggregate formation in degraded soils. Microbiol Res 183:26–41. https://doi.org/10.1016/j.micres.2015.11.007

Relucenti M, Familiari G, Donfrancesco O, Taurino M, Li X, Chen R, Artini M, Papa R, Selan L (2021) Microscopy methods for biofilm imaging: focus on SEM and VP-SEM pros and cons. Biology (Basel) 10(1):51. https://doi.org/10.3390/biology10010051

Samad A, Trognitz F, Compant S, Antonielli L, Sessitsch A (2017) Shared and host-specific microbiome diversity and functioning of grapevine and accompanying weed plants. Environ Microbiol 19(4):1407–1424. https://doi.org/10.1111/1462-2920.13618

Schoebitz M, López MD, Serrí H, Martínez O, Zagal E (2016) Combined application of microbial consortium and humic substances to improve the growth performance of blueberry seedlings. J Soil Sci Plant Nutr 16(4):1010–1023. https://doi.org/10.4067/S0718-95162016005000074

Sharma A, Saha TN, Arora A, Shah R, Nain L (2017) Efficient microorganism compost benefits plant growth and improves soil health in calendula and marigold. Hortic Plant J 3(2):67–72. https://doi.org/10.1016/j.hpj.2017.07.003

Sharma M, Sood G, Chauhan A (2023) Novel synergism of Cedecea lapagei KU14 and *Bacillus subtilis* KU21 for sustainable productivity of *Rosmarinus officinalis* in Northwest Himalayas. Rhizosphere 25:100683. https://doi.org/10.1016/j.rhisph.2023.100683

Singh A, Kumar M, Verma S, Choudhary P, Chakdar H (2020) Plant microbiome: trends and prospects for sustainable agriculture. In: Varma A, Tripathi S, Prasad R (eds) Plant Microbe Symbiosis. Springer, Cham https://doi.org/10.1007/978-3-030-36248-5_8

Sritongon N, Boonlue S, Mongkolthanaruk W, Jogloy S, Riddech N (2023) The combination of multiple plant growth promotion and hydrolytic enzyme producing rhizobacteria and their effect on Jerusalem artichoke growth improvement. Sci Rep 13:5917. https://doi.org/10.1038/s41598-023-33099-x

Srivastava P, Balhara M, Giri B (2020) Soil health in India: past history and future perspective. In: Soil health, pp 1–9. https://doi.org/10.1007/978-3-030-44364-1_1

Srivastava P, Sahgal M, Sharma K, Enshasy HAE, Gafur A, Alfarraj S, Ansari MJ, Sayyed RZ (2022) Optimization and identification of siderophores produced by *Pseudomonas monteilii* strain MN759447 and its antagonism toward fungi associated with mortality in *Dalbergia sissoo* plantation forests. Front Plant Sci 13:984522. https://doi.org/10.3389/fpls.2022.984522

Srivastava P, Chaudhary P, Parveen P (2023) Nanoparticles and their effect on plant health and productivity. In: Chaudhary P, Chaudhary A, Nadda AK, Khati P (eds) Advances in nanotechnology for smart agriculture: techniques and applications, 1st edn. CRC Press. https://doi.org/10.1201/9781003345565-6

Steffen W, Richardson K, Rockström J, Cornell SE, Fetzer I, Bennett EM, Biggs R, Carpenter SR, De Vries W, De Wit CA, Folke C (2015) Planetary boundaries: guiding human development on a changing planet. Science 347(6223):1259855. https://doi.org/10.1126/science.1259855

Sugiyama A, Ueda Y, Zushi T, Takase H, Yazaki K (2014) Changes in the bacterial community of soybean rhizospheres during growth in the field. PLoS One 9(6):e100709. https://doi.org/10.1371/journal.pone.0100709

Sultana F, Hossain MM (2022) Assessing the potentials of bacterial antagonists for plant growth promotion, nutrient acquisition, and biological control of Southern blight disease in tomato. PLoS One 17(6):e0267253. https://doi.org/10.1371/journal.pone.0267253

Tahat MM, Alananbeh MK, Othman YA, Leskovar ID (2020) Soil health and sustainable agriculture. Sustainability 12:4859. https://doi.org/10.3390/su12124859

Triveni S, Prasanna R, Kumar A, Bidyarani N, Singh R, Saxena AK (2015) Evaluating the promise of *Trichoderma* and *Anabaena* based biofilms as multifunctional agents in *Macrophomina phaseolina*-infected cotton crop. Biocontrol Sci Tech 25(6):656–670. https://doi.org/10.1080/09583157.2015.1006171

Tsoi R, Dai Z, You L (2019) Emerging strategies for engineering microbial communities. Biotechnol Adv 37(6):107372. https://doi.org/10.1016/j.biotechadv.2019.03.011

Tyagi J, Chaudhary P, Jyotsana, Bhagwati U, Bhandari G, Chaudhary A (2022a) Chapter 17: Impact of endophytic fungi in biotic stress management. In: Soni R, Suyal D, Goel R (eds) Plant protection: from chemicals to biologicals. De Gruyter, Berlin, Boston, pp 447–462. https://doi.org/10.1515/9783110771558-017

Tyagi J, Chaudhary P, Mishra A, Khatwani M, Dey S, Varma A (2022b) Role of endophytes in abiotic stress tolerance: with special emphasis on *Serendipita indica*. Int J Environ Res 16:62. https://doi.org/10.1007/s41742-022-00439-0

Venturi V, Fuqua C (2013) Chemical signaling between plants and plant-pathogenic bacteria. Annu Rev Phytopathol 51:17–37. https://doi.org/10.1146/annurev-phyto-082712-102239

Wang Z, Chen Z, Kowalchuk GA, Xu Z, Fu X, Kuramae EE (2021) Succession of the resident soil microbial community in response to periodic inoculations. Appl Environ Microbiol 87(9):e00046–e00021. https://doi.org/10.1128/AEM.00046-21

Xun W, Shao J, Shen Q, Zhang R (2021) Rhizosphere microbiome: functional compensatory assembly for plant fitness. Comput Struct Biotechnol J 19:5487–5493. https://doi.org/10.1016/j.csbj.2021.09.035

Yadav J, Verma JP, Jaiswal DK, Kumar A (2014) Evaluation of PGPR and different concentration of phosphorus level on plant growth, yield and nutrient content of rice (*Oryza sativa*). Ecol Eng 62:123–128. https://doi.org/10.1016/j.ecoleng.2013.10.013

Yaish MW, Al-Lawati A, Jana GA, Vishwas Patankar H, Glick BR (2016) Impact of soil salinity on the structure of the bacterial endophytic community identified from the roots of caliph medic (*Medicago truncatula*). PLoS One 11(7):e0159007. https://doi.org/10.1371/journal.pone.0159007

Zhang F, Huo Y, Cobb AB, Luo G, Zhou J, Yang G, Zhang Y (2018) *Trichoderma* biofertilizer links to altered soil chemistry, altered microbial communities, and improved grassland biomass. Front Microbiol 9:848. https://doi.org/10.3389/fmicb.2018.00848

Zhao X, Zhu D, Tan J, Wang R, Qi G (2023) Cooperative action of fulvic acid and *bacillus paralicheniformis* ferment in regulating soil microbiota and improving soil fertility and plant resistance to bacterial wilt disease. Microbiol Spectr 11(2):e04079–e04022. https://doi.org/10.1128/spectrum.04079-22

Zhou L, Liu W, Duan H, Dong H, Li J, Zhang S, Zhang J, Ding S, Xu T, Guo B (2023) Improved effects of combined application of nitrogen-fixing bacteria *Azotobacter beijerinckii* and microalgae *Chlorella pyrenoidosa* on wheat growth and saline-alkali soil quality. Chemosphere 313:137409. https://doi.org/10.1016/j.chemosphere.2022.137409

Zuluaga MYA, Lima Milani KM, Azeredo Gonçalves LS, Martinez de Oliveira AL (2020) Diversity and plant growth-promoting functions of diazotrophic/N-scavenging bacteria isolated from the soils and rhizospheres of two species of Solanum. PLoS One 15(1):e0227422. https://doi.org/10.1371/journal.pone.0227422

Microbial Inoculants and Soil Microbial Population

Hina Khatoon, Parul Chaudhary, and Anuj Chaudhary

Abstract The unintended consequences of introducing nontarget organisms into a new habitat are a major cause of worry as they can have a negative impact on biodiversity and ecosystem function. Microbial inoculants are typically expected to have targeted effect on specific functions but have less impacts on resident microbial population, whereas invasive macroorganisms are typically the focus of these studies. As microbial inoculants become increasingly important for attaining sustainable agricultural productivity, it is crucial to note that microbial inoculants often interact with native microbial communities, potentially impacting ecosystem processes. This chapter will give a broad view of the major influence of some microbial inoculants upon resident soil microbial population.

Keywords Agricultural productivity · Microbial inoculants · Microbial communities · Soil biodiversity

1 Introduction

Agricultural sustainability is facing a significant challenge due to increasing demands of food productivity with limited land resources. So, it becomes essential to utilize the limited resources more proficiently in order to get sustainable food

H. Khatoon (✉)
Department of Environmental Science, Dev Bhoomi Uttarakhand University, Dehradun, Uttarakhand, India

P. Chaudhary
School of Agriculture, Graphic Era Hill University, Dehradun, Uttarakhand, India

A. Chaudhary
School of Agriculture and Environmental Sciences, Shobhit University, Gangoh, Uttar Pradesh, India

P. Chaudhary, A. Chaudhary (eds.), *Microbial Inoculants*, Interdisciplinary Biotechnological Advances, https://doi.org/10.1007/978-981-97-0633-4_3

production (Owen et al. 2015). To solve this problem, different types of pesticides are used to improve food production and reduce plant diseases. However, the environmental effects of these agricultural practices decrease the soil fertility and must take advantage of fertile soil to maintain the soil's ability to be a healthy environment (Trivedi et al. 2012; Chaudhary et al. 2022a).

Microbial inoculants have already been utilized for nearly a century but have become increasingly popular in the past three decades, with the emergence of a number of commercial-grade inoculant products (Babalola and Glick 2012; Chaudhary et al. 2021a, b). Inoculation of microbes into the soil can lead to increased soil biological diversity by improving soil structural properties, thus providing an alternate approach by replacing the use of harmful chemical substances in agricultural land (Carvajal-Muñoz and Carmona-Garcia 2012; Chaudhary et al. 2021c, d).

Microbial inoculants can have a direct effect on atmospheric nitrogen fixation and mineral solubility (Babalola 2010; Agri et al. 2021). The interactions between plants and inoculants are unique to plant and soil bacterial genotypes, which may vary with types of microbes (Weiland-Bräuer et al. 2015; Manzar et al. 2022). Bacterial inoculation technologies have crucial role in the effectiveness of this process. In order to be effective, bacteria must survive in the soil, which can be facilitated by the type of applied inoculants (Jayaraman et al. 2014). A number of studies stated that microbial inoculants are increasingly being used to promote growth of the plants by enriching soil quality (Chaudhary et al. 2022; Chaudhary et al. 2023a, b, c). This chapter provides several major applications and impacts of inoculants on soil microbial population.

2 Microbial Inoculants

Microbial inoculants have caught the attention of many farmers and scientists, due to their ability to improve nutrient availability, absorption, and soil and plant health to increase sustainable yield (Chaudhary et al. 2021; O'Callaghan et al. 2022). Exotic macroorganisms have been demonstrated by ecologists to be a major threat to habitats, not only in terms of the extinction of native species but also in terms of the functioning of ecosystems in different environment (Roy et al. 2019). While some of these microorganisms are released naturally into the environment, the majority of planned introductions take place in the environment/agriculture sector, where the introduced microorganisms serve as soil bioremediators, biological control agents, and biofertilizers (Ahmad 2017; Kumari et al. 2021; Chaudhary et al. 2022b).

The purpose of soil microbial introductions is to regulate or enhance ecosystem functions and services, such as promoting plant growth, reducing litter, cycling nutrients, restoring biodiversity, and maintaining the fertility of soil (Tamayo-Velez and Osorio 2018; Sutherland et al. 2019).

While it is possible that the microbial community in the soil may be able to regroup and build itself back up following the disruption caused due to the inoculation, this finding highlights the environmental and developmental implications of inoculation due to microbes on soil-resident microbial communities that remain largely unexplored. Soil microbial communities' enormous diversity, as well as the challenges in characterizing this diversity, may make it difficult to understand how microbial inoculation may impact soil microbial communities (Jurburg and Salles 2015). However, as microbial inoculation expands with the implementation of sustainable agriculture practices, research must better assess the influences of microbe's introductions on resident soil population (Verma et al. 2019). In the agricultural industry, microbial inoculation is a widely used technique to acquire desirable properties in soil.

The term microbial inoculant is used to refer to a group of microorganisms that improve nutrient availability, nutrient absorption, and plant growth through their biophysical and biochemical processes in soil. Various terms are used to describe this group of microorganisms, including bio-stimulants, bioinoculants, and biological fertilizers (Ansari et al. 2014; Chaudhary et al. 2022c, d). The details about some microorganisms that had been extensively used as inoculants are explained in the following subheading as well as in Table 1 and Fig. 1.

2.1 Rhizobia *Inoculants*

The mechanisms by which rhizobia affects plant growth are as follows: (1) Fix nitrogen. (2) Help in the promotion of nitrogen-fixing free-living bacteria. (3) Increase nutrient uptake. (4) Increase plant hormone production rate. (5) Increase growth of useful microbes. (6) Manage microbial disease transmission. (7) Contribute to insect pest control (Alemneh et al. 2020).

Researchers used a cultivation-dependent method and a PCR-independent method (SSCP) to show that when field trial was done for a strain of *Sinorhizobium meliloti* strain L33, it had an effect on bacterial diversity in alfalfa rhizosphere. They found that it decreased the amount of γ-proteobacteria, but it also increased the

Table 1 Major impacts of microbial inoculants on soil microbes

Microbial inoculants	Observations	References
Free-living bacteria inoculants	Reported negative impacts on the microbial community in the rhizosphere	Piromyou et al. (2013)
Rhizobia inoculants	Field inoculation revealed substantial alterations in bacterial composition and biomass in the primary soil of common beans	Trabelsi et al. (2012)
Azospirillum inoculants	Inoculation altered the physiological properties of the roots of rice plant	de Salamone et al. (2012)
Arbuscular mycorrhizal fungi	Bacterial communities in tomato rhizosphere were significantly altered by inoculation	Lioussanne et al. (2010)
Biocontrol agent as inoculants	Inoculants had a short-term impact on microbial flora of cucumber plant, indicating that the biocontrol activity of this agent is of limited applicability	Gao and Chen (2012)
Inoculation of *B. velezensis* and *Brevundimonas diminuta*	Microbial inoculants had positive impact on microbial communities and improved rice plant health	Wang et al. (2023)

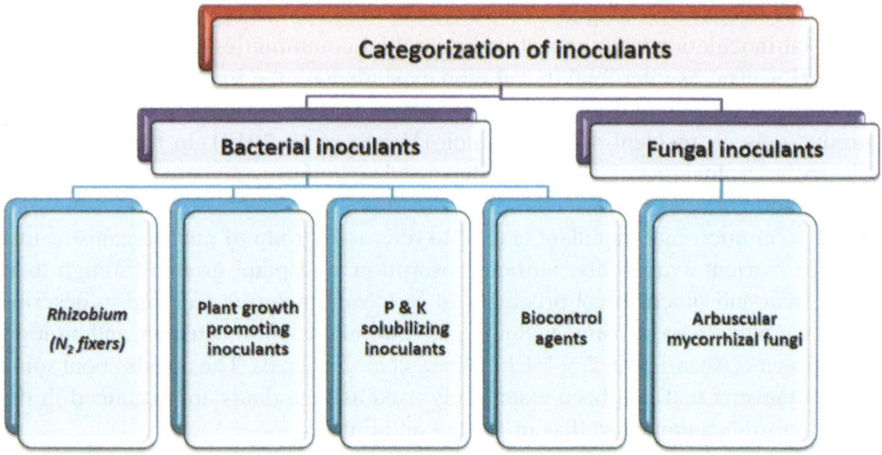

Fig. 1 Categorization of microbial inoculants into different groups

number of α-proteobacteria. The results of terminal restriction fragment analysis showed the effects of the intercropping, rhizobia inoculation, and growth stage on soybean endophytic bacteria diversity (Zhang et al. 2011). The number of roots was increased in the presence of introduced *Bradyrhizobium linconeense* and decreased in the presence of *Sinorhizzobium americaneum*, suggesting that the endogenous *Sinorhizobium* may be inhibited by competition. The presence of three different endophytic species in soybean roots may be related to the autoregulatory system of the leguminous organism, which is able to detect and address the presence of ineffective rhizobia with root nodules (Schumpp and Deakin 2010).

2.2 Azospirillum *Inoculants*

Azospirillum has been extensively studied for its agronomic advantages. *Azospirillum* inoculation has a big impact on root growth and alters the microbial population at rhizospheric zone (Yasuda et al. 2022).

ARISA (automated ribosomal intergenic spacer analysis) showed that maize inoculation with *Azospirillum lipoferum* CRT1 increased the genetic diversity of rhizobacterial populations (Schumpp and Deakin 2010). The physiology and surface area of the root changed as more root hairs were produced, resulting in increased uptake of nutrients. But, none of the bacterial community was affected significantly. De Salamone et al. (2012) suggested that the hormonal action of *Azospirillum* enhances the rate of nitrogen absorption resulting in higher biomass yields of rice crops as well as other microbial communities' growth in soil.

2.3 Pseudomonas *Inoculants*

Studies have been conducted on a variety of *Pseudomonas* species for their role in promoting plant growth. In the most recent study, the function of plant growth promotion of diazotrophs (cold adaptation) was identified, namely, the *P. jesenii* MP1 and *P. migulae* S10724. They are especially well-adapted to the ever-changing weather in the mountains and could be a great bioinoculants in high-altitude farming areas (Suyal et al. 2014; Kumar et al. 2014).

2.4 Phosphate Solubilizing Bacteria (PSB) Inoculants

Phosphorus plays an essential role in the metabolism of plants, which in turn has a direct impact on crop yield. Additionally, it plays a critical role in the functioning of essential enzymes that regulate metabolic pathways (Bechtaoui et al. 2021). Approximately 98% of Indian soils are estimated to insufficient phosphorus levels, which are necessary for the growth of plants. Numerous species of bacteria and fungi are reported to possess the capability of inorganic phosphate solubilization and are commonly referred to as phosphate solubilizing microorganisms (Ryan and Graham 2018; Chaudhary et al. 2021e, f). Some of the phosphorus-solubilizing microorganisms are *Arthrobacter*, *Bacillus*, *Delftia*, *Gordonia*, *Phyllobacterium*, *Pseudomonas*, *Rhodococcus* and *Serratia* sp., *Xanthomonas*, *Chryseobacterium*, *Xanthobacter agilis*, *Vibrio proteolyticus*, *Azotobacter*, *Enterobacter*, *Pantoea*, *Klebsiella*, *Pseodomonas* sp., and *Rhizobium leguminosarum* (Singh et al. 2012; Rani et al. 2013).

2.5 Plant Growth Promoting Bacteria (PGPB) Inoculants

The utilization of beneficial bacteria for plant growth, commonly referred to as plant growth promoting bacteria (PGPB), and their inoculation are essential approaches to sustainable agriculture, as the successful implementation of this practice allows for the reduction or elimination of the use of agrochemicals without affecting plant growth and yield (Poria et al. 2022). Microbial inoculants may be a cost-effective option for smallholder farmers, as they can be tested for their plant growth promotion ability (biofertilizers) and defense (biocontrol agents) with the aim of inoculant optimization (Ambrosini et al. 2012; Calvo et al. 2014). Gomez-Godinez et al. (2023) have reported the use of plant growth promoting bacteria as inoculants for crop yield enhancement.

2.6 Mycorrhizal Fungi Inoculants

Arbuscular mycorrhizal fungi (AMF) are the most common and beneficial mycorrhiza (Calvo et al. 2014). The main advantage of using mycorrhiza is its capability to increase the uptake and supply of nutrients such as nitrogen, phosphorus, potassium, zinc, iron, calcium, magnesium, and manganese to the roots of host plant (Begum et al. 2019). AMF are capable of forming symbiotic relationships with the majority of land plants and colonizing a larger area of soil. They absorb carbon from their hosts, which promotes plant growth by utilizing resources and returning minerals and water (Khaliq et al. 2022).

AMF interact with soil microorganisms that are connected to their extracellular matrix, creating a particular zone of the soil called mycorrhizosphere. Various bacterial species react to AMF, indicating a high level of specificity among AMF-associated bacteria (Emmett et al. 2021). *G. intraradices* that comes under *Arbuscular mycorrhizal* fungi can have an indirect effect on banana plants by stimulating the development of a systemic resistance to the two nematodes *Radopholus similis* and *Pratylenus coffeae* that feed on banana plants (Mandou et al. 2023). These studies demonstrated that AMF supports alterations in microbial community composition within the mycorrhizosphere. Identifying key microbial populations associated with increased biomass production will provide insight into their role in plant growth promotion and inhibition of plant pathogen emergence (Tyagi et al. 2022b).

2.7 Biocontrol Inoculants

The majority of commercial rhizobacterials have been developed as biological control inoculants for the purpose of increasing plant nutrition by reducing abiotic stress conditions such as plant pathogens (Sukweenadhia et al. 2015; Backer et al. 2018). Various kinds of microorganisms are utilized as biocontrol inoculants such as *Bacillus subtilis*, *Trichoderma harzianum*, and *P. fluorescens* (Mohiddin et al. 2010; Gao and Chen 2012).

3 Specific Features of Good Microbial Inoculants

Inoculants have been the focus of research due to their catabolic flexibility and superior root-colonization capabilities including motility, attachment, and growth (Mendoza-Suárez et al. 2021). They possess a broad range of enzyme and metabolite production capabilities, including the production of auxin, phosphate solubilization, siderophore production, and their ability to survive and proliferate in a competitive environment with microbial communities (Saeed et al. 2021). Microbial inoculation success is largely contingent upon the plant species, temperature and moisture of the soil, pathogens, and the preparation and application of inoculants (Saad et al. 2020).

4 Impacts of Inoculants on the Native Microbial Population

Microbes in soil are extremely composite and resilient to microbial inoculation (Mallon et al. 2015). They are created based on the availability of resources and the variety of soil microbes to enhance the nutrient utilization capability (Mallon et al. 2018). Agricultural practices around the world, like fertilizing and managing soil, have a big impact on biogeochemical cycles including carbon and nitrogen. It was observed that inoculated microbes rapidly colonize the host soil rhizosphere and alter the balance of soil microbial communities significantly. These alterations could be direct or indirect in terms of their interactions between the inoculants and soil microbes (Babalola 2014). Some microbial inoculants may alter the balance of soil microbial communities (e.g., decrease or increase), while others may have no effect at all. Piromyou et al. (2013) reported a nonsignificant impact on the microbial community in the rhizosphere in relation to free-living bacteria inoculants.

According to some studies, the soil rhizospheric community is highly resistant toward the introduced invasive bacterial species. The introduction of an unmodified microbial inoculum to the wheat-cropped soil impaired the natural flora of the actinobacteria endophyte, resulting in a decrease in diversity and level of colonization (Kandel et al. 2017). Conversely, the addition of one actinobacterium to the wheat plant increased the level of colonization and did not adversely affect the native flora (Yousaf et al. 2011). When N2 fixation is a property that is essential for the formation of symbiotic relationships between plants and bacteria, such as between leguminous plants and *Rhizobium*, the effects of the inoculum on microbial diversity are more significant than those of free-living bacteria (Bakhoum et al. 2012). Some potential impacts of microbial inoculants on soil microbial population are also presented in Fig. 2.

5 Major Significances of Microbial Inoculants

India is primarily an agricultural nation, and in order to feed its rapidly growing population, it is essential to enhance productivity. The FAO predicts that by 2030, there will be a 60% increase in the need for agricultural products (Mia and Shamsuddin 2010). In order to enhance the productivity of agricultural land to the desired level, a number of strategies like introduction of microbial inoculants to soil are becoming increasingly important, and some of the major significances of microbial inoculants are described in Table 2 and Fig. 3, as well as in the following subheading:

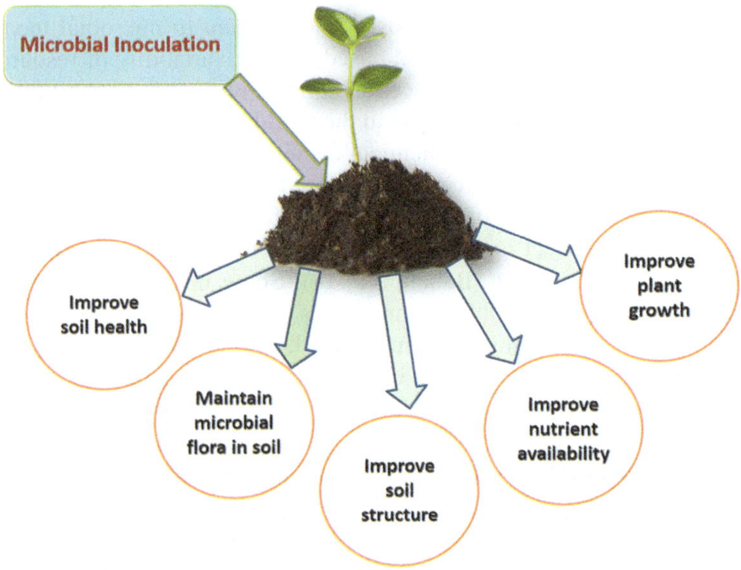

Fig. 2 Potential impacts of inoculation on native soil microbial population

Table 2 Beneficial properties and significances of some microbial inoculants

Microbial inoculants	Beneficial properties and significances	References
Rhizobacteria	Soil fertility improvement	Ambrosini et al. (2016)
Azospirillum	Nitrogen fixation	Cassan et al. (2020)
Phosphate solubilizers	Phosphorus solubilization	Raymond et al. (2021)
Arbuscular mycorrhizal fungi	Phosphorus solubilization	Zhang et al. (2019)
Rhizobacteria	Potassium solubilization	Meena et al. (2014)
Bacillus spp.	Phosphorus solubilization and improved soil health	Kumari et al. (2020)
Pseudomonas and *Pantoea* sp.	Mineral solubilization and improved plant and soil health parameters	Chaudhary and Sharma (2019); Kukreti et al. (2020)
Sporosarcina spp.	Soil structure improvement	Badiee et al. (2019)
Rhizobacteria	Bioremediation of contaminated soil	Guo et al. (2020)
Pseudomonas taiwanensis	Improved soil microbial population	Agri et al. (2021, 2022)
Bacillus spp.	Improved soil enzyme activities	Khati et al. (2017, 2018, 2019)

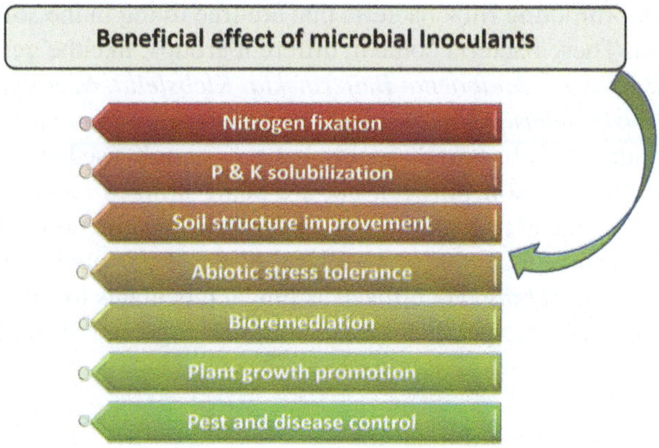

Fig. 3 Beneficial effect of microbial inoculants

5.1 Microbial Inoculants for Soil Fertility Improvement

Interactions between plants and inoculants with their effects on soil microbial communities suggested their contribution toward soil fertility (Ambrosini et al. 2016). The following are some nutrients that can enhance soil fertility by the application of microbial inoculants:

5.1.1 Nitrogen (N) Fixation

Nitrogen is the most important factor for plant growth. Mineral fertilizers and the fixation of nitrogen take place with the help of some microorganisms and are important for plant growth in agricultural soil (Grzyb et al. 2021). The role of microbes in N cycling is paramount. Bacteria possess a key enzyme, nitrogenase, which is responsible for exclusively fixing atmospheric N, either as symbiotic or asymbiotic (Mahmud et al. 2020). Soil N availability and supply can be improved in sustainable agriculture by using several microbial inoculants like PGPR, *rhizobia*, and *arbuscular mycorrhizal fungi* (Tyagi et al. 2022a, b).

The process of symbiotic N fixation is carried out by a variety of bacteria on both legumes and leafless plants. It's caused by certain bacteria, like *rhizobia* accountable for fixing legumes, while *Actinobacteria* and *Frankia* are responsible for fixing non-legumes (Mahmud et al. 2020). In order to attain the effective N fixation, inoculants such as legume and the rhizobium-based are taken into consideration to get compatibility and environmental adaptability (Lindström and Mousavi 2020). It was reported that rhizobia-inoculated legumes can boost the nitrogen content in soil. The fact that they can fix large quantity of nitrogen into the soil means they have a lot of capability to help farmers by replacing industrial N fertilizers. It is essential that the inoculum is capable of establishing, competing with, and maintaining an effective nodule formation in the environment by competing with others (Soumare et al. 2020).

Plant growth promoting rhizobacteria that are free living in the soil can fix nitrogen into the soil. These bacteria come in different groups, like the genera *Azoarcus*, *Pseudomonads*, *Nostoc*, *Anabaena*, *Beijerinckia*, *Klebsiella*, *Azotobacter*, *Pantoea*, *Azospirillum*, *Burkholderia*, *Bacillus*, and *Bradyrhizobium* (Cassan et al. 2020; Zilli et al. 2021). A number of bacterial species that are not only used alone as inoculants in order to enhance the soil nitrogen but are being utilized more efficiently with other microbes such as *Azospirillum* can be used in conjunction with rhizobia to improve soil N fertility (Cassan et al. 2020). Most of the fungal species such as *arbuscular mycorrhizal fungi* are utilized as fungal inoculants to enhance N fixation in association with plants, specifically with crops (Hodge and Storer 2015).

5.1.2 Phosphorus (P) Solubilization

Phosphorus is a micronutrient that makes up approximately 0.1% part of our Earth's crust and participates in plant growth development. It was determined that, under the artificial conditions, P-soluble bacteria are capable of dissolving complex phosphate compounds to simple organic ions (Barrow and Lambers 2022). Majority of microbial inoculants increase the sustainable supply of P to plants through P solubilization and P mobilization. Microorganisms have been identified to play an imperative function in the biogeochemical cycling of phosphorous in the rhizospheres by increasing the growth of microbial substrates based on P solubilization (Parnell et al. 2016; Rafi et al. 2019; Raymond et al. 2021).

PSMs are groups of certain bacteria and fungi that can be used to make insoluble form of phosphorus into soluble form and available to plants through the production of phosphatase and phytase enzymes. Different groups of microbes are responsible for P solubilization such as *Azospirillum*, *Azotobacter*, *Bacillus*, *Enterobacter*, *Streptomyces*, and *Pseudomonas* (Calvo et al. 2014; Owen et al. 2015; Hsu et al. 2015). The positive effects of *Arbuscular mycorrhizal fungi* was reported on crop growth and P nutrition (Zhang et al. 2019). Inorganic phosphate solubilization and mineralization by secretion of organic acids and phosphatase enzyme production are also performed by some groups of fungi including *Aspergillus* and *Penicillium* (Wang et al. 2015).

5.1.3 Potassium (K) Solubilization

Potassium is a micronutrient that is vital for the growth of plants. Certain groups of microorganisms are responsible for the solubilization of insoluble form of potassium in soil for plant growth (Meena et al. 2014). Rhizospheric microorganisms used as potassium solubilization inoculants include *Arthrobacter* sp., *Acidithiobacillus* sp., *Bacillus* sp., and *Paenibacillus* sp. (Zarjani et al. 2013). Fluctuating levels of K in plants can result in inadequate root development, slow growth, and reduced crop yields (White and Karley 2010). Soil and the rhizosphere contain K-soluble microorganisms such as PGPR and AMF that are involved in solubilization of potassium (Liu et al. 2012; Meena et al. 2014; Etesami et al. 2021).

5.2 *Changes in Soil Structure*

Soil structure is an essential factor in the sustainability of agriculture and the functioning of ecosystems, as it affects the soil features in relation to physical, biological, and chemical (Tahat et al. 2020). Several factors, including soil microorganisms and soil structure, are involved in formation and stabilization of soil masses (Helliwell et al. 2014; Badiee et al. 2019). Soil structure is affected by the size, type, and the mineralogy of the soil, which affects how well bacteria and fungi work as an inoculum. Microaggregate is classified into $<250\mu m$, and macroaggregate is classified as $>250\mu m$. This separation affects bacteria and fungi in different ways. Microaggregates are stabilized by fungi, while microaggregates are enhanced by bacteria. *Arbuscular mycorrhizal fungi* (AMF) take part in the soil structure enhancement process by maintaining soil aggregates (Leifheit et al. 2014).

5.3 *Microbial Inoculants for Abiotic Stress Tolerance*

Drought is becoming more common and severe in most parts of the world due to global climate change. Both fungi and bacteria can survive drought conditions in most ecosystems (Seleiman et al. 2021). However, fungi are more resistant to drought conditions than bacteria. In a study, it was found that when it was dry in forests and deserts, fungi stayed for longer time than bacteria (Yuste et al. 2011). AMF and saprophytic fungi are more resilient to heat and drought as compared to bacteria and actinomycetes (Leifheit et al. 2014).

Certain soil microorganisms are known to reduce salinity in plants and thereby enhance plant growth and productivity. Some of these soil microorganisms are *Agrobacteria*, *Azospirillum*, *Bacillus*, *Glomus*, *Gordonia*, and *Pseudomonas*, as well as *arbuscular mycorrhizal fungi* (Fukami et al. 2018; Jia et al. 2019). They are also eco-friendly and cost-effective. (Shilev et al. 2012; Sukweenadhia et al. 2015).

5.4 *Bioremediation of Contaminated Soil*

Microbial inoculants can be used in conjunction with plants to remove or convert contaminants into harmless substances and may also be employed for bioremediation of soil contaminants like organic and inorganic soil types (Kumar et al. 2021a, b, 2022; Singh et al. 2023). The microbial populations are affected by chelating gas release, acidification and phosphate solubilization, as well as redox changes. Heavy metals are also affected by plant accessibility. Bioinoculants are more cost-effective to phytoremediation polluted soil. Bioremediation strategies in soil remediation are safe and sustainable (Alori 2015; Kumar et al. 2021). In addition, other studies indicated that PGPB binds to metals, immobilizes them, and undergoes oxidation transformation to reduce metal toxicity (Guo et al. 2020; Zhang et al. 2020).

A number of bacterial species have been discovered to break down contaminants and soil pollutants including hydrocarbons (Afzal et al. 2012), dioxins, pesticides, neonicotinoids, and cadmium (Young et al. 2019; Pang et al. 2020).

5.5 *Microbial Inoculants as Biofertilizer*

Microbial inoculation is beneficial for the environment and does not pose any risk to plants, humans, or animals. Microbial inoculants improved the nutrient accessibility, yield, enhance root growth and exudation as biofertilizer (Babalola 2010; Ahemad and Kibret 2014; Babalola and Glick 2012).

The following are some of the most common microbial inoculants used as biofertilizers: *Azotobacter*, *Trichoderma*, *Azospirillum*, *Bradyrhizobium*, *Mycorrhiza*, phosphorus-solubilizing bacteria, *Rhizobium*, *Pseudomonas*, *Bacillus*, *Aspergillus*, *Cytophaga*, and phosphate solubilizing bacteria. Microbe-based fertilizers are cheaper and more effective than chemical fertilizers (Bargaz et al. 2018). They give off 25–30% of the nitrogen equivalent of chemical fertilizers. Also, they boost phosphorus and potassium levels, growth, and yield, make water more easily absorbed, and keep soil active (Mohammadi et al. 2011; N'Cho et al. 2013).

6 Conclusion

Microbes perform important function in the improvement of soil properties, plant growth, and yield by increasing plant formation, protecting soil from stress, enlightening soil structure, and increasing soil nutrient uptake. There is a lot of potential for microbial inoculants to become even more important in terms of both quality and quantity, as well as expanding their range of applications. A number of studies have reported change in soil microbial community due to introduction of inoculants. Therefore, these types of studies will lead to new practical applications with significant and integrative advancement of microbiology.

Microbial inoculants are promising and profitable, but large-scale commercial adoption has not been possible. This is due to various reasons, with identification, proliferation, supply chain, and usage by farmers. There are also some limitations in acceptance, such as the potential for easy contamination, limited farmer adoption, and absence of legislative policies and practices for management of soil population. To make sure the inoculants are consistent, the quality needs to be standardized, and there should be internationally recognized methods for quality testing. Application practice should be uniform for different types of farming and user-friendly, and the formulations should be steady and effective. As discussed in this chapter, microbial inoculant technologies could potentially improve bio-management of nutrients and plant growth. Microbial inoculants may also help reduce overuse and environmental impact of agrochemicals. Some additional researches are essential to comprehend the multifaceted relationship between soil and microorganisms, which will lead to replace the agrochemicals. Combining microorganisms in the form of inoculants has proven to be beneficial in many cases and can be further improved with research. In order to meet the objective of soil microbial community enhancement, the goal of using bioinoculant technology will lead to safer and eco-friendly microbial solutions.

References

Afzal M, Yousaf S, Reichenauer TG, Sessitsch A (2012) The inoculation method affects colonization and performance of bacterial inoculant strains in the phytoremediation of soil contaminated with diesel oil. Int J Phytoremediation 14:35–47. https://doi.org/10.1080/15226514.2011.552928

Agri U, Chaudhary P, Sharma A (2021) In vitro compatibility evaluation of agriusable nanochitosan on beneficial plant growth-promoting rhizobacteria and maize plant. Natl Acad Sci Lett 44:555–559. https://doi.org/10.1007/s40009-021-01047-w

Agri U, Chaudhary P, Sharma A, Kukreti B (2022) Physiological response of maize plants and its rhizospheric microbiome under the influence of potential bioinoculants and nanochitosan. Plant Soil 474:451–468. https://doi.org/10.1007/s11104-022-05351-2

Ahemad M, Kibret M (2014) Mechanisms and applications of plant growth promoting rhizobacteria: current perspective. J King Saud Univ Sci 26(1):1–20. https://doi.org/10.1016/j.jksus.2013.05.00

Ahmad J (2017) Bioremediation of petroleum sludge using effective microorganism (EM) technology. Pet Sci Technol 35(14):1515–1522. https://doi.org/10.1080/10916466.2017.1356850

Alemneh AA, Zhou Y, Ryder MH, Denton MD (2020) Mechanisms in plant growth-promoting rhizobacteria that enhance legume–rhizobial symbioses. J Appl Microbiol 129:1133–1156. https://doi.org/10.1111/jam.14754

Alori ET (2015) Phytoremediation using microbial community II. In: Ansari AA, Gill SS, Newman L, Lanza GR (eds) Phytoremediation: management of environmental contaminants, vol II. Springer, New York, pp 183–190. https://doi.org/10.1007/978-3-319-10969-5_15

Ambrosini A, Beneduzi A, Stefanski T, Pinheiro FG, Vargas LK, Passaglia LMP (2012) Screening of plant growth promoting Rhizobacteria isolated from sunflower (*Helianthus annuus* L.). Plant Soil 356:245–264. https://doi.org/10.1007/s11104-011-1079-1

Ambrosini A, de Souza R, Passaglia LMP (2016) Ecological role of bacterial inoculants and their potential impact on soil microbial diversity. Plant Soil 400:193–207. https://doi.org/10.1007/s11104-015-2727-7

Ansari MF, Tipre DR, Dave SR (2014) Efficiency evaluation of commercial liquid biofertilizers for growth of *Cicer aeritinum* (chickpea) in pot and field study. Biocatal Agric Biotechnol 4(1):17–24. https://doi.org/10.1016/j.bcab.2014.09.010

Babalola OO (2010) Beneficial bacteria of agricultural importance. Biotechnol Lett 32(11):1559–1570. https://doi.org/10.1007/s10529-010-0347-0

Babalola OO (2014) Does nature make provision for backups in the modification of bacterial community structures? Biotechnol Genet Eng Rev 30(1):31–48. https://doi.org/10.1080/02648725.2014.921497

Babalola OO, Glick BR (2012) Indigenous African agriculture and plant associated microbes: current practice and future transgenic prospects. Sci Res Essays 7(28):2431–2439. https://doi.org/10.5897/SRE11.1714

Backer R, Rokem JS, Ilangumaran G, Lamont J, Praslickova D, Ricci E, Subramanian S, Smith DL (2018) Plant growth-promoting rhizobacteria: context, mechanisms of action, and roadmap to commercialization of biostimulants for sustainable agriculture. Front Plant Sci 9:1473. https://doi.org/10.3389/fpls.2018.01473

Badiee H, Sabermahani M, Tabandeh F, Saeedi A (2019) Application of an indigenous bacterium in comparision with *Sporosarcina pasteurii* for improvement of fine granular soil. Int J Environ Sci Technol 16:8389–8400. https://doi.org/10.1007/s13762-019-02292-9

Bakhoum N, Ndoye F, Kane A (2012) Impact of rhizobial inoculation on *Acacia senegal* (L.) Willd. growth in greenhouse and soil functioning in relation to seed provenance and soil origin. World J Microbiol Biotechnol 28(7):2567–2579. https://doi.org/10.1007/s11274-012-1066-6

Bargaz A, Lyamlouli K, Chtouki M, Zeroual Y, Dhiba D (2018) Soil microbial resources for improving fertilizers efficiency in an integrated plant nutrient management system. Front Microbiol 9:1–25. https://doi.org/10.3389/fmicb.2018.01606

Barrow N, Lambers H (2022) Phosphate-solubilising microorganisms mainly increase plant phosphate uptake by effects of pH on root physiology. Plant Soil 476:397–402. https://doi.org/10.1007/s11104-021-05240-0

Bechtaoui N, Rabiu MK, Raklami A, Oufdou K, Hafidi M, Jemo M (2021) Phosphate-dependent regulation of growth and stresses management in plants. Front Plant Sci 12:679916. https://doi.org/10.3389/fpls.2021.679916

Begum N, Qin C, Ahanger MA, Raza S, Khan MI, Ashraf M, Ahmed N, Zhang L (2019) Role of arbuscular mycorrhizal fungi in plant growth regulation: implications in abiotic stress tolerance. Front Plant Sci 10:1068. https://doi.org/10.3389/fpls.2019.01068

Calvo P, Nelson L, Kloepper JW (2014) Agricultural uses of plant biostimulants. Plant Soil 383(1–2):3–41. https://doi.org/10.1007/s11104-014-2131-8

Carvajal-Muñoz JS, Carmona-Garcia CE (2012) Benefits and limitations of biofertilization in agricultural practices. Livest Res Rural Dev 24(3):1–8. http://www.lrrd.org/lrrd24/3/carv24043.htm

Cassan F, Coniglio A, Lopez G, Molina R, Nievas S, de Carlan C, Donadio F, Torres D, Rosas S, Pedrosa F, Mora V (2020) Everything you must know about Azospirillum and its impact on agriculture and beyond. Biol Fertil Soils 56:461–479. https://doi.org/10.1007/s00374-020-01463-y

Chaudhary P, Sharma A (2019) Response of nanogypsum on the performance of plant growth promotory bacteria recovered from nanocompound infested agriculture field. Environ Ecol 37:363–372

Chaudhary A, Parveen H, Chaudhary P, Bhatt P, Khatoon H (2021) Rhizospheric microbes and their mechanism. In: Bhatt P, Gangola S, Udayanga D, Kumar G (eds) Microbial technology for sustainable environment. Springer, Singapore. https://doi.org/10.1007/978-981-16-3840-4_6

Chaudhary P, Parveen H, Gangola S, Kumar G, Bhatt P, Chaudhary A (2021a) Plant Growth Promoting Rhizobacteria and their application in sustainable crop production. In: Bhatt P, Gangola S, Udayanga D, Kumar G (eds) Microbial technology for sustainable environment. Springer, Singapore. https://doi.org/10.1007/978-981-16-3840-4_13

Chaudhary P, Khati P, Chaudhary A, Gangola S, Kumar R, Sharma A (2021b) Bioinoculation using indigenous *Bacillus* spp. improves growth and yield of *Zea mays* under the influence of nanozeolite. 3 Biotech 11:11. https://doi.org/10.1007/s13205-020-02561-2

Chaudhary P, Khati P, Chaudhary A, Maithani D, Kumar G, Sharma A (2021c) Cultivable and metagenomic approach to study the combined impact of nanogypsum and *Pseudomonas taiwanensis* on maize plant health and its rhizospheric microbiome. PLoS One 16:e0250574. https://doi.org/10.1371/journal.pone.0250574

Chaudhary P, Khati P, Gangola S, Kumar A, Kumar R, Sharma A (2021d) Impact of nanochitosan and *Bacillus* spp. on health, productivity and defence response in *Zea mays* under field condition. 3 Biotech 11:237. https://doi.org/10.1007/s13205-021-02790-z

Chaudhary P, Chaudhary A, Parveen H, Rani A, Kumar G, Kumar A, Sharma A (2021e) Impact of nanophos in agriculture to improve functional bacterial community and crop productivity. BMC Plant Biol 21:519. https://doi.org/10.1186/s12870-021-03298-7

Chaudhary P, Sharma A, Chaudhary A, Khati P, Gangola S, Maithani D (2021f) Illumina based high throughput analysis of microbial diversity of maize rhizosphere treated with nanocompounds and *Bacillus* sp. Appl Soil Ecol 159:103836. https://doi.org/10.1016/j.apsoil.2020.103836

Chaudhary A, Chaudhary P, Upadhyay A, Kumar A, Singh A (2022) Effect of gypsum on plant growth promoting rhizobacteria. Environ Ecol 39(4A):1248–1256

Chaudhary P, Singh S, Chaudhary A, Agri U, Bhandari G (2022a) Chapter 3: Agrochemicals and their effects on soil microbial population. In: Soni R, Suyal D, Goel R (eds) Plant protection: from chemicals to biologicals. De Gruyter, Berlin, Boston, pp 45–58. https://doi.org/10.1515/9783110771558-003

Chaudhary P, Agri U, Chaudhary A, Kumar A, Kumar G (2022b) Endophytes and their potential in biotic stress management and crop production. Front Microbiol 13:933017. https://doi.org/10.3389/fmicb.2022.933017

Chaudhary P, Singh S, Chaudhary A, Sharma A, Kumar G (2022c) Overview of biofertilizers in crop production and stress management for sustainable agriculture. Front Plant Sci 13:930340. https://doi.org/10.3389/fpls.2022.930340

Chaudhary P, Chaudhary A, Agri U, Khatoon H, Singh A (2022d) Recent trends and advancements for agro-environmental sustainability at higher altitudes. In: Goel R, Soni R, Suyal DC, Khan M (eds) Survival strategies in cold-adapted microorganisms. Springer, Singapore. https://doi.org/10.1007/978-981-16-2625-8_19

Chaudhary P, Xu M, Ahamad L, Chaudhary A, Kumar G, Adeleke BS, Verma KK, Hu D-M, Širić I, Kumar P, Popescu SM, Abou Fayssal S (2023a) Application of synthetic consortia for improvement of soil fertility, pollution remediation, and agricultural productivity: a review. Agronomy 13:643. https://doi.org/10.3390/agronomy13030643

Chaudhary P, Chen S, Rajput VD, Jaiswal DK (2023b) Editorial: bioinoculants with nanocompounds to improve soil health: a step toward sustainable agriculture. Front Environ Sci 11:1270002. https://doi.org/10.3389/fenvs.2023.1270002

Chaudhary P, Chen S, Rajput VD, Jaiswal DK (eds) (2023c) Bioinoculants with nano-compounds to improve soil health: a step toward sustainable agriculture. Frontiers Media SA, Lausanne. https://doi.org/10.3389/978-2-8325-3287-4

de Salamone IEG, Funes JM, Di Salvo LP (2012) Inoculation of paddy rice with *Azospirillum brasilense* and *Pseudomonas fluorescens*: impact of plant genotypes on rhizosphere microbial communities and field crop production. Appl Soil Ecol 61:196–204. https://doi.org/10.1016/j.apsoil.2011.12.012

Emmett BD, Lévesque-Tremblay V, Harrison MJ (2021) Conserved and reproducible bacterial communities associate with extraradical hyphae of arbuscular mycorrhizal fungi. ISME J 15(8):2276–2288. https://doi.org/10.1038/s41396-021-00920-2

Etesami H, Jeong BR, Glick BR (2021) Contribution of arbuscular mycorrhizal fungi, phosphate–solubilizing bacteria, and silicon to P uptake by plant. Front Plant Sci 12:699618. https://doi.org/10.3389/fpls.2021.699618

Fukami J, Cerezini P, Hungria M (2018) *Azospirillum*: benefits that go far beyond biological nitrogen fixation. AMB Express 8(73):1–12. https://doi.org/10.1186/s13568-018-0608-1

Gao GD, Chen S (2012) Effect of biocontrol agent *Pseudomonas fluorescens* 2P24 on soil fungal community in cucumber rhizosphere using T-RFLP and DGGE. PLoS One 7:e31806. https://doi.org/10.1371/journal.pone.0031806

Gomez-Godinez LJ, Aguirre-Noyola JL, Martinez-Romero E, Arteaga-Garibay RI, Ireta-Moreno J, Ruvalcaba-Gomez JM (2023) A look at plant-growth-promoting bacteria. Plants 12:1668. https://doi.org/10.3390/plants12081668

Grzyb A, Wolna-Maruwka A, Niewiadomska A (2021) The significance of microbial transformation of nitrogen compounds in the light of integrated crop management. Agronomy 11(7):1415. https://doi.org/10.3390/agronomy11071415

Guo J, Muhammad H, Lv X, Wei T, Ren X, Jia H, Atif S, Hua L (2020) Prospects and applications of plant growth promoting rhizobacteria to mitigate soil metal contamination: a review. Chemosphere 246:125823. https://doi.org/10.1016/j.chemosphere.2020.125823

Helliwell JR, Miller AJ, Whalley WR, Mooney SJ, Sturrock CJ (2014) Quantifying the impact of microbes on structural development and behaviour in wet soil. Soil Biol Biochem 74:138–147. https://doi.org/10.1016/j.soilbio.2014.03.009

Hodge A, Storer K (2015) Arbuscular mycorrhiza and nitrogen: implications for individual plants through to ecosystems. Plant Soil 383(1–2):1–19. https://doi.org/10.1007/s11104-014-2162

Hsu PC, Condron LM, O'Callaghan M, Hurst M (2015) HemX is required for production of 2-ketogluconate, the predominant organic anion for inorganic phosphate solublization by *Burkholderia* sp. Ha185. Environ Microbiol Rep 7:918–928. https://doi.org/10.1111/1758- 2229.12326

Jayaraman D, Gilroy S, Ane JM (2014) Staying in touch: mechanical signals in plant–microbe interactions. Curr Opin Plant Biol 20:104–109. https://doi.org/10.1016/j.pbi.2014.05.003

Jia T, Wang J, Chang W, Fan X, Sui X, Song F (2019) Proteonomics analysis of E. angustifolia seedling inoculated mycorrhizal fungi under salt stress. Int J Mol Sci 20:788. https://doi.org/10.3390/ijms20030788

Jurburg SD, Salles JF (2015) Functional redundancy and ecosystem function—the soil microbiota as a case study. In: Biodiversity in ecosystems—linking structure and function. https://doi.org/10.5772/58981

Kandel SL, Joubert PM, Doty SL (2017) Bacterial endophyte colonization and distribution within plants. Microorganisms 5(4):77. https://doi.org/10.3390/microorganisms5040077

Khaliq A, Perveen S, Alamer KH, Zia Ul Haq M, Rafique Z, Alsudays IM, Althobaiti AT, Saleh MA, Hussain S, Attia H (2022) Arbuscular mycorrhizal fungi symbiosis to enhance plant–soil interaction. Sustainability 14(13):7840. https://doi.org/10.3390/su14137840

Khati P, Chaudhary P, Gangola S, Bhatt P, Sharma A (2017) Nanochitosan supports growth of *Zea mays* and also maintains soil health following growth. 3 Biotech 7:81. https://doi.org/10.1007/s13205-017-0668-y

Khati P, Parul, Bhatt P, Nisha Kumar R, Sharma A (2018) Effect of nanozeolite and plant growth promoting rhizobacteria on maize. 3 Biotech 8:141. https://doi.org/10.1007/s13205-018-1142-1

Khati P, Chaudhary P, Gangola S, Sharma P (2019) Influence of nanozeolite on plant growth promotory bacterial isolates recovered from nanocompound infested agriculture field. Environ Ecol 37:521–527

Kukreti B, Sharma A, Chaudhary P, Agri U, Maithani D (2020) Influence of nanosilicon dioxide along with bioinoculants on *Zea mays* and its rhizospheric soil. 3 Biotech 10:345. https://doi.org/10.1007/s13205-020-02329-8

Kumar S, Suyal DC, Dhauni N, Bhoriyal M, Goel R (2014) Relative plant growth promoting potential of himalayan psychrotolerant *Pseudomonas jesenii* strain MP1 against native *Cicer arietinum* L., *Vigna mungo* (L.) hepper; *Vigna radiata* (L.) Wilczek., *Cajanus cajan* (L.) Millsp. and *Eleusine coracana* (L.) Gaertn. Afr J Microbiol 8(50):3931–3943. https://doi.org/10.5897/AJMR2014.7035

Kumar A, Sharma A, Chaudhary P, Gangola S (2021) Chlorpyrifos degradation using binary fungal strains isolated from industrial waste soil. Biologia 10:3071–3081. https://doi.org/10.1007/s11756-021-00816-8

Kumar G, Lal S, Maurya SK, Bhattacherjee AK, Chaudhary P, Gangola S, Rajan S (2021a) Exploration of *Klebsiella pneumoniae* M6 for paclobutrazol degradation, plant growth attributes, and biocontrol action under subtropical ecosystem. PLoS One 16:e0261338. https://doi.org/10.1371/journal.pone.0261338

Kumar G, Suman A, Lal S, Ram RA, Bhatt P, Pandey G, Chaudhary P, Rajan S (2021b) Bacterial structure and dynamics in mango (*Mangifera indica*) orchards after long term organic and conventional treatments under subtropical ecosystem. Sci Rep 11:20554. https://doi.org/10.1038/s41598-021-00112-0

Kumar G, Lal S, Soni SK, Maurya SK, Shukla PK, Chaudhary P, Bhattacherjee AK, Garg N (2022) Mechanism and kinetics of chlorpyrifos co-metabolism by using environment restoring microbes isolated from rhizosphere of horticultural crops under subtropics. Front Microbiol 13:891870. https://doi.org/10.3389/fmicb.2022.891870

Kumari S, Sharma A, Chaudhary P, Khati P (2020) Management of plant vigor and soil health using two agriusable nanocompounds and plant growth promotory rhizobacteria in Fenugreek. 3 Biotech 10:1–11. https://doi.org/10.1007/s13205-020-02448-2

Kumari H, Khati P, Gangola S, Chaudhary P, Sharma A (2021) Performance of plant growth promotory rhizobacteria on Maize and soil characteristics under the influence of TiO_2 nanoparticles. Pantnagar J Res 19:28–39

Leifheit EF, Veresoglou SD, Lehmann A, Morris EK, Rillig MC (2014) Multiple factors influence the role of arbuscular mycorrhizal fungi in soil aggregation—a meta-analysis. Plant Soil 374:523–537. https://doi.org/10.1007/s11104-013-1899-2

Lindström K, Mousavi SA (2020) Effectiveness of nitrogen fixation in rhizobia. Microb Biotechnol 13(5):1314–1335. https://doi.org/10.1111/1751-7915.13517

Lioussanne L, Perreault F, Jolicoeur M, St-Arnaud M (2010) The bacterial community of tomato rhizosphere is modified by inoculation with arbuscular mycorrhizal fungi but unaffected by soil enrichment with mycorrhizal root exudates or inoculation with *Phytophthora nicotianae*. Soil Biol Biochem 42:473–483. https://doi.org/10.1016/j.soilbio.2009.11.034

Liu D, Lian B, Dong H (2012) Isolation of *Paenibacillus* sp. and assessment of its potential for enhancing mineral weathering. Geomicrobiol J 29(5):413–421. https://doi.org/10.1080/01490451.2011.576602

Mahmud K, Makaju S, Ibrahim R, Missaoui A (2020) Current progress in nitrogen fixing plants and microbiome research. Plants (Basel) 9(1):97. https://doi.org/10.3390/plants9010097

Mallon CA, Poly F, Le Roux X, Marring I, van Elsas J, Salles D, Falcao J (2015) Resource pulses can alleviate the biodiversity-invasion relationship in soil microbial communities. Ecology 96(4):915–926. https://doi.org/10.1890/14-1001.1

Mallon CA, Le Roux X, Van Doorn GS, Dini-Andreote F, Poly F, Salles JF (2018) The impact of failure: unsuccessful bacterial invasions steer the soil microbial community away from the invader's niche. ISME J 12(3):728–741. https://doi.org/10.1038/s41396-017-0003-y

Mandou MS, Souleymanou A, Chotangui AH (2023) Arbuscular mycorrhizal fungi inoculation and intercropping combine to control nematodes in bananas. Int J Plant Soil Sci 35(14):136–146. https://doi.org/10.9734/ijpss/2023/v35i143029

Manzar N, Jayaswal D, Kishan G, Chauhan S, Joshi D, Singh J, Chaudhary P, Jamali H, Kushwaha P, Parveen H, Kashyap AS, Sahu PK, Singh UB (2022) Application of molecular ecology approaches in sustainable agriculture for a better understanding of plant–microbiome interactions. In: Singh UB, Sahu PK, Singh HV, Sharma PK, Sharma SK (eds) Rhizosphere microbes. Microorganisms for sustainability, vol 40. Springer, Singapore. https://doi.org/10.1007/978-981-19-5872-4_3

Meena VS, Maurya BR, Verma JP (2014) Does a rhizospheric microorganism enhance K+ availability in agricultural soils? Microbiol Res 169(5–6):337–347. https://doi.org/10.1016/j.micres.2013.09.003

Mendoza-Suárez M, Andersen SU, Poole PS, Sánchez-Cañizares C (2021) Competition, nodule occupancy, and persistence of inoculant strains: key factors in the rhizobium-legume symbioses. Front Plant Sci 12:690567. https://doi.org/10.3389/fpls.2021.690567

Mia MAB, Shamsuddin ZH (2010) *Rhizobium* as a crop enhancer and biofertilizer for increased cereal production. Afr J Biotechnol 9(37):6001–6009. http://www.academicjournals.org/AJB

Mohammadi K, Khalesro S, Sohrabi Y, Heidari G (2011) A review: beneficial effects of the mycorrhizal fungi for plant growth. J Appl Environ Biol Sci 1(9):310–319

Mohiddin FA, Khan MR, Khan SM (2010) Why Trichoderma is considered super hero (super fungus) against the evil parasites? Plant Pathol J 9:1–11. https://doi.org/10.3923/ppj.2010.92.102

N'Cho CO, Yusuf AA, Ama-Abina JT, Jemo M, Abaidoo RC, Savane I (2013) Effects of commercial microbial inoculants and foliar fertilizers on soybean nodulation and yield in northern Guinea savannah of Nigeria. Int J Adv Agric Res 1:66–73. https://hdl.handle.net/10568/76651

O'Callaghan M, Ballard RA, Wright D (2022) Soil microbial inoculants for sustainable agriculture: limitations and opportunities. Soil Use Manag 38:1340–1369. https://doi.org/10.1111/sum.12811

Owen D, Williams AP, Griffith GW, Withers PJA (2015) Use of commercial bio-inoculants to increase agricultural production through improved phosphorus acquisition. Appl Soil Ecol 86:41–54. https://doi.org/10.1016/j.apsoil.2014.09.012

Pang S, Lin Z, Zhang W, Mishra S, Bhatt P, Chen S (2020) Insights into the microbial degradation and biochemical mechanisms of neonicotinoids. Front Microbiol 11:868. https://doi.org/10.3389/fmicb.2020.00868

Parnell JJ, Berka R, Young HA, Sturino JM, Kang Y, Barnhart DM, DiLeo MV (2016) From the lab to the farm: an industrial perspective of plant beneficial microorganisms. Front Plant Sci 7:1110. https://doi.org/10.3389/fpls.2016.01110

Piromyou P, Noisangiam R, Uchiyama H, Tittabutr P, Boonkerd N, Teaumroong N (2013) Indigenous microbial community structure in rhizosphere of Chinese kale as affected by plant growth-promoting rhizobacteria inoculation. Pedosphere 23(5):577–592. https://doi.org/10.1016/S1002-0160(13)60051-X

Poria V, Dębiec-Andrzejewska K, Fiodor A, Lyzohub M, Ajijah N, Singh S, Pranaw K (2022) Plant Growth-Promoting Bacteria (PGPB) integrated phytotechnology: a sustainable approach for remediation of marginal lands. Front Plant Sci 13:999866. https://doi.org/10.3389/fpls.2022.999866

Rafi MM, Krishnaveni M, Charyulu P (2019) Phosphate-solubilizing microorganisms and their emerging role in sustainable agriculture. In: Buddolla V (ed) Recent developments in applied microbiology and biochemistry. Elsevier, pp 223–233. https://doi.org/10.1016/B978-0-12-816328-3.00017-9

Rani A, Shouche Y, Goel R (2013) Comparative *in situ* remediation potential of *Pseudomonas putida* 710A and *Commamonas aquatica* 710B using plant (*Vigna radiata* (L.) wilczek) assay. Ann Microbiol 63(3):923–928. https://doi.org/10.1007/s13213-012-0545-1

Raymond NS, Gomez B, Bom FJT, Nybroe O, Jensen LS, Muller-stover DS, Oberson A, Richardson A (2021) Phosphate-solubilizing microorganisms for improved crop productivity: a critical assessment. New Phytol 229:1268–1277. https://doi.org/10.1111/nph.16924

Roy HE, Bacher S, Essl F, Adriaens T, Aldridge DC, Bishop JDD (2019) Developing a list of invasive alien species likely to threaten biodiversity and ecosystems in the European Union. Glob Chang Biol 25(3):1032–1048. https://doi.org/10.1111/gcb.14527

Ryan MH, Graham JH (2018) Little evidence that farmers should consider abundance or diversity of arbuscular mycorrhizal fungi when managing crops. New Phytol 220:1092–1107. https://doi.org/10.1111/nph.15308

Saad MM, Eida AA, Hirt H (2020) Tailoring plant-associated microbial inoculants in agriculture: a roadmap for successful application. J Exp Bot 71(13):3878–3901. https://doi.org/10.1093/jxb/eraa111

Saeed Q, Xiukang W, Haider FU, Kučerik J, Mumtaz MZ, Holatko J, Naseem M, Kintl A, Ejaz M, Naveed M, Brtnicky M, Mustafa A (2021) Rhizosphere bacteria in plant growth promotion, biocontrol, and bioremediation of contaminated sites: a comprehensive review of effects and mechanisms. Int J Mol Sci 22(19):10529. https://doi.org/10.3390/ijms221910529

Schumpp O, Deakin WJ (2010) How inefficient rhizobia prolong their existence within nodules. Trends Plant Sci 15:189–195. https://doi.org/10.1016/j.tplants.2010.01.001

Seleiman MF, Al-Suhaibani N, Ali N, Akmal M, Alotaibi M, Refay Y, Dindaroglu T, Abdul-Wajid HH, Battaglia ML (2021) Drought stress impacts on plants and different approaches to alleviate its adverse effects. Plants (Basel) 10(2):259. https://doi.org/10.3390/plants10020259

Shilev S, Sancho ED, Benlloch-González M (2012) Rhizospheric bacteria alleviate salt-produced stress in sunflower. J Environ Manag 95(Suppl):S37–S41. https://doi.org/10.1016/j.jenvman.2010.07.019

Singh AV, Chandra R, Goel R (2012) Phosphate solubilization by *Chryseobacterium* sp. and their combined effect with N and P fertilizers on plant growth promotion. Arch Agron Soil Sci 59(5):641–651. https://doi.org/10.1080/03650340.2012.664767

Singh S, Chaudhary P, Sharma A (2023) Efficient decolourization of toluidine blue and methylene blue using *Bacillus cereus*. Vegetos. https://doi.org/10.1007/s42535-023-00747-6

Soumare A, Diedhiou AG, Thuita M, Hafidi M, Ouhdouch Y, Gopalakrishnan S, Kouisni L (2020) Exploiting biological nitrogen fixation: a route towards a sustainable agriculture. Plants 9(8):1011. https://doi.org/10.3390/plants9081011

Sukweenadhia J, Kima Y, Choib E, Kohc S, Leed S, Kima Y, Yanga DC (2015) *Paenibacillus yonginensis* DCY84Tinduces changes in *Arabidopsis thaliana* gene expression against aluminum, drought, and salt stress. Microbiol Res 172:7–15. https://doi.org/10.1016/j.micres.2015.01.007

Sutherland WJ, Broad S, Butchart SHM, Clarke SJ, Ockendon N (2019) A horizon scan of emerging issues for global conservation in 2019. Trends Ecol Evol 34(1):83–94. https://doi.org/10.1016/j.tree.2018.11.001

Suyal DC, Shukla A, Goel R (2014) Growth promotory potential of the psychrophilic diazotroph *Pseudmonas migulae* S10724 against native *Vigna radiata* (L.) Wilczek. 3 Biotech 4:665–668. https://doi.org/10.1007/s13205-014-0259-0

Tahat MM, Alananbeh KM, Othman YA, Leskovar DI (2020) Soil health and sustainable agriculture. Sustainability 12(12):4859. https://doi.org/10.3390/su12124859

Tamayo-Velez A, Osorio NW (2018) Soil fertility improvement by litter decomposition and inoculation with the fungus *Mortierella* sp. in avocado plantations of Colombia. Commun Soil Sci Plant Anal 49(2):139–147. https://doi.org/10.1080/00103624.2017.1417420

Trabelsi D, Ben Ammar H, Mengoni A, Mhamdi R (2012) Appraisal of the crop-rotation effect of rhizobial inoculation on potato cropping systems in relation to soil bacterial communities. Soil Biol Biochem 54:1–6. https://doi.org/10.1016/j.soilbio.2012.05.013

Trivedi P, Pandey A, Palni LMS (2012) Bacterial inoculants for field applications under mountain ecosystem: present initiatives and future prospects. In: Maheshwari DK (ed) Bacteria in agrobiology: plant probiotics. Springer, Berlin, Heidelberg, pp 15–44. https://doi.org/10.1007/978-3-642-27515-9_2

Tyagi J, Chaudhary P, Jyotsana, Bhagwati U, Bhandari G, Chaudhary A (2022a) Chapter 17: Impact of endophytic fungi in biotic stress management. In: Soni R, Suyal D, Goel R (eds) Plant protection: from chemicals to biologicals. De Gruyter, Berlin, Boston, pp 447–462. https://doi.org/10.1515/9783110771558-017

Tyagi J, Chaudhary P, Mishra A, Khatwani M, Dey S, Varma A (2022b) Role of endophytes in abiotic stress tolerance: with special emphasis on *Serendipita indica*. Int J Environ Res 16:62. https://doi.org/10.1007/s41742-022-00439-0

Verma M, Mishra J, Arora NK (2019) Plant growth-promoting rhizobacteria: diversity and applications. Environmental biotechnology: for sustainable future. Springer, pp 129–173. https://doi.org/10.1007/978-981-10-7284-0_6

Wang H, Liu S, Zhal L, Zhang J, Ren T, Fan B, Liu H (2015) Preparation and utilization of phosphate biofertilizers using agricultural waste. J Integr Agric 14(1):158–167. https://doi.org/10.1016/S2095-3119(14)60760-7

Wang J, Zhao S, Xu S, Zhao W, Zhang X, Lei Y, Zhai H, Huang Z (2023) Co-inoculation of antagonistic *Bacillus velezensis* FH-1 and *Brevundimonas diminuta* NYM3 promotes rice growth by regulating the structure and nitrification function of rhizosphere microbiome. Front Microbiol 14:1101773. https://doi.org/10.3389/fmicb.2023.1101773

Weiland-Bräuer N, Pinnow N, Schmitz RA (2015) Novel reporter for identification of interference with acyl homoserine lactone and autoinducer-2 quorum sensing. Appl Environ Microbiol 81(4):1477–1489. https://doi.org/10.1128/AEM.03290-14

White PJ, Karley AJ (2010) Potassium. In: Hell R, Mendel RR (eds) Cell biology of metals and nutrients, plant cell monographs, vol 17. Springer, Berlin, pp 199–224. https://doi.org/10.12691/wjar-5-5-1

Yasuda M, Dastogeer KMG, Sarkodee-Addo E, Tokiwa C, Isawa T, Shinozaki S, Okazaki S (2022) Impact of *Azospirillum* sp. B510 on the rhizosphere microbiome of rice under field conditions. Agronomy 12(6):1367. https://doi.org/10.3390/agronomy12061367

Young SD, Koten C, Gray CW, Cavanagh JAE, Wakelin S (2019) Symbiosis between *Rhizobium leguminosarum* by *trifolii* strain TA1 and a white clover cultivar benefits clover tolerance to cadmium toxicity. N Z J Agric Res 63:353–364. https://doi.org/10.1080/00288233.2019.1680394

Yousaf S, Afzal M, Reichenauer TG, Brady CL, Sessitsch A (2011) Hydrocarbon degradation, plant colonization and gene expression of alkane degradation genes by endophytic *Enterobacter ludwigii* strains. Environ Pollut 159(10):2675–2683. https://doi.org/10.1016/j.envpol.2011.05.031

Yuste JC, Peñuelas J, Estiarte M, Garcia-mas J, Mattana S, Ogaya R, Pujol M, Sardan J (2011) Drought-resistant fungi control soil organic matter decomposition and its response to temperature. Glob Chang Biol 17:1475–1486. https://doi.org/10.1111/j.1365-2486.2010.02300.x

Zarjani JK, Aliasgharzad N, Oustan S, Emadi M, Ahmadi A (2013) Isolation and characterization of potassium solubilizing bacteria in some Iranian soils. Arch Agron Soil Sci 59(12):1713–1723. https://doi.org/10.1080/03650340.2012.756977

Zhang YZ, Wang ET, Li M (2011) Effects of rhizobial inoculation, cropping systems and growth stages on endophytic bacterial community of soybean roots. Plant Soil 347:147–161. https://doi.org/10.1007/s11104-011-0835-6

Zhang S, Lehmann A, Zheng W, You Z, Rillig MC (2019) Arbuscular mycorrhizal fungi increase grain yields: a meta-analysis. New Phytol 222:543–555. https://doi.org/10.1111/nph.15570

Zhang H, Yuan X, Xiong T, Wang H, Jiang L (2020) Bioremediation of co-contaminated soil with heavy metals and pesticides: influence factors, mechanisms and evaluation methods. Chem Eng J 398:2567. https://doi.org/10.1016/j.cej.2020.125657

Zilli JE, Pacheco RS, Gianluppi V, Smiderle OJ, Urquiaga S, Hungaria M (2021) Biological N2 Fixation and yield performance of soybean inoculated with *Bradyrhizobium*. Nutr Cycl Agroecosyst 119:323–336. https://doi.org/10.1007/s10705-021-10128-7

Plant-Microbe Interactions and Crop Production

Neha Bisht and Parul Chaudhary

Abstract Plants are living entities that provide food, oxygen, and habitat for other organisms in the environment. Plants, like all living things, require adequate nutrition, water, and sunlight to develop and thrive. Plants are continually interacting with their surroundings, including the microbes that live in the soil, roots, and leaves. These interactions may have a significant impact on plant health and growth, as well as the surrounding ecosystem. Microbes have specialized roles in the environment, including managing toxicity, suppressing illness, promoting plant development and cycling, and utilizing nutrients. The interactions between microbes and plants can be pathogenic, symbiotic, and associative that have different impacts on plant health. Keeping up the research to better understand the dynamics of rhizosphere communities would provide sustainable farming approaches with much-needed momentum. Plant microbe interactions alter immunity, physiological functions, and soil quality, ultimately leading to sustainable agriculture. This chapter focuses on the plant-microbe communications in terms of nutrient uptake, stress tolerance, and crop production.

Keywords Root exudates · Stress · Nutrient acquisition · Plant-microbe interactions

1 Introduction

Increasing population in India is facing tremendous obstacles in feeding the people and sustainable agricultural production as well as human health concerns. This loss is associated with the use of chemical pesticides and fertilizers and adaption of poor farming practices. Diversity of natural microflora and evolution of naive species

N. Bisht (✉)
Department of Microbiology, Govind Ballabh Pant University of Agriculture and Technology, Pantnagar, Uttarakhand, India

P. Chaudhary
School of Agriculture, Graphic Era Hill University, Dehradun, Uttarakhand, India

P. Chaudhary, A. Chaudhary (eds.), *Microbial Inoculants*, Interdisciplinary Biotechnological Advances, https://doi.org/10.1007/978-981-97-0633-4_4

through mutation in soil environment will eventually increase while avoiding the use of chemicals in soil (Das et al. 2022; Chaudhary et al. 2022a). The use of microbial inoculants is one of the greatest solutions to this problem that provide nutrients to the plant and are involved in degradation of complex compounds in a sustainable way (Liu et al. 2018; Chaudhary and Sharma 2019; Kumari et al. 2020; Kumari et al. 2021; Parveen et al. 2022). The interaction between plant and microbe is fruitful for each other. Plants act as a home for the microbial species, which reside as an endophyte in either rhizosphere zone, plant surfaces, intercellular surfaces, and apoplastic surfaces. In turn, microbial cells provide food or nutrients to plants, for instance, nitrogen, phosphates, phenols, hydroxyl ions and acids, etc., which aid plant development and resistance to stress situations (Schirawski and Perlin 2018; Nadarajah and Abdul Rahman 2021).

Microbes and plants have a complicated interaction. While some microorganisms are beneficial to plants, some, like those that cause plant diseases, can be harmful. Microbes can help plants absorb nutrients, defend against pathogens, and tolerate environmental stresses (Tyagi et al. 2022a, b). In turn, plants can also provide nutrients and a habitat for microbes. One important example of the interaction is the relationship between plants and mycorrhizal fungi. Mycorrhizal fungi are a type of beneficial fungus that forms a symbiotic relationship with plant roots. The fungus helps the plant absorb nutrients from the soil, such as phosphorus, in exchange for carbohydrates produced by the plant (Bonfante and Genre 2010). This relationship is especially important in nutrient-poor soils, where plants may struggle to obtain the nutrients they need to grow and thrive. Another example is the interaction between plants and nitrogen-fixing bacteria. *Rhizobium* are able to convert atmospheric nitrogen into a form that plants can use for growth. In exchange, the bacteria receive carbohydrates produced by the plant. This relationship is particularly important for legumes, such as beans and peas, which rely on nitrogen-fixing bacteria to grow (Wang et al. 2018).

Phytoremediation in plants is improved by microbes associated with them as they help in breakdown of organic contaminants (Jambon et al. 2018). Crop varieties with genetic alterations having greater disease and stress tolerance are produced (Kumar and Verma 2018). But plant growth promoting microorganisms (PGPM) serve as a promising alternate for sustainable crop production as they help in providing phytohormones and nutrition and eliminate phytopathogens and toxic pollutants (Spence and Bais 2015; Chaudhary et al. 2022a, b, c; Singh et al. 2023). PGPR (plant growth promoting rhizobacteria) and mycorrhizal fungi are involved in production of ionophores, siderophores, antioxidants, and nutrients in stress conditions. These microbes benefit the plant and soil environment by regulating soil fertility and pH and alleviating phytopathogens to make a plant disease-free (Agri et al. 2021, 2022). Micro- and macronutrients are some of the important key factors in maintaining the overall plant growth. Some of the essential macronutrients such as N, P, S, K, C, H, O, Ca, Mg, and S and micronutrients such as Fe, Mn, B, Mo, Cu, Zn, Cl, Ni, Co, S, and Si are required for plant growth. Some examples of nutrient deficiency in plants causing disease are Ca deficiency such as grape tomato leading to blossom end rot, Mg deficiency in *Frangula alnus* causing yellowing of leaves and interveinal chlorosis in sweetgum leaf, and K deficiency in palm causing yellowing and falling of leaves (Broadley et al. 2017).

Root exudates secreted by root tips of plants are involved in attracting soil microbes which in turn help in nutrient sequestration from insoluble complexes. Figure 1 shows the interaction of microbes with plant affecting the overall activity of plant health. The majority of plant species related with PGPR are primarily from the following genera such as *Agrobacterium*, *Azoarcus*, *Bacillus*, *Alcaligens*, *Bradyrhizobium*, *Enterobacter*, *Beijerinckia*, *Frankia*, and *Rhizobium* (Habibi et al. 2014; Kukreti et al. 2020; Agri et al. 2022). Interaction between plants and microbes is essential for the establishment and maintenance of plant health, growth, and nutrient acquisition (Manzar et al. 2022). Beneficial microbes, PGPR, mycorrhizal fungi, and endophytes can enhance plant growth and health by providing essential nutrients, suppressing pathogens, and inducing systemic resistance against biotic and abiotic stresses (Chaudhary et al. 2021a, 2022d, e). Pathogenic microbes can cause significant damage to plant tissues and reduce plant growth and yield. Thus, understanding the dynamics of plant-microbe interactions is crucial for developing sustainable agriculture practices and maintaining plant health and productivity. In this book chapter, we will present an overview of the current understanding of plant-microbe interactions and their role in maintaining plant health and the various types of plant-microbe interactions, their mechanisms, and their impact on plant growth and health. We will also highlight the emerging trends and challenges in the field of plant-microbe interactions and their implications for sustainable agriculture.

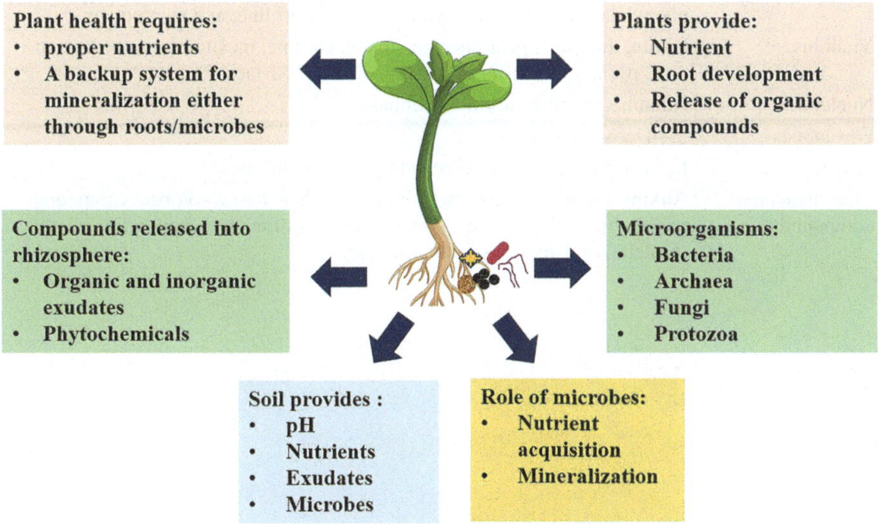

Fig. 1 Interaction of plant and microbes affecting plant health activity

2 The Plant's Rhizospheric Region

Plant roots consisting of a region where the population of microbes is maximum number are known as the rhizospheric region. The microbes present in the rhizosphere are known as PGPR. Microbial activity can be calculated by plating technique known as "rhizosphere effect" or rhizosphere/soil ratio (Mhlongo et al. 2018). This ratio is calculated by dividing the microbial population present in rhizospheric area by microbial population in the non-rhizospheric area (Sukumaran et al. 2015). Microbial population present in rhizosphere affects plant growth. The microbial population is feasible to regulate soil organic matter and mineralization for sustaining quality of plant and soil (Chaudhary et al. 2023a, b, c). Exudates released from plant roots, litter, and organic matter present in soil are the three main sources of carbon which are shared among plant and microbes (Nadarajah et al. 2021). Carbohydrates, organic acids, and a number of other complex compounds are known as root exudates that feed the PGPR in the rhizosphere (Hacquard 2016). Different root exudates secreted by the plant roots are shown in Table 1. Plants inturn get benefited by microbes providing them nutrients and water, organic matter degradation and CO_2 for photosynthesis. The rhizosphere therefore plays a vital role in maintaining nutrients and water content in plants and soil (Bhaduri et al. 2015).

Table 1 Compounds found in root exudates (Nadarajah et al. 2021)

Compounds	Exudates
Carbohydrates	Sucrose, maltose, rhamnose, fructose, arabinose, glucose, and raffinose
Organic acids	Tartaric, malic, oxalic, acetic, citric, and fumaric acid
Amino acid	Leucine, valine, glutamine, asparagines, serine, glycine, glutamic acid, phenylalanine, threonine, tyrosine, lysine, proline, and tryptophan
Vitamins	Biotin, thiamine, pentothenate, niacin, choline, inositol, p-amino bezoic acid, pyridoxine, m-methyl nicotinic acid, "M-factor"
Nucleotide	Adenine, guanine, uridine, citidine
Flavanones	3,4-Deoxy flavanones
Enzymes	Invertase, amylase, protease, and polygalacturonase
Miscellaneous compounds	Auxins, organic phosphorus compounds, benetonite coloring compounds, nodulating inhibitors, *Azotobacter* growth stimulator, scopoletin (6-methoxy-7-hydroxycoumarin), etc.

3 Mechanisms Adopted by PGPR

PGPR adapt direct and indirect mechanisms to enhance growth of plants to cope under stressful environmental conditions. Figure 2 highlights some of the direct and indirect mechanisms.

3.1 *Direct Mechanisms*

PGPRs are involved in plant growth and development through improving uptake and mineralization of nutrients, as well as utilization of organic molecules.

3.1.1 Phytohormone Production

Phytohormones are known to be produced by plants to induce plant growth, immunity, stress tolerance, and maturity. PGPR also produce phytohormones in stress conditions. They activate the cellular machinery of plant through the activation of transcription factors. Auxin, gibberellins, cytokinins, jasmonic acid, and brassinosteroids modulate plant growth and response to stress (Maheshwari et al. 2015).

(a) **Auxins/Indole Acetic Acid (IAA):** IAA is responsible for improving plant health and development through expansion, division, and differentiation of plant cells, promotion of root and shoot elongation, apical dominance, etc. IAA are produced by various rhizobacteria, e.g., *Azospirillum*, *Agrobacterium*, *Bacillus*, *Paenibacillus*, *Pseudomonas*, and *Erwinia* (Chandra et al. 2018; Khati et al. 2017, 2018). Production of IAA can be regulated by environmental factors like pH and presence of tryptophan secreted by plant roots (Jaiswal et al. 2018). L-tryptophan is the precursor of IAA which is used by PGPR for synthesis of IAA by the process of deamination, decarboxylation, and/or hydrolysis reac-

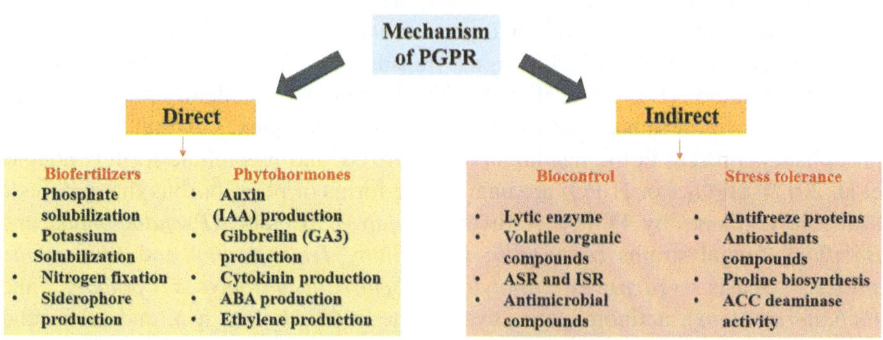

Fig. 2 Direct and indirect mechanisms of plant growth promotion by PGPR

tions. IAA help plants to resist adverse conditions and lead to proper growth of plant (Duca et al. 2018; Lebrazi et al. 2020).

(b) **Gibberellin:** Stem elongation, dormancy, flowering, germination, sex expression, leaf senescence, and fruit ripening are controlled by phytohormone gibberellins (GAs). Many PGPR also produce GA at adverse situations. The bacteria that produce GA are mainly sp. of *Burkholderia cepacian*, *Acinetobacter*, *Bacillus pumilus*, and *Promicromonospora* (Spence and Bais 2015). The production of GA results in positive feedback for production of endogenous of plant. Positive effects in growth of banana plants were seen when inoculated with *Bacillus siamensis* that is able to produce GA (Ambawade and Pathade 2015). GA3 was reported to be a major GA present in *Azospirillum* that promote the growth and development of plant (Pii et al. 2015).

(c) **Cytokinins:** Cytokinins promote cell growth, proliferation, and differentiation in plants. They influence apical dominance, inhibit aging of tissues, and thus promote leaf growth. Cytokinins are also known to inhibit premature leaf senescence. Cytokinins are involved in negative feedback on apical dominance which is known to be upregulated by auxins. Cytokinins which can be produced by PGPR are an alternative for enhancing plant growth to improve crop yield and quality (Akhtar et al. 2020).

(d) **Ethylene and Abscisic Acid:** Ethylene promotes fruit ripening. Root exudates contain 1-aminocyclopropane-1-carboxylate (ACC) which is precursor for the formation of ethylene. Ammonium and α-ketobutyrate are formed from ACC via ACC deaminase produced by PGPR, therefore decreasing ethylene concentration (Spence and Bais 2015). At low concentrations, ethylene mitigates detrimental effects of biotic and abiotic stresses. Abscisic acid, like ethylene, affects plant development under adverse conditions. Some of the ABA producing bacteria are *B. pumilus*, *B. subtilis*, *Pseudomonas putida*, and *Achromobacter xylosoxidans* that positively influence plant homeostasis.

3.1.2 Phosphate Solubilization

Phosphorus is a second essential element required by plants after nitrogen. Phosphate is required in various processes like photosynthesis, carbon sequestration, membrane formation, energy transfer, signal transduction, disease resistance capacity, and seed formation (Lebrazi et al. 2020). The P content available in soil has not been utilized by plants because of its fixation to insoluble complexes. Around 90–95% of phosphate is present in the insoluble, immobilized, and precipitate form (Cabugao et al. 2017). HPO_4^{-2} or H_2PO_4 are unavailable forms of phosphorous that are made available to plants by PGPR. A diverse group of bacteria (*Pseudomonads* and *Bacillus*), fungal strains (*Aspergilli*, *Penicillium*, *Trichoderma*, and *Rhizoctonia solani*), yeasts (*Yarrowia lipolytica*, *Schizosaccharomyces pombe*, and *Pichiafermentans*), actinomycetes, cyanobacteria (*Anabeana* sp.), and arbuscular mycorrhizal fungi (*Glomus fasciculatum*) are known to be involved in phosphate solubilization. Acidification of the medium, chelation (Ca, Al, Fe), exchange reactions and production of various acids (citric acid, glycolic acid, succinic acid, gluconic acid, oxalic acid, lactic acid, ascorbic acid, etc.) and enzymes (acid phosphatases, phytases, phosphonatases, and C-P lyases), and production of H_2S are well-known mechanisms of P-solubilization. Application of microorganisms as phosphate solubilizing microorganisms (PSM) improves plant growth and yield of crop (Radhakrishnan et al. 2015; Chaudhary et al. 2021b, c, d, e, f).

3.1.3 Siderophore Production

Siderophores are iron/ion chelating molecules. Among micronutrients, iron is mostly unavailable because at higher pH the easily available form of iron, ferrous ion (Fe2+), makes insoluble complexes with other metal ions (Kumar et al. 2017). Iron has essential role in many cellular reactions regarding growth and development; therefore, it is regarded as most vital nutrient. As huge amount of chemical form of iron is provided to plants, it has adverse effect on plant health. Therefore, as an alternative, several soil microorganisms that play a significant role in solubilization of these complexes by secreting certain bioactive molecule and thereby enhancing nutrient availability in soil for plant growth are used (Singh et al. 2022). Among several microbial mechanisms which can alleviate the stressful condition, siderophore production is most eloquent. Siderophores are low molecular weight organic molecules synthesized by several soil bacteria and fungi (Timofeeva et al. 2022). Ionophores are specific types of ion chelating molecules which chelate other ions like Mn, B, Cl, Ni, Co, S, Si, Zn, K, Na, Mg, Cu, and Al including iron that are present in insoluble form. Siderophores chelate ferric ions from insoluble complexes and convert them to soluble ferrous forms and provide them to plants. Siderophore or ionophores sequester Fe^{3+} ions and provide them to plants. PGPR inhibit the growth of phytopathogens by competing for available iron present in the soil (Bisht et al. 2018).

3.1.4 Biological Nitrogen Fixation

Nitrogen assimilating by plant is the most limiting factor for plant growth (Courty et al. 2015). Nitrogen (78%) present in atmosphere still is not available to plants. There is no plant yet known that fixes atmospheric nitrogen into utilizable form. However, biological nitrogen fixation is known to change nitrogen to ammonia by nitrogen fixer using nitrogenase enzymes. This enzyme complex is composed of two-component dinitrogenase reductase which is an iron protein and dinitrogenase that is a metal cofactor. The first component provides electrons, and the second component utilizes the electron to reduce nitrogen to ammonia. Mo-nitrogenase, V-nitrogenase, and Fe-nitrogenase are identified as the main metal cofactors. Symbiotic and nonsymbiotic associations lead to nitrogen fixation with the aid of nitrogen-fixing organisms. *Azospirillum*, *Gluconacetobacter diazotrophicus*, and cyanobacteria are some nonsymbiotic nitrogen-fixing bacteria. *Rhizobiaceae* family members are nitrogen-fixing bacteria. A very little amount of fixed nitrogen is supplied to plant by nonsymbiotic bacteria. Symbiotic associations by bacteria lead to colonization inside the roots of establishment of leguminous plants. Nodule formation is observed in intracellular parts of plant cell where rhizobia cell colonizes (Böhme and Masepohl 2018). Biological nitrogen-fixing organisms are proven to be promising alternative for balancing nitrogen status in plants. Worldwide, significant consideration has been given to the limitation of nitrogen in agricultural soils, and new tools and techniques are developed to prevent the application of chemical fertilizers for crop improvement. A combination of a free-living diazotrophic nitrogen fixer bacterium *Anabaena* and *Azolla* in a waterlogged rice field compensates the shortage of nitrogen. Therefore, *Anabaena* is considered as a "biofertilizer" (Fosu-Mensah et al. 2015).

3.1.5 Potassium Assimilation

Potassium is also considered as a major element for plant growth and development. Potassium present in atmosphere that is in available form is only 10%; the remaining 90% is unavailable. Plants deficient with potassium are not able to develop roots properly, seeds produced are very small, and yellowing of leaves and lower yield are observed. Therefore, as an alternative, PGPR are inoculated to the plant roots to sustain the potassium in soil and plant. PGPR are known to solubilize rock phosphate via organic acid production and make them accessible to plants (Ansari et al. 2017).

3.2 Indirect Mechanisms

Indirect mechanism of plant growth promotion by PGPR promotes the plant growth by production of different compounds and is a good approach to alleviate soilborne diseases.

3.2.1 Biocontrols

PGPR can produce a wide variety of antimicrobial compounds for defense such as broad-spectrum antibiotics, lactic acid produced by *Lactobacillus lactis*, hydrolytic enzymes like lysozymes, chitinase, proteases, lipases and β,1-3, and glucanases that play a prominent role in eliminating phytopathogens by either lysing through lytic or lysogenic pathways. There are different kinds of exotoxins and bacteriocins that consist of bactericidal activity.

3.2.2 Antimicrobial Compounds

Antibiotics are considered as organic compounds that have a low molecular weight. These compounds play a negative role in growth and metabolism of other microbial population (de Jesus Sousa and Olivares 2016). Densely populated bacterial cells produce biofilm that releases numerous metabolites like antibiotics and toxins in their surroundings. The presence of the toxins and antibiotics inhibits the phytopathogens present in periphery. Biosurfactants, produced by *Bacillus subtilis* in the biofilm, are inhibitory to other pathogens (Quadros et al. 2011). Biosurfactants have antifungal and antibacterial activity and are cyclic in nature containing amino acid and lipids. Hence, *B. subtilis* biofilm induces resistance in plants. On the other hand, bacteriocins differ from the traditional antibiotics in relatively narrow bactericidal spectrum, i.e., toxic for only closely related to the producing strains (Simons et al. 2020). Strains of *Leuconostoc mesenteroides* produce bacteriocins, but the compounds involved have not been completely purified and characterized yet. In general, bacteriocins produced by *Leuconostoc* are active against *Listeria monocytogenes*, not against lactic acid bacteria. *Leuconostoc mesenteroides* isolated from goats' milk produces an entirely heat labile antilisterial bacteriocin, while *Leuconostoc mesenteroides* inaccessible from cheddar cheese produces a heat stable (100 °C, 30 min) antilisterial bacteriocin (Goswami et al. 2016).

3.2.3 ASR (Acquired Systemic Resistance) and ISR (Induced Systemic Resistance)

ASR and ISR are the mechanisms adopted by plants against phytopathogen attack (Kamle et al. 2020). When a pathogenic microbe comes in contact with the plant host. The plant disease response is triggered and ASR is activated. The plant in response to the elicitor shows some signs of infection such as necrosis. During this pathway of plant immune response, plant is protected against a second invasion of pathogen. Concentration of salicylic acid and proteins related to pathogenesis (PRPs) is accumulated in higher amount during ASR pathway (Romera et al. 2019). Nonpathogenic organisms or abiotic factors trigger the ISR pathway. This immune system of plant defense is activated through resistance cascade of jasmonic acid and ethylene (Zamioudis et al. 2015; Verbon et al. 2017; Martínez-Medina et al. 2017).

3.2.4 Stress Tolerance

PGPR are involved in activation of stress tolerance genes. Biotic and abiotic stresses lead to alteration in plant growth (Mangal et al. 2023). Osmo-protectants like proline maintain water balance under deficient conditions and facilitate water uptake from soil (Sharma et al. 2023). Various antioxidant enzymes such as catalases, peroxidases, and superoxide dismutase are produced by plants that capture free radicals and prevent cells from damage (González 2023). PGPR inoculation to plant greatly influences the antioxidant formation in stress conditions. For instance, an increase in antioxidant peroxidase was observed when tomato plant was inoculated with *Bacillus subtilis* (Nadarajah 2020). Likewise, corn plants inoculated with PGPR *Piriformospora indica* showed a negative impact on biotic stress. The activity of catalase and superoxide dismutase was increased (Gusain et al. 2023). Free radicals produced during normal metabolic functions in cells are unstable molecules and can damage other molecules and are either neutralized directly or indirectly by antioxidant molecules. Antioxidants can participate in enzymatic systems involved in this function indirectly. Flavonoids are the main antioxidant compounds having neutralistic properties. Flavonoids are potent metal capturing compounds and also consist of antioxidant characters. They are well known for their antagonistic property for inflammation, allergy, and viral infection (Manna et al. 2021).

3.2.5 Quorum Quenching

Certain bacterial cells inhibit the growth of a plant pathogen by attacking its cellular machinery. Quorum sensing is a property of phytopathogens by which they release chemical signals or messengers to aggregate other pathogenic microbes to infect the plant. Once the cell attains the cellular maturity (cell density), it starts to sense the presence of bacterial population. The bacterial cell starts to change their metabolism by turning on different sets of genes. Bacterial cells secrete low molecular compounds called autoinducers. Autoinducers bind bacterial cellular receptors and send signals to activate transduction cascade and change the bacterial gene expression. The bacterial pathogen becomes more virulent. To inhibit the action of pathogenic microbes, PGPRs play a vital role in quenching the quorum sensing. PGPRs produce enzyme known as lactonase, it degrades autoinducer, and plant seedlings are rescued. This strategy is well known in lab but yet not tested in field (Olanrewaju et al. 2017).

3.2.6 Bacteriophages

Bacteriophages or bacterial viruses are known to lyse phytopathogens. The isolation and characterization of bacteriophage to strain level eventually leads to lyse target pathogen without affecting other strains. The bacteriophages are sprayed on the infected areas of plant. A mixture of bacteriophages is generally applied on the plants to ensure that mutants do not develop because various sites for bacteriophages binding are present on pathogenic microbe. UV light sensitivity is generally observed in bacteriophages; therefore, they are applied on plants at dark (Buttimer et al. 2017). Some bacteriophages are applied weekly or daily to be effective. Some bacteriophages act as biocontrol agent against bacterial pathogen such as *Xanthomonas campestris* and *Pseudomonas syringae* bacteria that causes bacterial spot and canker disease of tomatoes (Yu et al. 2016; Peitl et al. 2017). Bacteriophages undergo lytic and lysogenic life cycle. Bacteria are treated with phages to rescue with diseases, and the dealing of bacterial infections by phages is identified as phage therapy (Doss et al. 2017; O'Brien 2017; Buttimer et al. 2017).

4 Microbial Interactions in Nutrient Acquisition

Nutrient acquisition is regarded as one of the most limiting factors to crop yield. PGPRs play a vital role in acquisition of nutrients from soil to plants either by mineralization or solubilization of ions (Rashid et al. 2016). PGPR can be inoculated with plants or sprayed in form of fertilizers. Microbe and plant associated exudates are released into environment, and the process of nodulation initiates. Bacteria bind the Nod factors present in plant transcription machinery to receptor kinases containing lysin motif (LysM RKLs). Meanwhile, bacteria get carbon in exchange of activating Nod factor binding (Vishwakarma et al. 2020). Iron deficiency is one of the most important aspects of plant growth. Iron deficiency transcription factor is activated by bacterial volatile compounds. PGPRs secrete ionophores that solubilize or chelate nutrient ions in the nutrient deficient conditions. Ionophores act as carrier molecules or channels that can bind ion and transport into cell or allow many ions to pass across plant cell. Any kind of stress in soil reduces the capacity to acquire metal due to formation of insoluble complexes with carbonates (Bisht et al. 2018). The form of phosphorus which is utilized by the plants for their growth and development is present in insoluble form and made available to plant by concentrating it into soluble forms by phosphate solubilizing bacteria. The most efficient PSM belongs to genera *Bacillus*, *Pseudomonas*, *Rhizobium*, and *Trichoderma* (Freitas et al. 2015). These bacteria in phosphorous limiting conditions produce ionophores that bind the insoluble form of phosphorous and transport them across cell membrane so that it is available to plants. Ionophores also bind essential ions (Zn, K, Na, Mg, Cu, Al) that are unavailable to plant under saline conditions. One of the specific ionophore is siderophores that specifically bind iron under iron-limiting conditions. In alkaline conditions, iron in form of ferric ion is not available to be absorbed by plants. Therefore, to make this form of iron available to plants, PGPRs produce a specific type of compound to make iron accessible to plants. These compounds are known as siderophores that have high affinity to bind ferric iron. Siderophores bind to ferric iron and make a complex with them, and then this complex is transported to cytosol (Rosier et al. 2018). Reduction of the ferric iron bound to siderophore occurs inside the cytosol, and ferric iron is converted to ferrous iron. The ferrous iron form is readily absorbable by plants. A variety of siderophores can be synthesized by bacteria to chelate diverse metal ions. Ionophores are also a specific type of metal chelating compound (Agrahari et al. 2020). They either bind to metal compounds or act as a channel to pass the metal ions. Some plants such as monocots have also been reported to release phytosiderophores in order to chelate metal ions in stress conditions. The affinity of phytosiderophore to chelate metal ions in comparison with microbe-associated siderophore is less.

5 Interactions Among Plant-Microbe and Their Impact on Plant Health

The interactions among plant-microbe and their environment directly influence plant heath. The interactions between plant and neighboring plant, plant and microbe, and microbe and microbe in vicinity are mechanisms that enhance crop production in a sustainable way.

5.1 Root-Root Interaction

The interaction of plant roots with roots of another plant in vicinity is either influenced in a positive or negative way. Positive interactions among roots lead to acquisition of nutrients, defense, and stress tolerance genes are passed. If negative interaction occurs, diseases related genes are passed. For instance, during positive interaction, when mixed cropping is done between leguminous and nonleguminous plants, P levels in plants increase. This is because leguminous plants release carboxylate enzymes which are involved in uptake of P from soil in soluble form and, thus, proliferate growth in neighboring plants (Meller et al. 2020). Various mechanisms are involved in interaction between roots of different plants such as quorum sensing, chemical messengers, or the exudate release by the plants in vicinity. Signal transformation occurs, and the roots are attracted toward the roots of another plant for exchange of materials (Canarini et al. 2019). Nutrient acquisition in plants occurs via presence or absence of certain acids and environmental conditions. When acids like phosphatases and carboxylase are present, nutrients like P and C are acquired, while N acquisition is achieved in absence of these acids (Jin et al. 2015). A specific group of fungus known as *Arbuscular mycorrhiza*, which make symbiotic association with leguminous plants produce carboxylate exudates that help to maintain phosphorous levels in plant roots (Li et al. 2021). Citric acid and fumaric acid are involved in production of biofilm to attract other bacteria. This citric acid is released from plants like cucumber and citrus plants to which bacteria like *Bacillus amyloliquefaciens* is attracted. Fumaric acid released from roots of banana fruit is able to captivate bacterium *Bacillus subtilis* (Zhang et al. 2014). Another mechanism by which the roots of neighboring plants are placed in vicinity is allelopathy. Allelopathy is a mechanism by which the plant growth is enhanced by the growth of neighboring plants.

There are several plants that release phytotoxins, and these compounds inhibit the growth of other plants and reduce the competition for nutrients and resources. For instance, catechin toxin released from plants inhibits growth of plant, viz., roots, seeds, leaf, etc. (Mommer et al. 2016). Some volatile compounds act as allelochemicals and regulate signaling in fungus mycorrhiza. These volatile compounds are organic compounds and regulate the functioning in the rhizosphere. Some plants show inhibitory effect toward the phytotoxin released by their own. These plants are allelopathic and act on other plant species and regulate self-defense mechanisms

(Kong et al. 2019). However, certain non-allelopathic plants also exist and can be resistant. Rhizobiome controls the exudates present around the roots that help maintain a feedback regulation of interactions between plants and microbes. Root system is very competitive; deep root system is less competitive in comparison with wide root system. Hormones are known to induce growth of roots in a positive way (Semchenko et al. 2018). Meanwhile, mycorrhiza fungi play a vital role in water and nutrient absorption. Further, pathogenic microorganisms and application of fungicide present in soil render negative impact on plant growth and the soil microbial population (Berruti et al. 2016).

5.2 Interactions Between Root and Microbe

Interaction among root and microbe is more interesting than between root and root. Root-microbe interactions draw a timeline figure among the relationship between biotic and abiotic factors. The interactions may be symbiotic and nonsymbiotic. Either can be beneficial, neutral, parasitic, and competitive. The symbiotic interactions can be established between rhizobacteria and legume, actinobacteria and root, and mycorrhiza and root (Thapa and Prasanna 2018). The most important relationship is between leguminous crops and microbes (Dong et al. 2019; Sivakumar et al. 2020). Rhizobacteria produce Nod factors that are received by the leguminous plant receptors. Nod factors act as inducing signals, and formation of nodule takes place which promotes cell growth (Vishwakarma et al. 2020). Atmospheric nitrogen is then processed into ammonia in this organ and further used in protein synthesis (Igiehon and Babalola 2018). This process is self-regulatory through feedback mechanism (Pankievicz et al. 2019). The process of nodulation is studied in different symbionts, viz., *Sinorhizobium meliloti*, *Mesorhizobium loti*, and *Bradyrhizobium japonicum* (Lorite et al. 2018). Flavonoids that are released into soil by legume plants activate the protein such as NodD, and this protein in turn activates nodulation genes which are encoded by Sym plasmids (Larrainzar and Wienkoop 2017). A change in protein content within the nodule was observed in *Medicago truncatula* during the formation of leghaemoglobin and enolase isoforms (Trouvelot et al. 2014). Ethylene was reported to be an important factor responsible for symbioses and nodulation of the host. In ethylene-insensitive mutant of *M. truncatula* (skl) that had a defective ethylene pathway, hyper nodulation was observed. Nod factors are also detected to be present in soybean while inoculated with *B. Japonicum*. Calcium-dependent protein kinase (CDPK) activates symbioses process. The presence of antioxidant enzymes such as lectins is involved in attachment of rhizobacteria in infection process. Pathogenesis related protein (PR10) isoforms were identified in *M. truncatula* roots occupied by *S. meliloti* while doing its proteomic analysis. Nutrient exchange influences symbiosis, and bacteroid present in roots helps in nutrient exchange (Nadarajah 2017). The membranes of bacteroid contain proteins, nod factors, receptors, and protein digesting enzymes (Liu et al. 2018). The nodule itself contains enzymes for synthesis of C, N, and Fe, transport, and defense (Wang et al. 2016). ABC transporters existing in symbiotic bacteria are responsible for nutrient transfer and nodulation (Baslam et al. 2020).

Actinomycetes are also reported to form interactions with plants. They form symbiotic relationship between angiosperms, for instance, *Frankia alni* and *Alnus* sp. These species secrete a set of hydrolytic enzymes that mediate interaction among these species (Begum et al. 2019). Mycorrhizal fungi are also known to form a symbiotic relationship. The fungus infects the leguminous crops and colonizes the roots. Mycorrhiza lives separately in plant membrane and does not harm the plant exchange system. The fungus provides P to plant and takes carbohydrates and lipids from plant (Nouioui et al. 2019). *M. truncatula* inoculated with *Glomus mosseae* showed better nutrition and stress tolerance capacity (Nadarajah and Kumar 2019). *Trichoderma* spp. that are used as biocontrol agents also make symbiotic associations with leguminous plants. The species of *T. harzianum* is known to produce cell wall degrading enzyme proteases that destroy fungal cell wall (Slama et al. 2019). Another species *T. asperellum* produces PR proteins for defense (Nadarajah 2018). Table 2 provides the plant-microbe interactions and their mechanism of action and significance.

5.3 *Interaction Between Microbe and Microbe*

Microbe-microbe interaction can be pathogenic and competitive. These interactions can occur within pathogen, endophytes, succession by microbes, and biotic and abiotic factors. Pathogens present in phyllosphere influence the microbial community (Nouioui et al. 2019). Figure 3 shows beneficial, neutral, and non-beneficial interaction among the microbes present in plant surrounding. Microbial community is affected when infected by SBL in maize. When there is pathogen infection in plant, it is also open for other infections. Mildew pathogen infects Brassicaceae, when it is infected with white rust. *A. thaliana* and *Albugo laibachii* have more susceptibility toward *Phytophthora infestans* pathogen (Prince et al. 2017). Quorum sensing (QS) and biofilm production by bacterial population are utilized for beneficial interactions (Chaudhry et al. 2021). QS mediates increase in pathogenicity and

Table 2 PGPR used in different crops and their mechanism of action and significance

Crop	PGPR	Mechanisms of action	Significance	References
Maize	*Bacillus subtilis*	Production of plant growth regulators (auxins, cytokinins, and gibberellins)	Enhance root development, increase plant growth and yield	Kumar et al. (2020)
Wheat	*Azospirillum* spp.	Nitrogen fixation and synthesis of phytohormones	Improve nutrient uptake, plant growth, and yield	Maksimov et al. (2011)
Tomato	*Bacillus amyloliquefaciens*	Production of antimicrobial compounds	Control of fungal pathogens; improve plant growth and yield	Singh et al. (2016)
Soybean	*Rhizobium* spp.	Plant growth regulator production and N2 fixation	Improve nutrient uptake, plant growth, and yield	Summuna et al. (2019)

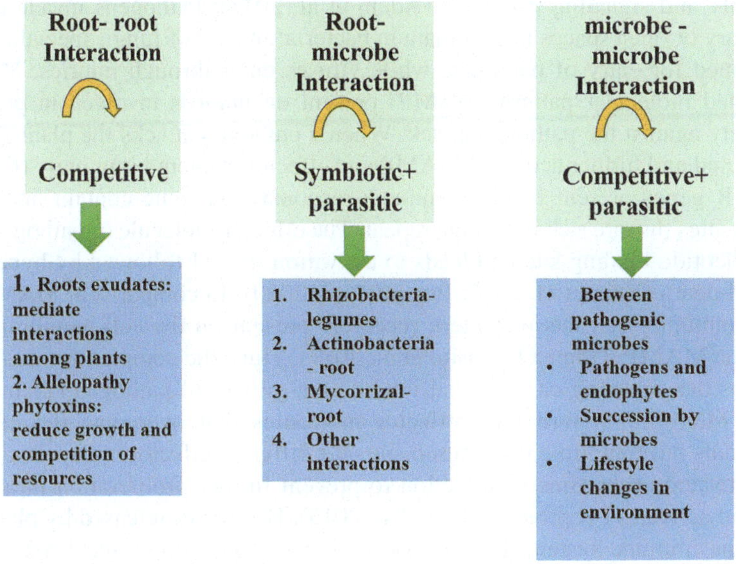

Fig. 3 Various interactions among plant-microbe in phyllosphere and rhizosphere region

virulence in pathogenic microbe by producing signals and effect plant transcriptome and proteome (Rajesh and Ravishankar Rai 2014). Hyperparasitism, competition, and/or antibiosis by endophytes inhibits foliar pathogens. Powdery mildew of *Erysiphe* sp. is inhibited by fungal endophytes (Ludlow et al. 2019). Some endophytes act as biocontrol agents by producing antibiotics such as polyketide synthase (Jakuschkin et al. 2016). *Ulocladium atrum*, *Truncaztella angustata*, and *T. atroviride* showed resistance to infection in *P. trichocarpa* in contrast to rust pathogens (Nadarajah 2016). Inhibitors and enzymes of quorum sensing and quenching mechanism are used by endophytes to inhibit the signaling process. AHL lactonase enzyme inhibits *Erwinia carotovora*, *Bacillus thuringiensis*, and *Enterobacter asburiae*. Trehalose enzyme not only stabilizes membranes and enzymes in drought conditions but raises disease resistance by activating systemic resistance (ISR) (Li et al. 2021). Co-inoculation of *Fusarium verticillioides* with the biotrophic pathogen *Ustilago maydis* inhibited its growth. *A. laibachii* is inhibited by biocontrol agents *Moesziomyces* sp. and *Ustilaginales* through the secretion of hydrolase (Eitzen et al. 2021).

6 Genetic Level Regulatory Pathway Adapted by PGPR for Defense Mechanisms Against Phytopathogens

Plants and microorganisms show beneficial effects toward each other. But when the interaction is harmful for the host, immune system is triggered (Yang et al. 2017). The plant like human and animal consists of immune system containing physiological and biochemical structural components (Yu et al. 2018). Plant has innate immunity and signaling pathways (Ádám et al. 2018). Pathogens invade through any injury or open spaces like stomata in bacteria. In case of fungi, specialized sacs are formed for entry of pathogen, while viruses enter through injuries. Microbe-associated molecular pattern (MAMP) present on plant is involved in providing immunity against the pathogen attack. When a pathogen attacks the plant, MAMP is activated and further activated MAMP and effector triggered immunity (Cui et al. 2015). R genes present in plant immune responsive genome contain nucleotide-binding sites that are rich in leucine repeat. The effector molecule or pathogen binds the nucleotide-binding site and leads to activation of ETI followed by hypersensitive cell death (Neik et al. 2017; Begum et al. 2019). In comparison, MAMP triggered immunity consists of pattern receptors present on the cell membranes and binds the MAMP (Gómez Expósito et al. 2017). Thus, the complex formed in turn activates the signaling cascade that inhibits the growth of pathogen (Zhang et al. 2014). Microbes in turn have effector molecules that overcome the effect of PTI. Plants have multiple recognition sites for different effector molecules. SAR is also activated at the time of infection to prevent further proliferation into distant cells and activates PR proteins (Lee et al. 2015). Hormones released by plants also affect the immune system. Induced systemic resistance (ISR) and SAR are well known to be activated through these hormones (Mhlongo et al. 2018). SAR is responsible to produce signals after infection in plant (Fiorilli et al. 2020). Different chemicals such as methyl salicylate, glycerol-3-phosphate, azelaic acid, and pipecolic acid are involved in signal transfer (Avila-Quezada et al. 2018; Nadarajah et al. 2021; Chadha 2019).

7 Harmful Interactions

Plant-microbe interaction is beneficial as well as harmful for plant growth and development. PGPRs are found in closeness with plant roots and improve plant health (Thynne et al. 2015). Meanwhile, evolving pathogenic microorganisms evade host's innate immune system. The pathogenic microorganisms evolve their host range and virulence and cause disease in same or new ones (Goss 2015). Better approaches to understand the mechanisms behind pathogen invasion lead to disease control through genetic and defense strategies (Masri et al. 2015). There are several mechanisms known for emerging pathogens: (a) novel host found for endemic bacteria; (b) microorganisms turn to be pathogenic after being endemic; (c) microbe becomes pathogenic to novel plants after being introduced to new environment; and (d) insect vector when feed on new host spread harmful microbes to plants (Singh et al. 2021). Under abiotic stress conditions (drought, heat, saline, alkaline, etc.), plants become more susceptible to pathogenic attack. Different tools and techniques are known to obtain information on emerging population. Databases of phytopathogens are also available to study plant-microbe interactions. Species of *Xanthomonas*, *Ralstonia solanacearum*, *Xylella fastidiosa*, and *Pseudomonas syringae* are known to cause disease in plants. Exposed plant parts such as natural openings and wounds are attacked by the pathogenic microbe and trigger the immune response (Zoledowska et al. 2019). Abiotic factors that regulate interaction of plant with pathogens are host vulnerability, population size, and climate (Brader et al. 2017). Few nonpathogenic microbes behave as pathogenic with alterations in host or change in size of microbial population. A number of plant-microbiota members are known for increasing plant resistance from phytopathogenic microbial species, known as biocontrol agents (Brader et al. 2017). Hence, modern tools can help understand strange behavior of plant-microbe interactions (Shelake et al. 2019).

8 Conclusion

Plant-microbe interactions can also have important implications for agriculture. Plant health is inclined by different reasons, including the interactions between plants and microbes. While some microorganism can be disadvantageous to plants, many others are beneficial and can help plants grow and thrive. Understanding these interactions is important for maintaining healthy plant populations and promoting sustainable agriculture practices. Soil microbial structure provides insight on plant microbe-associated beneficial interactions that play an important role in nutrient acquisition, stress response, and disease resistance. Rather than identifying microbes in environment, it is beneficial to know the roles of microbes that singly and in combination beneficial for plant health.

References

Ádám AL, Nagy ZÁ, Kátay G, Mergenthaler E, Viczián O (2018) Signals of systemic immunity in plants: progress and open questions. Int J Mol Sci 19(4):1146. https://doi.org/10.3390/ijms19041146

Agrahari RK, Singh P, Koyama H, Panda SK (2020) Plant-microbe interactions for sustainable agriculture in the postgenomic era. Curr Genomics 21(3):168–178. https://doi.org/10.2174/1389202921999200505082116

Agri U, Chaudhary P, Sharma A (2021) In vitro compatibility evaluation of agriusable nanochitosan on beneficial plant growth-promoting rhizobacteria and maize plant. Natl Acad Sci Lett 44:555–559. https://doi.org/10.1007/s40009-021-01047-w

Agri U, Chaudhary P, Sharma A, Kukreti B (2022) Physiological response of maize plants and its rhizospheric microbiome under the influence of potential bioinoculants and nanochitosan. Plant Soil 474:451–468. https://doi.org/10.1007/s11104-022-05351-2

Akhtar SS, Mekureyaw MF, Pandey C, Roitsch T (2020) Role of cytokinins for interactions of plants with microbial pathogens and pest insects. Front Plant Sci 10:1777. https://doi.org/10.3389/fpls.2019.01777

Ambawade MS, Pathade GR (2015) Production of gibberellic acid by Bacillus siamensis BE 76 isolated from banana plant (Musa spp.). Int J Sci Res 4(7):394–398

Ansari A, Ali A, Asif M (2017) Biologically active pyrazole derivatives. New J Chem 41(1):16–41

Avila-Quezada GD, Esquivel JF, Silva-Rojas HV, Leyva-Mir SG, Garcia-Avila CJ, Quezada-Salinas A, Noriega-Orozco L, Rivas-Valencia P, Ojeda-Barrios D, Melgoza-Castillo A (2018)

Emerging plant diseases under a changing climate scenario: threats to our global food supply. Emirates J Food Agric 30:443–450. https://doi.org/10.9755/ejfa.2018.v30.i6.1715

Baslam M, Mitsui T, Sueyoshi K, Ohyama T (2020) Recent advances in carbon and nitrogen metabolism in C3 plants. Int J Mol Sci 22(1):318. https://doi.org/10.3390/ijms22010318

Begum N, Qin C, Ahanger MA, Raza S, Khan MI, Ashraf M, Ahmed N, Zhang L (2019) Role of arbuscular mycorrhizal fungi in plant growth regulation: implications in abiotic stress tolerance. Front Plant Sci 10:1068. https://doi.org/10.3389/fpls.2019.01068

Berruti A, Lumini E, Balestrini R, Bianciotto V (2016) Arbuscular mycorrhizal fungi as natural biofertilizers: let's benefit from past successes. Front Microbiol 6:1559. https://doi.org/10.3389/fmicb.2015.01559

Bhaduri D, Pal S, Purakayastha TJ, Chakraborty K, Yadav RS, Akhtar MS (2015) Soil quality and plant-microbe interactions in the rhizosphere. Sustain Agric Rev 17:307–335. https://doi.org/10.1007/978-3-319-16742-8_9

Bisht N, Sharma B, Joshi N, Tewari L (2018) Role of microbial siderophore in Iron acquisition and amelioration of plant stress in saline soil. Biotech Today 8(2):7–15. https://doi.org/10.5958/2322-0996.2018.00012.1

Böhme H, Masepohl B (2018) Differentiation of Vegetative Anabaena Cells into Nitrogen-Fixing Heterocysts. Routledge, In Plant responses to environmental stresses, pp 91–110

Bonfante P, Genre A (2010) Mechanisms underlying beneficial plant–fungus interactions in mycorrhizal symbiosis. Nat Commun 1:48. https://doi.org/10.1038/ncomms1046

Brader G, Compant S, Vescio K, Mitter B, Trognitz F, Ma LJ, Sessitsch A (2017) Ecology and genomic insights into plant-pathogenic and plant-nonpathogenic endophytes. Annu Rev Phytopathol 55:61–83. https://doi.org/10.1146/annurev-phyto-080516-035641

Broadley MR, White PJ, Hammond JP, Zelko I, Lux A (2017) Zinc in plants. New Phytol 219(4):1127–1142. https://doi.org/10.1111/nph.14726

Buttimer C, McAuliffe O, Ross RP, Hill C, O'Mahony J, Coffey A (2017) Bacteriophages and bacterial plant diseases. Front Microbiol 8:34. https://doi.org/10.3389/fmicb.2017.00034

Cabugao KG, Timm CM, Carrell AA, Childs J, Lu TS, Pelletier DA, Weston DJ, Norby RJ (2017) Root and rhizosphere bacterial phosphatase activity varies with tree species and soil phosphorus availability in Puerto Rico tropical Forest. Front Plant Sci 8:1834. https://doi.org/10.3389/fpls.2017.01834

Canarini A, Kaiser C, Merchant A, Richter A, Wanek W (2019) Root exudation of primary metabolites: mechanisms and their roles in plant responses to environmental stimuli. Front Plant Sci 10:157. https://doi.org/10.3389/fpls.2019.00157

Chadha S (2019) Plant–microbe interaction: gene-to-metabolite network. In: Bioactive molecules in plant defense: signaling in growth and stress, pp 75–100. https://doi.org/10.1007/978-3-030-27165-75

Chandra S, Askari K, Kumari M (2018) Optimization of indole acetic acid production by isolated bacteria from *Stevia rebaudiana* rhizosphere and its effects on plant growth. J Genet Eng Biotechnol 16(2):581–586. https://doi.org/10.1016/j.jgeb.2018.09.001

Chaudhary P, Sharma A (2019) Response of nanogypsum on the performance of plant growth promotory bacteria recovered from nanocompound infested agriculture field. Environ Ecol 37:363–372

Chaudhary A, Parveen H, Chaudhary P, Bhatt P, Khatoon H (2021a) Rhizospheric microbes and their mechanism. In: Bhatt P, Gangola S, Udayanga D, Kumar G (eds) Microbial technology for sustainable environment. Springer, Singapore. https://doi.org/10.1007/978-981-16-3840-4_6

Chaudhary P, Chaudhary A, Parveen H, Rani A, Kumar G, Kumar A, Sharma A (2021b) Impact of nanophos in agriculture to improve functional bacterial community and crop productivity. BMC Plant Biol 21:519. https://doi.org/10.1186/s12870-021-03298-7

Chaudhary P, Khati P, Chaudhary A, Gangola S, Kumar R, Sharma A (2021c) Bioinoculation using indigenous *Bacillus* spp. improves growth and yield of *Zea mays* under the influence of nanozeolite. 3 Biotech 11:11. https://doi.org/10.1007/s13205-020-02561-2

Chaudhary P, Khati P, Chaudhary A, Maithani D, Kumar G, Sharma A (2021d) Cultivable and metagenomic approach to study the combined impact of nanogypsum and *Pseudomonas taiwanensis* on maize plant health and its rhizospheric microbiome. PLoS One 16:e0250574. https://doi.org/10.1371/journal.pone.0250574

Chaudhary P, Khati P, Gangola S, Kumar A, Kumar R, Sharma A (2021e) Impact of nanochitosan and *Bacillus* spp. on health, productivity and defence response in *Zea mays* under field condition. 3 Biotech 11:237. https://doi.org/10.1007/s13205-021-02790-z

Chaudhary P, Parveen H, Gangola S, Kumar G, Bhatt P, Chaudhary A (2021f) Plant Growth Promoting Rhizobacteria and their application in sustainable crop production. In: Bhatt P, Gangola S, Udayanga D, Kumar G (eds) Microbial technology for sustainable environment. Springer, Singapore. https://doi.org/10.1007/978-981-16-3840-4_13

Chaudhary A, Chaudhary P, Upadhyay A, Kumar A, Singh A (2022a) Effect of gypsum on plant growth promoting rhizobacteria. Environ Ecol 39(4A):1248–1256

Chaudhary P, Agri U, Chaudhary A, Kumar A, Kumar G (2022b) Endophytes and their potential in biotic stress management and crop production. Front Microbiol 13:933017. https://doi.org/10.3389/fmicb.2022.933017

Chaudhary P, Singh S, Chaudhary A, Agri U, Bhandari G (2022c) Chapter 3: Agrochemicals and their effects on soil microbial population. In: Soni R, Suyal D, Goel R (eds) Plant protection: from chemicals to biologicals. De Gruyter, Berlin, Boston, pp 45–58. https://doi.org/10.1515/9783110771558-003

Chaudhary P, Singh S, Chaudhary A, Sharma A, Kumar G (2022d) Overview of biofertilizers in crop production and stress management for sustainable agriculture. Front Plant Sci 13:930340. https://doi.org/10.3389/fpls.2022.930340

Chaudhary P, Chaudhary A, Agri U, Khatoon H, Singh A (2022e) Recent trends and advancements for agro-environmental sustainability at higher altitudes. In: Goel R, Soni R, Suyal DC, Khan M (eds) Survival strategies in cold-adapted microorganisms. Springer, Singapore. https://doi.org/10.1007/978-981-16-2625-8_19

Chaudhary P, Xu M, Ahamad L, Chaudhary A, Kumar G, Adeleke BS, Verma KK, Hu D-M, Širić I, Kumar P, Popescu SM, Abou Fayssal S (2023a) Application of synthetic consortia for improvement of soil fertility, pollution remediation, and agricultural productivity: a review. Agronomy 13:643. https://doi.org/10.3390/agronomy13030643

Chaudhary P, Chen S, Rajput VD, Jaiswal DK (2023b) Editorial: bioinoculants with nano-compounds to improve soil health: a step toward sustainable agriculture. Front Environ Sci 11:1270002. https://doi.org/10.3389/fenvs.2023.1270002

Chaudhary P, Chen S, Rajput VD, Jaiswal DK (eds) (2023c) Bioinoculants with nano-compounds to improve soil health: a step toward sustainable agriculture. Frontiers Media SA, Lausanne. https://doi.org/10.3389/978-2-8325-3287-4

Chaudhry V, Runge P, Sengupta P, Doehlemann G, Parker JE, Kemen E (2021) Shaping the leaf microbiota: plant-microbe-microbe interactions. J Exp Bot 72(1):36–56. https://doi.org/10.1093/jxb/eraa417

Courty PE, Smith P, Koegel S, Redecker D, Wipf D (2015) Inorganic nitrogen uptake and transport in beneficial plant root-microbe interactions. Crit Rev Plant Sci 34(1–3):4–16. https://doi.org/10.1080/07352689.2014.897897

Cui H, Tsuda K, Parker JE (2015) Effector-triggered immunity: from pathogen perception to robust defense. Annu Rev Plant Biol 66:487–511. https://doi.org/10.1146/annurev-arplant-050213-040012

Das PP, Singh KR, Nagpure G, Mansoori A, Singh RP, Ghazi IA, Kumar A, Singh J (2022) Plant-soil-microbes: a tripartite interaction for nutrient acquisition and better plant growth for sustainable agricultural practices. Environ Res 214:113821. https://doi.org/10.1016/j.envres.2022.113821

de Jesus Sousa JA, Olivares FL (2016) Plant growth promotion by Streptomycetes: ecophysiology, mechanisms and applications. Chem Biol Technol Agric 3:24. https://doi.org/10.1186/s40538-016-0073-5

Dong CJ, Wang LL, Li Q, Shang QM (2019) Bacterial communities in the rhizosphere, phyllosphere and endosphere of tomato plants. PLoS One 14(11):e0223847. https://doi.org/10.1371/journal.pone.0223847

Doss J, Culbertson K, Hahn D, Camacho J, Barekzi N (2017) A review of phage therapy against bacterial pathogens of aquatic and terrestrial organisms. Viruses 9(3):50. https://doi.org/10.3390/v9030050

Duca DR, Rose DR, Glick BR (2018) Indole acetic acid overproduction transformants of the rhizobacterium Pseudomonas sp. UW4. Antonie Van Leeuwenhoek 111:1645–1660. https://doi.org/10.1007/s10482-018-1051-7

Eitzen K, Sengupta P, Kroll S, Kemen E, Doehlemann G (2021) A fungal member of the *Arabidopsis thaliana* phyllosphere antagonizes *Albugo laibachii* via a GH25 lysozyme. Elife 10:e65306. https://doi.org/10.7554/eLife.65306

Fiorilli V, Catoni M, Lanfranco L, Zabet NR (2020) Editorial: interactions of plants with bacteria and fungi: molecular and epigenetic plasticity of the host. Front Plant Sci 11:274. https://doi.org/10.3389/fpls.2020.00274

Fosu-Mensah BY, Vlek PL, Manske G, Mensah M (2015) The influenec of Azolla pinnata on floodwater chemistry, grain yield and nitrogen uptake of rice in dano, southwestern Burkina Faso. J Agric Sci 7(8):118. https://doi.org/10.5539/jas.v7n8p118

Freitas MA, Medeiros FH, Carvalho SP, Guilherme LR, Teixeira WD, Zhang H, Paré PW (2015) Augmenting iron accumulation in cassava by the beneficial soil bacterium Bacillus subtilis (GBO3). Front Plant Sci 6:596. https://doi.org/10.3389/fpls.2015.00596

Gómez Expósito R, de Bruijn I, Postma J, Raaijmakers JM (2017) Current insights into the role of rhizosphere bacteria in disease suppressive soils. Front Microbiol 8:2529. https://doi.org/10.3389/fmicb.2017.02529

González EM (2023) Drought stress tolerance in plants. Int J Mol Sci 24(7):6562. https://doi.org/10.3390/ijms24076562

Goss EM (2015) Genome-enabled analysis of plant-pathogen migration. Annu Rev Phytopathol 53:121–135. https://doi.org/10.1146/annurev-phyto-080614-115936

Goswami D, Thakker JN, Dhandhukia PC (2016) Portraying mechanics of plant growth promoting rhizobacteria (PGPR): a review. Cogent Food Agric 2(1):1127500. https://doi.org/10.1080/23311932.2015.1127500

Gusain S, Joshi S, Joshi R (2023) Sensing, signalling, and regulatory mechanism of cold-stress tolerance in plants. Plant Physiol Biochem 197:107646. https://doi.org/10.1016/j.plaphy.2023.107646

Habibi S, Djedidi S, Prongjunthuek K, Mortuza MF, Ohkama-Ohtsu N, Sekimoto H et al (2014) Physiological and genetic characterization of rice nitrogen fixer PGPR isolated from rhizosphere soils of different crops. Plant Soil 379(1–2):51–66. https://doi.org/10.1007/s11104-014-2035-7

Hacquard S (2016) Disentangling the factors shaping microbiota composition across the plant holobiont. New Phytol 209(2):454–457. https://doi.org/10.1111/nph.13760

Igiehon NO, Babalola OO (2018) Rhizosphere microbiome modulators: contributions of nitrogen fixing bacteria towards sustainable agriculture. Int J Environ Res Public Health 15(4):574. https://doi.org/10.3390/ijerph15040574

Jaiswal A, Das K, Koli DK, Pabbi S (2018) Characterization of cyanobacteria for IAA and siderophore production and their effect on rice seed germination. Int J Curr Microbiol App Sci 5:212–222

Jakuschkin B, Fievet V, Schwaller L, Fort T, Robin C, Vacher C (2016) Deciphering the Pathobiome: intra- and interkingdom interactions involving the pathogen Erysiphe alphitoides. Microb Ecol 72(4):870–880. https://doi.org/10.1007/s00248-016-0777-x

Jambon I, Thijs S, Weyens N, Vangronsveld J (2018) Harnessing plant-bacteria-fungi interactions to improve plant growth and degradation of organic pollutants. J Plant Interact 13(1):119–130. https://doi.org/10.1080/17429145.2018.1441450

Jin J, Tang C, Sale P (2015) The impact of elevated carbon dioxide on the phosphorus nutrition of plants: a review. Ann Bot 116(6):987–999. https://doi.org/10.1093/aob/mcv088

Kamle M, Borah R, Bora H, Jaiswal AK, Singh RK, Kumar P (2020) Systemic acquired resistance (SAR) and induced systemic resistance (ISR): role and mechanism of action against phytopathogens. In: Fungal biotechnology and bioengineering, pp 457–470. https://doi.org/10.1007/978-3-030-41870-0_20

Khati P, Chaudhary P, Gangola S, Bhatt P, Sharma A (2017) Nanochitosan supports growth of *Zea mays* and also maintains soil health following growth. 3 Biotech 7:81. https://doi.org/10.1007/s13205-017-0668-y

Khati P, Parul, Bhatt P, Nisha Kumar R, Sharma A (2018) Effect of nanozeolite and plant growth promoting rhizobacteria on maize. 3 Biotech 8:141. https://doi.org/10.1007/s13205-018-1142-1

Kong CH, Xuan TD, Khanh TD, Tran HD, Trung NT (2019) Allelochemicals and signaling chemicals in plants. Molecules 24(15):2737. https://doi.org/10.3390/molecules24152737

Kukreti B, Sharma A, Chaudhary P, Agri U, Maithani D (2020) Influence of nanosilicon dioxide along with bioinoculants on *Zea mays* and its rhizospheric soil. 3 Biotech 10:345. https://doi.org/10.1007/s13205-020-02329-8

Kumar V, Menon S, Agarwal H, Gopalakrishnan D (2017) Characterization and optimization of bacterium isolated from soil samples for the production of siderophores. Resour-EfficTechnol 3(4):434–439. https://doi.org/10.1016/j.reffit.2017.04.004

Kumar A, Verma JP (2018) Does plant—microbe interaction confer stress tolerance in plants: a review? Microbiol Res 207:41–52. https://doi.org/10.1016/j.micres.2017.11.004

Kumari S, Sharma A, Chaudhary P, Khati P (2020) Management of plant vigor and soil health using two agriusable nanocompounds and plant growth promotory rhizobacteria in Fenugreek. 3 Biotech 10:1–11. https://doi.org/10.1007/s13205-020-02448-2

Kumar V, Jain L, Jain SK, Chaturvedi S, Kaushal P (2020) Bacterial endophytes of rice (Oryza sativa L.) and their potential for plant growth promotion and antagonistic activities. South African J Bot 134:50–63. https://doi.org/10.1016/j.sajb.2020.02.017

Kumari H, Khati P, Gangola S, Chaudhary P, Sharma A (2021) Performance of plant growth promotory rhizobacteria on Maize and soil characteristics under the influence of TiO_2 nanoparticles. Pantnagar J Res 19:28–39

Larrainzar E, Wienkoop S (2017) A proteomic view on the role of legume symbiotic interactions. Front Plant Sci 8:1267. https://doi.org/10.3389/fpls.2017.01267

Lebrazi S, Fadil M, Chraibi M, Fikri-Benbrahim K (2020) Screening and optimization of indole-3-acetic acid production by *Rhizobium* sp. strain using response surface methodology. J Genet Eng Biotechnol 18(1):1–10. https://doi.org/10.1186/s43141-020-00035-9

Lee HJ, Park YJ, Seo PJ, Kim JH, Sim HJ, Kim SG, Park CM (2015) Systemic immunity requires SnRK2.8-mediated nuclear import of NPR1 in arabidopsis. Plant Cell 27(12):3425–3438. https://doi.org/10.1105/tpc.15.00371

Li T, Wu S, Yang W, Selosse MA, Gao J (2021) How mycorrhizal associations influence orchid distribution and population dynamics. Front Plant Sci 12:647114. https://doi.org/10.3389/fpls.2021.647114

Liu A, Contador CA, Fan K, Lam HM (2018) Interaction and regulation of carbon, nitrogen, and phosphorus metabolisms in root nodules of legumes. Front Plant Sci 9:1860. https://doi.org/10.3389/fpls.2018.01860

Lorite MJ, Estrella MJ, Escaray FJ, Sannazzaro A, Videira e Castro IM, Monza J, León-Barrios M (2018) The Rhizobia-Lotus symbioses: deeply specific and widely diverse. Front Microbiol 9:2055. https://doi.org/10.3389/fmicb.2018.02055

Ludlow EJ, Vassiliadis S, Ekanayake PN, Hettiarachchige IK, Reddy P, Sawbridge TI, Rochfort SJ, Spangenberg GC, Guthridge KM (2019) Analysis of the indole diterpene gene cluster for biosynthesis of the epoxy-Janthitrems in *Epichloë* endophytes. Microorganisms 7(11):560. https://doi.org/10.3390/microorganisms7110560

Maheshwari DK, Dheeman S, Agarwal M (2015) Phytohormone-producing PGPR for sustainable agriculture. In: Bacterial metabolites in sustainable agroecosystem, vol 12, pp 159–182. https://doi.org/10.1007/978-3-319-24654-3_7

Maksimov IV, Abizgildina RR, Pusenkova LI, Bulgakov VP (2011) Effect of bacterial phytohormones on plant growth stimulating activity of rhizospheric pseudomonads. Microbiology 80(6):732–739. https://doi.org/10.1134/S0026261711060136

Mangal V, Lal MK, Tiwari RK, Altaf MA, Sood S, Kumar D, Aftab T (2023) Molecular insights into the role of reactive oxygen, nitrogen and sulphur species in conferring salinity stress tolerance in plants. J Plant Growth Regul 42(2):554–574. https://doi.org/10.1007/s00344-022-10591-8

Manna M, Thakur T, Chirom O, Mandlik R, Deshmukh R, Salvi P (2021) Transcription factors as key molecular target to strengthen the drought stress tolerance in plants. Physiol Plant 172(2):847–868. https://doi.org/10.1111/ppl.13268

Manzar N, Jayaswal D, Kishan G, Chauhan S, Joshi D, Singh J, Chaudhary P, Jamali H, Kushwaha P, Parveen H, Kashyap AS, Sahu PK, Singh UB (2022) Application of molecular ecology approaches in sustainable agriculture for a better understanding of plant–microbiome interactions. In: Singh UB, Sahu PK, Singh HV, Sharma PK, Sharma SK (eds) Rhizosphere microbes. Microorganisms for sustainability, vol 40. Springer, Singapore. https://doi.org/10.1007/978-981-19-5872-4_3

Martínez-Medina A, Van Wees SCM, Pieterse CMJ (2017) Airborne signals from *Trichoderma* fungi stimulate iron uptake responses in roots resulting in priming of jasmonic acid dependent defences in shoots of *Arabidopsis thaliana* and *Solanum lycopersicum*. Plant Cell Environ 40:2691–2705. https://doi.org/10.1111/pce.13016

Masri L, Branca A, Sheppard AE, Papkou A, Laehnemann D, Guenther PS, Prahl S, Saebelfeld M, Hollensteiner J, Liesegang H, Brzuszkiewicz E, Daniel R, Michiels NK, Schulte RD, Kurtz J, Rosenstiel P, Telschow A, Bornberg-Bauer E, Schulenburg H (2015) Host-pathogen coevolution: the selective advantage of *Bacillus thuringiensis* virulence and its cry toxin genes. PLoS Biol 13(6):e1002169. https://doi.org/10.1371/journal.pbio.1002169

Meller S, Frossard E, Spohn M, Luster J (2020) Plant nutritional status explains the modifying effect of provenance on the response of beech sapling root traits to differences in soil nutrient supply. Front For Glob Change 3:535117. https://doi.org/10.3389/ffgc.2020.535117

Mhlongo MI, Piater LA, Madala NE, Labuschagne N, Dubery IA (2018) The chemistry of plant-microbe interactions in the rhizosphere and the potential for metabolomics to reveal signaling related to defense priming and induced systemic resistance. Front Plant Sci 9:112. https://doi.org/10.3389/fpls.2018.00112

Mommer L, Hinsinger P, Prigent-Combaret C, Visser EJ (2016) Advances in the rhizosphere: stretching the interface of life. Plant Soil 407:1–8. https://doi.org/10.1007/s11104-016-3040-9

Nadarajah KK (2016) Rhizosphere interactions: life below ground. In: Plant-microbe interaction: an approach to sustainable agriculture, pp 3–23. https://doi.org/10.1007/978-981-10-2854-0_1

Nadarajah KK (2017) *Rhizobium* in rice yield and growth enhancement. In: *Rhizobium* biology and biotechnology, pp 83–103. https://doi.org/10.1007/978-3-319-64982-5_6

Nadarajah KK (2018) In silico identification of plant-derived secondary metabolites in defense. In: In silico approach for sustainable agriculture, pp 275–293. https://doi.org/10.1007/978-981-13-0347-0_16

Nadarajah K, Kumar IS (2019) Drought response in rice: The miRNA story. Int J Mol Sci 20(15):3766. https://doi.org/10.3390/ijms20153766

Nadarajah KK (2020) ROS homeostasis in abiotic stress tolerance in plants. Int J Mol Sci 21(15):5208. https://doi.org/10.3390/ijms21155208

Nadarajah K, Hamid ANW, Rahman ANSN (2021) SA-mediated regulation and control of abiotic stress tolerance in rice. Int J Mol Sci 22(11):5591. https://doi.org/10.3390/ijms22115591

Nadarajah K, Abdul Rahman NSN (2021) Plant–microbe interaction: aboveground to belowground, from the good to the bad. Int J Mol Sci 22(19):10388. https://doi.org/10.3390/ijms221910388

Neik TX, Barbetti MJ, Batley J (2017) Current status and challenges in identifying disease resistance genes in *Brassica napus*. Front Plant Sci 8:1788. https://doi.org/10.3389/fpls.2017.01788

Nouioui I, Cortés-Albayay C, Carro L, Castro JF, Gtari M, Ghodhbane-Gtari F, Klenk HP, Tisa LS, Sangal V, Goodfellow M (2019) Genomic insights into plant-growth-promoting potentialities of the genus *Frankia*. Front Microbiol 10:1457. https://doi.org/10.3389/fmicb.2019.01457

O'Brien PA (2017) Biological control of plant diseases. Australas Plant Pathol 44:1–12. https://doi.org/10.1007/s13313-017-0481-4

Olanrewaju OS, Glick BR, Babalola OO (2017) Mechanisms of action of plant growth promoting bacteria. World J Microbiol Biotechnol 33:1–16. 2364-9. https://doi.org/10.1007/s11274-017-

Pankievicz V, Irving TB, Maia LG, Ané JM (2019) Are we there yet? The long walk towards the development of efficient symbiotic associations between nitrogen-fixing bacteria and non-leguminous crops. BMC Biol 17(1):1–17. https://doi.org/10.1186/s12915-019-0710-0

Parveen H, Chaudhary A, Chaudhary P, Sultana R, Khati P, Kumar G, Khati P, Rana M, Bhatt P (2022) Omics approaches for the production of the microbial enzymes and applications. In: Bhatt P (ed) Industrial applications of microbial enzymes, 1st edn. CRC Press. https://doi.org/10.1201/9781003202998-16

Peitl V, Štefanović M, Karlović D (2017) Depressive symptoms in schizophrenia and dopamine and serotonin gene polymorphisms. Prog Neuro-Psychopharmacol Biol Psychiatry 77:209–215. https://doi.org/10.1016/j.pnpbp.2017.04.011

Pii Y, Mimmo T, Tomasi N, Terzano R, Cesco S, Crecchio C (2015) Microbial interactions in the rhizosphere: beneficial influences of plant growth-promoting rhizobacteria on nutrient acquisition process: a review. Biol Fertil Soils 51(4):403–415. https://doi.org/10.1007/s00374-015-0996-1

Prince DC, Rallapalli G, Xu D, Schoonbeek HJ, Çevik V, Asai S, Jones JD (2017) Albugo-imposed changes to tryptophan-derived antimicrobial metabolite biosynthesis may contribute to suppression of non-host resistance to *Phytophthora infestans* in *Arabidopsis thaliana*. BMC Biol 15:1–22. https://doi.org/10.1186/s12915-017-0360-z

Quadros PC, Duarte CTM, Pastore MG (2011) Biological activities of a mixture of biosurfactant from Bacillus subtilis and alkaline lipase from Fusarium oxysporum. Braz J Microbiol 42(1):354–361. https://doi.org/10.1590/S1517-83822011000100045

Radhakrishnan R, Khan AL, Kang SM, Lee I-J (2015) A comparative study of phosphate solubilization and the host plant growth promotion ability of *Fusarium verticillioides* rk01 and *Humicola* sp. Knu01 under salt stress. Ann Microbiol 65:585–593. https://doi.org/10.1007/s13213-014-0894-z

Rajesh PS, Ravishankar Rai V (2014) Quorum quenching activity in cell-free lysate of endophytic bacteria isolated from *Pterocarpus santalinus* Linn., and its effect on quorum sensing regulated biofilm in *Pseudomonas aeruginosa* PAO1. Microbiol Res 169(7–8):561–569. https://doi.org/10.1016/j.micres.2013.10.005

Rashid MI, Mujawar LH, Shahzad T, Almeelbi T, Ismail IM, Oves M (2016) Bacteria and fungi can contribute to nutrients bioavailability and aggregate formation in degraded soils. Microbiol Res 183:26–41. https://doi.org/10.1016/j.micres.2015.11.007

Romera FJ, García MJ, Lucena C, Martínez-Medina A, Aparicio MA, Ramos J, Pérez-Vicente R (2019) Induced systemic resistance (ISR) and Fe deficiency responses in dicot plants. Front Plant Sci 10:287. https://doi.org/10.3389/fpls.2019.00287

Rosier A, Medeiros FH, Bais HP (2018) Defining plant growth promoting rhizobacteria molecular and biochemical networks in beneficial plant-microbe interactions. Plant Soil 428:35–55. https://doi.org/10.1007/s11104-018-3679-5

Schirawski J, Perlin MH (2018) Plant-microbe interaction 2017-the good, the bad and the diverse. Int J Mol Sci 19(5):1374. https://doi.org/10.3390/ijms19051374

Semchenko M, Lepik A, Abakumova M, Zobel K (2018) Different sets of belowground traits predict the ability of plant species to suppress and tolerate their competitors. Plant Soil 424:157–169. https://doi.org/10.1007/s11104-017-3282-1

Sharma A, Kaur A, Tyagi S, Upadhyay SK (2023) Glutathione peroxidases in plants: innumerable role in abiotic stress tolerance and plant development. J Plant Growth Regul 42(2):598–613. https://doi.org/10.1007/s00344-022-10601-9

Shelake RM, Pramanik D, Kim JY (2019) Exploration of plant-microbe interactions for sustainable agriculture in CRISPR era. Microorganisms 7:269. https://doi.org/10.3390/microorganisms7080269

Simons A, Alhanout K, Duval RE (2020) Bacteriocins, antimicrobial peptides from bacterial origin: overview of their biology and their impact against multidrug-resistant bacteria. Microorganisms 8(5):639. https://doi.org/10.3390/microorganisms8050639

Singh RP, Jha PN, Jha PN (2016) The plant-growth-promoting bacterium *Bacillus amyloliquefaciens* Bk6 stimulates the growth and expression of antioxidant enzymes in tomato (*Solanum lycopersicum* L.) under long-term salinity stress. J Plant Growth Regul 35:427–440. https://doi.org/10.1007/s00344-015-9539-7

Singh KP, Kumari P, Rai PK (2021) Current status of the disease-resistant gene(s)/QTLs, and strategies for improvement in *Brassica juncea*. Front Plant Sci 12:617405. https://doi.org/10.3389/fpls.2021.617405

Singh SK, Wu X, Shao C, Zhang H (2022) Microbial enhancement of plant nutrient acquisition. Stress Biol 2:3. https://doi.org/10.1007/s44154-021-00027-w

Singh S, Chaudhary P, Sharma A (2023) Efficient decolourization of toluidine blue and methylene blue using *Bacillus cereus*. Vegetos. https://doi.org/10.1007/s42535-023-00747-6

Sivakumar N, Saarthishkum R, Selvakumar G, Shyamkumar R, Arjunekumar K (2020) Phyllospheric microbiomes: diversity, ecological significance, and biotechnological applications. In: Plant microbiomes for sustainable agriculture, pp 113–172. https://doi.org/10.1007/978-3-030-38453-1_5

Slama HB, Cherif-Silini H, Chenari Bouket A, Qader M, Silini A, Yahiaoui B, Alenezi FN, Luptakova L, Triki MA, Vallat A, Oszako T, Rateb ME, Belbahri L (2019) Screening for *fusarium* antagonistic bacteria from contrasting niches designated the endophyte *bacillus halotolerans* as plant Warden against *fusarium*. Front Microbiol 9:3236. https://doi.org/10.3389/fmicb.2018.03236

Spence C, Bais H (2015) Role of plant growth regulators as chemical signals in plant-microbe interactions: a double edged sword. Curr Opin Plant Biol 27:52–58. https://doi.org/10.1016/j.pbi.2015.05.028

Sukumaran S, Dreisigacker S, Lopes M, Chavez P, Reynolds MP (2015) Genome-wide association study for grain yield and related traits in an elite spring wheat population grown in temperate irrigated environments. Theor Appl Genet 128:353–363. https://doi.org/10.1007/s00122-014-2435-3

Summuna B, Gupta S, Sheikh PA (2019) Plant growth and health promoting plant-microbe interactions. In: Plant health under biotic stress: volume 2: microbial interactions, pp 253–260. https://doi.org/10.1007/978-981-13-6040-4_13

Thapa S, Prasanna R (2018) Prospecting the characteristics and significance of the phyllosphere microbiome. Ann Microbiol 68:229–245. https://doi.org/10.1007/s13213-018-1331-5

Thynne E, McDonald MC, Solomon PS (2015) Phytopathogen emergence in the genomics era. Trends Plant Sci 20(4):246–255. https://doi.org/10.1016/j.tplants.2015.01.009

Timofeeva AM, Galyamova MR, Sedykh SE (2022) Bacterial siderophores: classification, biosynthesis, perspectives of use in agriculture. Plants (Basel) 11(22):3065. https://doi.org/10.3390/plants11223065

Trouvelot S, Héloir M-C, Poinssot B, Gauthier A, Paris F, Guillier C, Combier M, Trdá L, Daire X, Adrian M (2014) Carbohydrates in plant immunity and plant protection: roles and potential application as foliar sprays. Front Plant Sci 5:592. https://doi.org/10.3389/fpls.2014.00592

Tyagi J, Chaudhary P, Jyotsana, Bhagwati U, Bhandari G, Chaudhary A (2022a) Chapter 17: Impact of endophytic fungi in biotic stress management. In: Soni R, Suyal D, Goel R (eds) Plant protection: from chemicals to biologicals. De Gruyter, Berlin, Boston, pp 447–462. https://doi.org/10.1515/9783110771558-017

Tyagi J, Chaudhary P, Mishra A, Khatwani M, Dey S, Varma A (2022b) Role of endophytes in abiotic stress tolerance: with special emphasis on *Serendipita indica*. Int J Environ Res 16:62. https://doi.org/10.1007/s41742-022-00439-0

Verbon EH, Trapet PL, Stringlis IA, Kruijs S, Bakker PAHM, Pieterse CMJ (2017) Iron and immunity. Annu Rev Phytopathol 55:355–375. https://doi.org/10.1146/annurev-phyto-080516-035537

Vishwakarma K, Kumar N, Shandilya C, Mohapatra S, Bhayana S, Varma A (2020) Revisiting plant-microbe interactions and microbial consortia application for enhancing sustainable agriculture: a review. Front Microbiol 11:560406. https://doi.org/10.3389/fmicb.2020.560406

Wang M, Ding L, Gao L, Li Y, Shen Q, Guo S (2016) The interactions of aquaporins and mineral nutrients in higher plants. Int J Mol Sci 17(8):1229. https://doi.org/10.3390/ijms17081229

Wang Q, Liu J, Zhu H (2018) Genetic and molecular mechanisms underlying symbiotic specificity in legume-*Rhizobium* interactions. Front Plant Sci 9:313. https://doi.org/10.3389/fpls.2018.00313

Yang Q, Balint-Kurti P, Xu M (2017) Quantitative disease resistance: dissection and adoption in maize. Mol Plant 10(3):402–413. https://doi.org/10.1016/j.molp.2017.02.004

Yu JG, Lim JA, Song YR, Heu S, Kim GH, Koh YJ, Oh CS (2016) Isolation and characterization of bacteriophages against *Pseudomonas syringae* pv. actinidiae causing bacterial canker disease in kiwifruit. J Microbiol Biotechnol 26(2):385–393. https://doi.org/10.4014/jmb.1509.09012

Yu MH, Zhao ZZ, He JX (2018) Brassinosteroid signaling in plant-microbe interactions. Int J Mol Sci 19(12):4091. https://doi.org/10.3390/ijms19124091

Zamioudis C, Korteland J, Van Pelt JA, van Hamersveld M, Dombrowski N, Bai Y (2015) Rhizobacterial volatiles and photosynthesis-related signals coordinate MYB72 expression in *Arabidopsis* roots during onset of induced systemic resistance and iron-deficiency responses. Plant J 84:309–322. https://doi.org/10.1111/tpj.12995

Zhang N, Wang D, Liu Y, Li S, Shen Q, Zhang R (2014) Effects of different plant root exudates and their organic acid components on chemotaxis, biofilm formation and colonization by beneficial rhizosphere-associated bacterial strains. Plant Soil 374:689–700. https://doi.org/10.1007/s11104-013-1915-6

Zoledowska S, Presta L, Fondi M, Decorosi F, Giovannetti L, Mengoni A, Lojkowska E (2019) Metabolic modeling of Pectobacterium parmentieri SCC3193 provides insights into metabolic pathways of plant pathogenic bacteria. Microorganisms 7(4):101. https://doi.org/10.3390/microorganisms7040101

Microbial Advancement in Agriculture

Viabhav Kumar Upadhayay, Sergio de los Santos Villalobos, S. T. M. Aravindharajan, Bharti Kukreti, Manoj Kumar Chitara, Vandana Jaggi, Adita Sharma, and Ajay Veer Singh

Abstract The inexorable expansion in the global demand for food, coupled with the need to reduce the environmental impact of conventional farming practices, has created the need for innovative and eco-friendly solutions. Among the myriad of promising alternatives, agriculturally significant microorganisms have emerged as effective tools in the fight against this challenge. Such microorganisms, replete with diverse plant growth-stimulating traits, have been harnessed to fortify soil health, enhance crop growth, and augment yield. Biofilm-forming microbial inoculants, for example, have shown remarkable efficacy in crop production, while microbial-assisted biofortification of food crops has emerged as an increasingly popular prac-

V. K. Upadhayay (✉)
Department of Microbiology, College of Basic Sciences and Humanities, Dr. Rajendra Prasad Central Agricultural University, Pusa, Samastipur, Bihar, India

S. de los Santos Villalobos
Laboratorio de Biotecnología del Recurso Microbiano, Instituto Tecnológico de Sonora, Ciudad Obregón, Sonora, Mexico

S. T. M. Aravindharajan
Division of Microbiology, ICAR-Indian Agricultural Research Institute, New Delhi, India

B. Kukreti · A. V. Singh
Department of Microbiology, CBSH, Govind Ballabh Pant University of Agriculture and Technology, Pantnagar, Uttarakhand, India

M. K. Chitara
Department of Plant Pathology, College of Agriculture, Chandra Shekhar Azad University of Agriculture and Technology, Kanpur, Uttar Pradesh, India

V. Jaggi
Department of Microbiology, CBSH, Govind Ballabh Pant University of Agriculture and Technology, Pantnagar, Uttarakhand, India

Department of Pediatrics and Human Development, College of Human Medicine, Michigan State University, Grand Rapids, Michigan, United States

A. Sharma
College of Fisheries, Dholi, Dr. Rajendra Prasad Central Agricultural University, Muzaffarpur, Bihar, India

P. Chaudhary, A. Chaudhary (eds.), *Microbial Inoculants*, Interdisciplinary Biotechnological Advances, https://doi.org/10.1007/978-981-97-0633-4_5

tice. Moreover, microbial inoculants can be effective tools to mitigate the consequences of climate change by curtailing greenhouse gas emissions, sequestration of carbon, etc. The synergistic application of nanomaterials and microbes has also been found to significantly increase crop output. A recent and particularly promising approach is "rhizosphere microbiome transplantation," which involves the transfer of beneficial microorganisms from healthy soil to affected crop roots, thereby conferring disease resistance and boosting crop vitality. This chapter emphasizes the significant and game-changing role of microbes and microbial based strategies in agriculture, which holds the key to unlocking a transformative and enduring future for crop-based food production.

Keywords Biofilm · Biofortification · Climate change · Crop growth · Rhizosphere microbiome transplantation · Nanomaterials · Soil health

1 Introduction

In the contemporary world, a pivotal challenge lies in the provision of a constant supply of sustenance for the escalating human population. This task is further exacerbated by the persistent onslaught of climate change, which has deleterious effects on agricultural productivity. For several decades, the widespread use of chemical-based fertilizers, pesticides, and herbicides has resulted in significant environmental degradation, affecting human health and soil quality (Parveen et al. 2022; Chaudhary et al. 2022b). Fortunately, the use of agriculturally important microorganisms serves as a beacon of hope that mitigates the negative effects associated with our overreliance on agrochemicals (Chaudhary et al. 2023a). The role of microorganisms in agriculture is of paramount importance, and their impact on soil fertility and plant growth has been the subject of extensive scientific study (Agri et al. 2022). Microbes show remarkable competence in nutrient management due to their promising role in nutrient cycling and providing various nutrients to plants (Agri et al. 2021; Kumari et al. 2021). The conversion of atmospheric nitrogen into a plant-available form is a critical ecosystem service provided by nitrogen-fixing bacteria, a certain class of bacteria. These microorganisms have the capacity to convert atmospheric nitrogen gas into ammonia, which can subsequently be assimilated by plants to foster growth. Similarly, phosphate-solubilizing bacteria can convert insoluble phosphorus into plant-accessible forms, thus augmenting plant growth and productivity. The primary objective of utilizing plant growth-promoting microorganisms is to increase crop production while minimizing the dependency on chemical-based fertilizers (Khan et al. 2023a). Furthermore, microbes have a significant impact on soil structure and stability. They secrete extracellular polymeric substances (EPS) that facilitate the formation of soil aggregates, thereby improving soil structure and stability (Kaviya et al. 2019). This, in turn, enhances soil water retention, mitigating erosion and leading to better crop growth and higher yields. It is pertinent to note that microorganisms that inhabit the rhizosphere are commonly referred to as "rhizospheric bacteria," while those that reside within plant tissues are known as "endophytes" (Chaudhary et al. 2022c). Under in vitro conditions, isolated microbes are evaluated for various plant growth-promoting traits, and highly efficient strains can be selected based on rigorous field-based trials (Upadhayay et al. 2022c). In modern times, a variety of novel approaches can be incorporated to amplify microbial advances in the contemporary agricultural landscape (Chaudhary et al. 2022d). These approaches could include (a) the utilization of agriculturally important microorganisms and biofilm-forming bacteria to boost crop productivity; (b) the exploration of microbial roles in soil fertility, climate-smart agriculture, and biofortification; (c) the transplanting of rhizospheric microbiomes; and (d) the combined impact of microbial strains as potential inoculants with select nanomaterials (Chaudhary et al. 2021b, c, d, e) (Fig. 1).

The use of biofilms in agriculture is an eminent microbial advancement that has shown immense promises. Biofilms are microbial communities that form a protective layer around plant roots, providing a physical barrier against pathogens while promoting nutrient uptake. Biofilms can enhance plant growth and productivity, leading to increased yield and better-quality produce (Ansari et al. 2021; Taktek

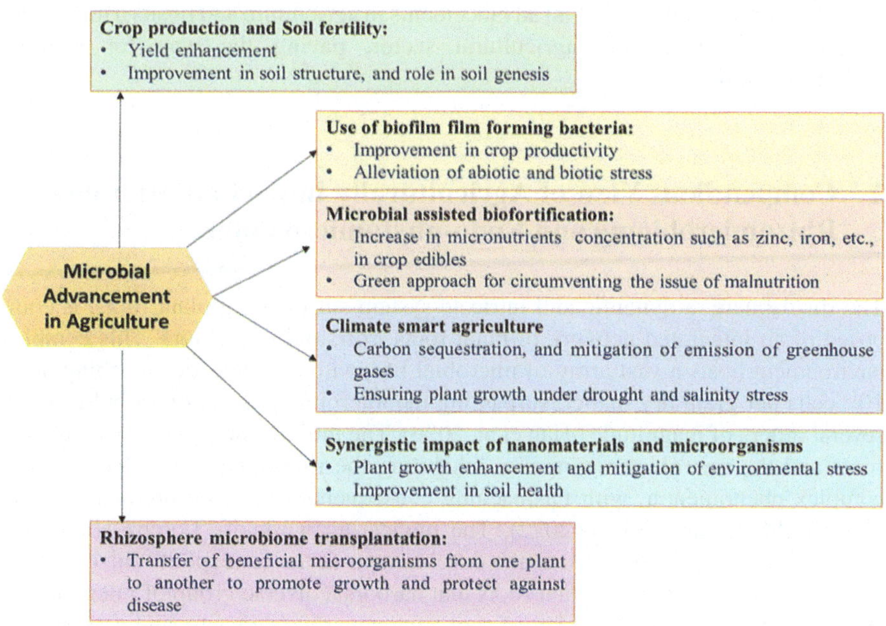

Fig. 1 Flowchart illustrating microbial advancement in agriculture

et al. 2017). Biofortification, the process of improving the nutrient content of crops through microbial interventions (Upadhayay et al. 2022a), is another promising strategy for crop improvement (Upadhayay et al. 2021). Microbes can be used to increase the bioavailability of essential nutrients such as iron, zinc, and selenium, which are often deficient in diets in many parts of the world (Upadhayay et al. 2018; Khati et al. 2019; Chaudhary et al. 2022e). This can improve the nutritional value of crops and contribute to better health outcomes. Additionally, microbes have the ability to promote carbon sequestration in soil, reducing greenhouse gas emissions and mitigating climate change (Ahmed et al. 2020).

Rhizosphere microbiome transplantation is another technique used in microbial advancements for crop improvement. This involves the transfer of beneficial microorganisms from one plant to another to promote growth and protect against disease (Jiang et al. 2022). This technique can be used to establish beneficial microbial communities that can enhance plant growth and protect against pathogens, reducing the need for synthetic inputs such as pesticides.

Furthermore, the synergistic application of nanomaterials and plant growth-promoting microbes has recently garnered attention for its superior effectiveness in improving crop productivity (Khati et al. 2018; Kumari et al. 2020; Upadhayay et al. 2022c), mitigating abiotic stress (Muhammad et al. 2022), and promoting soil health (Kukreti et al. 2020; Chaudhary et al. 2021f, g). The integration of microbial technologies in agriculture can foster more sustainable and environmentally friendly practices while diminishing reliance on synthetic inputs. Thus, this chapter illustrates the cutting-edge microbial advancements in agriculture and posits a microbial-driven revolution in the agricultural sector, paving the way for a more sustainable future.

2 Compendious View of Agriculturally Important Microbes: Rhizomicrobiome and Endo-phytomicrobiome

The rhizosphere, a dynamic and intricate system surrounding plant roots, is composed of an integrated network of plant roots, soil, and microbiota. This complex environment hosts a vast array of microbial life, with cell densities reaching up to 10^{11} cells per gram dry matter, surpassing the microbial population of bulk soil by several orders of magnitude (Tian et al. 2020). The intricate interplay between plant roots and the microbial community inhabiting the rhizosphere is a dynamic and complex phenomenon, with far-reaching consequences for crop productivity and plant health (Manzar et al. 2022). The pioneering work of "Lorenz Hiltner," a revered plant physiologist, led to the development of the concept of "rhizosphere," a narrow zone surrounding plant roots that harbors a diverse group of microbes, all shaped by the intricate interplay of root exudates, rhizodeposits, and a myriad of chemical compounds. Plants are known to actively engage in a complex and sophisticated system of communication with soil microbiota, resulting in the establishment of a diverse and dynamic microbial community that exerts a profound influence on plant growth and health. These microbes colonize various parts of the plant, including the phyllosphere, endophytic tissues, and the rhizosphere, and contribute to the modulation of the host plant's genomic and metabolic capabilities. The intricate network of beneficial interactions between the plant and its associated microbes involves a range of functions, such as biocontrol, nutrient acquisition, phytohormone production, and immune response against diverse biotic and abiotic stressors, ultimately leading to a boost in plant growth and productivity (Kaul et al. 2018; Dini-Andreote 2020; Chaudhary et al. 2021a, 2022a). The rhizosphere, the region of soil surrounding plant roots, harbors a diverse and dynamic microbial community that together constitute the rhizomicrobiome. This intricate web of microorganisms illustrates a prolific role in shaping the well-being and vitality of plants, even under challenging environmental conditions (Tian et al. 2020). Remarkably, scientific research has demonstrated that certain members of the rhizosphere microbial community can serve as potent biofertilizers, providing numerous benefits to plant growth and development. Numerous reports have highlighted the successful exploitation of rhizosphere-inhabiting microbes for their biofertilizer properties. Among the most notable are bacterial genera such as *Azospirillum*, *Bacillus*, *Pseudomonas*, *Serratia*, *Streptomyces*, and *Stenotrophomonas*, which have been revealed to exert positive effects on plant growth and development. The diversity and richness of rhizospheric microflora have been shown to be largely contingent upon the particular plant species under consideration and are indicative of the intimate and tightly regulated relationships between plants and their microbial symbionts. Intriguingly, recent research has revealed that the root microbiota of *Hordeum vulgare* exhibits a striking preference for *Comamonadaceae*, *Flavobacteriaceae*, and *Rhizobiaceae* bacterial families, which are known to play key roles in plant nutrient acquisition *and* stress tolerance (Bulgarelli et al. 2015). In stark contrast, the root microbiota of *Thymus zygis* is dominated by *Bradyrhizobiaceae*, *Nocardioidaceae*, and *Geodermatophilaceae*, highlighting the exceptional diversity and specificity of microbial communities associated with different plant species (Pascual et al. 2016). The plant growth-promoting bacteria (PGPB) or plant growth-promoting rhizobacteria (PGPR), as they are commonly known, represent a highly diverse group of microbes that possess remarkable abilities to enhance plant growth and development (Satyaprakash et al. 2017). These remarkable organisms have garnered significant interest due to their ability to establish mutually beneficial relationships with plants, which can lead to substantial improvements in crop yields and sustainability. Notably, a wide array of PGPR genera have been identified as effective plant growth enhancers, including *Acinetobacter*, *Azospirillum*, *Azotobacter*, *Bacillus*, *Bradyrhizobium*, *Burkholderia*, *Erwinia*, *Klebsiella*, *Mesorhizobium*, *Micrococcus*, *Nitrobacter*, *Pseudomonas*, *Rhizobium*, and many others (Di Benedetto et al. 2017; Moustaine et al. 2017). On the other hand, some microbes are known to reside in the internal parts of plants, which are referred to as endophytes. They are capable of colonizing the tissues of vascular plants (Brader et al. 2017; Fadiji and Babalola 2020). Endophytes can colonize various plant parts such as the stem, roots, petioles, leaf segments, inflorescences of weeds, fruit, buds, seeds, and even the dead and hollow hyaline cells of plants (Gouda et al. 2016; Yadav 2018). Endophytism is a fascinating phenomenon whereby microbes establish a symbiotic relationship with plants without causing any harm. The entry of endophytes into their host plants is a complex process that involves various mechanisms and pathways. Among these, the most common entry points for endophytes are root cracks, including primary and lateral roots, root hair cells, and tissue wounds caused by plant development, which provide an ideal entry point for these microbes to access the plant interior (Santoyo et al. 2016). Initially, endophytes were thought to be harmless to the host plants, and their association with plants can be either obligate or facultative (Nair and Padmavathy 2014). The population of endophytes in a plant species is highly variable and depends on various factors such as host species and its developmental stage, inoculum density, and environmental condition (Dudeja and Giri 2014). Endophytes occupy an ecological niche similar to that of phytopathogens, making them suitable as biocontrol agents (Tyagi et al. 2022a). Several endophytic microbes possess the ability to carry out different activities within their hosts. Studies unveiled the significance of endophytes in enhancing plant growth and providing protection to their host plants by diverse mechanisms such as providing nutrients and producing alkaloids (Fadiji and Babalola 2020). They modulate plant activities in many dimensions by altering the production of endogenous phytohormones (including auxin, cytokinin, and gibberellic acid) and producing bioactive compounds (including terpenoids and flavonoids) and chitinase enzymes to suppress plant pathogens (Christina et al. 2013; Wang et al. 2014; Tyagi et al. 2022b). The intricate relationship between endophytes and their host plants goes beyond mere growth promotion. Endophytes are known to elicit a remarkable plant defense mechanism, called induced systemic resistance (ISR), which enhances the plant's protection against a broad spectrum of pathogens (Pieterse et al. 2012). This unique capability of endophytes has garnered significant attention in recent years, with researchers exploring their potential applications in biotechnology, agriculture, and medicine. The identification of major endophyte genera has been a subject of intense research, involving various methods such as "16S rRNA gene sequencing," "fatty acid analysis," and "carbon source utilization analysis." Some of the most prominent endophyte genera identified through these methods include *Cellulomonas*, *Clavibacter*, *Curtobacterium*, *Pseudomonas*, and *Microbacterium*. The diversity of endophytic microflora and their intricate interactions with their host plants continue to fascinate scientists and pave the way for new discoveries in plant-microbe interactions. As we deepen our understanding of the role of endophytes in plant health and protection, we can exploit their potential to improve crop productivity and reduce the use of harmful pesticides. The remarkable ability of endophytes to stimulate plant defense responses and promote growth highlights their enormous potential as a sustainable and eco-friendly alternative to conventional agricultural practices.

3 Microbial Role in Crop Growth

Maize, rice, and wheat are globally significant cereal crops that play a critical role in ensuring food security. Additionally, leguminous crops such as pulses serve as essential sources of nutrients and protein. However, climatic variations and human activities, including the use of high levels of agrochemicals, have led to soil fertility imbalances, resulting in reduced crop yields. Fortunately, the use of agriculturally important microorganisms provides an eco-friendly solution to combat this issue, offering a promising avenue for sustainable agriculture with reduced dependence on chemical-based fertilizers and pesticides. These microorganisms, known as "plant growth-promoting microbes," primarily reside in the soil and colonize the rhizosphere of plants, depicting crop growth via a range of several mechanisms, including nutrient solubilization (P, K, Zn, etc.), fixation of N_2, and production of siderophores, HCN, exopolysaccharides, phytohormones (IAA), and antipathogenic compounds (Khan et al. 2020, 2023b; Singh et al. 2020). Moreover, microbial inoculants have been found to improve crop tolerance to abiotic stresses such as drought, salinity, and metal stress, making them an important area of research. Multiple bacterial strains have been investigated for their capability to mitigate abiotic stresses in crops (Upadhayay et al. 2023a). The effectiveness of biofertilizer application in enhancing crop production depends on various factors, including soil conditions, crop type, and the specific biofertilizer used (Khan et al. 2023a). Research studies have shown that the use of biofertilizers can increase crop yields by an average of 10–40% (Nosheen et al. 2021), although this percentage may vary depending on the crop type and environmental conditions. Parameters such as plant height, dry matter, chlorophyll content, number of tillers, thousand grain weight, grain yield, biological yield, etc. are determined to illustrate the impact of microbial inoculation on crop growth. A recent study by Khan et al. (2022) demonstrated the effect of PGPR inoculants on oat productivity under field trial conditions and showed significant improvement in agronomical parameters, especially grain yield, in an eco-friendly manner. Table 1 illustrates the improvement in crop-based plants in response to agriculturally important microbial inoculants.

4 Importance in Soil Fertility and Soil Genesis

The beneficial microbial population found on the upper surface of the Earth has played a significant role in modern agriculture. These are certain kinds of macrofauna and microflora, also considered natural soil engineers due to their significant role in soil genesis processes and enhanced soil fertility (Kaviya et al. 2019). The root-associated microorganisms are involved in the biodegradation process of hard rock into soil formation commonly known as weathering (Gregory 2022). The weathering of rocks and mineralization by microbial communities are key processes that facilitate mineral dissolution and increase the release of elements in the

Table 1 The role of beneficial microorganisms in enhancing crop performance

Microbial strains	Crop	Beneficial attributes leading to crop improvement	References
Azospirillum brasilense	*Zea mays*	• Enhancement in agronomic efficiency, nitrogen uptake in plants • Augmentation in grain yield (by 7%)	Galindo et al. (2022)
Bacillus sp. (*B. simplex and B. pumilus*)	Wheat	• Significant enhancement in yield (between 9.6 and 29.29%)	Öksel et al. (2022)
Bacterial consortium (*Bacillus* and *Acinetobacter*)	Maize	• Enhancement in N content in shoot and grain and P content in straw and grain • Improvements in plant height, photosynthetic rates, length and weight of cob, grain yield under salinity stress condition	Shabaan et al. (2022)
Pseudomonas jesenii MP1	Kidney bean (*Phaseolus vulgaris*)	• Improvement in grain yield and nutrient status • Enhancement in soil health	Khan et al. (2023b)
Combination of *Bacillus amyloliquefaciens* and biochar	Wheat	• Improvement in grain yield, straw yield, and biological yield in drought stress	Zafar-ul-Hye et al. (2019)
A. brasilense and *R. pisi*	Wheat	• Increment in grain yield, number of grains per plant, and thousand grain weight • Significant increment in leaf epicuticular wax and relative water content	Zaheer et al. (2019)
Acidovorax delafieldii + half dose of inorganic fertilizer	Rice	• Increment in plant height, weight of both root and shoot, and grain yield • Improvement in NPK uptake	Cavite et al. (2021)
Serratia spp., strain BRM 32114	Rice	• Enhancement in stomatal conductance and nutrient content (N, Ca, and Mg) • Improvement in dry matter, number of grain per plant, weight of 1000 grains, and grain yield	Nascente et al. (2019)
Arbuscular mycorrhizal fungi (AMF) (*Funneliformis mosseae*, *Rhizophagus intraradices*)	Soybean	• Increment in the number and dry weight of nodules • Enhancement in the number of pods, dry weight of shoot, and seed yield	Adeyemi et al. (2021)
Azotobacter + mycorrhizal fungi (*Glomus intraradices*)	Lentil	• Maximum increment in chlorophyll content, biomass yield, seed yield, and protein content of seed	Amirnia et al. (2019)

biogeochemical cycle of important nutrients such as carbon, nitrogen, sulfur, phosphorus, and others and the conversion of the element (Sathya et al. 2016; Samuels et al. 2020; Singh et al. 2023). Macrofauna including earthworms help in enhancing soil fertility by mixing soil layers and binding the soil with organic matter resulting in improving the soil's physical, chemical, and biological features (Ahmed and Al-Mutairi 2022). The microflora including symbiotic, viz., *Rhizobium* spp., and nonsymbiotic, viz., *Azotobacter*, *Azospirillum*, *Bacillus*, and *Klebsiella* spp., could be created a mutual link between the soil and roots of the plants and perform various kinds of a biogeochemical process such as nutrient recycling and nutrient decomposition (Toor and Adnan 2020). The diversity of the microorganisms in the soil could be decided by the soil type and associated host crop secretion such as root exudates (Chitara et al. 2022). The diversity of microorganisms in soil is affected by the types of amendments added to it. In North China, researchers conducted a 4-year experiment to investigate the impact of soil amendments on microbial diversity. Different types of fertilizers, including nitrogen alone and a combination of "NPK," "organic manure," "nitrogen plus organic manure," and "NPK plus organic manure," were applied in an apple orchard. The results exhibited that the maximum microbial diversity was detected in "NPK plus organic manure treatment," followed by "organic manure" and a combination of "NPK and nitrogen plus manure." The diverse microbial populations detected included actinomycetes, bacteria, and fungi (Zhu et al. 2020). A 40-year balanced application of fertilizer and organic manure to the red paddy soil improved the fertility status of the soil and sustained the archaeal and bacterial abundance and bacterial community structure, which can help in improving rice productivity (Wang et al. 2021). A reduction in the chemical fertilizer application with an organic fertilizer in maize crops significantly improved the yield, and microbial fertility status can help in ensuring food security and long-term sustainability to the environment (Han et al. 2021).

5 Microbial Role in Climate-Smart Agriculture

The escalating demand for crop-based sustenance has precipitated an overwhelming reliance on agrochemicals and the extensive exploitation of farmland, ultimately contributing to the proliferation of greenhouse gases that are fundamentally responsible for the intensification of global warming. Specifically, "carbon dioxide (CO_2)," "methane (CH_4)," and "nitrous oxide (N_2O)" are among the most prominent greenhouse gases that are responsible for the exacerbation of climate change. CO_2, in particular, constitutes a substantial proportion of greenhouse gases, with 65% of its emissions originating from fossil-based fuels and the remaining 11% being sourced from other activities, including forestry. The other constituents of greenhouse gases, such as CH_4, N_2O, and fluorinated gases, account for 16%, 6%, and 2%, respectively (Malhi et al. 2021). These gases are instrumental in driving up atmospheric temperatures, rendering agriculture highly vulnerable to the far-reaching effects of climate change, including shifts in temperature patterns, changes in precipitation, and an increased incidence of catastrophic events, such as drought, heat stress, salinity, flooding, etc. (Shahzad et al. 2021). Drought, salinity, heat stress, etc. are examples of abiotic stress that have affected global agriculture and pose an existential threat to the global food supply (Upadhayay et al. 2023a). The contemporary world is faced with unprecedented challenges arising from the effects of climate change, and as such, the adoption of climate-smart agriculture practices has become an essential undertaking. Achieving objectives such as adaptation enhancement, minimizing greenhouse gas emissions, and the attainment of national food security goals necessitates the widespread adoption of these practices (Katel et al. 2022; Mohanty and Swain 2018). Microbes, owing to their survival strategies in the face of various stressful conditions posed by global climate change, have emerged as "intelligent players" in the quest for mitigating the impact of climate change on sustainable agriculture. The application of microbial inoculants, possessing multiple plant growth-stimulating effects and survival mechanisms under diverse abiotic stresses, provides a promising approach to supporting host plants under stressful conditions (Upadhayay et al. 2023a, c). The selection of microbial inoculants based on their particular abiotic stress-tolerant traits, such as drought and salinity tolerance, allows for the attribution of specific microorganisms as drought-tolerant or salinity-tolerant microorganisms.

The benefits of microorganisms as stress mitigating agents in plants have been well-documented in literature. Through the production of hormones, synthesis of ACC deaminase, accumulation of osmo-protectants, and the regulation of the expression of "stress-responsive genes," certain microbes can confer stress tolerance in plants, providing a promising avenue for sustainable agriculture. In a study conducted by Kour et al. (2020b), the drought mitigation effects of two microbial strains (*A. calcoaceticus* and *Penicillium* sp.) were evaluated in foxtail millet. The results showed that the plants inoculated with these strains had enhanced accumulation of substances that provide stress relief, such as proline, carbohydrates, and glycine betaine, indicating a drought stress survival strategy. Similarly, Devarajan et al. (2021) demonstrated the effectiveness of *Bacillus megaterium* PB50 in increasing drought stress tolerance in plants. The study evaluated several functional attributes, including the increase in relative water content, levels of sugars, proline, proteins, IAA, and significantly higher expression of stress-responsive genes (such as LEA, HSP70, RAB16B). Furthermore, Haroon et al. (2022b) demonstrated the potential of certain bacterial strains, including *B. megaterium*, *B. tequilensis*, and *P. putida*, as plant growth promoters, inducing salt stress tolerance in wheat through the production of ACC deaminase and EPS.

Carbon sequestration is an essential process for capturing atmospheric carbon dioxide (CO2) and storing it in the soil for prolonged periods. This process is facilitated by various mechanisms, with the intricate role of soil microorganisms in the carbon (C) cycle being well-established. Their activity has been recognized as a major driver of variations in soil C storage capacity (Bhattacharyya et al. 2022). Soil microorganisms significantly contribute to soil carbon sequestration by controlling diverse pathways for CO2 capture, enhancing the capacity for sedimentation of carbonates, and facilitating the development of stable carbon forms that protect soil carbon (Ahmed et al. 2020). Moreover, soil bacteria play a crucial role in mitigating the effects of climate change and in carbon fixation. Bacteria that promote plant growth also support mycorrhizal colonization in the rhizosphere, resulting in increased production of glomalin, an indispensable carbon and nitrogen storage in the soil system (Nautiyal et al. 2019). Furthermore, microbes have been demonstrated to enhance plant growth in various literature (Upadhayay et al. 2022b, d), which indirectly leads to more carbon allocation in plant biomass and presents a prominent option for carbon recycling (Gougoulias et al. 2014). Furthermore, plant probiotics influence soil quality, ultimately regulating carbon content. Overall, the important role of soil microorganisms in carbon sequestration underscores the need for their continued study and management in promoting sustainable soil and environmental practices.

Ralstonia eutropha, a chemotrophic bacterium, represents a noteworthy example of this phenomenon, as it employs its metabolic machinery to produce polyhydroxyalkanoates (PHAs), a biopolymer with tremendous potential for various novel applications (Rossi et al. 2015). The research conducted by Liu et al. (2018) has also shed light on the remarkable ability of soil bacteria to harness atmospheric carbon dioxide by transforming it into calcium carbonate precipitates, thereby facilitating biogenic deposition of CO_2 in drylands. *Arthrobacter* sp., *Microbacterium* sp., *Streptomyces* sp., etc. are among the soil microorganisms that were found to exhibit a noteworthy role in this carbon sequestration process, exemplifying the untapped potential of microorganisms in tackling climate change. In addition to their crucial role in the carbon cycle, microorganisms also play a significant role in mitigating greenhouse gas emissions, specifically methane and nitrous oxide. Methanotrophs and nitrous oxide respiring bacteria are highly effective in reducing methane and nitrous oxide emissions, respectively, and hold great potential in combating the greenhouse effect (Mohanty and Swain 2018). Recent research has shown that certain bacterial strains, such as *Methylobacterium oryzae* and *Paenibacillus polymyxa*, can be utilized as competent inoculants to reduce methane emissions (Rani et al. 2021). Moreover, a study by Davamani et al. (2020) illustrated the remarkable ability of a consortium of methanotrophs to significantly reduce methane emissions by 60%, although it was found to decrease paddy grain yield by 35%. Furthermore, research conducted by Wu et al. (2018) demonstrated that *B. amyloliquefaciens* significantly promoted plant growth while also mitigating 50% of N_2O emissions in a greenhouse experiment. These findings highlight the promising potential of microorganisms in mitigating greenhouse gas emissions and supporting sustainable agriculture practices.

6 Potential of Microbial Biofilm in Agricultural Application

Bacteria typically exist in a free-living state but can form biofilms under specific environmental conditions (Omar et al. 2017). Biofilms have been observed since the 1680s, when Leeuwenhoek studied microbial biofilms on teeth using a primitive light microscope (De Vos 2015). "Biofilms" are considered as multicellular, cooperative groups of microbes that attach to solid surface using self-produced extracellular polymeric substances (EPS). Biofilms are well-documented and can be classified based on appearance (simple to complex, flat to tower, clustered film to featureless) and diversity (Koo et al. 2017). The process of biofilm formation begins with initial adsorption to a surface, followed by intercellular attachment through quorum sensing, microcolony formation, biofilm maturation, and dispersion (Achinas et al. 2019; Pandit et al. 2020). Biofilm production by various microbiota is influenced by several factors, including cell surface structure and charge, bacterial metabolites, substratum surface roughness, temperature, pH, nutrient and water availability, and heavy metals (Ansari et al. 2017). The biofilm density is directly proportional to the cell number, ranging from 10^8 to 10^{11} cells/g wet weight, with 90% dry weight accounting for the gelatinous matrix layer and 10% dry weight for microorganisms (Flemming et al. 2016). The rhizosphere and rhizoplane are major hot spots for various plant growth-promoting bacteria (PGPB) that uptake nutrients secreted by plants (Saleh et al. 2020). However, successfully enhancing crop yields with PGPB requires their successful establishment against native microbes through competition. To avoid this cumbersome situation, application of beneficial biofilm bacteria is a potential option as they secure themselves against adverse stress conditions as well as competition from native microbiota in their niches (Triveni et al. 2012). Biofilms are composed of a diverse group of microorganisms including bacteria, fungi, and cyanobacteria such as *Bacillus*, *Rhizobium*, *Azospirillum*, *Herbaspirillum*, *Bradyrhizobium*, *Gluconacetobacter*, *Burkholderia*, *Azotobacter*, *Anabaena*, *Erwinia*, *Paenibacillus*, *Mesorhizobium*, *Pseudomonas*, *Xyella*, and *Trichoderma* (Velmourougane et al. 2017). Biofilm bacteria not only have a specific relationship with plants, but their interaction with arbuscular mycorrhizal fungi (AMF) creates a "tripartite association" (Ghignone et al. 2012).

Moreover, agriculturally important biofilm biofertilizers (BFBF) were first introduced by Gamini Seneviratne from Sri Lanka, which cut down chemical fertilizer use by 25% and increased yields up to 20–30% in paddy, tea, and vegetables (Biofilm 2023). From a plant growth perspective, BFBF has extensively helped in the nutrient cycle from soil to plant. It has influenced plant growth by solubilizing different minerals such as insoluble phosphate (Yadav et al. 2015), potassium (Verma et al. 2017), and zinc (Verma et al. 2015). Nitrogen fixation also gets increased by high nitrogenase activity (Rubin et al. 2017), and BFBF enhances the production of growth hormones such as indole acetic acid (IAA). BFBF also produces hydrolytic enzymes, HCN, volatile organic compounds, iron chelating agents (siderophores), ammonia, and ACC deaminase (Rubin et al. 2017; Kour et al. 2020a). Apart from this, BFBF plays a crucial role in reinstating soil health by increasing soil organic carbon (OC) and microbial biomass carbon (Seneviratne et al. 2011; Zakeel and Safeena 2019). The co-inoculation of *Stenotrophomonas* sp. K96 and *Pseudomonas* sp. M45 expresses strong biofilm, which facilitates significant vegetative growth attributes in coffee plantation (Hazarika et al. 2021). According to Taktek et al. (2017), the inoculation of *Rhizobium miluonense* Rm3 and *Burkholderia anthina* Ba8 biofilms can dissolve inorganic phosphate minerals, release organic acid, and lower pH. Prasanna et al. (2015) reported that inoculation of a cyanobacteria biofilm formulation (*Anabaena-Azotobacter* biofilm) in maize crops significantly increased Zn2+ concentration, plant defense enzyme activity, soil health parameters, and yields.

The use of biofilm bacteria has been shown to aid in the resistance of salinity stress by improving relative water content (RWC) and mineral transfer to plants through the aggregation of soil macrospores and binding of Na^{2+} (Qurashi and Sabri 2012; Mohammed 2018; Zakeel and Safeena 2019). A study by Kasim et al. (2016) revealed that increasing salt concentration (0, 250, 500, and 1000 mM NaCl) in vitro resulted in an increase in biofilm production by 20 PGPB. After screening 9 efficient strains from the 20 PGPB, treating barley grain under salt stress showed a higher proportion of shoot length (10–15%), fresh biomass of roots and shoots (35 and 64%), and relative water content (RWC) of roots and shoots (12–15% and 7%), respectively. Additionally, similar results were obtained from a study by Mohammed (2018), which reported that *Pseudomonas anguilliseptica* SAW 24 improved the shoot height and fresh biomass of *faba bean* during salinity stress.

Biofilm bacteria play a crucial role in relieving drought stress as compared to non-biofilm PGPB. These bacteria produce a significant amount of aminocyclopropane-1-carboxylate deaminase (ACC) and osmoprotectants like proline, trehalose, glycine betaine, amino acids, ectoine, and glutamine under water deficit conditions, which helps to sustain and multiply PGPB under severe drought conditions (Timmusk et al. 2014). Wang et al. (2019) reported that inoculating biofilm-producing *Bacillus amyloliquefaciens* 54 in tomato plants increased the relative water content, abscisic acid, and antioxidant enzymes and decreased the level of malondialdehyde during drought stress, as compared to the non-biofilm mutant. Ansari et al. (2021) also reported similar findings, where the inoculation of *Pseudomonas azotoformans* FAP5 significantly enhanced the production of biofilms under drought stress, which helped to improve the growth attributes of wheat plants. Biofilm bacteria also play a crucial role in controlling disease-causing plant pathogens in agricultural systems. These bacteria encounter plant pathogens through competition for colonization, triggering host defensive mechanisms (ISR and SAR), and producing antimicrobials (Ajijah et al. 2023). Most of the biocontrol studies of biofilm microbes focus on *Bacillus* and *Pseudomonas*. The biofilm formation of *Bacillus* sp. is triggered by the lipopeptide molecule "surfactin," which enters the Spo0A pathway to activate Histidine Kinase KinC (Chen et al. 2022). Zeriouh et al. (2014) concluded that surfactin of *B. subtilis* UMAF6614 synthesized biofilm that actively controlled *Xanthomonas campestris*, *Pectobacterium carotovorum*, and *Podosphaera fusca* on melon phylloplane. Cao et al. (2018) found that biofilm formation of *Bacillus velezensis* (Y6 and F7) was stimulated by surfactin, iturin, and fengycin against *Ralstonia solanacearum* and *Fusarium oxysporum* in tomato. Moreover, multispecies biofilms play a simultaneous role in pathogen suppression and cycling the unavailable nutrients to the host.

7 Microbial-Assisted Biofortification

Micronutrient deficiencies in food crops are a significant global issue that can negatively impact both humans and plants. "Biofortification," a strategy that involves adding essential nutrients to various parts of plants, can help address this problem (Koç and Karayiğit 2022). Unfortunately, environmental and soil factors often prevent the uptake of micronutrients from soil to plants, leading to insufficient amounts of these nutrients and hindering plant growth and development (Masunaga and Marques Fong 2018; Upadhayay et al. 2018). Micronutrient deficiencies, which include a lack of zinc, iron, selenium, copper, manganese, and vitamins, are a significant global health concern that can lead to malnutrition. Zinc, iron, and selenium are crucial micronutrients that are essential in appropriate amounts through a routine diet to maintain several life processes. Zinc, in particular, is considered the most vital micronutrient for all organisms, including humans, as it plays essential structural roles in numerous proteins, enzymes, and transcription factors that regulate gene expression (Roohani et al. 2013).

Zinc insufficiency is a pressing global health issue and is linked to a plethora of human health complications, ranging from stunted physical growth, elevated susceptibility to diverse infections, compromised wound healing, chronic diarrhea, skeletal malformations, to an augmented risk of miscarriage (Khan et al. 2019). Iron deficiency can have devastating effects on both plant and human health. Plants with insufficient iron levels may suffer from chlorosis, which results in stunted growth and reduced crop yield, thereby impacting global food security (Li et al. 2021). Meanwhile, iron deficiency in humans can lead to nutritional anemia and impaired immune function, particularly in children, as well as impaired neurocognitive development (Mitterstiller et al. 2022). Additionally, selenium is a crucial micronutrient with roles in a wide range of metabolic pathways, including antioxidant defense and thyroid hormone metabolism (Mojadadi et al. 2021). The dearth of Se is related with a plethora of afflictions, ranging from cardiovascular disorders, reduced male fertility, and hypothyroidism to compromised immune response, augmented vulnerability to infections, cancer, oxidative stress-related disorders, and even epilepsy (Genchi et al. 2023; Upadhayay et al. 2018). The battle against the detrimental effects of micronutrient deficiencies is being waged on the front lines of agriculture, where the pursuit of producing safe and nutritious food crops has become a quest for elevating the levels of vital micronutrients within the edible portions of plants. As we seek to address the critical issue of biofortification, it is important to acknowledge that agronomic and genetic methods may not be suitable for developing nations with predominantly rural populations (Upadhayay et al. 2018, 2019, 2022b). This is because the usage of chemical-based fertilizers, which is often integral to these approaches, can cause significant harm to both soil health and human well-being (Upadhayay et al. 2022c). Furthermore, genetic and transgenic methods for biofortification can prove to be time-consuming, labor-intensive, and economically impracticable (Upadhayay et al. 2019, 2022d).

Therefore, it is imperative that we explore alternative approaches that prioritize the safety and sustainability of agricultural practices, especially in areas where traditional farming techniques are still prevalent. Only by embracing innovative solutions that prioritize the health of both the environment and the people who rely on it can we hope to achieve lasting success in the pursuit of biofortification. As a cost-effective and sustainable alternative, plant growth-promoting microorganisms can be used to increase micronutrient concentrations in plants and can be proficiently practiced as "microbial-assisted biofortification" (Fig. 2). Plant growth-promoting bacteria and rhizobacteria, for example, can enhance crop production through nitrogen fixation, mineral solubilization, phytohormone production, and biocontrol (Upadhayay et al. 2023a, c). The rhizosphere is an active and dynamic environment where various bacterial species interact with plants to support their growth and development. One crucial role of these bacteria is to help mobilize micronutrients like zinc and iron, making them more easily assimilated by plants (Upadhayay et al. 2018, 2022c). Soil microorganisms achieve this through mechanisms such as organic acid production and siderophore synthesis (Upadhayay et al. 2019, 2022b). By producing organic acids, bacteria can make zinc more available to plants by sequestering cations and dropping the pH of the rhizosphere (Haroon et al. 2022a). Meanwhile, bacteria that produce siderophores are effective at chelating iron, which increases its availability to plants. Siderophores are organic compounds that not only help with iron uptake but may also contribute in the mobilization of other metals, including zinc (Timofeeva et al. 2022).

The agroecosystem contains both organic and inorganic forms of selenium, including four oxidation states of inorganic selenium: Se^{2-} (selenide), Se^{0} (elemental Se), Se^{4+} (selenite), and Se^{6+} (selenate) (Hossain et al. 2021). Bacteria have the ability to reduce oxidized and methylated selenium oxyanions to different selenium

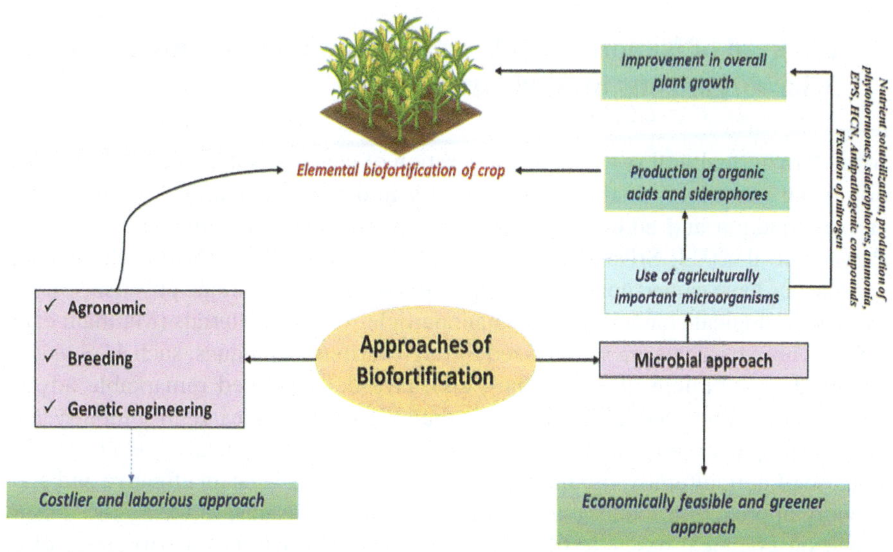

Fig. 2 Strategies for biofortifying crops: a comparison of different approaches, with emphasis on microbial-assisted biofortification

compounds. This reduction process allows these microorganisms to convert Se^{6+} and Se^{4+} to Se^{0}, generating Se methylated compounds and organic selenium compounds such as SeMet and SeMeSeCys (Mora et al. 2015). By doing so, bacteria make selenium available for use in agriculture systems. Soil- and rhizosphere-associated microorganisms can be used to improve nutrient acquisition of essential micronutrients from soil to various parts of plants. This microbial-assisted precise nutrient acquisition approach is a suitable method for crop biofortification with required nutrient elements like zinc, iron, and selenium (Upadhayay et al. 2018). Table 2 illustrates the use of agriculturally important microorganism as natural biofortifying agents in biofortification of different crops.

The benefits of biofortification can be duly appreciated by considering the following salient factors:

- Firstly, the efficacy of bacterial strains in the solubilization of micronutrients should be rigorously validated via in vitro assays to ensure optimal results.
- Secondly, the potential micronutrient-solubilizing bacteria must exhibit a diverse range of plant growth-promoting properties to effectively augment the growth and development of crops.
- Thirdly, bacterial inoculants must demonstrate a tangible and significant increase in the micronutrient content of the edible portions of food crops to justify their use in biofortification.
- Lastly, the micronutrient content can be accurately estimated using advanced techniques such as "atomic absorption spectroscopy" or "inductively coupled plasma-mass spectrometry (ICP-MS)," providing valuable insights into the efficacy of the biofortification process.

8 Cocktail of Nanomaterials and Agriculturally Important Microorganisms in Agriculture

In modern agricultural systems, the integration of nanotechnology has emerged as a promising and effective strategy, particularly in developing nations, for enhancing crop production and addressing food security challenges (Dhaigude et al. 2023; Parveen et al. 2023; Srivastava et al. 2023; Rani et al. 2023). Nanotechnological approaches have gained considerable attention owing to the unique physical, chemical, and biological characteristics of nanoparticles or nanomaterials (Maithani et al. 2023). These particles are synthesized through various techniques, such as physical, chemical, and biological approaches, and have demonstrated remarkable advantages over conventional fertilizers due to their high surface area, high solubility, and other novel physicochemical properties (Priya et al. 2023; Upadhayay et al. 2023b). The use of agriculturally significant microorganisms has been an effective and eco-friendly approach in modern agriculture for several decades, reducing the dependence on chemical-based fertilizers and pesticides. Bacteria, in particular, offer unique plant growth-promoting properties, including nutrient solubilization, nitrogen fixation, phytohormone production, EPS, ammonia, iron chelating agents, and

Table 2 Micronutrient biofortification in crops through microbial treatments

Microbial inoculant(s)	Plant	Plant part	Element(s)	References
Bacillus cereus (ZnSB13)	Chickpea	Grain	Zinc	Batool et al. (2021)
Microbial consortium (*Paenibacillus polymyxa* ZM27 + *B. subtilis* ZM63 + *B. aryabhattai* S10)	Maize	Grain	Iron and zinc	Ahmad et al. (2022)
Co-inoculation of *Piriformospora indica* and *Azotobacter chroococcum* WR5 (AzWR5)	*Triticum aestivum* L.	Grain	Iron and zinc	Singh et al. (2022b)
Bacillus altitudinis WR10	*Triticum aestivum*	Grain	Iron	Sun et al. (2021)
Rhizobium radiobacter LB2	Lettuce	Leaf	Zinc	Verma et al. (2020)
Selenobacteria mixture and *G. claroideum*	*Wheat*	Grain	Selenium	Durán et al. (2013)
Pseudomonas sp. VBZ4	*Solanum lycopersicum* L.	Fruit	Zinc	Karnwal (2021a)
Burkholderia cepacian + ZnO	*Rice*	Grain	Zinc	Upadhayay et al. (2022c)
Pseudomonas spp.	*Okra*	Fruit	Zinc	Karnwal (2021b)
Bacillus biofertilizer	*Maize*	Grain	Iron and zinc	Mumtaz et al. (2022)
Consortium of siderophore producing bacterial strains + Fe	Groundnut (*Arachis hypogea* L.)	Kernel	Iron and zinc	Sarwar et al. (2022)

other attributes (Roshani et al. 2020; Upadhayay et al. 2022b). Similarly, fungi, particularly mycorrhizal fungi, enhance nutrient uptake efficiency in host plants, thereby contributing to plant growth and development (Upadhayay et al. 2019). Furthermore, scientific literature has demonstrated the potential of plant growth-promoting microorganisms in alleviating abiotic and biotic stress, thus improving plant resilience to environmental stressors (Singh et al. 2022a; Upadhayay et al. 2023a). The application of a judicious combination of selective plant growth-promoting microbial strains, which have demonstrated compatibility with the nanoparticles of interest, has been established in numerous studies (Agri et al. 2021; Chaudhary et al. 2023b, c, d). The co-utilization of nanomaterials and agriculturally important microorganisms exhibits significant potentiality for sustainable agriculture, representing an advanced alternative to agrochemicals, and is rapidly gaining recognition in the agricultural domain (Bhandari et al. 2023; Chaudhary et al. 2023e, f). The concept of proficient fertilization, which this revolutionary approach embodies, is overwhelmingly preferred over traditional chemical-based fertilization techniques due to its unprecedented effectiveness in resource consumption, steady and sustained release of nutrients, significant enhancement in crop production with the use of minimal doses of fertilizer, and negligible deleterious effects on soil. The novel concept of "nanobiofertilizer" presents a cutting-edge approach that amalgamates nanomaterials and biofertilizers to enhance crop yields (Upadhayay et al. 2023c). Recent breakthroughs in nanotechnology have paved the way for the encapsulation of agriculturally important microbial strains in appropriate nanomaterials (Panichikkal and Krishnankutty 2022; Akhtar et al. 2022) or the simultaneous encapsulation of both nanomaterials and microbial strains in an appropriate carrier (Moradi Pour et al. 2022; Upadhayay et al. 2023c), based on the design of experiment. This innovative approach not only ensures the effectiveness and shelf life of the microbial element but also provides a controlled and sustained supply of both nanomaterials and probiotic strains. Study by Khati et al. (2017) reported that application of "nanochitosan" and *Bacillus* spp. improved the health of maize plants. Table 3 illustrates the combined impact of nanomaterials and potential microbial strains on crop productivity, representing a new strategy that fosters hope for

Table 3 Combined impact of nanomaterials and agriculturally important microorganisms

Combination of nanomaterials and microbial strain (s)	Crop	Beneficial attributes	References
Bacillus mycoides PM35 + titanium dioxide (TiO_2)—nanoparticles	Barley (*Hordeum vulgare* L.)	• Improvement in plant growth and biomass under cadmium stress	Ma et al. (2023)
Bacillus pumilus + nanoparticles (ZnO and TiO_2)	Maize	• Improvement in agronomical parameters against cadmium stress • Improvement in phytohormones, proline, and nutritional content	Shafiq et al. (2022)
ZnO nanoparticles + *Pseudomonas* sp.	Wheat	• Enhancement in plant growth, biomass, and photosynthetic pigments • Protection of plant from heat and drought stress	Azmat et al. (2022)
Bacillus sp. + nanozeolite	Maize	• Increment in plant height, fresh and dry weight of root and shoot, photosynthetic pigments, total sugars, etc. • 29.80% increment in maize productivity	Chaudhary et al. (2021c)
Nanogypsum + *Pseudomonas taiwanensis*	Maize	• Improvement in agronomical parameters of plants • Improvement in structure and function of soil	Chaudhary et al. (2021d)
Nano-zinc oxide (1 g/L) + *Azosprillium*	*Triticale*	• Enhancement in seed quality, increasing of grain filling period, zinc and protein content	Kamari and Sharifi (2017)
ZnO nanoparticles + *Azospirillum brasilense*	Wheat	• Improvement in growth-yield parameters and nutrient uptake under drought condition	Muhammad et al. (2022)

sustainable agriculture. Besides, the combined use of nanocompounds and bioinoculants also augments the soil fertility status (Agri et al. 2022).

9 Rhizosphere Microbiome Transplant: A New-Fangled Tactic in Plant Disease Management

To achieve disease suppression and confer other beneficial properties to their host, probiotics for humans and plant growth-promoting microbes for plants are vital. However, their effectiveness can often be impeded by the instability of their colonization within their respective host and aggressive competition with other microbes residing within the host's microbiome (Bziuk et al. 2022; Schlatter et al. 2017). To transcend the boundaries imposed by single-strain or synthetic community inoculations and achieve optimal outcomes, "fecal microbiome transplantation" has emerged as a promising solution. This avant-garde approach has demonstrated remarkable success in treating a range of clinical ailments, including, but not limited to, *Clostridium difficile* infections (Kassam et al. 2013) and cancer (Baruch et al. 2021). The notion of "rhizosphere microbiome transplantation (RMT)" in the context of the plant rhizosphere has been somewhat neglected, likely due to the challenge of implementing this strategy on a large scale while ensuring its efficacy (Bziuk et al. 2022). However, the transplantation of "protective" microbiomes from resilient to vulnerable plants presents a unique opportunity to gain a more profound comprehension of the various factors that govern the assembly of the rhizosphere microbiome (Jiang et al. 2022).

A holistic approach toward community coalescence presents a promising perspective for developing effective strategies to mitigate plant diseases (Gopal et al. 2013; Jiang et al. 2022). Previous studies utilizing cultivation-dependent methods have revealed the significance of certain bacterial species, such as *Pseudomonas*, *Bacillus*, and others, in exerting disease suppression in plants (Kyselková and Moënne-Loccoz 2012; Gómez Expósito et al. 2017). Recent research utilizing cultivation-independent approaches has provided a comprehensive understanding of the role of soil microbiomes in suppressing plant diseases (Schlatter et al. 2017). In the classical approach, it has been empirically observed that the introduction of a disease-suppressive soil to a conducive soil (at a ratio of 1:9 (w/w)) is a highly effective means of enhancing plant resistance against *Rhizoctonia solani* infections. This method has been further substantiated and elucidated, revealing a complex interplay between soil microbiomes and plant health. Specifically, it has been shown to engender a significant enrichment of 17 distinct bacterial populations, spanning the phyla *Proteobacteria*, *Firmicutes*, and *Actinobacteria*, underscoring the intricate and dynamic nature of soil-plant interactions (Mendes et al. 2011). Recently, studies have shown that the use of RMT can reduce the development of bacterial wilt disease and decrease its incidence. This phenomenon has been observed in soils derived from healthy plants, which are enriched with *Firmicutes* and *Proteobacteria* and associated with the enrichment of *Flavobacterium* in the rhizosphere microbiome (Kwak et al. 2018; Wei et al. 2019). In a recent study by Jiang et al. (2022), the rhizosphere microbiomes of 12 eggplant varieties were evaluated for their level of resistance against bacterial wilt disease caused by the pathogen *R. solanacearum*. Transplanting rhizosphere microbiomes from distinct, resistant donors led to varying degrees of wilt disease suppression, with a maximum reduction in disease incidence of up to 47%. In a separate study, Bziuk et al. (2022) reported that transplanting a rhizosphere microbiome from field-grown barley to potting soil-grown barley seedlings decreased susceptibility to powdery mildew caused by *Blumeria graminis* f. sp. *hordei*, a common fungal pathogen of barley. These studies provide convincing indication to support the use of rhizosphere microbiome transplants as a promising approach to protect plants against pathogen infections.

10 Conclusion

The utilization of microorganisms in agriculture gives a beacon of hope amidst the widespread level contamination of the environment caused by the disproportionate use of chemical-based fertilizers, pesticides, and herbicides. The application of agriculturally important microorganisms exhibits outstanding proficiency in nutrient management, crop growth promotion, and maintaining of soil health. Some cutting-edge approaches include application of biofilm-forming bacteria, microbial-assisted biofortification, and rhizospheric microbiome transplantation, and the combined use of nanomaterials and plant growth-promoting microbial strain show tremendous potential for the development of sustainable and eco-friendly agricultural practices. It is also important to explore the pivotal contribution of microbes as an approach for practicing climate-smart agriculture. By harnessing the power of microbes, it is imperative to bring this harmonious approach to agriculture that is in tune with the natural world, minimizing our dependency on synthetic inputs and representing a transformative shift toward a more sustainable and equitable future.

References

Achinas S, Charalampogiannis N, Euverink GJW (2019) A brief recap of microbial adhesion and biofilms. Appl Sci 9:2801. https://doi.org/10.3390/app9142801

Adeyemi NO, Atayese MO, Sakariyawo OS, Azeez JO, Olubode AA, Ridwan M, Adebiyi A, Oni O, Ibrahim I (2021) Influence of different arbuscular mycorrhizal fungi isolates in enhancing growth, phosphorus uptake and grain yield of soybean in a phosphorus deficient soil under field conditions. Commun Soil Sci Plant Anal 52:1171–1183. https://doi.org/10.1080/00103624.2021.1879117

Agri U, Chaudhary P, Sharma A (2021) In vitro compatibility evaluation of agriusable nanochitosan on beneficial plant growth-promoting rhizobacteria and maize plant. Natl Acad Sci Lett 44:555–559. https://doi.org/10.1007/s40009-021-01047-w

Agri U, Chaudhary P, Sharma A, Kukreti B (2022) Physiological response of maize plants and its rhizospheric microbiome under the influence of potential bioinoculants and nanochitosan. Plant Soil 474:451–468. https://doi.org/10.1007/s11104-022-05351-2

Ahmad M, Hussain A, Dar A, Luqman M, Ditta A, Iqbal Z, Ahmad HT, Nazli F, Soufan W, Almutairi K, Sabagh AE (2022) Combating iron and zinc malnutrition through mineral biofortification in maize through plant growth promoting *Bacillus* and *Paenibacillus* species. Front Plant Sci 13:1094551. https://doi.org/10.3389/fpls.2022.1094551

Ahmed N, Al-Mutairi KA (2022) Earthworms effect on microbial population and soil fertility as well as their interaction with agriculture practices. Sustainability 14:7803. https://doi.org/10.3390/su14137803

Ahmed AAQ, Odelade KA, Babalola OO (2020) Microbial inoculants for improving carbon sequestration in agroecosystems to mitigate climate change. In: Leal Filho W (ed) Handbook of climate change resilience. Springer, Cham. https://doi.org/10.1007/978-3-319-93336-8_119

Ajijah N, Fiodor A, Pandey AK, Rana A, Pranaw K (2023) Plant growth-promoting bacteria (PGPB) with biofilm-forming ability: a multifaceted agent for sustainable agriculture. Diversity 15:112. https://doi.org/10.3390/d15010112

Akhtar N, Ilyas N, Meraj TA, Pour-Aboughadareh A, Sayyed RZ, Mashwani ZU, Poczai P (2022) Improvement of plant responses by nanobiofertilizer: a step towards sustainable agriculture. Nanomaterials (Basel) 12:965. https://doi.org/10.3390/nano12060965

Amirnia R, Ghiyasi M, Siavash Moghaddam S, Rahimi A, Damalas CA, Heydarzadeh S (2019) Nitrogen-fixing soil bacteria plus mycorrhizal fungi improve seed yield and quality traits of lentil (lens culinaris Medik). J Soil Sci Plant Nutr 19:592–602. https://doi.org/10.1007/s42729-019-00058-3

Ansari FA, Jafri H, Ahmad I, Abulreesh HH (2017) Factors affecting biofilm formation in vitro and in the rhizosphere. In: Biofilms in plant and soil health. Wiley, pp 275–290

Ansari FA, Jabeen M, Ahmad I (2021) *Pseudomonas azotoformans* FAP5, a novel biofilm-forming PGPR strain, alleviates drought stress in wheat plant. Int J Environ Sci Technol 18(12):3855–3870. https://doi.org/10.1007/s13762-020-03045-9

Azmat A, Tanveer Y, Yasmin H, Hassan MN, Shahzad A, Reddy M, Ahmad A (2022) Coactive role of zinc oxide nanoparticles and plant growth promoting rhizobacteria for mitigation of synchronized effects of heat and drought stress in wheat plants. Chemosphere 297:133982. https://doi.org/10.1016/j.chemosphere.2022.133982

Baruch EN, Youngster I, Ben-Betzalel G, Ortenberg R, Lahat A, Katz L (2021) Fecal microbiota transplant promotes response in immunotherapy-refractory melanoma patients. Science 371:602–609. https://doi.org/10.1126/science.abb5920

Batool S, Asghar HN, Shehzad MA, Yasin S, Sohaib M, Nawaz F (2021) Zinc-solubilizing bacteria-mediated enzymatic and physiological regulations confer zinc biofortification in chickpea (*Cicer arietinum* L.). J Soil Sci Plant Nutr 21:2456–2471. https://doi.org/10.1007/s42729-021-00537-6

Bhandari G, Dhasmana A, Chaudhary P, Gupta S, Gangola S, Gupta A, Rustagi S, Shende SS, Rajput VD, Minkina T, Malik S, Slama P (2023) A perspective review on green nanotechnology in agro-ecosystems: opportunities for sustainable agricultural practices & environmental remediation. Agriculture 13:668. https://doi.org/10.3390/agriculture13030668

Bhattacharyya SS, Ros GH, Furtak K, Iqbal HMN, Parra-Saldívar R (2022) Soil carbon sequestration—an interplay between soil microbial community and soil organic matter dynamics. Sci Total Environ 815:152928. https://doi.org/10.1016/j.scitotenv.2022.152928

Brader G, Compant S, Vescio K, Mitter B, Trognitz F, Ma LJ, Sessitsch A (2017) Ecology and genomic insights into plant-pathogenic and plant-nonpathogenic endophytes. Annu Rev Phytopathol 55:61–83. https://doi.org/10.1146/annurev-phyto-080516-035641

Bulgarelli D, Garrido-Oter R, Münch PC, Weiman A, Dröge J, Pan Y (2015) Structure and function of the bacterial root microbiota in wild and domesticated barley. Cell Host Microbe 17:392–403. https://doi.org/10.1016/j.chom.2015.01.011

Bziuk N, Maccario L, Sørensen SJ, Schikora A, Smalla K (2022) Barley rhizosphere microbiome transplantation—a strategy to decrease susceptibility of barley grown in soils with low

microbial diversity to powdery mildew. Front Microbiol 13:830905. https://doi.org/10.3389/fmicb.2022.830905
Cao Y, Pi H, Chandrangsu P, Li Y, Wang Y, Zhou H (2018) Antagonism of two plant-growth promoting *Bacillus velezensis* isolates against *Ralstonia solanacearum* and *Fusarium oxysporum*. Sci Rep 8:4360. https://doi.org/10.1038/s41598-018-22782-z
Cavite HJM, Mactal AG, Evangelista EV, Cruz JA (2021) Growth and yield response of upland rice to application of plant growth-promoting rhizobacteria. J Plant Growth Regul 40:494–508. https://doi.org/10.1007/s00344-020-10114-3
Chaudhary A, Parveen H, Chaudhary P, Bhatt P, Khatoon H (2021a) Rhizospheric microbes and their mechanism. In: Bhatt P, Gangola S, Udayanga D, Kumar G (eds) Microbial technology for sustainable environment. Springer, Singapore. https://doi.org/10.1007/978-981-16-3840-4_6
Chaudhary P, Chaudhary A, Parveen H, Rani A, Kumar G, Kumar A, Sharma A (2021b) Impact of nanophos in agriculture to improve functional bacterial community and crop productivity. BMC Plant Biol 21:519. https://doi.org/10.1186/s12870-021-03298-7
Chaudhary P, Khati P, Chaudhary A, Gangola S, Kumar R, Sharma A (2021c) Bioinoculation using indigenous *Bacillus* spp. improves growth and yield of *Zea mays* under the influence of nanozeolite. 3 Biotech 11:11. https://doi.org/10.1007/s13205-020-02561-2
Chaudhary P, Khati P, Chaudhary A, Maithani D, Kumar G, Sharma A (2021d) Cultivable and metagenomic approach to study the combined impact of nanogypsum and *Pseudomonas taiwanensis* on maize plant health and its rhizospheric microbiome. PLoS One 16:e0250574. https://doi.org/10.1371/journal.pone.0250574
Chaudhary P, Khati P, Gangola S, Kumar A, Kumar R, Sharma A (2021e) Impact of nanochitosan and *Bacillus* spp. on health, productivity and defence response in *Zea mays* under field condition. 3 Biotech 11:237. https://doi.org/10.1007/s13205-021-02790-z
Chaudhary P, Parveen H, Gangola S, Kumar G, Bhatt P, Chaudhary A (2021f) Plant Growth Promoting Rhizobacteria and their application in sustainable crop production. In: Bhatt P, Gangola S, Udayanga D, Kumar G (eds) Microbial technology for sustainable environment. Springer, Singapore. https://doi.org/10.1007/978-981-16-3840-4_13
Chaudhary P, Sharma A, Chaudhary A, Khati P, Gangola S, Maithani D (2021g) Illumina based high throughput analysis of microbial diversity of maize rhizosphere treated with nanocompounds and *Bacillus* sp. Appl Soil Ecol 159:103836. https://doi.org/10.1016/j.apsoil.2020.103836
Chaudhary A, Chaudhary P, Upadhyay A, Kumar A, Singh A (2022a) Effect of gypsum on plant growth promoting rhizobacteria. Environ Ecol 39(4A):1248–1256
Chaudhary P, Singh S, Chaudhary A, Agri U, Bhandari G (2022b) Chapter 3: Agrochemicals and their effects on soil microbial population. In: Soni R, Suyal D, Goel R (eds) Plant protection: from chemicals to biologicals. De Gruyter, Berlin, Boston, pp 45–58. https://doi.org/10.1515/9783110771558-003
Chaudhary P, Agri U, Chaudhary A, Kumar A, Kumar G (2022c) Endophytes and their potential in biotic stress management and crop production. Front Microbiol 13:933017. https://doi.org/10.3389/fmicb.2022.933017
Chaudhary P, Chaudhary A, Agri U, Khatoon H, Singh A (2022d) Recent trends and advancements for agro-environmental sustainability at higher altitudes. In: Goel R, Soni R, Suyal DC, Khan M (eds) Survival strategies in cold-adapted microorganisms. Springer, Singapore. https://doi.org/10.1007/978-981-16-2625-8_19
Chaudhary P, Singh S, Chaudhary A, Sharma A, Kumar G (2022e) Overview of biofertilizers in crop production and stress management for sustainable agriculture. Front Plant Sci 13:930340. https://doi.org/10.3389/fpls.2022.930340
Chaudhary P, Xu M, Ahamad L, Chaudhary A, Kumar G, Adeleke BS, Verma KK, Hu D-M, Širić I, Kumar P, Popescu SM, Abou Fayssal S (2023a) Application of synthetic consortia for improvement of soil fertility, pollution remediation, and agricultural productivity: a review. Agronomy 13:643. https://doi.org/10.3390/agronomy13030643
Chaudhary P, Chen S, Rajput VD, Jaiswal DK (2023b) Editorial: bioinoculants with nanocompounds to improve soil health: a step toward sustainable agriculture. Front Environ Sci 11:1270002. https://doi.org/10.3389/fenvs.2023.1270002

Chaudhary P, Chaudhary A, Nadda AK, Khati P (eds) (2023c) Advances in nanotechnology for smart agriculture: techniques and applications, 1st edn. CRC Press. https://doi.org/10.1201/9781003345565

Chaudhary P, Chaudhary A, Dhaigude V, Fayssal SA, Kumar P, Sharma A, Chen S (2023d) Environmental-friendly nanoparticles in agriculture. In: Chaudhary P, Chaudhary A, Nadda AK, Khati P (eds) Advances in nanotechnology for smart agriculture: techniques and applications, 1st edn. CRC Press. https://doi.org/10.1201/9781003345565-13

Chaudhary P, Chaudhary A, Khati P, Kumar G, Tyagi J, Behera M (2023e) Nanotechnology and omics approach in agrobiotechnology. In: Fernandez-Luqueno F, Patra JK (eds) Agricultural and environmental nanotechnology. Interdisciplinary biotechnological advances. Springer, Singapore. https://doi.org/10.1007/978-981-19-5454-2_12

Chaudhary P, Chen S, Rajput VD, Jaiswal DK (eds) (2023f) Bioinoculants with nano-compounds to improve soil health: a step toward sustainable agriculture. Frontiers Media SA, Lausanne. https://doi.org/10.3389/978-2-8325-3287-4

Chen Z, Srivastava P, Zarazúa-Osorio B, Marathe A, Fujita M, Igoshin OA (2022) *Bacillus subtilis* histidine kinase KinC activates biofilm formation by controlling heterogeneity of single-cell responses. mBio 13:e0169421. https://doi.org/10.1128/mbio.01694-21. Pubmed: 35012345

Chitara MK, Sharma S, Parihar M, Jeevan B, Sangannavar PA, Kumar A (2022) The structure, function, and utility of the rhizosphere microbiome of cereal crops. In: Microbial resource technologies for sustainable development. Elsevier, pp 77–111

Christina A, Christapher V, Bhore SJ (2013) Endophytic bacteria as a source of novel antibiotics: an overview. Pharmacogn Rev 7:11–16. https://doi.org/10.4103/0973-7847.112833

Davamani V, Parameswari E, Arulmani S (2020) Mitigation of methane gas emissions in flooded paddy soil through the utilization of methanotrophs. Sci Total Environ 726:138570. https://doi.org/10.1016/j.scitotenv.2020.138570

de Vos WM (2015) Microbial biofilms and the human intestinal microbiome. npj Biofilms Microbiomes 1:15005. https://doi.org/10.1038/npjbiofilms.2015.5

Devarajan AK, Muthukrishanan G, Truu J, Truu M, Ostonen I, Kizhaeral SS (2021) The foliar application of rice phyllosphere bacteria induces drought-stress tolerance in *Oryza sativa* (L.). Plants (Basel) 10:387. https://doi.org/10.3390/plants10020387

Dhaigude V, Naliyapara HB, Chaudhary P, Fayssal SA, Nadda AK, Minkina T, Rajput VD (2023) Role of nanobiosensors in agriculture advancement. In: Chaudhary P, Chaudhary A, Nadda AK, Khati P (eds) Advances in nanotechnology for smart agriculture: techniques and applications, 1st edn. CRC Press. https://doi.org/10.1201/9781003345565-9

Di Benedetto NA, Corbo MR, Campaniello D, Cataldi MP, Bevilacqua A, Sinigaglia M, Flagella Z (2017) The role of plant growth promoting bacteria in improving nitrogen use efficiency for sustainable crop production: a focus on wheat. AIMS Microbiol 3:413–434. https://doi.org/10.3934/microbiol.2017.3.413

Dini-Andreote F (2020) Endophytes: the second layer of plant defense. Trends Plant Sci 25:319–322. https://doi.org/10.1016/j.tplants.2020.01.007

Dudeja SS, Giri R (2014) Beneficial properties, colonization, establishment and molecular diversity of endophytic bacteria in legume and nonlegume. Afr J Microbiol Res 8:1562–1572. https://doi.org/10.5897/AJMR2013.6541

Durán P, Acuña JJ, Jorquera MA, Azcón R, Borie F, Cornejo P, Mora ML (2013) Enhanced selenium content in wheat grain by co-inoculation of selenobacteria and arbuscular mycorrhizal fungi: a preliminary study as a potential Se biofortification strategy. J Cereal Sci 57:275–280. https://doi.org/10.1016/j.jcs.2012.11.012

Fadiji AE, Babalola OO (2020) Elucidating mechanisms of endophytes used in plant protection and other bioactivities with multifunctional prospects. Front Bioeng Biotechnol 8:467. https://doi.org/10.3389/fbioe.2020.00467

Flemming HC, Wingender J, Szewzyk U, Steinberg P, Rice SA, Kjelleberg S (2016) Biofilms: an emergent form of bacterial life. Nat Rev Microbiol 14:563–575. https://doi.org/10.1038/nrmicro.2016.94

Galindo FS, Rodrigues WL, Fernandes GC, Boleta EHM, Jalal A, Rosa P (2022) Enhancing agronomic efficiency and maize grain yield with *Azospirillum brasilense* inoculation under Brazilian savannah conditions. Front Environ Sci 10:821628. https://doi.org/10.3389/fenvs.2022.821628

Genchi G, Lauria G, Catalano A, Sinicropi MS, Carocci A (2023) Biological activity of selenium and its impact on human health. Int J Mol Sci 24:2633. https://doi.org/10.3390/ijms24032633

Ghignone S, Salvioli A, Anca I, Lumini E, Ortu G, Petiti L (2012) The genome of the obligate endobacterium of an AM fungus reveals an interphylum network of nutritional interactions. ISME J 6:136–145. https://doi.org/10.1038/ismej.2011.110

Gómez Expósito R, de Bruijn I, Postma J, Raaijmakers JM (2017) Current insights into the role of rhizosphere bacteria in disease suppressive soils. Front Microbiol 8:2529. https://doi.org/10.3389/fmicb.2017.02529

Gopal M, Gupta A, Thomas GV (2013) Bespoke microbiome therapy to manage plant diseases. Front Microbiol 4:355. https://doi.org/10.3389/fmicb.2013.00355

Gouda S, Das G, Sen SK, Shin HS, Patra JK (2016) Endophytes: a treasure house of bioactive compounds of medicinal importance. Front Microbiol 7:1538. https://doi.org/10.3389/fmicb.2016.01538

Gougoulias C, Clark JM, Shaw LJ (2014) The role of soil microbes in the global carbon cycle: tracking the below-ground microbial processing of plant-derived carbon for manipulating carbon dynamics in agricultural systems. J Sci Food Agric 94:2362–2371. https://doi.org/10.1002/jsfa.6577

Gregory PJ (2022) Russell REVIEW Are plant roots only "in" soil or are they "of" it? Roots, soil formation and function. Eur J Soil Sci 73:e13219. https://doi.org/10.1111/ejss.13219

Han J, Dong Y, Zhang M (2021) Chemical fertilizer reduction with organic fertilizer effectively improve soil fertility and microbial community from newly cultivated land in the Loess Plateau of China. Appl Soil Ecol 165:103966. https://doi.org/10.1016/j.apsoil.2021.103966

Haroon M, Khan ST, Malik A (2022a) Zinc-solubilizing bacteria: an option to increase zinc uptake by plants. In: Khan ST, Malik A (eds) Microbial biofertilizers and micronutrient availability. Springer, Cham. https://doi.org/10.1007/978-3-030-76609-2_11

Haroon U, Khizar M, Liaquat F, Ali M, Akbar M, Tahir K (2022b) Halotolerant plant growth-promoting rhizobacteria induce salinity tolerance in wheat by enhancing the expression of SOS genes. J Plant Growth Regul 41(6):2435–2448. https://doi.org/10.1007/s00344-021-10457-5

Hazarika SN, Saikia K, Borah A, Thakur D (2021) Prospecting endophytic bacteria endowed with plant growth promoting potential isolated from *Camellia sinensis*. Front Microbiol 12:738058. https://doi.org/10.3389/fmicb.2021.738058

Hossain A, Skalicky M, Brestic M, Maitra S, Sarkar S, Ahmad Z (2021) Selenium biofortification: roles, mechanisms, responses and prospects. Molecules 26:881. https://doi.org/10.3390/molecules26040881

Jiang G, Zhang Y, Gan G, Li W, Wan W, Jiang Y (2022) Exploring rhizo-microbiome transplants as a tool for protective plant-microbiome manipulation. ISME Commun 2:1–10. https://doi.org/10.1038/s43705-022-00094-8

Kamari H, Sharifi R (2017) Effects of Nano-Zinc oxide and plant growth promoting rhizobacteria on zinc and protein content, phosphatase activity and related traits to grain growth of Triticale. Iran J Seed Sci Technol 6:151–163. https://doi.org/10.22034/ijsst.2017.113549

Karnwal A (2021a) *Pseudomonas* spp., a zinc-solubilizing vermicompost bacteria with plant growth-promoting activity moderates zinc biofortification in tomato. Int J Veg Sci 27:398–412. https://doi.org/10.1080/19315260.2020.1812143

Karnwal A (2021b) Zinc solubilizing *Pseudomonas* spp. from vermicompost bestowed with multifaceted plant growth promoting properties and having prospective modulation of zinc biofortification in *Abelmoschus esculentus* L. J Plant Nutr 44:1023–1038. https://doi.org/10.1080/01904167.2020.1862199

Kasim WA, Gaafar RM, Abou-Ali RM, Omar MN, Hewait HM (2016) Effect of biofilm forming plant growth promoting rhizobacteria on salinity tolerance in barley. Ann Agric Sci 61:217–227

Kassam Z, Lee CH, Yuan Y, Hunt RH (2013) Fecal microbiota transplantation for *Clostridium difficile* infection: systematic review and meta-analysis. Am J Gastroenterol 108:500–508. https://doi.org/10.1038/ajg.2013.59

Katel S, Mandal HR, Subedi D, Koirala S, Timsina S, Poudel A (2022) Climate Smart Agriculture for food security, adaptation, and migration: a review. Turk J Agric Food Sci Technol 10:1558–1564. https://doi.org/10.24925/turjaf.v10i8.1558-1564.5162

Kaul S, Gupta S, Sharma T, Dhar MK (2018) Unfolding the role of rhizomicrobiome toward sustainable agriculture. In: Giri B, Prasad R, Varma A (eds) Root biology. Soil biology, vol 52. Springer, Cham. https://doi.org/10.1007/978-3-319-75910-4_14

Kaviya N, Upadhayay VK, Singh J, Khan A, Panwar M, Singh AV (2019) Role of microorganisms in soil genesis and functions. In: Varma A, Choudhary D (eds) Mycorrhizosphere and pedogenesis. Springer, Singapore. https://doi.org/10.1007/978-981-13-6480-8_2

Khan A, Singh J, Upadhayay VK, Singh AV, Shah S (2019) Microbial biofortification: a green technology through plant growth promoting microorganisms. In: Sustainable green technologies for environmental management. Springer, Singapore, pp 255–269. https://doi.org/10.1007/978-981-13-2772-8_13

Khan A, Upadhayay VK, Panwar M, Singh AV (2020) Soil microbiota: a key bioagent for revitalization of soil health in hilly regions. In: Goel R, Soni R, Suyal D (eds) Microbiological advancements for higher altitude agro-ecosystems & sustainability. Rhizosphere biology. Springer, Singapore. https://doi.org/10.1007/978-981-15-1902-4_10

Khan A, Singh AV, Upadhayay VK, Ballabh A, Prasad B (2022) Influence of PGPR on growth and yield of oat (Avena sativa L.) under field conditions. Indian J Ecol 49:1351–1356

Khan A, Panthari D, Sharma RS, Punetha A, Singh AV, Upadhayay VK (2023a) Biofertilizers: a microbial-assisted strategy to improve plant growth and soil health. In: Advanced microbial techniques in agriculture, environment, and health management. Elsevier, pp 97–118. https://doi.org/10.1016/B978-0-323-91643-1.00007-7

Khan A, Singh AV, Pareek N, Arya P, Upadhayay VK, Kumar Jugran A, Mishra PK, Goel R (2023b) Credibility assessment of cold adaptive *Pseudomonas jesenni* MP1 and *P. palleroniana* N26 on growth, rhizosphere dynamics, nutrient status, and yield of the kidney bean cultivated in Indian Central Himalaya. Frontiers. Plant Sci 14:1042053. https://doi.org/10.3389/fpls.2023.1042053

Khati P, Chaudhary P, Gangola S, Bhatt P, Sharma A (2017) Nanochitosan supports growth of *Zea mays* and also maintains soil health following growth. 3 Biotech 7:81. https://doi.org/10.1007/s13205-017-0668-y

Khati P, Parul, Bhatt P, Nisha Kumar R, Sharma A (2018) Effect of nanozeolite and plant growth promoting rhizobacteria on maize. 3 Biotech 8:141. https://doi.org/10.1007/s13205-018-1142-1

Khati P, Chaudhary P, Gangola S, Sharma P (2019) Influence of nanozeolite on plant growth promotory bacterial isolates recovered from nanocompound infested agriculture field. Environ Ecol 37:521–527

Koç E, Karayiğit B (2022) Assessment of biofortification approaches used to improve micronutrient-dense plants that are a sustainable solution to combat hidden hunger. J Soil Sci Plant Nutr 22:475–500. https://doi.org/10.1007/s42729-021-00663-1

Koo H, Allan RN, Howlin RP, Stoodley P, Hall-Stoodley L (2017) Targeting microbial biofilms: current and prospective therapeutic strategies. Nat Rev Microbiol 15:740–755. https://doi.org/10.1038/nrmicro.2017.99

Kour D, Rana KL, Kaur T, Yadav N, Yadav AN, Rastegari AA, Saxena AK (2020a) Microbial biofilms: functional annotation and potential applications in agriculture and allied sectors. In: New and future developments in microbial biotechnology and bioengineering: microbial biofilms. Elsevier, pp 283–301

Kour D, Rana KL, Yadav AN, Sheikh I, Kumar V, Dhaliwal HS, Saxena AK (2020b) Amelioration of drought stress in foxtail millet (*Setaria italica* L.) by P-solubilizing drought-tolerant microbes with multifarious plant growth promoting attributes. Environ Sustain 3:23–34. https://doi.org/10.1007/s42398-020-00094-1

Kukreti B, Sharma A, Chaudhary P, Agri U, Maithani D (2020) Influence of nanosilicon dioxide along with bioinoculants on *Zea mays* and its rhizospheric soil. 3 Biotech 10:345. https://doi.org/10.1007/s13205-020-02329-8

Kumari S, Sharma A, Chaudhary P, Khati P (2020) Management of plant vigor and soil health using two agriusable nanocompounds and plant growth promotory rhizobacteria in Fenugreek. 3 Biotech 10:1–11. https://doi.org/10.1007/s13205-020-02448-2

Kumari H, Khati P, Gangola S, Chaudhary P, Sharma A (2021) Performance of plant growth promotory rhizobacteria on Maize and soil characteristics under the influence of TiO_2 nanoparticles. Pantnagar J Res 19:28–39

Kwak MJ, Kong HG, Choi K, Kwon SK, Song JY, Lee J (2018) Author correction: rhizosphere microbiome structure alters to enable wilt resistance in tomato. Nat Biotechnol 36:1117. https://doi.org/10.1038/nbt1118-1117

Kyselková M, Moënne-Loccoz Y (2012) *Pseudomonas* and other microbes in disease-suppressive soils. In: Sustainable agriculture. Springer, Dordrecht, pp 93–140. https://doi.org/10.1007/978-94-007-4113-3_5

Li J, Cao X, Jia X, Liu L, Cao H, Qin W, Li M (2021) Iron deficiency leads to chlorosis through impacting chlorophyll synthesis and nitrogen metabolism in *Areca catechu* L. Front Plant Sci 12:710093. https://doi.org/10.3389/fpls.2021.710093

Liu Z, Zhang Y, Fa K, Zhao H, Qin S, Yan R, Wu B (2018) Desert soil bacteria deposit atmospheric carbon dioxide in carbonate precipitates. Catena 170:64–72. https://doi.org/10.1016/j.catena.2018.06.001

Ma J, Li Y, Chen F, Sun Y, Zhu Y, Wang L (2023) *Bacillus mycoides* PM35 in combination with titanium dioxide (TiO2)-nanoparticles enhanced morpho-physio-biochemical attributes in Barley (*Hordeum vulgare* L.) under cadmium stress. Chemosphere 323:138224. https://doi.org/10.1016/j.chemosphere.2023.138224

Maithani D, Sharma A, Dasila H, Tiwari A, Upadhayay VK (2023) Nanotechnology for crop improvement and sustainable agriculture. In: Advances in nanotechnology for smart agriculture. CRC Press, pp 323–339

Malhi GS, Kaur M, Kaushik P (2021) Impact of climate change on agriculture and its mitigation strategies: a review. Sustainability 13:1318. https://doi.org/10.3390/su13031318

Manzar N, Jayaswal D, Kishan G, Chauhan S, Joshi D, Singh J, Chaudhary P, Jamali H, Kushwaha P, Parveen H, Kashyap AS, Sahu PK, Singh UB (2022) Application of molecular ecology approaches in sustainable agriculture for a better understanding of plant–microbiome interactions. In: Singh UB, Sahu PK, Singh HV, Sharma PK, Sharma SK (eds) Rhizosphere microbes. Microorganisms for sustainability, vol 40. Springer, Singapore. https://doi.org/10.1007/978-981-19-5872-4_3

Masunaga T, Marques Fong JD (2018) Strategies for increasing micronutrient availability in soil for plant uptake. In: Plant micronutrient use efficiency. Elsevier, pp 195–208

Mendes R, Kruijt M, de Bruijn I, Dekkers E, van der Voort M, Schneider JH (2011) Deciphering the rhizosphere microbiome for disease-suppressive bacteria. Science 332:1097–1100. https://doi.org/10.1126/science.1203980

Mitterstiller A-M, von Raffay L, Nairz M (2022) Iron deficiency, anemia, and the immune system. In: Nutrition and health. Springer International Publishing, pp 235–248

Mohammed AF (2018) Effectiveness of exopolysaccharides and biofilm forming plant growth promoting rhizobacteria on salinity tolerance of faba bean (Vicia faba L.). Afr J Microbiol Res 12:399–404

Mohanty S, Swain CK (2018) Role of microbes in climate smart agriculture. In: Panpatte D, Jhala Y, Shelat H, Vyas R (eds) Microorganisms for green revolution. Microorganisms for sustainability, vol 7. Springer, Singapore. https://doi.org/10.1007/978-981-10-7146-1_7

Mojadadi A, Au A, Salah W, Witting P, Ahmad G (2021) Role for selenium in metabolic homeostasis and human reproduction. Nutrients 13:3256. https://doi.org/10.3390/nu13093256

Mora ML, Durán P, Acuña J, Cartes P, Demanet R, Gianfreda L (2015) Improving selenium status in plant nutrition and quality. J Soil Sci Plant Nutr 15:486–503. https://doi.org/10.4067/S0718-95162015005000041

Moradi Pour M, Saberi Riseh R, Skorik YA (2022) Sodium alginate-gelatin nanoformulations for encapsulation of *Bacillus velezensis* and their use for biological control of pistachio gummosis. Materials (Basel) 15:2114. https://doi.org/10.3390/ma15062114. Pubmed: 35329566

Moustaine M, Elkahkahi R, Benbouazza A, Benkirane R, Achbani EH (2017) Effect of plant growth promoting rhizobacterial (PGPR) inoculation on growth in tomato (*Solanum lycopersicum* L.) and characterization for direct PGP abilities in Morocco. Int J Environ Agric Biotechnol 2:238708. https://doi.org/10.22161/ijeab/2.2.5

Muhammad F, Raza MAS, Iqbal R, Zulfiqar F, Aslam MU, Yong JWH (2022) Ameliorating drought effects in wheat using an exclusive or co-applied rhizobacteria and ZnO nanoparticles. Biology 11:1564. https://doi.org/10.3390/biology11111564

Mumtaz MZ, Ahmad M, Zafar-ul-Hye M, Saqib M, Akhtar MFUZ, Zaheer MS (2022) Seed-applied zinc-solubilising. Crop Pasture Sci 73:503–514. https://doi.org/10.1071/CP21415

Nair DN, Padmavathy S (2014) Impact of endophytic microorganisms on plants, environment and humans. Sci World J 2014:250693. https://doi.org/10.1155/2014/250693

Nascente AS, Lanna AC, de Sousa TP, Chaibub AA, de Souza ACA, de Filippi MCC (2019) N fertilizer dose-dependent efficiency of *Serratia* spp. for improving growth and yield of upland rice (*Oryza sativa* L.). Int J Plant Prod 13:217–226. https://doi.org/10.1007/s42106-019-00049-5

Nautiyal P, Rajput R, Pandey D, Arunachalam K, Arunachalam A (2019) Role of glomalin in soil carbon storage and its variation across land uses in temperate Himalayan regime. Biocatal Agric Biotechnol 21:101311. https://doi.org/10.1016/j.bcab.2019.101311

Nosheen S, Ajmal I, Song Y (2021) Microbes as biofertilizers, a potential approach for sustainable crop production. Sustainability 13:1868. https://doi.org/10.3390/su13041868

Öksel C, Balkan A, Bilgin O, Mirik M, Başer İ (2022) Investigation of the effect of PGPR on yield and some yield components in winter wheat (*Triticum aestivum* L.). Turk J Field Crop 27:127–133. https://doi.org/10.17557/tjfc.1019160

Omar A, Wright JB, Schultz G, Burrell R, Nadworny P (2017) Microbial biofilms and chronic wounds. Microorganism 5:9. https://doi.org/10.3390/microorganisms5010009

Pandit A, Adholeya A, Cahill D, Brau L, Kochar M (2020) Microbial biofilms in nature: unlocking their potential for agricultural applications. J Appl Microbiol 129:199–211. https://doi.org/10.1111/jam.14609

Panichikkal J, Krishnankutty RE (2022) Chitosan and gold nanoparticles supplementation for augmentation of indole-3-acetic acid production by rhizospheric *Pseudomonas aeruginosa* and plant growth enhancement. Curr Microbiol 79:185. https://doi.org/10.1007/s00284-022-02850-4

Parveen H, Chaudhary A, Chaudhary P, Sultana R, Khati P, Kumar G, Khati P, Rana M, Bhatt P (2022) Omics approaches for the production of the microbial enzymes and applications. In: Bhatt P (ed) Industrial applications of microbial enzymes, 1st edn. CRC Press. https://doi.org/10.1201/9781003202998-16

Parveen H, Chaudhary P, Srivastava P, Priya S, Bhandari G, Chaudhary A (2023) Role of nanoparticles in agriculture. In: Chaudhary P, Chaudhary A, Nadda AK, Khati P (eds) Advances in nanotechnology for smart agriculture: techniques and applications, 1st edn. CRC Press. https://doi.org/10.1201/9781003345565-1

Pascual J, Blanco S, García-López M, García-Salamanca A, Bursakov SA, Genilloud O (2016) Assessing bacterial diversity in the rhizosphere of Thymus zygis growing in the Sierra Nevada National Park (Spain) through culture-dependent and independent approaches. PLoS One 11:e0146558. https://doi.org/10.1371/journal.pone.0146558

Pieterse CM, Van der Does D, Zamioudis C, Leon-Reyes A, Van Wees SC (2012) Hormonal modulation of plant immunity. Annu Rev Cell Dev Biol 28:489–521. https://doi.org/10.1146/annurev-cellbio-092910-154055

Prasanna R, Bidyarani N, Babu S, Hossain F, Shivay YS, Nain L (2015) Cyanobacterial inoculation elicits plant defense response and enhanced Zn mobilization in maize hybrids. Cogent Food Agric 1:998507. https://doi.org/10.1080/23311932.2014.998507

Priya S, Kumar A, Upadhayay VK, Chaudhary A, Parveen H, Kumar G (2023) Impact of nanoparticles on biotic stress tolerance. In: Advances in nanotechnology for smart agriculture. CRC Press, pp 197–220

Qurashi AW, Sabri AN (2012) Biofilm formation in moderately halophilic bacteria is influenced by varying salinity levels. J Basic Microbiol 52:566–572. https://doi.org/10.1002/jobm.201100253

Rani V, Bhatia A, Kaushik R (2021) Inoculation of plant growth promoting-methane utilizing bacteria in different N-fertilizer regime influences methane emission and crop growth of flooded paddy. Sci Total Environ 775:145826. https://doi.org/10.1016/j.scitotenv.2021.145826

Rani A, Kumar T, Trivedi B, Chaudhary P (2023) Nanotechnology in climate smart agriculture. In: Chaudhary P, Chaudhary A, Nadda AK, Khati P (eds) Advances in nanotechnology for smart agriculture: techniques and applications, 1st edn. CRC Press. https://doi.org/10.1201/9781003345565-14

Roohani N, Hurrell R, Kelishadi R, Schulin R (2013) Zinc and its importance for human health: an integrative review. J Res Med Sci 18:144–157

Roshani KA, Singh AV, Upadhayay VU, Prasad B (2020) Development of potential microbial consortia and their assessment on wheat (*Triticum aestivum*) seed germination. Environ Ecol 38:6–16

Rossi F, Olguín EJ, Diels L, De Philippis R (2015) Microbial fixation of CO2 in water bodies and in drylands to combat climate change, soil loss and desertification. New Biotechnol 32:109–120. https://doi.org/10.1016/j.nbt.2013.12.002

Rubin RL, van Groenigen KJ, Hungate BA (2017) Plant growth promoting rhizobacteria are more effective under drought: a meta-analysis. Plant Soil 416:309–323. https://doi.org/10.1007/s11104-017-3199-8

Saleh D, Sharma M, Seguin P, Jabaji S (2020) Organic acids and root exudates of *Brachypodium distachyon*: effects on chemotaxis and biofilm formation of endophytic bacteria. Can J Microbiol 66:562–575. https://doi.org/10.1139/cjm-2020-0041

Samuels T, Bryce C, Landenmark H, Marie-Loudon C, Nicholson N, Stevens AH, Cockell C (2020) Microbial weathering of minerals and rocks in natural environments. In: Biogeochemical cycles: ecological drivers and environmental impact, pp 59–79. https://doi.org/10.1002/9781119413332.ch3

Santoyo G, Moreno-Hagelsieb G, Orozco-Mosqueda M, Glick BR (2016) Plant growth-promoting bacterial endophytes. Microbiol Res 183:92–99. https://doi.org/10.1016/j.micres.2015.11.008

Sarwar S, Khaliq A, Yousra M, Sultan T (2022) Iron biofortification potential of siderophore producing rhizobacterial strains for improving growth, yield and iron contents of groundnut. J Plant Nutr 45:2332–2347. https://doi.org/10.1080/01904167.2022.2063733

Sathya A, Vijayabharathi R, Gopalakrishnan S (2016) Soil microbes: the invisible managers of soil fertility. In: Microbial inoculants in sustainable agricultural productivity: Vol. 2: functional applications, pp 1–16

Satyaprakash MS, Nikitha T, Vani SS, Sadhana B, Vani SS (2017) Phosphorous and phosphate solubilising bacteria and their role in plant nutrition. Int J Curr Microbiol Appl Sci 6:2133–2144. https://doi.org/10.20546/ijcmas.2017.604.251

Schlatter D, Kinkel L, Thomashow L, Weller D, Paulitz T (2017) Disease suppressive soils: new insights from the soil microbiome. Phytopathology 107:1284–1297. https://doi.org/10.1094/PHYTO-03-17-0111-RVW

Seneviratne G, Jayasekara APDA, De Silva MSDL, Abeysekera UP (2011) Developed microbial biofilms can restore deteriorated conventional agricultural soils. Soil Biol Biochem 43:1059–1062. https://doi.org/10.1016/j.soilbio.2011.01.026

Shabaan M, Asghar HN, Zahir ZA, Sardar MF, Parveen R, Faiza AQ (2022) Halotolerant rhizobacterial consortium confers salt tolerance to maize under naturally salt-affected soil. Soil Sci Soc Am J 86:1264–1279. https://doi.org/10.1002/saj2.20438

Shafiq T, Yasmin H, Shah ZA, Nosheen A, Ahmad P, Kaushik P, Ahmad A (2022) Titanium oxide and zinc oxide nanoparticles in combination with cadmium tolerant *Bacillus pumilus* ameliorates the cadmium toxicity in maize. Antioxidants (Basel) 11:2156. https://doi.org/10.3390/antiox11112156

Shahzad A, Ullah S, Dar AA, Sardar MF, Mehmood T, Tufail MA (2021) Nexus on climate change: agriculture and possible solution to cope future climate change stresses. Environ Sci Pollut Res Int 28:14211–14232. https://doi.org/10.1007/s11356-021-12649-8

Singh J, Singh AV, Upadhayay VK, Khan A (2020) Comparative evaluation of developed carrier based bio-formulations bearing multifarious PGP properties and their effect on shelf life under different storage conditions. Environ Ecol 38:96–103

Singh J, Singh AV, Upadhayay VK, Khan A, Chandra R (2022a) Prolific contribution of *Pseudomonas protegens* in Zn biofortification of wheat by modulating multifaceted physiological response under saline and non-saline conditions. World J Microbiol Biotechnol 38:227. https://doi.org/10.1007/s11274-022-03411-4

Singh S, Sakshi, Annapurna K, Shrivastava N, Varma A (2022b) Symbiotic interplay of *Piriformospora indica* and *Azotobacter chroococcum* augments crop productivity and biofortification of Zinc and Iron. Microbiol Res 262:127075. https://doi.org/10.1016/j.micres.2022.127075

Singh S, Chaudhary P, Sharma A (2023) Efficient decolourization of toluidine blue and methylene blue using *Bacillus cereus*. Vegetos. https://doi.org/10.1007/s42535-023-00747-6

Srivastava P, Chaudhary P, Parveen P (2023) Nanoparticles and their effect on plant health and productivity. In: Chaudhary P, Chaudhary A, Nadda AK, Khati P (eds) Advances in nanotechnology for smart agriculture: techniques and applications, 1st edn. CRC Press. https://doi.org/10.1201/9781003345565-6

Sun Z, Yue Z, Liu H, Ma K, Li C (2021) Microbial-assisted wheat iron biofortification using endophytic *Bacillus altitudinis* WR10. Front Nutr 8:704030. https://doi.org/10.3389/fnut.2021.704030

Taktek S, St-Arnaud M, Piché Y, Fortin JA, Antoun H (2017) Igneous phosphate rock solubilization by biofilm-forming mycorrhizobacteria and hyphobacteria associated with *Rhizoglomus irregulare* DAOM 197198. Mycorrhiza 27:13–22. https://doi.org/10.1007/s00572-016-0726-z

Tian T, Reverdy A, She Q, Sun B, Chai (2020) The role of rhizodeposits in shaping rhizomicrobiome. Environ Microbiol Rep 12:160–172. https://doi.org/10.1111/1758-2229.12816

Timmusk S, Abd El-Daim IA, Copolovici L, Tanilas T, Kännaste A, Behers L et al (2014) Drought-tolerance of wheat improved by rhizosphere bacteria from harsh environments: enhanced biomass production and reduced emissions of stress volatiles. PLoS One 9:e96086. https://doi.org/10.1371/journal.pone.0096086

Timofeeva AM, Galyamova MR, Sedykh SE (2022) Bacterial siderophores: classification, biosynthesis, perspectives of use in agriculture. Plants (Basel) 11:3065. https://doi.org/10.3390/plants11223065

Toor MD, Adnan M (2020) Role of soil microbes in agriculture; a review. Open Access J Biogener Sci Res 4:1–5. https://doi.org/10.46718/JBGSR.2020.04.000091

Triveni S, Prasanna R, Saxena AK (2012) Optimization of conditions for in vitro development of Trichoderma viride-based biofilms as potential inoculants. Folia Microbiol 57:431–437. https://doi.org/10.1007/s12223-012-0154-1

Tyagi J, Chaudhary P, Jyotsana, Bhagwati U, Bhandari G, Chaudhary A (2022a) Chapter 17: Impact of endophytic fungi in biotic stress management. In: Soni R, Suyal D, Goel R (eds) Plant protection: from chemicals to biologicals. De Gruyter, Berlin, Boston, pp 447–462. https://doi.org/10.1515/9783110771558-017

Tyagi J, Chaudhary P, Mishra A, Khatwani M, Dey S, Varma A (2022b) Role of endophytes in abiotic stress tolerance: with special emphasis on *Serendipita indica*. Int J Environ Res 16:62. https://doi.org/10.1007/s41742-022-00439-0

Upadhayay VK, Singh AV, Pareek N (2018) An insight in decoding the multifarious and splendid role of microorganisms in crop biofortification. Int J Curr Microbiol Appl Sci 7:2407–2418. https://doi.org/10.20546/ijcmas.2018.706.286

Upadhayay VK, Singh J, Khan A, Lohani S, Singh AV (2019) Mycorrhizal mediated micronutrients transportation in food based plants: a biofortification strategy. In: Varma A, Choudhary D (eds) Mycorrhizosphere and pedogenesis. Springer, Singapore. https://doi.org/10.1007/978-981-13-6480-8_1

Upadhayay VK, Singh AV, Khan A, Pareek N (2021) Influence of zinc solubilizing bacterial co-inoculation with zinc oxide supplement on rice plant growth and Zn uptake. J Pharm Innov 10:113–116

Upadhayay VK, Khan A, Singh J, Singh AV (2022a) Bacterial assisted improved Zn consignment in root and shoot of rice plant by zinc solubilizing Serratia marcescence bearing plant probiotic traits. Adv Biol Res 13:01–08

Upadhayay VK, Singh AV, Khan A (2022b) Cross talk between zinc-solubilizing bacteria and plants: a short tale of bacterial-assisted zinc biofortification. Front Soil Sci 1:788170. https://doi.org/10.3389/fsoil.2021.788170

Upadhayay VK, Singh AV, Khan A, Singh J, Pareek N, Raghav A (2022c) FE-SEM/EDX based zinc mobilization analysis of *Burkholderia cepacia* and *Pantoea rodasii* and their functional annotation in crop productivity, soil quality, and zinc biofortification of paddy. Front Microbiol 13:852192. https://doi.org/10.3389/fmicb.2022.852192

Upadhayay VK, Singh AV, Khan A, Sharma A (2022d) Contemplating the role of zinc-solubilizing bacteria in crop biofortification: an approach for sustainable bioeconomy. Front Agron 4:903321. https://doi.org/10.3389/fagro.2022.903321

Upadhayay VK, Maithani D, Dasila H, Taj G, Singh AV (2023a) Microbial services for mitigation of biotic and abiotic stresses in plants. In: Advanced microbial techniques in agriculture, environment, and health management. Elsevier, pp 67–81

Upadhayay VK, Naitam MG, Ghosh S, Patel VK, Sharma A, Chandra AK, Taj G (2023b) Techniques for integrating nanotechnology in agriculture. In: Advances in nanotechnology for smart agriculture. CRC Press, pp 19–40

Upadhayay VK, Chitara MK, Mishra D, Jha MN, Jaiswal A, Kumari G (2023c) Synergistic impact of nanomaterials and plant probiotics in agriculture: a tale of two-way strategy for long-term sustainability. Front Microbiol 14:1133968. https://doi.org/10.3389/fmicb.2023.1133968

Velmourougane K, Prasanna R, Saxena AK (2017) Agriculturally important microbial biofilms: present status and future prospects. J Basic Microbiol 57:548–573. https://doi.org/10.1002/jobm.201700046

Verma P, Yadav AN, Khannam KS, Panjiar N, Kumar S, Saxena AK, Suman A (2015) Assessment of genetic diversity and plant growth promoting attributes of psychrotolerant bacteria allied with wheat (Triticum aestivum) from the northern hills zone of India. Ann Microbiol 65:1885–1899. https://doi.org/10.1007/s13213-014-1027-4

Verma P, Yadav AN, Kumar V, Singh DP, Saxena AK (2017) Beneficial plant-microbes interactions: biodiversity of microbes from diverse extreme environments and its impact for crop improvement. In: Singh D, Singh H, Prabha R (eds) Plant-microbe interactions in agro-ecological perspectives. Springer, Singapore. https://doi.org/10.1007/978-981-10-6593-4_22

Verma M, Singh A, Dwivedi DH, Arora NK (2020) Zinc and phosphate solubilizing Rhizobium radiobacter (LB2) for enhancing quality and yield of loose leaf lettuce in saline soil. Environ Sustain 3:209–218. https://doi.org/10.1007/s42398-020-00110-4

Wang M, Xing Y, Wang J, Xu Y, Wang G (2014) The role of the chi1 gene from the endophytic bacteria *Serratia proteamaculans* 336x in the biological control of wheat take-all. Can J Microbiol 60:533–540. https://doi.org/10.1139/cjm-2014-0212

Wang DC, Jiang CH, Zhang LN, Chen L, Zhang XY, Guo JH (2019) Biofilms positively contribute to *Bacillus amyloliquefaciens* 54-induced drought tolerance in tomato plants. Int J Mol Sci 20:6271. https://doi.org/10.3390/ijms20246271

Wang JL, Liu KL, Zhao XQ, Zhang HQ, Li D, Li JJ, Shen RF (2021) Balanced fertilization over four decades has sustained soil microbial communities and improved soil fertility and rice productivity in red paddy soil. Sci Total Environ 793:148664. https://doi.org/10.1016/j.scitotenv.2021.148664

Wei Z, Gu Y, Friman VP, Kowalchuk GA, Xu Y, Shen Q, Jousset A (2019) Initial soil microbiome composition and functioning predetermine future plant health. Science. Advances 5:eaaw0759. https://doi.org/10.1126/sciadv.aaw0759

Wu S, Zhuang G, Bai Z, Cen Y, Xu S, Sun H (2018) Mitigation of nitrous oxide emissions from acidic soils by *Bacillus amyloliquefaciens*, a plant growth-promoting bacterium. Glob Chang Biol 24:2352–2365. https://doi.org/10.1111/gcb.14025

Yadav AN (2018) Biodiversity and biotechnological applications of host specific endophytic fungi for sustainable agriculture and allied sectors. Acta Sci Microbiol 1:44

Yadav AN, Sharma D, Gulati S, Singh S, Dey R, Pal KK (2015) Haloarchaea endowed with phosphorus solubilization attribute implicated in phosphorus cycle. Sci Rep 5:12293. https://doi.org/10.1038/srep12293

Zafar-ul-Hye M, Danish S, Abbas M, Ahmad M, Munir TM (2019) ACC deaminase producing PGPR *Bacillus amyloliquefaciens* and *Agrobacterium fabrum* along with biochar improve wheat productivity under drought stress. Agronomy 9:343. https://doi.org/10.3390/agronomy9070343

Zaheer MS, Raza MAS, Saleem MF, Khan IH, Ahmad S, Iqbal R, Manevski K (2019) Investigating the effect of *Azospirillum brasilense* and *Rhizobium pisi* on agronomic traits of wheat (*Triticum aestivum* L.). Arch Agron Soil Sci 65:1554–1564. https://doi.org/10.1080/03650340.2019.1566954

Zakeel MCM, Safeena MIS (2019) Biofilmed biofertilizer for sustainable agriculture. In: Ansari R, Mahmood I (eds) Plant health under biotic stress. Springer, Singapore. https://doi.org/10.1007/978-981-13-6040-4_3

Zeriouh H, de Vicente A, Pérez-García A, Romero D (2014) Surfactin triggers biofilm formation of *Bacillus subtilis* in melon phylloplane and contributes to the biocontrol activity. Environ Microbiol 16:2196–2211. https://doi.org/10.1111/1462-2920.12271

Zhu Z, Bai Y, Lv M, Tian G, Zhang X, Li L (2020) Soil fertility, microbial biomass, and microbial functional diversity responses to four years fertilization in an apple orchard in North China. Hortic Plant J 6:223–230. https://doi.org/10.1016/j.hpj.2020.06.003

Application of Omics Approaches to Improve Bioinoculant Performance

Saurabh Gangola, Samiksha Joshi, Rekha Goswami, and Geeta Bhandari

Abstract Bioinoculants are regarded as ecologically friendly microorganisms that are used to enhance soil quality and supply host plants with vital nutrients. This alternative to chemical fertilizers has resulted in increased agricultural yields, prompting interest in studying the innovation process. Gene-editing tools can modify the desired gene product in designed microbial inoculants. Furthermore, contemporary biotechnological methods such as OptKnock, flux balance analysis, constraint-based modeling, and modeling of biological networks have been discussed to improve the capabilities of bioinoculants. While these methods provide a fascinating perspective of the metabolic network, detailed kinetic data is required to better understand these fluxes. Additionally, by incorporating enzyme-coding genes, these techniques can help stimulate metabolic networks. This review explains how systems biology and gene-editing techniques can be used to enhance the potential of bioinoculants while also acknowledging the current challenges and multifaceted future of systems biology.

Keywords Bioinoculants · Omics approach · Enzymes · Crop productivity · Tools

---

S. Gangola · S. Joshi (✉)
School of Agriculture, Graphic Era Hill University, Bhimtal Campus, Bhimtal, Uttarakhand, India

R. Goswami
Department of Environmental Sciences, Graphic Era Hill University, Dehradun, Uttarakhand, India

G. Bhandari
Department of Biosciences, Himalayan School of Biosciences, Swami Rama Himalayan University, Dehradun, Uttarakhand, India

P. Chaudhary, A. Chaudhary (eds.), *Microbial Inoculants*, Interdisciplinary Biotechnological Advances, https://doi.org/10.1007/978-981-97-0633-4_6

1 Introduction

Utilizing some beneficial microbial species such as bacteria, algae, and fungi in soil amendments, also known as bioinoculants, can aid in the growth and development of plants. These bioinoculants are composed of active or inactive bacterial cells that are capable of fixing nitrogen, showing solubilization properties for phosphorus, potassium, and zinc, and also produce HCN and siderophores; it also shows positive impact when colonized in the rhizosphere roots of the host plant (Patil and Solanki 2016; Chaudhary et al. 2021; Kumari et al. 2021; Agri et al. 2021, 2022). Recently, various genera of *Rhizobium*, *Azospirillum*, *Bacillus*, *Pseudomonas*, and *Lactobacillus* of rhizobacteria class have been recognized as one of the promoting plant developments and are used as bioinoculants (Ahemad and Kibret 2014; Kumari et al. 2020; Kukreti et al. 2020). They are essential to the ecosystem's ability to survive. Understanding the role and relationship of microbial communities depends on how microorganisms interact with other biotic and abiotic variables (Imam et al. 2016; Dangi et al. 2018; Manzar et al. 2022). The bacterial microbiome, out of all the microbiomes, interacts well with the agricultural field (Gilbert et al. 2010). Bioinoculants are currently used as a key strategy for enhancing crop productivity through the integration of nutrient supply systems. This approach aids in maintaining agricultural output and environmental preservation (Jat et al. 2015; Frew et al. 2018; Chaudhary et al. 2022a, b, 2023a, b). Numerous modern formulation techniques are employed, including formulations based on metabolites, liquids, solid carriers (both inorganic and organic), and synthetic polymers (Arora et al. 2017). The prevalent method for cultivating bioinoculants involves using a fed-batch approach to create a dense culture of microorganisms, specifically *Pseudomonas* spp. (Mutturi et al. 2017). For producing biofertilizers, microbe inoculants like *Frankia*, *Azospirillum*, *Pseudomonas*, and *Dyadobacter* are grown in larger quantities using liquid or solid nutritional media through a scale-up process (Tatsukami 2017). To speed up the colonization process, these microbial communities synthesize a variety of inorganic and organic substances, including tartaric, acetic, succinic, oxalic, gluconic, 2-keto-gluconic, and lactic acids. Plants are protected by various beneficial microbes that function as bioinoculants in a variety of ways, including phytostimulation and biological control agents (Bhattacharyya and Jha 2012; Tyagi et al. 2022a, b). The process of rhizospheric colonization between plants and bacteria is shown to produce and exchange many types of signals, which may be advantageous, harmful, or neutral for the plants. According to a recent investigation, several volatile chemicals generated by the microbial community can alter how plants grow and develop (Tyagi et al. 2018). A variety of techniques, including genome engineering and gene editing, are being widely studied for improving the accuracy of predicting microbial interactions, as documented in (Esvelt and Wang 2013). In addition to these methodologies, in silico approaches are also used to examine such interactions. Metabolomics techniques have shown promise in uncovering complex metabolic pathways using electronic ionization (EI), mass spectrometry (MS), gas chromatography (GC), and compound annotation in databases. Principal component analysis (PCA) and independent component analysis (ICA) are frequently employed to identify patterns in nontargeted metabolomics research. It has been discovered that metabolites are more similar to the phenotype than mRNA transcripts and proteins are. This is likely due to the fact that metabolites are able to combine protein interactions, gene expression, and regulatory processes. The use of computational modeling and metabolic engineering as novel techniques for anticipating genetic mutations within metabolic networks has become increasingly common in recent years. Biotechnologies such as constraint-based modeling (CBM), OptKnock, flux balance analysis (FBA), and biological network modeling are utilized in the process of predicting the capability of bioinoculants. This review focuses on systems biology concepts and gene-editing technologies to enhance the effectiveness of bioinoculants, as highlighted in (Basu et al. 2018). Systems biology has evolved as a result of mathematical and computational modeling, and numerous systems biology technologies, such as genomics, transcriptomics, metabolomics, and proteomics, are being used to research microbial inoculants that benefit host plants. Finally, this study discusses the challenges and prospects for future multidisciplinary approaches and the development of new biotechnological instruments.

2 Application of Systems Biology Approaches

Several disciplines, including industrial microbiology, medical microbiology, and agricultural microbiology, have employed a systems biology approach to study the microbial community (Belibasakis et al. 2018). Systems biology methods can be used to alter the metabolic network of bacteria. Although the hazy goal of metabolic engineering may change, the technique always stays the same or doesn't change. The emergence of computational modeling has brought about a change in the strategies used in metabolic engineering, as it enables the prediction of the consequences of genetic modifications on the metabolic network (Baart and Martens 2012; Suyal et al. 2016) (Table 1).

Direct DNA separation from environmental materials including sludge, soil, and water allows for transcriptomics to be used in the study of gene expression in the microbial-plant community. The analysis of transcriptomes has helped scientists identify beneficial microbial interactions, as they found. For instance, increased root growth was seen in rhizosphere cultures isolated from 40 to 45% of 150 *Pseudomonas* spp. (Etalo et al. 2018; Papadopoulou et al. 2018). The use of metatranscriptomics and metagenomics in studying microbial communities may have relevance to the rhizospheric microbial population. In soil with disease-suppressive properties, researchers have uncovered around 33,000 bacterial and archaeal species using culture-dependent functional analyses and PhyloChip metagenomics (Igiehon and Babalola 2018). Microarray or transcriptome analysis can be used to probe potentially dangerous microbial interactions. After the discovery of pathogen-associated molecular patterns and defensive signaling molecules like jasmonic acid (JA),

Table 1 Different databases used for the improvement of bioinoculant performance

Techniques	Database name	Website (URL)	Function	References
Proteomics	UniProtKB	www.uniprot.org/	Extract information of protein	Kodati et al. (2021)
	MaxQuant	https://www.maxquant.org/	Functional interpretation	Kim et al. (2015)
	X!tandem	https://www.thegpm.org/tandem/	Functional interpretation	Kim et al. (2013)
Metabolomics	MetaFluxNet	http://mbel.kaist.ac.kr/lab/mfn/	Metabolic flux analysis	Litsios et al. (2018)
	BRENDA	http://www.brenda-enzymes.info/	Give information on molecular and biochemical enzyme	Kim et al. (2018)
	Reactome	http://signal.salk.edu/interactome.html	Pathway analysis	Kumari et al. (2015)
	BioCarta	http://www.biocarta.com/	Pathway analysis	Lai et al. (2012)
Genomics	PANTHER	http://www.pantherdb.org/	Signal transduction processes	Laserson et al. (2011)
	GenMAPP	http://www.genmapp.org/	Signal transduction processes	Lee et al. (2021)
Molecular interaction	MINT (The Molecular INTeraction Database)	https://mint.bio.uniroma2.it/	Molecular interaction analysis	Litvinov et al. (2021)
	IntAct	https://www.ebi.ac.uk/intact/home	Molecular interaction analysis	Liu et al. (2015)

ethylene (ET), abscisic acid (ABA), and salicylic acid (SA), researchers used microarray technology to study a network of signal transduction pathways (Simons-Senftle et al. 2018). The discovery of regulatory and resistance genes was the end outcome of this process. The identification of microbial genes engaged in interaction and strategies for disease-resistance genes as a result of multispecies transcriptome research has also provided insight into the gene activation mechanism for disease resistance (Chaudhary and Shukla 2019). The R-gene is known to play a significant role in the defense mechanisms of tomatoes, while the OxO gene can enhance or reduce the symptoms of late blight in *Sclerotinia sclerotiorum* and *Botrytis cinerea*. Additionally, the pepper ferredoxin I protein has been found to outperform resistance in *Ralstonia solanacearum*. To acquire an in-depth comprehension of systems biology, it is necessary to implement a methodological plan that takes into account findings from research on metabolomics, proteomics, transcriptomics, and genomes, as well as methods from the fields of statistics and bioinformatics. Recent studies have revealed that gel-based and liquid chromatography-mass spectrometry (LC MS/MS) proteomics techniques can be utilized to analyze a vast collection of soybean bacteroid proteins. This was discovered by researchers in the field of soybean research (Thiele and Palsson 2010). As a result, *Bradyrhizobium japonicum* has been found to express nearly 44% of its genome's coding potential during symbiosis, thanks to the use of proteomics techniques. This finding has uncovered new information about bacterial metabolism in the relationship between *Glycine max* and *B. japonicum*. Furthermore, it has the potential to improve symbiotic metabolism by influencing the expression and activity of the enzymes involved. In addition, research in metabolomics, proteomics, genomics, and transcriptomics can enhance the capacity of bacteria that serve as bioinoculants.

3 Utilization of Metabolomic Strategies and Methods

Metabolomics is the study and analysis of the diverse metabolites of a cell, tissue, or its adjacent environment, especially when describing various microbiological processes (Alamgir 2018). Recent research has shown how genetic changes can affect the intracellular metabolome of microorganisms in its entirety. Additionally, lipidomics is employed to determine the composition of proteins and lipids in membranes under stress circumstances (Du et al. 2018). During the development phase, fluxomic techniques have been employed to investigate the carbon metabolism when either glucose or succinate is the sole source of carbon (Litsios et al. 2018). By identifying different substances or metabolites, metabolomic methods significantly advance research. As per the reports, a number of metabolites exhibit diverse biological and physiochemical traits. For instance, *Pseudomonas putida* generates putidacin, while *Bacillus subtilis* produces bacteriocin, which are secondary metabolites possessing antifungal and antibacterial properties. Numerous bacterial species, including *Bacillus*, *Pseudomonas*, *Actinobacteria*, and *Myxococcus*, create polyketides, which are classified under a distinct class of secondary metabolites (Cimermancic et al. 2014). It is well established that pyrazines, which are volatile secondary metabolites generated by bacteria such as *Chondromyces*, *Streptomyces*, *Pseudomonas*, and *Bacillus*, exhibit antibacterial characteristics (Fig. 1). These metabolites offer a far better view of sustainable agricultural productivity and the dwindling of pathogenic species. The results of gene expression can be used to categorize the biochemical components and phenotype of a tissue or cell by identifying numerous types of metabolites (Buck et al. 2018; Lakshmanan et al. 2018). Measuring metabolites qualitatively and quantitatively provides a broad and in-depth perspective of the metabolic components of microorganisms that may be used to assess gene activity (Wang et al. 2013). The alterations in the transcriptome and proteome can be compared to further investigate the identified metabolites. Current metabolomics applications and their methods are also covered in this study. Metabolite profiling and metabolite fingerprinting are utilized in the rapidly developing field of metabolomics (Azad and Shulaev 2018). Liquid chromatography is used to separate secondary metabolites, capillary electrophoresis is used to separate ionic metabolites, and gas chromatography is used to separate primary metabolites and volatile substances in metabolomics. When compared to HPLC,

Fig. 1 The formation process of secondary metabolites and their analytical methods for identification

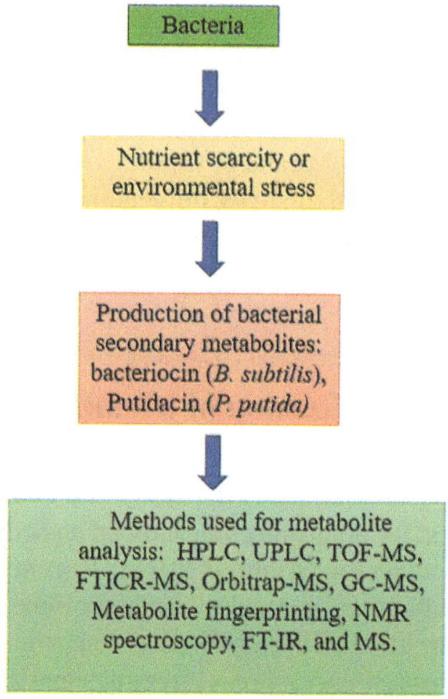

ultrahigh-performance liquid chromatography (UPLC) offers superior high sensitivity and resolution, making it an effective approach in the field of metabolomics. Mass analyzer techniques, such as TOF-MS, FTICR-MS, and Orbitrap-MS, are frequently used for metabolite profiling due to their quick scan times (Rustagi et al. 2018). GC-MS has been extensively utilized in both metabolomics and electron impact. Metabolite fingerprinting is a technique that identifies metabolic signatures of various metabolites under stress conditions (Van Der Oost et al. 2014). For analysis of statistics and patterns, numerous multivariate, such as self-organizing maps, principal component analysis, discriminant function analysis, analysis of variance, and hierarchical clustering, can be used to study metabolite fingerprinting. Metabolite fingerprinting uses a variety of analytical techniques, including NMR spectroscopy, FT-IR, and MS (Wang et al. 2013). Since NMR spectroscopy is less sensitive than MS and can only detect a small number of cellular metabolites, MS has greater advantages. NMR, on the other hand, has less resolution than MS, but one of MS's side effects is that it produces a lot of complicated spectra, which makes it difficult to justify data. Additionally, MS can be used to recognize a sizable number of secondary metabolites. There are a variety of methods for metabolite fingerprinting, which is used to identify metabolites, according to reports (Yang and Long 2016). It has been stated that metabolome samples with low molecular weight were photographed using infrared (IR) spectroscopy. Forage proteins quickly ferment because of metabolic substances as well (Martinez-Fernandez et al. 2018). In this investigation, PCA was used to determine how the cultivars varied and how much the metabolite fingerprinting between spring growth and second regrowth changed. Seasonal variations have also been linked to fluctuation and deviation in metabolic components such glucose, choline, and malic acid (Jafari 2018). The distinction between metabolites activated by data processing, metabolic pathways, and metabolic models can be made using metabolomics (Khan et al. 2019). Understanding metabolic engineering in microbial species, which enables identification and detection of numerous metabolites, is aided by this. The metabolome exposes the biological mechanism involved in phenotypic control in a cell, tissue, or organism. By using metabolomic analysis, it has recently been possible to monitor the gene function by looking at the microorganism's metabolic state.

4 Applications of Proteomics Approaches

Proteomic methods, which are connected to metabolomics and transcriptomics analysis, provide crucial information on the gene expression of functionally translated areas and modifications in proteomes at the protein level (Ramalingam et al. 2015). By utilizing these methods in specific ways, the capacity of bioinoculants can be improved, leading to an increase in crop productivity (Prasad et al. 2018). Proteomic methods can be implemented in various ways, including translational proteomics, protein-protein interaction analysis to detect protein complexes, post-translational modification analysis to identify how a specific protein has been altered, and comparative proteomics studies for comparing protein profiles across stress responses, developmental stages, cells, genotypes, and other factors. The results of the bioinformatics analysis and all the data from the aforementioned methods provide important information about the biological processes that may be used to improve the bioinoculant's capacity and, consequently, the quality of the plants (Schenkluhn et al. 2010). A comprehensive investigation of the cell proteome in a complex system through differential proteomics involves the development of novel technology (Imam et al. 2017). Various techniques are employed for proteomics research, including electrophoresis, chromatography, metabolic labeling, and MS (Sander and Joung 2014). Gel-based systems and protein profiling are particularly useful tools for studying microbial communities under stress conditions. When working with gel-based systems, the methods of two-dimensional gel electrophoresis and two-dimensional differential gel electrophoresis, respectively, are utilized for the purposes of measuring and identifying proteins. While 2D-GE can visualize and resolve multiple forms of the same protein, its low sensitivity makes it unsuitable for protein quantification. According to Beer and Speicher (2018), it is possible to improve the compatibility and sensitivity of MS testing by switching from utilizing Coomassie dye to fluorescent dyes. Therefore, the improved 2D-DIGE gel electrophoresis method, which labels protein mixtures with fluorescence, is increasingly being used due to its ability to detect proteins at sub picomolar levels. This approach is highly sensitive and has been used to study leguminous crops' early responses to disease and symbiosis (Sharma et al. 2013). Proteomics has also been used to research SWItch/Sucrose Non-Fermentable (SNF) and to differentiate between the roles of various proteins and noncoding RNAs (diCenzo et al. 2018).

5 Application of Transcriptomics and Genomics Approaches

Transcriptomic techniques, including RNA deep sequencing and microarrays, are employed to study transcriptional responses via RNA-based regulation and significant transcriptional regulators. Notably, FixL/FixK/FixJ and NodD are exceptional regulators that activate symbiotic genes and induce nodulation genes, respectively (D Arrigo 2016; Poole et al. 2018). These investigations covered a wide range of topics, including root exudate-induced adaptability and rhizosphere colonization (Compant et al. 2010). Numerous studies on transcriptomics have recently looked at the effects of finding protein-encoding genes and flavonoids. In order to identify the gene network, *Sinorhizobium meliloti* protein-coding genes' transcript levels were controlled and tracked in *Medicago sativa*. The investigation focused on the expression of genes and the regulation of transcription during the cell cycle of *S. meliloti*. FixJ acts as a key regulator of gene expression under oxic environments. With the help of the genes bacA and fixJ, it has been possible to decipher the function of regulators, certain genes, and other metabolic processes. Numerous studies have looked at the differences in the transcriptomics of bacteroides with distinct morphologies (D'Alessio et al. 2017). Numerous studies have recently looked at the genetic differences in growth-promoting microorganisms and how they affect the bacteroid and nodule transcriptomes. Numerous researchers used laser-capture microdissection to individually photograph each zone of the nodule in order to determine the spatial expression patterns (Schaum et al. 2018).

Genomics, like transcriptomics, is a multidisciplinary field that delves into the study of genomes, encompassing their structure, function, evolution, mapping, and manipulation. To examine the structure and functionality of whole genomes, it entails the analysis and sequencing of genomes utilizing bioinformatics and DNA sequencing. The Signature-tagged Mutant Library (STM) was purportedly used in the genome-scale study of the genes involved in symbiosis. The genomes of multiple strains were inserted with STMs from a *S. meliloti* library. Finally, an STM library of *S. fredii* was reported, as was an STM library of *Mesorhizobium loti*, which comprised 13 Nod mutants. The STM technique created transposon sequencing, also known as insertion sequencing (Inseq), for the purpose of evaluating the activities of genes quantitatively. Hundreds of genomes are identified across the genome using single insertion and next-generation sequencing. Numerous research teams modified transposon sequencing (Tn-seq) for microbes that promote plant growth. In order to analyze the core genome of *R. leguminosarum*, a mariner transposon was used to investigate the oxygen and carbon levels in the fitness of damaged genes (diCenzo et al. 2018). Using Tn5-derived transposons and mariner transposons, a nodule-specific cysteine-rich (NCR) peptide was used to examine *S. meliloti* sensitivity to the whole genome (diCenzo et al. 2018). This approach is expected to be very beneficial in elucidating genes at the genome scale at different stages of microbial life patterns.

6 Application of CRISPR/Cas Gene-Editing Method

The most modern method, known as gene editing, is utilized to modify the desired gene product through genetic engineering (Wang et al. 2017). To boost the effectiveness of these methods, microbial cells' genetic material is manipulated (Fig. 2). Genetic engineering includes recombinant DNA technologies, genetic modification, gene cloning, and genetic engineering (Kumar et al. 2016; Basu et al. 2018). Microorganisms that have been genetically engineered hold tremendous promise for enhancing crop protection under stressful conditions (Rosier et al. 2018). Through the alteration of enzyme specificity and pathway design, these methods have recently been employed to create novel strains with desirable features.

The emergence of recombinant technology tools and gene editing enabled agricultural genetic engineering to be successful (Bhardwaj et al. 2014). TALENs, Zinc finger proteins, and, more recently, CRISPR/Cas9 systems are major gene-editing technologies used to improve the genome changes of bacteria used as bioinoculants (Wawrzyniak et al. 2018). The CRISPR/Cas system may be utilized to alter microorganisms for agricultural applications (Caplan et al. 2015). Engineered nucleases, or molecular scissors, are employed in this procedure. The CRISPR/Cas system, sometimes referred to as third-generation nucleases, is crucial for the development

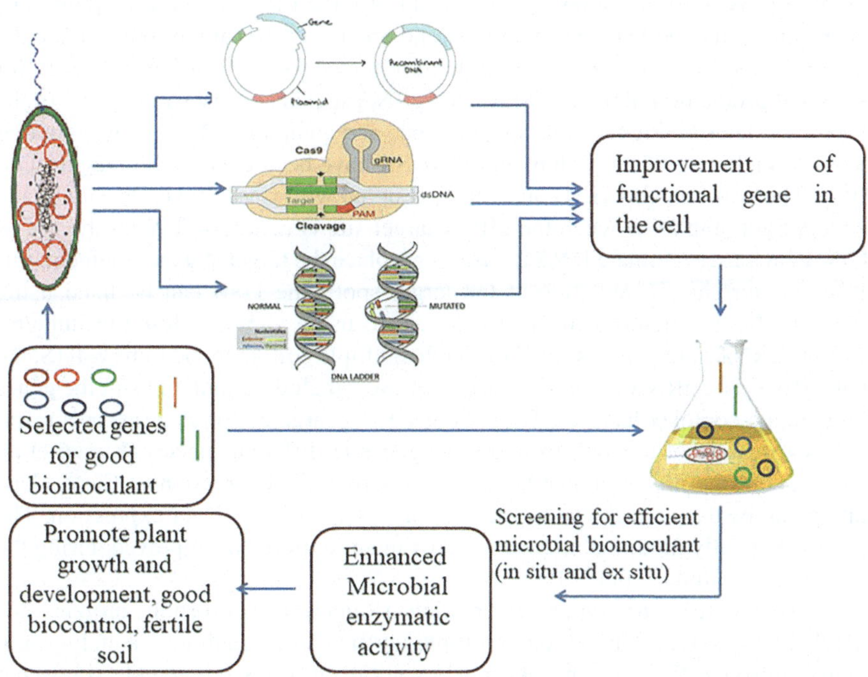

Fig. 2 Improvement of bioinoculant performance by different gene-editing strategies using the recombinant DNA technique, CRISPR/Cas, and gene mutation technology

of bacteria that can be employed as bioinoculants (Farrar et al. 2014). The efficacy and simplicity of the CRISPR/Cas system make it more effective for gene editing. More than ten CRISPR/Cas systems reportedly functioned properly. Three categories and 11 subtypes are used to categories them. Each form of CRISPR/Cas has a distinct and particular Cas protein connected with it. A particular targeted piece of foreign DNA is where RNA directs the DNA endonuclease Cas9 for the inhibitory process (Kumar et al. 2016). CRISPRs were also used to create gRNAs, which are composed of a tandem array of direct repeat sequences that match with unique sequences and are separated by spacer sequences. These sequences range in length from 35 to 45 base pairs. Following this, CRISPR can be processed and translated to create mature, well-defined CRISPR RNAs (crRNAs). The Cas protein and CRISPR RNAs work together to create crRNP (ribonucleoprotein) (Verma et al. 2018). Along with gRNA, the most astonishing and beneficial aspect of CRISPR is its specificity. When used in conjunction with gRNA, CRISPR enables precise and targeted binding to target DNA that is contained within the gRNA's modified genetic engineering (Turi et al. 2018). The twofold nickase fusion system and Cas9 nuclease aid in the binding specificity. In order to prevent off-target binding and the binding of two gRNAs toward both downstream and upstream targets, the double nickase system is helpful (Chuai et al. 2017). This method of binding, specifically Cas9 via fusion with restriction enzymes, can be improved without nuclease activity. According to reports, the nuclease activity of the restriction enzyme is stimulated by the close proximity of binding. With the aid of CRISPR/Cas9, an important gene can be removed or added when double-strand breaks (DSBs) are introduced into the target region (Hsu et al. 2014). The codon-optimized form of Cas9 is necessary for the proper production of the CRISPR/Cas sgRNA sequence to be successful. For the purpose of determining the cleavage sequence and creating gRNA, many computer methods, including CHOPCHOP and ECRISP, have been created (Montague et al. 2014). The two domains HNH and RucV, coupled with Cas9, break the strand and create a blunt end DSB when the gRNA target site is recognized. With the use of HDR, which uses template DNA to insert or replace the targeted gene fragment, this DSB can introduce a mutation at the target spot. The DSB can be fixed using NHEJ. Different microorganisms can be edited using various editing techniques. Multiple gRNAs are demonstrated to enable multiple editing in the same cell (Satge et al. 2016). The gRNA exhibits a variety of uses, including mutations in the genes that regulate complex features. The GFP gene was targeted for cleavage as a result of the expression of the sgRNA and Cas9 genes in different species (Kumar et al. 2016). Codon optimization can be used to improve the expression of Cas9. Gene editing can be used to achieve the desired need for Cas9 promoter expression. The use of genetically modified microorganisms in agriculture has improved during the past 10 years (Shukla et al. 2018).

The successful achievement of creating a microbial consortium utilizing the CRISPR/Cas system that exhibits multiple advantageous qualities, including biocontrol, nitrogen fixation, and phosphate solubilization, is remarkable (Pretscher et al. 2018). The production of different crops can be increased by using polymicrobial inoculants as a good consortium. With the development of proteomics and microbial genomes, more study is being done in this field to better understand the interactions between soil bacteria, plants, and microbes.

7 Advanced Biotechnological Approaches

Analysis of the metabolic pathway systems is aided by metabolic engineering and its use in biotechnology. According to reports, a variety of approaches are employed in signaling, network route modeling, and stimulation. The various cellular processes can be better understood and seen in a dynamic and quantitative light, thanks to systems-level modeling methodologies. Particularly, CBA (constraint-based analysis) achieved a success for the activation of the metabolic network by integrating with FBA. The flux balance analysis (FBA) method follows the principle of mass conservation while identifying the distribution of fluxes in a metabolic network using a stoichiometric matrix and an objective function that is relevant to biology. FBA is a useful tool for studying genome-scale reconstructions of various microorganisms, as it enables the in silico examination of the effects of gene insertions and deletions (Kim et al. 2018). Metabolic engineering uses a variety of biological data from FBA for diverse goals. The models that are utilized are distinct in every method that can have an impact on the dependability of stimulation. In this study, the different modeling approaches are separated into two distinct groups: (1) CBM and interaction-based modeling with stoichiometry (static models) and (2) mechanistic modeling with stoichiometry and kinetic parameters.

8 Application of Constraint-Based Modeling

Performing metabolic reconstruction is a time-consuming procedure that requires locating, classifying, and connecting many components of a metabolic network. These components include metabolites, proteins, and genes. This process is essential for understanding the numerous metabolic processes that occur within a cell. As a cell's systems continuously expand through genetic sequencing, reconstruction, and integration into various metabolic processes, the process of metabolic reconstruction remains crucial.

According to reports, CBM monitored the activity of each metabolic network under chemical-physical inhibition. The topology of the network can be fully described by a metabolic reconstruction, which can then be utilized as the foundation for the creation of genome-scale models that can be applied to the modeling of yet another distinct metabolic network of bacteria. (Tsai and Joung 2016). The discovery of novel techniques to control the numerous illnesses caused by genetically modified organisms (GMOs) in plants may also be facilitated by mathematical modeling and systems biology approaches. A specific microbial genomic network is said to benefit from the addition of novel enzyme-coding genes from various species, thanks to OptStrain. In order to manipulate the up- and downregulation of metabolic enzymes as well as gene knockouts to achieve the required metabolite synthesis, OptReg constraint-based modeling was recently created (Thal et al. 2018). By conducting bioinformatics investigations, it is possible to obtain gene annotations of protein sequences for numerous organisms, which in turn facilitates the reconstruction of a metabolic network. Flux-based analysis can be used to analyze metabolic networks in addition to CBA. In the absence of thorough kinetic data, several researchers claimed that studying luxes provides motivating insights (Azhar et al. 2016).

9 Conclusion

The health of plants and the sustainability of many ecosystems depend heavily on bioinoculants. The present review has provided a description of the techniques that are presently accessible for understanding systems biology, including metabolomics, transcriptomics, proteomics, and genomics. The diverse metabolic networks used by bacteria have been described using systems biology technologies including CBM, OptKnock, and FBA. The interactions between the pathogen and the host plant can also be managed using other gene-editing techniques like TALENs and CRISPR/Cas. The abilities needed to switch from conventional agricultural equipment to contemporary agricultural biotechnology are made possible by these systems biology approaches. Knowing molecular biology and systems biology can facilitate the progress of sustainable agriculture based on microbes. These methods assist farmers in comprehending the positive impact of soil microbial populations. These methods undoubtedly create new possibilities for agricultural advancement in the future that are based on advantageous microbial amenities to reduce the need for synthetic chemicals and pesticides. These methods allow for the achievement of both economic objectives and environmentally sound practices. By employing genomic techniques and computational biology, researchers have shown that it is possible to improve the metabolic and communication pathways responsible for promoting positive effects on plants while maintaining ecological sustainability.

References

Agri U, Chaudhary P, Sharma A (2021) In vitro compatibility evaluation of agriusable nanochitosan on beneficial plant growth-promoting rhizobacteria and maize plant. Natl Acad Sci Lett 44:555–559. https://doi.org/10.1007/s40009-021-01047-w

Agri U, Chaudhary P, Sharma A, Kukreti B (2022) Physiological response of maize plants and its rhizospheric microbiome under the influence of potential bioinoculants and nanochitosan. Plant Soil 474:451–468. https://doi.org/10.1007/s11104-022-05351-2

Ahemad M, Kibret M (2014) Mechanisms and applications of plant growth promoting rhizobacteria: current perspective. J King Saud Univ Sci 26:1–20. https://doi.org/10.1016/j.jksus.2013.05.001

Alamgir ANM (2018) Introduction. Biotechnology, in vitro production of natural bioactive compounds, herbal preparation, and disease management (treatment and prevention). In: Alamgir ANM (ed) Therapeutic use of medicinal plants and their extracts, vol 2. Springer, Cham, pp 585–664. https://doi.org/10.1007/978-3-319-92387-1_7

Arora NK, Verma M, Mishra J (2017) Rhizobial bio-formulations: past, present and future. In: Rhizotrophs: plant growth promotion to bioremediation. Springer, Singapore, pp 69–99. https://doi.org/10.1007/978-981-10-4862-3_4

Azad RK, Shulaev V (2018) Metabolomics technology and bioinformatics for precision medicine. Brief Bioinform 20:1957. https://doi.org/10.1093/bib/bbx170

Azhar AH, Mohamad MS, Deris S, Napis S, Sinnott RA (2016) A hybrid of ant colony optimization and f lux variability analysis to improve the production of L-phenylalanine and biohydrogen. Int J Adv Soft Comput Appl 8(2)

Baart GJ, Martens DE (2012) Genome-scale metabolic models: reconstruction and analysis. In: Neisseria meningitidis. Humana Press, Springer, pp 107–126. https://doi.org/10.1007/978-1-61779-346-2_7

Basu S, Rabara RC, Negi S, Shukla P (2018) Engineering PGPMOs through gene editing and systems biology: a solution for phytoremediation? Trends Biotechnol 36(5):499–510. https://doi.org/10.1016/j.tibtech.2018.01.011

Beer LA, Speicher DW (2018) Protein detection in gels using fixation. Curr Protoc Protein Sci 91(1):10–15. https://doi.org/10.1002/0471140864.ps1005s29

Belibasakis GN, Curtis MA, Hajishengallis G, Zaura E (2018) Meeting report: the 12th European oral microbiology workshop (EOMW) 2018 in Stockholm, Sweden, pp 64–69. https://doi.org/10.1080/21505594.2017.1376147

Bhardwaj D, Ansari MW, Sahoo RK, Tuteja N (2014) Biofertilizers function as key player in sustainable agriculture by improving soil fertility, plant tolerance and crop productivity. Microb Cell Factories 13:66. https://doi.org/10.1186/1475-2859-13-66

Bhattacharyya PN, Jha DK (2012) Plant growth-promoting rhizobacteria (PGPR): emergence in agriculture. World J Microbiol Biotechnol 28:1327–1350. https://doi.org/10.1007/s11274-011-0979-9

Buck LT, Burness G, Campbell KL, Darveau CA, Driedzic W, Guderley H, McClelland GB, Moon TW, Moyes CD, Schulte PM (2018) 50 years of comparative biochemistry: the legacy of Peter Hochachka. Comp Biochem Physiol B: Biochem Mol Biol 224:1–11. https://doi.org/10.1016/j.cbpb.2018.02.002

Caplan AL, Parent B, Shen M, Plunkett C (2015) No time to waste—the ethical challenges created by CRISPR: CRISPR/Cas, being an efficient, simple, and cheap technology to edit the genome of any organism, raises many ethical and regulatory issues beyond the use to manipulate human germ line cells. EMBO Rep 16:1421–1426. https://doi.org/10.15252/embr.201541337

Chaudhary T, Shukla P (2019) Bioinoculants for bioremediation applications and disease resistance: innovative perspectives. Indian J Microbiol 59(2):1–8. https://doi.org/10.1007/s12088-019-00783-4

Chaudhary P, Parveen H, Gangola S, Kumar G, Bhatt P, Chaudhary A (2021) Plant Growth Promoting Rhizobacteria and their application in sustainable crop production. In: Bhatt P, Gangola S, Udayanga D, Kumar G (eds) Microbial technology for sustainable environment. Springer, Singapore. https://doi.org/10.1007/978-981-16-3840-4_13

Chaudhary P, Agri U, Chaudhary A, Kumar A, Kumar G (2022a) Endophytes and their potential in biotic stress management and crop production. Front Microbiol 13:933017. https://doi.org/10.3389/fmicb.2022.933017

Chaudhary P, Singh S, Chaudhary A, Sharma A, Kumar G (2022b) Overview of biofertilizers in crop production and stress management for sustainable agriculture. Front Plant Sci 13:930340. https://doi.org/10.3389/fpls.2022.930340

Chaudhary P, Xu M, Ahamad L, Chaudhary A, Kumar G, Adeleke BS, Verma KK, Hu D-M, Širić I, Kumar P, Popescu SM, Abou Fayssal S (2023a) Application of synthetic consortia for improvement of soil fertility, pollution remediation, and agricultural productivity: a review. Agronomy 13:643. https://doi.org/10.3390/agronomy13030643

Chaudhary P, Chen S, Rajput VD, Jaiswal DK (eds) (2023b) Bioinoculants with nano-compounds to improve soil health: a step toward sustainable agriculture. Frontiers Media SA, Lausanne. https://doi.org/10.3389/978-2-8325-3287-4

Chuai G, Wang QL, Liu Q (2017) In silico meets in vivo: towards computational CRISPR-based sgRNA design. Trends Biotechnol 35:12–21. https://doi.org/10.1016/j.tibtech.2016.06.008

Cimermancic P, Medema MH, Claesen J, Kurita K, Brown LC, Mavrommatis K, Pati A, Godfrey PA, Koehrsen M, Clardy J, Birren BW (2014) Insights into secondary metabolism from a global analysis of prokaryotic biosynthetic gene clusters. Cell 158(2):412–421. https://doi.org/10.1016/j.cell.2014.06.034

Compant S, Clement C, Sessitsch A (2010) Plant growth-promoting bacteria in the rhizo- and endosphere of plants: their role, colonization, mechanisms involved and prospects for utilization. Soil Biol Biochem 42:669–678. https://doi.org/10.1016/j.soilbio.2009.11.024

D Arrigo I (2016) The transcriptional landscape of the production organism *Pseudomonas putida*. PhD diss.

D'Alessio M, Nordeste R, Doxey AC, Charles TC (2017) Transcriptome analysis of polyhydroxybutyrate cycle mutants reveals discrete loci connecting nitrogen utilization and carbon storage in *Sinorhizobium meliloti*. mSystems 2:e00035–e00017. https://doi.org/10.1128/mSystems.00035-17

Dangi AK, Sharma B, Khangwal I, Shukla P (2018) Combinatorial interactions of biotic and abiotic stresses in plants and their molecular mechanisms: systems biology approach. Mol Biotechnol 60(8):636–650

diCenzo GC, Zamani M, Checcucci A, Fondi M, Griffitts JS, Finan TM, Mengoni A (2018) Multi-disciplinary approaches for studying rhizobium–legume symbioses. Can J Microbiol 65(1):1–33. https://doi.org/10.1139/cjm-2018-0377

Du ZY, Lucker BF, Zienkiewicz K, Miller TE, Zienkiewicz A, Sears BB, Kramer DM, Benning C (2018) Galactoglycerolipid lipase PGD1 is involved in thylakoid membrane remodeling in response to adverse environmental conditions in *Chlamydomonas*. Plant Cell 30(2):447–465. https://doi.org/10.1105/tpc.17.00446

Esvelt KM, Wang HH (2013) Genome-scale engineering for systems and synthetic biology. Mol Syst Biol 9:641. https://doi.org/10.1038/msb.2012.66

Etalo DW, Jeon JS, Raaijmakers JM (2018) Modulation of plant chemistry by beneficial root microbiota. Nat Prod Rep 35:398. https://doi.org/10.1039/C7NP00057J

Farrar K, Bryant D, Cope Selby N (2014) Understanding and engineering beneficial plant–microbe interactions: plant growth promotion in energy crops. Plant Biotechnol J 12:1193–1206. https://doi.org/10.1111/pbi.12279

Frew A, Powell JR, Glauser G, Bennett AE, Johnson SN (2018) Mycorrhizal fungi enhance nutrient uptake but disarm defenses in plant roots, promoting plant-parasitic nematode populations. Soil Biol Biochem 126:123–132. https://doi.org/10.1016/j.soilbio.2018.08.019

Gilbert JA, Meyer F, Antonopoulos D, Balaji P, Brown CT, Brown CT, Desai N, Eisen JA, Evers D, Field D, Feng W (2010) Meeting report: the tera base metagenomics workshop and the vision of an Earth microbiome project. Stand Genomic Sci 3:243. https://doi.org/10.4056/sigs.1433550

Hsu PD, Lander ES, Zhang F (2014) Development and applications of CRISPR–Cas9 for genome engineering. Cell 157:1262–1278. https://doi.org/10.1016/j.cell.2014.05.010

Igiehon NO, Babalola OO (2018) Below-ground-above-ground plant–microbial interactions: focusing on soybean, rhizobacteria and mycorrhizal fungi. Open Microbiol J 12:261. https://doi.org/10.2174/1874285801812010261

Imam J, Singh PK, Shukla P (2016) Plant microbe interactions in post genomic era: perspectives and applications. Front Microbiol 7:1488. https://doi.org/10.3389/fmicb.2016.01488

Imam J, Shukla P, Prasad Mandal N, Variar M (2017) Microbial interactions in plants: perspectives and applications of proteomics. Curr Protein Pept Sci 18(9):956–965. https://doi.org/10.2174/1389203718666161122103731

Jafari T (2018) Application of NMR metabolomics profiling on questions arising from molecular biology. PhD diss., University of Turku

Jat LK, Singh YV, Meena SK, Meena SK, Parihar M, Jatav HS, Meena RK, Meena VS (2015) Does integrated nutrient management enhance agricultural productivity? J Pure Appl Microbiol 9(2):1211–1221

Khan N, Bano A, Babar MA (2019) Metabolic and physiological changes induced by plant growth regulators and plant growth promoting rhizobacteria and their impact on drought tolerance in Cicer arietinum L. PLoS One 14(3):0213040. https://doi.org/10.1371/journal.pone.0213040

Kim M, Lee KH, Yoon SW, Kim BS, Chun J, Yi H (2013) Analytical tools and databases for metagenomics in the next-generation sequencing era. Genomics Inform 11(3):102–113. https://doi.org/10.5808/GI.2013.11.3.102

Kim D, Langmead B, Salzberg SL (2015) HISAT: a fast spliced aligner with low memory requirements. Nat Methods 12(4):357–360. https://doi.org/10.1038/nmeth.3317

Kim DG, Seo SW, Cho BK, Lohumi S, Hong SJ, Lee WH (2018) Review of current approaches for implementing metabolic reconstruction. J Biosyst Eng 43(1):45–58. https://doi.org/10.5307/JBE.2018.43.1.045

Kodati S, Adesemoye AO, Yuen GY, Volesky JD, Everhart SE (2021) Origin of agricultural plant pathogens: diversity and pathogenicity of Rhizoctonia fungi associated with native prairie grasses in the Sandhills of Nebraska. PLoS One 16:e0249335. https://doi.org/10.1371/journal.pone.0249335

Kukreti B, Sharma A, Chaudhary P, Agri U, Maithani D (2020) Influence of nanosilicon dioxide along with bioinoculants on *Zea mays* and its rhizospheric soil. 3 Biotech 10:345. https://doi.org/10.1007/s13205-020-02329-8

Kumar P, Kumar V, Chattopadhyay T (2016) Gene cloning, recombinant expression and purification of a germin-like protein from tomato (Solanum lycopersicum L.). J Crop Improv 30(5):595–607

Kumari A, Das P, Parida AK (2015) Proteomics, metabolomics, and ionomics perspectives of salinity tolerance in halophytes. Front Plant Sci 6:537. https://doi.org/10.3389/fpls.2015.00537

Kumari S, Sharma A, Chaudhary P, Khati P (2020) Management of plant vigor and soil health using two agriusable nanocompounds and plant growth promotory rhizobacteria in Fenugreek. 3 Biotech 10:1–11. https://doi.org/10.1007/s13205-020-02448-2

Kumari H, Khati P, Gangola S, Chaudhary P, Sharma A (2021) Performance of plant growth promotory rhizobacteria on Maize and soil characteristics under the influence of TiO_2 nanoparticles. Pantnagar J Res 19:28–39

Lai B, Ding R, Li Y, Duan L, Zhu H (2012) A de novo metagenomic assembly program for shotgun DNA reads. Bioinformatics 28:1455–1462. https://doi.org/10.1093/bioinformatics/bts162

Lakshmanan M, Koduru L, Lee DY (2018) Software applications for phenotype analysis and strain design of cellular systems. Emerg Areas Bioeng 2:771–791. https://doi.org/10.1002/9783527803293.ch44

Laserson J, Jojic V, Koller D (2011) Genovo: de novo assembly for metagenomes. J Comput Biol 18:429–443. https://doi.org/10.1089/cmb.2010.0244

Lee SS, Joo YH, Choi JS, Jeong SM, Paradhipta DH, Noh HT, Han OK, Kim SC (2021) Effects of ensiling period and bacterial inoculants on chemical compositions and fermentation characteristics of Rye silage. J Korean Soc Grassland Forage Sci 41(4):259–266

Litsios A, Ortega AD, Wit EC, Heinemann M (2018) Metabolic-flux dependent regulation of microbial physiology. Curr Opin Microbiol 42:71–78. https://doi.org/10.1016/j.mib.2017.10.029

Litvinov DY, Karlov GI, Divashuk MG (2021) Metabolomics for crop breeding: general considerations. Genes 12(10):1602. https://doi.org/10.3390/genes12101602

Liu B, Zhang N, Zhao S, Chang J, Wang Z, Zhang G, Si H, Wang D (2015) Proteomic changes during tuber dormancy release process revealed by iTRAQ quantitative proteomics in potato. Plant Physiol Biochem 86:181–190. https://doi.org/10.1016/j.plaphy.2014.12.003

Manzar N, Jayaswal D, Kishan G, Chauhan S, Joshi D, Singh J, Chaudhary P, Jamali H, Kushwaha P, Parveen H, Kashyap AS, Sahu PK, Singh UB (2022) Application of molecular ecology approaches in sustainable agriculture for a better understanding of plant–microbiome interactions. In: Singh UB, Sahu PK, Singh HV, Sharma PK, Sharma SK (eds) Rhizosphere microbes. Microorganisms for sustainability, vol 40. Springer, Singapore. https://doi.org/10.1007/978-981-19-5872-4_3

Martinez-Fernandez G, Duval S, Kindermann M, Schirra HJ, Denman SE, McSweeney CS (2018) 3-NOP vs. halogenated compound: methane production, ruminal fermentation and microbial community response in forage fed cattle. Front Microbiol 9:1582. https://doi.org/10.3389/fmicb.2018.01582

Montague TG, Cruz JM, Gagnon JA, Church GM, Valen E (2014) CHOPCHOP: a CRISPR/Cas9 and TALEN web tool for genome editing. Nucleic Acids Res 42:W401–W407. https://doi.org/10.1093/nar/gku410

Mutturi S, Sahai V, Bisaria VS (2017) Fed-batch cultivation for high-density culture of Pseudomonas spp. for bioinoculant preparation. In: Varma A (ed) Modern tools and techniques to understand microbes. Springer, Cham, pp 381–400. https://doi.org/10.1007/978-3-319-49197-4_24

Papadopoulou ES, Perruchon C, Vasileiadis S, Rousidou C, Tanou G, Samiotaki M, Molassiotis A, Karpouzas DG (2018) Metabolic and evolutionary insights in the transformation of diphenylamine by a *Pseudomonas putida* strain unravelled by genomic, proteomic, and transcription analysis. Front Microbiol 9:676. https://doi.org/10.3389/fmicb.2018.00676

Patil HJ, Solanki MK (2016) Microbial inoculant: modern era of fertilizers and pesticides. In: Pratap Singh D (ed) Microbial inoculants in sustainable agricultural productivity. Springer, New Delhi, pp 319–343. https://doi.org/10.1007/978-81-322-2647-5_19

Poole P, Ramaschandran V, Terpolilli J (2018) Rhizobia: from saprophytes to endosymbionts. Nat Rev Microbiol 16:291. https://doi.org/10.1038/nrmicro.2017.171

Prasad R, Gill SS, Tuteja N (2018) New and future developments in microbial biotechnology and bioengineering: crop improvement through microbial biotechnology. Elsevier, USA

Pretscher J, Fischkal T, Branscheidt S, Jäger L, Kahl S, Schlander M, Thines E, Claus H (2018) Yeasts from different habitats and their potential as biocontrol agents. Fermentation 4:31. https://doi.org/10.3390/fermentation4020031

Ramalingam A, Kudapa H, Pazhamala LT, Weckwerth W, Varshney RK (2015) Proteomics and metabolomics: two emerging areas for legume improvement. Front Plant Sci 6:1116. https://doi.org/10.3389/fpls.2015.01116

Rosier A, Medeiros FHV, Bais HP (2018) Defining plant growth promoting rhizobacteria molecular and biochemical networks in beneficial plant–microbe interactions. Plant Soil 428(1–2):1–21. https://doi.org/10.1007/s11104-018-3679-5

Rustagi A, Singh G, Agrawal S, Gupta PK (2018) Proteomic studies revealing enigma of plant–pathogen interaction. In: Rustagi A (ed) Molecular aspects of plant–pathogen interaction. Springer, pp 239–264. https://doi.org/10.1007/978-981-10-7371-7_11

Sander JD, Joung JK (2014) CRISPR–Cas systems for editing, regulating and targeting genomes. Nat Biotechnol 32:347. https://doi.org/10.1038/nbt.2842

Satge C, Moreau S, Sallet E, Lefort G, Auriac MC, Rembliere C, Cottret L, Gallardo K, Noirot C, Jardinaud MF, Gamas P (2016) Reprogramming of DNA methylation is critical for nodule development in Medicago truncatula. Nat Plants 2:16166. https://doi.org/10.1038/nplants.2016.166

Schaum CE, Buckling A, Smirnoff N, Studholme DJ, Yvon-Durocher G (2018) Environmental fluctuations accelerate molecular evolution of thermal tolerance in a marine diatom. Nat Commun 9:1719. https://doi.org/10.1038/s41467-018-03906-5

Schenkluhn L, Hohnjec N, Niehaus K, Schmitz U, Colditz F (2010) Differential gel electrophoresis (DIGE) to quantitatively monitor early symbiosis- and pathogenesis-induced changes of the Medicago truncatula root proteome. J Proteome 73:753–768. https://doi.org/10.1016/j.jprot.2009.10.009

Sharma SB, Sayyed RZ, Trivedi MH, Gobi TA (2013) Phosphate solubilizing microbes: sustainable approaches for managing phosphorus deficiency in agricultural soils. Springer Plus 2:587. https://doi.org/10.1186/2193-1801-2-587

Shukla E, Thorat L, Bendre AD, Jadhav S, Pal JK, Nath BB, Gaikwad SM (2018) Cloning and characterization of trehalase: a conserved glycosidase from oriental midge, Chironomus ramosus 3 Biotech, 8:1–7

Simons-Senftle MN, Sarkar D, Maranas CD (2018) Modeling plant metabolism: advancements and future capabilities. In: Engineering nitrogen utilization in crop plants. Springer International Publishing, pp 57–76. https://doi.org/10.1007/978-3-319-92958-3_4

Suyal DC, Soni R, Sai S, Goel R (2016) Microbial inoculants as biofertilizer. In: Suyal Chandra D (ed) Microbial inoculants in sustainable agricultural productivity. Springer, New Delhi, pp 311–318. https://doi.org/10.1007/978-81-322-2647-5_18

Tatsukami Y (2017) Studies on symbiosis-specific phenotype of *Mesorhizobium loti* and its function to host plant. PhD diss., Kyoto University

Thal B, Braun HP, Eubel H (2018) Proteomic analysis dissects the impact of nodulation and biological nitrogen fixation on Vicia faba root nodule physiology. Plant Mol Biol 97:233–251. https://doi.org/10.1007/s11103-018-0736-7

Thiele I, Palsson BQ (2010) A protocol for generating a high-quality genome-scale metabolic reconstruction. Nat Protoc 5:93. https://doi.org/10.1038/nprot.2009.203

Tsai SQ, Joung JK (2016) Defining and improving the genome wide specificities of CRISPR–Cas9 nucleases. Nat Rev Genet 17:300. https://doi.org/10.1038/nrg.2016.28

Turi KN, Romick-Rosendale L, Ryckman KK, Hartert TV (2018) A review of metabolomics approaches and their application in identifying causal pathways of childhood asthma. J Allergy Clin Immunol 141:1191–1201. https://doi.org/10.1016/j.jaci.2017.04.021

Tyagi S, Mulla SI, Lee KJ, Chae JC, Shukla P (2018) VOCs-mediated hormonal signaling and crosstalk with plant growth promoting microbes. Crit Rev Biotechnol 38(8):1277–1296. https://doi.org/10.1080/07388551.2018

Tyagi J, Chaudhary P, Jyotsana, Bhagwati U, Bhandari G, Chaudhary A (2022a) Chapter 17: Impact of endophytic fungi in biotic stress management. In: Soni R, Suyal D, Goel R (eds) Plant protection: from chemicals to biologicals. De Gruyter, Berlin, Boston, pp 447–462. https://doi.org/10.1515/9783110771558-017

Tyagi J, Chaudhary P, Mishra A, Khatwani M, Dey S, Varma A (2022b) Role of endophytes in abiotic stress tolerance: with special emphasis on *Serendipita indica*. Int J Environ Res 16:62. https://doi.org/10.1007/s41742-022-00439-0

Van Der Oost J, Westra ER, Jackson RN, Wiedenheft B (2014) Unravelling the structural and mechanistic basis of CRISPR–Cas systems. Nat Rev Microbiol 12:479. https://doi.org/10.1038/nrmicro3279

Verma A, Kumar S, Kumar G, Saini JK, Agrawal R, Satlewal A, Ansari MW (2018) Rhizosphere metabolite profiling: an opportunity to understand plant–microbe interactions for crop improvement. In: Prasad R, Gill SS, Tuteja N (eds) Crop improvement through microbial biotechnology. Elsevier, pp 343–361. https://doi.org/10.1016/B978-0-444-63987-5.00017-7

Wang H, Yang H, Shivalila CS, Dawlaty MM, Cheng AW, Zhang F, Jaenisch R (2013) One-step generation of mice carrying mutations in multiple genes by CRISPR/ Cas-mediated genome engineering. Cell 153:910–918. https://doi.org/10.1016/j.cell.2013.04.025

Wang L, Chen L, Li R, Zhao R, Yang M, Sheng J, Shen L (2017) Reduced drought tolerance by CRISPR/Cas9-mediated SlMAPK3 mutagenesis in tomato plants. J Agric Food Chem 65:8674–8682. https://doi.org/10.1021/acs.jafc.7b02745

Wawrzyniak R, Kosnowska A, Macioszek S, Bartoszewski R, Jan Markuszewski M (2018) New plasma preparation approaches to enrich metabolome coverage in untargeted metabolomics: plasma protein bound hydrophobic metabolite release with proteinase K. Sci Rep 8:9541. https://doi.org/10.1038/s41598-018-27983-0

Yang X, Long K (2016) Novel small protein identification and quantitative proteomic analysis in *Pseudomonas putida* KT-2440 (Doctoral dissertation, Novo Nordisk Foundation Center for Biosustainability).

Role of Microbes in Plant Biotic Stress Management

Bartholomew Saanu Adeleke, Modupe Stella Ayilara, Uswat T. Adeyemi, Qudus O. Uthman, Anuj Chaudhary, and Sami Abou Fayssal

Abstract For ages, biotic stress (e.g., pathogens, weeds, and insects) has been considered as a major threat for plants causing yield reductions. Synthetic chemicals have been used for control, but due to their ecotoxicity, a safer and more sustainable alternative tool such as biopesticides is being encouraged. Biopesticides encompass the use of natural agents, such as plants, microbes, and natural enemies (animals), to control plant pests. Among all these agents, microbes are more easily manipulated and their preference relies on their hasty growth rate. Bacteria, fungi, and viruses, when maximally explored, can be efficiently used to control plant pathogen via a multitude of mechanisms. This includes a direct interference with the normal metabolism of the pathogen; an enhancement of plants root physiology; an induc-

B. S. Adeleke (✉)
Food Security and Safety Focus Area, Faculty of Natural and Agricultural Sciences, North-West University, Mmabatho, South Africa

Department of Biological Sciences, Microbiology Unit, School of Science, Olusegun Agagu University of Science and Technology, Okitipupa, Nigeria
e-mail: bs.adeleke@oaustech.edu.ng

M. S. Ayilara
Food Security and Safety Focus Area, Faculty of Natural and Agricultural Sciences, North-West University, Private Bag X2046, Mmabatho, South Africa

Department of Biological Sciences, Kings University, Ode-Omu, Nigeria

U. T. Adeyemi
Department of Agricultural Economics and Farm Management, Faculty of Agriculture, University of Ilorin, Ilorin, Nigeria

Q. O. Uthman
Soil, Water and Ecosystem Sciences, University of Florida, Gainesville, FL, USA

A. Chaudhary
School of Agriculture and Environmental Sciences, Shobhit University, Gangoh, Uttar Pradesh, India

S. Abou Fayssal
Department of Agronomy, Faculty of Agronomy, University of Forestry, Sofia, Bulgaria

Department of Plant Production, Faculty of Agriculture, Lebanese University, Beirut, Lebanon

P. Chaudhary, A. Chaudhary (eds.), *Microbial Inoculants*, Interdisciplinary Biotechnological Advances, https://doi.org/10.1007/978-981-97-0633-4_7

tion of the host plant systemic resistance, competition, and antagonism; an alteration of the plant-pathogen signaling and parasitism; and the production of different metabolites, such as organic acids, enzymes, and hormones. This chapter summarizes the microbe species involved in biotic stress management and their mechanism (mode of action) and the factors affecting their survival.

Keywords Food insecurity · Host-microbe interactions · Microbial functionalities · Phytopathogens · Plant growth stimulators

1 Introduction

Biotic stress refers to the negative impact of living organisms, such as pests, diseases, and weeds on the growth and development of plants. In agriculture, biotic stress can cause significant economic losses by reducing crop yields and/or quality, thus leading to food insecurity (Gull et al. 2019). For instance, herbivorous insects feeding on plant tissue can make plants more susceptible to diseases and cause direct damage (Garzo et al. 2020). Plant pathogens, such as fungi, bacteria, viruses, and nematodes can induce plant disease phenomenon's, such as wilting, root rot, and other deformities (Adnan et al. 2019; Tyagi et al. 2022a). Weeds can also compete with crops for nutrients, light, and water, which results in crop yield reduction (Ekwealor et al. 2019).

To mitigate the effects of biotic stress, farmers and researchers can employ various methods, e.g., crop rotation, breeding with resistant species, implementation of biological control agents, and applied pesticides (Kumar et al. 2020; Chaudhary et al. 2022a). However, these methods have varied effectiveness levels and may have unintended consequences, such as the emergence of pesticide-resistant pests (Teshome et al. 2020). Therefore, sustainable and integrated approaches are needed to manage biotic stress in agriculture systems. Microbes play an important role in biotic stress management, which is caused by ordinary living organisms, such as pests, pathogens, and weeds, and improves plant health as well as productivity (Moustafa-Farag et al. 2019; Chaudhary et al. 2023a, b, c). Some microbes can act as biocontrol agents against plant pests or diseases. For instance, *Bacillus thuringiensis* (Bt) is a bacterium that produces a toxin that kills certain insect pests and thereby can be used as a biocontrol agent. This bacterium has been explored in the production of biopesticides against some insects. Other examples of biocontrol agents include fungi like *Beauveria bassiana* and *Metarhizium anisopliae* for the control of a variety of insect pests (Liu et al. 2022). Microbes can induce systemic resistance in plants, which can be defined as a plant's ability to resist or tolerate biotic stress (Chaudhary et al. 2021, 2022). Some bacteria and fungi can trigger the plant's defense system, mostly pronounced by the production of phytohormones and other compounds that help the plant to resist and mitigate the occurring stress (Iqbal et al. 2021; Kumari et al. 2021). Conceivably, *Pseudomonas fluorescens* induced systemic resistance in tomato plants against *Fusarium oxysporum* (Fatima and Anjum 2017). Some nitrogen-fixing bacteria from the genera *Rhizobium* and *Azotobacter* can fix atmospheric nitrogen into a usable form for plants. This can help improve plant growth and reduce dependence on chemical fertilizers (Ayilara et al. 2022a). Soil and plant microbes compete with pests and pathogens for resources, such as nutrients and space. For instance, some beneficial soil bacteria potentially outcompete pathogens for nutrients and space, thus reducing their impact on plants (Ayilara et al. 2022b). Notably, microbes can play an important role in biotic stress management by providing biocontrol, inducing systemic resistance, fixing nitrogen, and competing with other pathogens for limited resources available in the plant environment (Khati et al. 2019; Chaudhary et al. 2022b).

2 Microbes Involved in Plant Biotic Stress Management

Different groups of microbes have been reported to alleviate plant biotic stress, and these include those that reside in the rhizosphere, rhizoplane, bulk soil, as well as the endophytes, which reside in the root, seeds, leaves, bark, and stem of different plants (Smercina et al. 2019; Adeleke et al. 2022). Figures 1 and 2 outline the main plant biotic stressors faced and microbes and their role in biotic stress management, respectively.

The exudates released by different plants favor the survival of different rhizosphere microbes; this creates permissible zones for the abundance of microbes

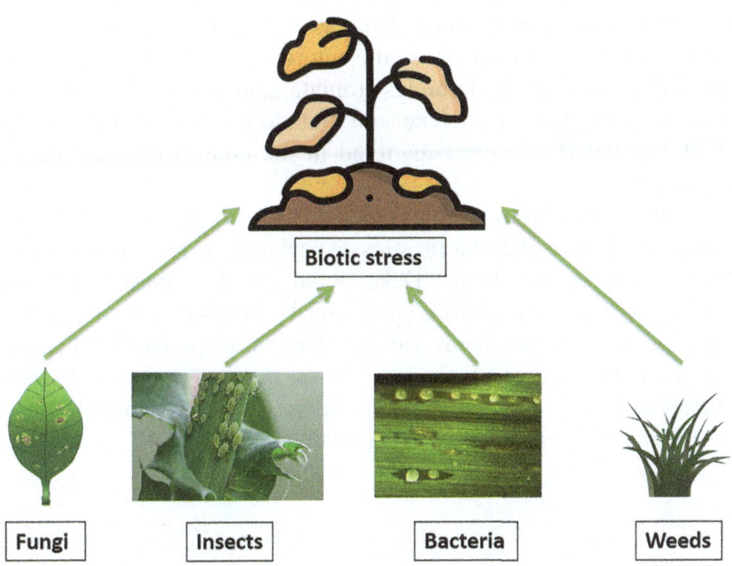

Fig. 1 Main plant biotic stressors

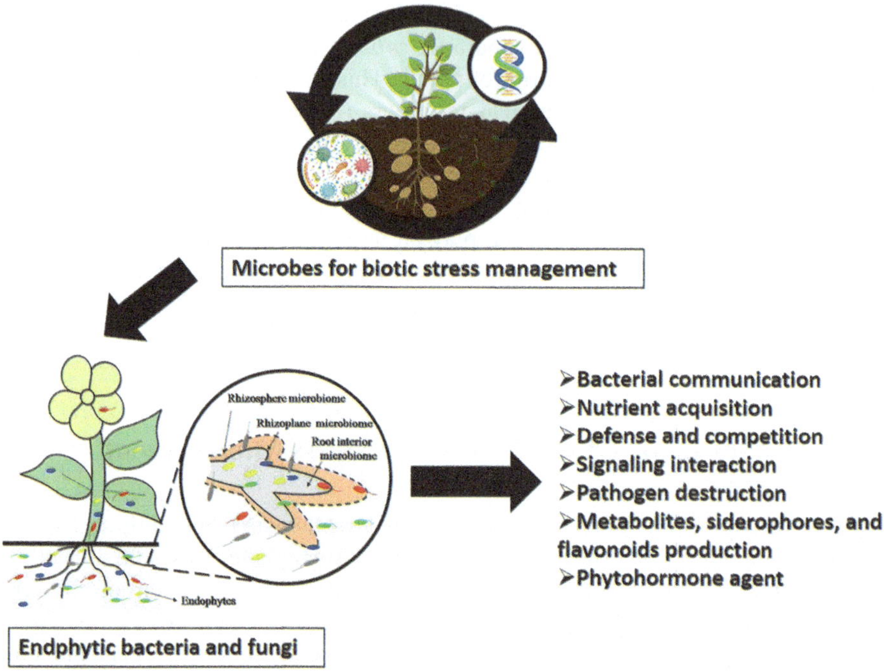

Fig. 2 Microbes and their role in biotic stress management

in the rhizosphere (Chaudhary et al. 2022c), even though there can be a variation in their diversity owing to different factors, i.e., plant genotype, environment, and plant development stage (Chagas et al. 2018). In some cases, these rhizosphere microbes gain access to the interior portion of the plants through injury openings (wounds), hydathodes stomata, and pores, where they establish themselves as endophytes (Adeleke and Babalola 2021; Pathak et al. 2022). Some plant-associated microbes involved in the control of plant diseases are listed in Table 1.

Asides from controlling microbial pathogens, microorganisms have also been reported to control other plant biotic stresses including insects and weeds (Mishra et al. 2013; Verma et al. 2020). Deka et al. (2021) reported the activity of *Paenibacillus popilliae*, *Bacillus papillae,* and *Verticillium lecanii* against insects i.e., *Popillia japonica*, Coleoptera, and Lepidoptera, respectively. Moreover, Usta (2013) reported the biocontrol potential of *B. thuringiensis* and *B. sphaericus* against different insects. *Pseudomonas fluorescens* was very active against weeds, *Bromus tectorum,* and *Setaria viridis* (Verma et al. 2020).

Table 1 Microbes associated with plant disease control

Microbes	Species	Stress agent	References
Bacteria	*Lactobacillus plantarum*	*Botrytis cinerea*	Chen et al. (2020)
	Bacillus sp., *Sphingomonas* sp., *Pantoea* sp., *Curtobacterium* sp., *Pseudomonas* sp.	*Colletotrichum gloeosporioides, Botrytis cinerea, Sclerotinia sclerotiorum*	Xu et al. (2019)
	Bacillus tequilensis	*Magnaporthe oryzae*	Li et al. (2018)
	Acinetobacter calcoaceticus, Bacillus thuringiensis, Bacillus subtilis, Bacillus siamensis, Bacillus amyloliquefaciens	*Fusarium oxysporum*	Khalil et al. (2021)
	Chitinophaga sp., *Pedobacter* sp., *Rhizobium* sp., *Pseudomonas* sp., *Chryseobacterium* sp., *Flavobacterium* sp., *Variovorax* sp.	*Pythium ultimum, Rhizopus oryzae*	Yin et al. (2021)
	Bacillus velezensis, Bacillus subtilis, Bacillus amyloliquefaciens	*Phytophthora* sp., *Fusarium* sp.	Nguyen et al. (2020)
	Streptomyces sp.	*Macrophomina phaseolina, Rhizoctonia solani*	Díaz-Díaz et al. (2022)
	Bacillus sp., *Pseudomonas* sp.	*Fusarium oxysporum*	Mon et al. (2021)
Fungi	*Geotrichum* sp., *Schwanniomyces vanrijiae, P. kudriavzevii*	*Monilinia fructigena*	Madbouly et al. (2020)
	Trichoderma sp.	*Phaeoacremonium minimum*	Carro-Huerga et al. (2020)
	Trichoderma sp.	*Fusarium oxysporum*	Mon et al. (2021)
	Trichoderma virens	*Rhizoctonia* sp., *Pythium* sp.	Thambugala et al. (2020)
	Paraphaeosphaeria minitans	*Sclerotinia sclerotiorum, Sclerotinia minor*	Thambugala et al. (2020)
	Piriformospora indica	*Fusarium oxysporum*	Fontana et al. (2021)
Viruses	Potato tuber moth GV *(PhopGV)*	*Phthorimaea operculella*	Singh et al. (2019)
	Araujia mosaic virus	*Araujia hortorum*	Verma et al. (2020)
	Lymantria dispar multicapsid nucleopolyhedrovirus	*Lepidopteran caterpillars*	Ruiu 2018
	Nucleopolyhedrovirus	*Spodoptera frugiperda*	Cuartas-Otálora et al. (2019)
Nematode	*Steinernema carpocapsae*	*Rhynchophorus ferrugineus*	Ruiu 2018
	Heterorhabditis downesi	*Otiorhynchus sulcatus*	Ruiu 2018

3 Factors Affecting Microbial Survival in Biotic Stress Management

There are several factors affecting microbial survival in biotic stress management; these include temperature, humidity, and pH, all being environmental factors. These factors affect the microbial survival and maximal use in agricultural systems (Gupta et al. 2017). Microbes have optimal temperature and pH ranges for growth and survival, and any deviation from these ranges can be lethal. High temperatures, for instance, cause microbial inactivation due to desiccation and protein denaturation, whereas low temperatures promote higher survival rates. On the other hand, some microbes lose viability at very low temperatures due to the ice crystal formation on their surface (Gupta et al. 2017).

Humidity is another important environmental factor that influences microbial survival. High humidity levels can encourage the growth of microbes; however, each microorganism has a certain relative humidity required for survival that should be taken into consideration (Zoz et al. 2016). For instance, the lowest limit of relative humidity for fungal growth is about 65%, while bacterial species require a much higher value to proliferate, develop, and multiplicate. Nutrient availability is critical for microbial growth and survival and is mainly affected by a variety of factors, i.e., soil type, organic matter content, and fertilizer application. Oxygen is also required for microbial survival; however, each microbe requires a certain oxygen level based on its metabolic needs (Jacoby et al. 2017). Soil moisture is also related to the availability of oxygen in soil, which is critical for the survival of aerobic microorganisms. Excessive soil moisture can cause anaerobic conditions in soil, thus preventing aerobic microbes from surviving, while a very low soil moisture content inhibits microbial survival (Abdul Rahman et al. 2021).

Microbial interactions are also an important factor in biotic stress management that affects microbial survival. In the agricultural ecosystem, microbes interact with each other, and these interactions can affect their survival (Chaudhary et al. 2022d). Microbial interactions can be mutualistic or antagonistic. When one microbe inhibits the growth or survival of another, this is referred to as an antagonistic interaction, whereas when microbes form symbiotic relationships that benefit their growth and survival, mutualistic interactions occur (Manzar et al. 2022).

Furthermore, antagonistic interactions can affect the survival of biocontrol agents in the agricultural ecosystem, while mutualistic interactions between microbes and plants can enhance the survival of biocontrol agents (Kohl et al. 2019). Rhizobia, for instance, are bacteria that form a mutualistic relationship with legume plants; they fix nitrogen and provide needed nutrients for the plant, while the latter provides them energy sources for their survival, growth, and development. This mutualistic relationship benefits both parties by improving their survival in the agricultural ecosystem.

Predation is an additional important interaction observed between microbes; it can seriously affect the survival of potential pathogens (Moënne-Loccoz et al. 2015). Bacteriophages, for instance, are viruses that infect and kill bacteria and can be also used as biocontrol agents against plant pathogens. Similarly, protozoa and nematodes can feed on bacteria and fungi and can be also used as biocontrol agents in biotic stress management (Lahlali et al. 2022).

Microbial survival in the agricultural ecosystem can be also influenced by host plant factors. Plants produce defense compounds that prevent pests and pathogens from growing and surviving. These defense compounds, like biocontrol agents and plant growth-promoting rhizobacteria (PGPR), can also inhibit the growth and survival of beneficial microbes (El-Saadony et al. 2022). Plant species and cultivars differ in their susceptibility to pests and pathogens and production of defense compounds (Kaur et al. 2022). Some plants produce high levels of defense compounds that inhibit the growth and survival of biocontrol agents and PGPR, while others produce low levels of such compounds which make them more susceptible to pests and pathogens infestation. Figure 3 summarizes the main factors affecting the microbial survival under biotic stress management.

All factors influencing the microbial survival in biotic stress management have important implications for agricultural sustainability. Sustainable agriculture is a farming system that prioritizes environmental health, economic profitability, and social and economic equity. Microbes contribute to sustainable agriculture by acting as biocontrol agents and contributing to enhanced soil health and fertility (Khati et al. 2018; Agri et al. 2021, 2022; Chaudhary et al. 2021a, b).

The need for integrated pest management (IPM) practices in agriculture is among the implications affecting microbial survival. IPM is a pest management system that employs a variety of methods, including biological, cultural, and chemical methods,

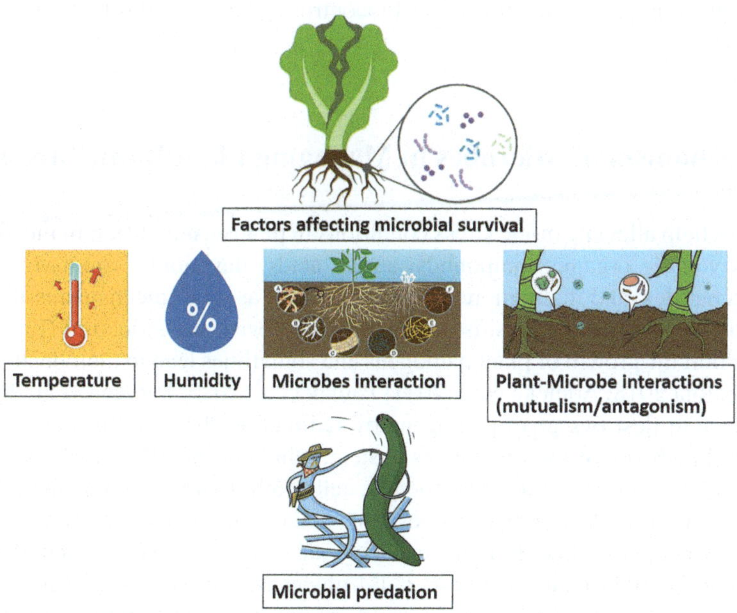

Fig. 3 Factors affecting microbial survival in biotic stress management

to control pests while minimizing the environmental impact. For instance, beneficial microbes are an important IPM component used as biocontrol agents. Therefore, there is a critical need to understanding the factors that influence microbial survival to reach a successful IPM approach.

Adding to that, there is a need for agricultural soil health management practices. Microbes play an important role in the maintenance of soil health and fertility, which is critical for sustainable agriculture (Kumari et al. 2020; Chaudhary et al. 2021c, d, e). Soil health management practices, such as cover cropping, crop rotation, organic matter management, and beneficial microbes, can therefore improve microbial survival and activity in soil, resulting in improved soil health and fertility (Kukreti et al. 2020; Chaudhary et al. 2021f, g).

Moreover, site-specific management strategies are in a major need. Environmental factors such as temperature and moisture can differ significantly between agricultural sites, thus affecting the survival of biocontrol agents and PGPR. In result, effective biotic stress management requires site-specific management strategies that take into account the environmental conditions of the site. The selection of biocontrol agents and PGPR that are adapted to the site's environmental conditions is also important. Biocontrol agents and PGPR that are adapted to the site's environmental conditions have a better chance to survive and establish in the soil, resulting in effective biotic stress management.

Host plant factors have important implications for sustainable agriculture as well. The susceptibility of host plants to defense compounds can affect the survival and effectiveness of biocontrol agents and PGPR in biotic stress management. Choosing plant species and cultivars that are less susceptible to defense compounds can therefore improve the efficacy of biocontrol agents and PGPR in biotic stress management.

4 Mechanism of Microbes in Managing Plant Biotic Stress

Microbes help alleviate biotic stress conditions in plants using different metabolites, such as volatile organic compounds, amino acids, flavonoids, coumarins, lipids, polysaccharides, and many more (Table 2), as well as other mechanisms including competition, antibiosis, parasitism, and systemic resistance induction (Fig. 4). All these inhibit the growth of plant pathogens and sometimes lead to their death (Enebe and Babalola 2018; Adeleke et al. 2021; Pathak et al. 2022; Parveen et al. 2022).

Pyocyanin destroys phytopathogens by redirecting the electron flow from the biological pathway of the organism's system, which eventually increases the oxygen reduction products and, consequently, cell death. However, some microbes are resistant to pyocyanin, owing to the level of catalase and superoxide dismutase they possess, which make them immune to such toxic compound (Marrez and Mohamad 2020). On the other hand, chitinase, β-1,4 glucanases, proteases, and β-1,3 glucanases produced by *Trichoderma* sp. are important in the degradation of pathogens cell wall, and sometimes, they interfere with their cell wall synthesis, leading to cell

Table 2 Metabolites used by microbes in the control of plant pathogens

Metabolite	Producing microbe	Pathogen targeted	Plant involved	References
Phenazine-1-carboxylic acid	*Burkholderia* sp.	*Fusarium oxysporum*	Banana	Xu et al. (2020)
Pyocyanin	*Pseudomonas aeruginosa*	*Botrytis cinerea*	Tomato	Marrez and Mohamad (2020)
2,4-diacetylphloroglucinol	*Pseudomonas fluorescens*	*Rhizoctonia solani, Ralstonia solanacearum, Macrophomina phaseolina, Sclerotium rolfsii, Fusarium oxysporum*	Tomato	Suresh et al. (2021)
Pyrrolnitrin and pyoluteorin	*Pseudomonas protegens*	*Heterobasidion* sp.	Stone pine	Pellicciaro et al. (2022)
Hydrogen cyanide	*Pseudomonas japonica, Bacillus megaterium, Pseudomonas tolaasii*	*Meloidogyne incognita, Agrobacterium tumefaciens*	Tomato	El-Rahman et al. (2019)
Siderophores	*Penicillium chrysogenum*	*Ralstonia solanacearum*	*Cymbidium aloifolium*	Chowdappa et al. (2020)
N-acyl homoserine lactonase	*Bacillus thuringiensis*	*Pseudomonas syringae*	Plum	Janakiev et al. (2020)
Kurstakin	*Bacillus* sp.	*Rhizoctonia solani, Fusarium oxysporum, Fusarium graminearum, Ascochyta citrullina*	Rice	Yu et al. (2022a)
Chitinase, ACC deaminase, and ammonia	*Pseudomonas geniculata, Burkholderia territorii, Bacillus siamensis, Burkholderia ubonensis, Brevibacillus gelatini, Pseudomonas beteli, Bacillus subtilis*	*Fusarium solani*	Black pepper	Lau et al. (2020)
Wuyiencin	*Streptomyces albulus*	*Sphaerotheca fuliginea*	Cucumber	Yang et al. (2021)
Viridin	*Trichoderma virens*	*Fusarium* sp.	Tomato	Singh et al. (2022)

(continued)

Table 2 (continued)

Metabolite	Producing microbe	Pathogen targeted	Plant involved	References
Trichodermin	*Trichoderma* sp.	*Phytophthora infestans*, *Alternaria solani*	Tomato	Borzykh et al. (2022)
Gliotoxin	*Trichoderma virens*	*Sclerotium rolfsii*	Peanut	Hua et al. (2021)
Trichodermester A	*Trichoderma atroviride*	*Pestalotiopsis theae*	Tea	Tang et al. (2021)
Harzianopyridone	*Trichoderma* sp.	*Ganoderma boninense*	Oil palm	Nusaibah and Musa (2019)
6-pentyl-αpyrone (6PP)	*Trichoderma asperellum*	*Botrytis cinerea*, *Alternaria brassicicola*	Arabidopsis	Alfiky and Weisskopf (2021)
Epl1-Tas	*Trichoderma asperellum*	*Alternaria alternata*	*Populus davidiana*, *Populus alba*	Yu et al. (2018)
Massoilactone	*Trichoderma harzianum*	*Botrytis cinerea*	Tomato	Nusaibah and Musa (2019)
Koningin	*Trichoderma* sp.	*Fusarium* spp.	Pulse plants	Mishra et al. (2020)

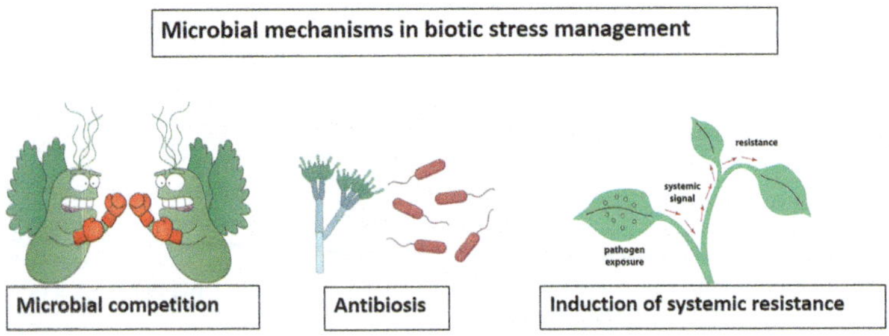

Fig. 4 Microbial mechanisms in biotic stress management

death (Shahriar et al. 2022; Tyagi et al. 2022b). Some microbes are capable of fostering plant resistance to pathogenic organisms, and this helps to promote the plant's healthy status. In this case, the pathogen is not the target but the metabolism and immune system of the plant is facilitated. For instance, exopolysaccharides produced by *Lactobacillus planetarium* have been reported to destroy *Xanthomonas gardneri* pathogen. However, this caused bacteria spot infection in tomato plant, which reduced the water movement toward the leaf stomata pore (Blainski et al. 2018).

Colletotrichum falcatum, a pathogen infection sugarcane, produces proteins that helps the plant to defend itself against the adverse effects of the metabolite. This turns possible by the production of hydrogen peroxide and callose at the infection spot, which prevents cellular lesion on the plant (Ashwin et al. 2017). Similarly, *Pseudomonas* sp. found in the tomato plant rhizosphere helps induce plant resistance by enhancing its intracellular organic compound assimilation. This prevents tomato plant from pathogens responsible of root and shoot diseases (Enebe and Babalola 2018).

Microbes produce antibiotics (low-molecular-weight organic compounds), which are harmful to plant pathogens; a reduction of plant diseases incidence is a direct result. Six major groups of antibiotic compounds are produced and utilized by microorganisms, i.e., hydrogen cyanide, phenazines, pyoluteorin, cyclic lipopeptides, phloroglucinols, and pyrrolnitrin (Saeed et al. 2021). The mode of action of the antibiotics includes the inhibition of the cell wall synthesis, prevention of the complexes on the small subunits of the ribosome, and the alteration of the cell membrane structure (Saeed et al. 2021). Microbes induce a systemic response in the plant to prevent the invasion of different pathogens and is usually undergone through the production of jasmonic acid and ethylene (Yu et al. 2022b). Jasmonates are derived from fatty acids that regulate the resistance level in plants. They can also modify the rhizosphere microbial diversity and population, whereas salicylic acid is an essential signaling molecule that increases in case of plants infection to help the latter defend against biotic stress (Yu et al. 2022b).

Furthermore, microbes can control weeds that compete with the host plant for nutrients, space, and other resources. In this regard, the mode of action of microbes includes direct interference with weeds normal metabolism through parasitism, production of metabolites, enhancement of root physiology of the host plant, increase in nutrient acquisition pattern of the host plant, alteration of the plant-pathogen signaling, and induction of the host plant systemic resistance (which lead to an improved immune system) (Masteling et al. 2019). *Pseudomonas aeruginosa* from *Momordica charantia* has been reported to control *Pennisetum purpureum, Amaranthus spinosum, Pisum sativa,* and *Oryza sativa* weeds due to their secretion of quinoline derivatives (Lawrance et al. 2019). Similarly, Bo et al. (2019) reported the biocontrol of *D. sanguinalis* weed using metabolites produced by *Streptomyces* sp. Microbes also help in the protection of plants against insect pathogens through the production of protective biofilms and metabolites and the competition with insects for resources. Particularly, *Streptomyces* sp., *Bacillus* sp., *Burkholderia* sp., *Paenibacillus* sp., *Serratia* sp., *Brevibacillus* sp., and *Pseudomonas* sp. are the major species involved in plant protection against insects (Ruiu 2020). Moreover, Kousar et al. (2020) revealed the ability of *Pseudomonas aeruginosa* and *Bacillus endophyticus* to control *Spodoptera litura.* This was actually traced to their ability to produce certain metabolites, including abscisic acid, *p*-kaempferol, salicylic acid, rutin, indole-3-acetic acid, caffeic acid, flavonoid glycoside, and *p*-coumaric acid.

5 Conclusion

Microbes play a critical role in the management of biotic stress in agriculture, and their survival in the agricultural ecosystem can be influenced by environmental factors, microbial interactions, and host plant factors. Understanding the factors affecting microbial survival and possible mechanisms is essential to better explore them for the development of sustainable agricultural systems that rely on biocontrol agents and functionality of microbes in biotic stress management. However, there is still certain gap in microbial stress management that research should fill. For that, it is urged to engineer robust microbial consortia as being more effective in the mitigation and management of different biotic stressors. Such step can help fighting biotic stresses nevertheless the environmental factors (and even abiotic stresses) and thus an improved balance of biotic/abiotic stress management.

References

Abdul Rahman NSN, Abdul Hamid NW, Nadarajah K (2021) Effects of abiotic stress on soil microbiome. Int J Mol Sci 22:9036. https://doi.org/10.3390/ijms22169036

Adeleke BS, Babalola OO (2021) The endosphere microbial communities, a great promise in agriculture. Int Microbiol 24:1–17. https://doi.org/10.1007/s10123-020-00140-2

Adeleke BS, Ayangbenro AS, Babalola OO (2021) Genomic analysis of endophytic Bacillus cereus T4S and its plant growth-promoting traits. Plants 10:1776. https://doi.org/10.3390/plants10091776

Adeleke B, Fadiji A, Ayilara M, Igiehon O, Nwachukwu B, Babalola O (2022) Strategies to enhance the use of endophytes as bioinoculants in agriculture. Horticulturae 8:498. https://doi.org/10.3390/horticulturae8060498

Adnan M, Islam W, Shabbir A, Khan KA, Ghramh HA, Huang Z, Chen HY, Lu G-D (2019) Plant defense against fungal pathogens by antagonistic fungi with Trichoderma in focus. Microb Pathog 129:7–18. https://doi.org/10.1016/j.micpath.2019.01.042

Agri U, Chaudhary P, Sharma A (2021) In vitro compatibility evaluation of agriusable nanochitosan on beneficial plant growth-promoting rhizobacteria and maize plant. Natl Acad Sci Lett 44:555–559. https://doi.org/10.1007/s40009-021-01047-w

Agri U, Chaudhary P, Sharma A, Kukreti B (2022) Physiological response of maize plants and its rhizospheric microbiome under the influence of potential bioinoculants and nanochitosan. Plant Soil 474:451–468. https://doi.org/10.1007/s11104-022-05351-2

Alfiky A, Weisskopf L (2021) Deciphering Trichoderma–plant–pathogen interactions for better development of biocontrol applications. J Fungi 7:61. https://doi.org/10.3390/jof7010061

Ashwin N, Barnabas EL, Ramesh Sundar A, Muthumeena M, Malathi P, Viswanathan R (2017) Disease suppressive effects of resistance-inducing agents against red rot of sugarcane. Eur J Plant Pathol 149:285–297. https://doi.org/10.1007/s10658-017-1181-1

Ayilara MS, Abberton M, Oyatomi OA, Babalola OO, Odeyemi O (2022a) Potentials of underutilized legumes in food security. Front Soil Sci 2:1020193. https://doi.org/10.3389/fsoil.2022.1020193

Ayilara MS, Adeleke BS, Babalola OO (2022b) Bioprospecting and challenges of plant microbiome research for sustainable agriculture, a review on soybean endophytic bacteria. Microb Ecol 85:1113–1135. https://doi.org/10.1007/s00248-022-02136-z

Blainski JML, Da Rocha Neto AC, Schimidt EC, Voltolini JA, Rossi MJ, Di Piero RM (2018) Exopolysaccharides from Lactobacillus plantarum induce biochemical and physiological alterations in tomato plant against bacterial spot. Appl Microbiol Biotechnol 102:4741–4753. https://doi.org/10.1007/s00253-018-8946-0

Bo AB, Kim JD, Kim YS, Sin HT, Kim HJ, Khaitov B, Ko YK, Park KW, Choi JS (2019) Isolation, identification and characterization of *Streptomyces* metabolites as a potential bioherbicide. PLoS One 14:e0222933. https://doi.org/10.1371/journal.pone.0222933

Borzykh OI, Sergiienko VG, Tytova LV, Biliavska LO, Boroday VV, Tkalenko GM, Balan GO (2022) Potential of some bioagents in fungal diseases controlling and productivity enhancement of tomatoes. Arch Phytopathol Plant Protect 55:1750–1765. https://doi.org/10.1080/03235408.2022.2116685

Carro-Huerga G, Compant S, Gorfer M, Cardoza RE, Schmoll M, Gutiérrez S, Casquero PA (2020) Colonization of Vitis vinifera L. by the endophyte Trichoderma sp. strain T154: biocontrol activity against Phaeoacremonium minimum. Front Plant Sci 11:1170. https://doi.org/10.3389/fpls.2020.01170

Chagas FO, De Cassia Pessotti R, Caraballo-Rodriguez AM, Pupo MT (2018) Chemical signaling involved in plant–microbe interactions. Chem Soc Rev 47:1652–1704. https://doi.org/10.1039/c7cs00343a

Chaudhary A, Parveen H, Chaudhary P, Bhatt P, Khatoon H (2021) Rhizospheric microbes and their mechanism. In: Bhatt P, Gangola S, Udayanga D, Kumar G (eds) Microbial technology for sustainable environment. Springer, Singapore. https://doi.org/10.1007/978-981-16-3840-4_6

Chaudhary P, Chaudhary A, Parveen H, Rani A, Kumar G, Kumar A, Sharma A (2021a) Impact of nanophos in agriculture to improve functional bacterial community and crop productivity. BMC Plant Biol 21:519. https://doi.org/10.1186/s12870-021-03298-7

Chaudhary P, Khati P, Chaudhary A, Gangola S, Kumar R, Sharma A (2021b) Bioinoculation using indigenous *Bacillus* spp. improves growth and yield of *Zea mays* under the influence of nanozeolite. 3 Biotech 11:11. https://doi.org/10.1007/s13205-020-02561-2

Chaudhary P, Khati P, Chaudhary A, Maithani D, Kumar G, Sharma A (2021c) Cultivable and metagenomic approach to study the combined impact of nanogypsum and *Pseudomonas taiwanensis* on maize plant health and its rhizospheric microbiome. PLoS One 16:e0250574. https://doi.org/10.1371/journal.pone.0250574

Chaudhary P, Khati P, Gangola S, Kumar A, Kumar R, Sharma A (2021d) Impact of nanochitosan and *Bacillus* spp. on health, productivity and defence response in *Zea mays* under field condition. 3 Biotech 11:237. https://doi.org/10.1007/s13205-021-02790-z

Chaudhary P, Parveen H, Gangola S, Kumar G, Bhatt P, Chaudhary A (2021e) Plant Growth Promoting Rhizobacteria and their application in sustainable crop production. In: Bhatt P, Gangola S, Udayanga D, Kumar G (eds) Microbial technology for sustainable environment. Springer, Singapore. https://doi.org/10.1007/978-981-16-3840-4_13

Chaudhary P, Sharma A, Chaudhary A, Khati P, Gangola S, Maithani D (2021f) Illumina based high throughput analysis of microbial diversity of maize rhizosphere treated with nanocompounds and *Bacillus* sp. Appl Soil Ecol 159:103836. https://doi.org/10.1016/j.apsoil.2020.103836

Chaudhary A, Chaudhary P, Upadhyay A, Kumar A, Singh A (2022) Effect of gypsum on plant growth promoting rhizobacteria. Environ Ecol 39(4A):1248–1256

Chaudhary P, Singh S, Chaudhary A, Agri U, Bhandari G (2022a) Chapter 3: Agrochemicals and their effects on soil microbial population. In: Soni R, Suyal D, Goel R (eds) Plant protection: from chemicals to biologicals. De Gruyter, Berlin, Boston, pp 45–58. https://doi.org/10.1515/9783110771558-003

Chaudhary P, Agri U, Chaudhary A, Kumar A, Kumar G (2022b) Endophytes and their potential in biotic stress management and crop production. Front Microbiol 13:933017. https://doi.org/10.3389/fmicb.2022.933017

Chaudhary P, Chaudhary A, Agri U, Khatoon H, Singh A (2022c) Recent trends and advancements for agro-environmental sustainability at higher altitudes. In: Goel R, Soni R, Suyal DC, Khan

M (eds) Survival strategies in cold-adapted microorganisms. Springer, Singapore. https://doi.org/10.1007/978-981-16-2625-8_19

Chaudhary P, Singh S, Chaudhary A, Sharma A, Kumar G (2022d) Overview of biofertilizers in crop production and stress management for sustainable agriculture. Front Plant Sci 13:930340. https://doi.org/10.3389/fpls.2022.930340

Chaudhary P, Xu M, Ahamad L, Chaudhary A, Kumar G, Adeleke BS, Verma KK, Hu D-M, Širić I, Kumar P, Popescu SM, Abou Fayssal S (2023a) Application of synthetic consortia for improvement of soil fertility, pollution remediation, and agricultural productivity: a review. Agronomy 13:643. https://doi.org/10.3390/agronomy13030643

Chaudhary P, Chen S, Rajput VD, Jaiswal DK (2023b) Editorial: bioinoculants with nano-compounds to improve soil health: a step toward sustainable agriculture. Front Environ Sci 11:1270002. https://doi.org/10.3389/fenvs.2023.1270002

Chaudhary P, Chen S, Rajput VD, Jaiswal DK (eds) (2023c) Bioinoculants with nano-compounds to improve soil health: a step toward sustainable agriculture. Frontiers Media SA, Lausanne. https://doi.org/10.3389/978-2-8325-3287-4

Chen C, Cao Z, Li J, Tao C, Feng Y, Han Y (2020) A novel endophytic strain of Lactobacillus plantarum CM-3 with antagonistic activity against *Botrytis cinerea* on strawberry fruit. Biol Control 148:104306. https://doi.org/10.1016/j.biocontrol.2020.104306

Chowdappa S, Jagannath S, Konappa N, Udayashankar AC, Jogaiah S (2020) Detection and characterization of antibacterial siderophores secreted by endophytic fungi from *Cymbidium aloifolium*. Biomolecules 10:1412. https://doi.org/10.3390/biom10101412

Cuartas-Otálora PE, Gómez-Valderrama JA, Ramos AE, Barrera-Cubillos GP, Villamizar-Rivero LF (2019) Bio-insecticidal potential of nucleopolyhedrovirus and granulovirus mixtures to control the fall armyworm Spodoptera frugiperda (JE Smith, 1797) (Lepidoptera: Noctuidae). Viruses 11:684. https://doi.org/10.3390/v11080684

Deka B, Baruah C, Babu A (2021) Entomopathogenic microorganisms: their role in insect pest management. Egypt J Biol Pest Control 31:121. https://doi.org/10.1186/s41938-021-00466-7

Díaz-Díaz M, Bernal-Cabrera A, Trapero A, Medina-Marrero R, Sifontes-Rodríguez S, Cupull-Santana RD, García-Bernal M, Agustí-Brisach C (2022) Characterization of actinobacterial strains as potential biocontrol agents against Macrophomina phaseolina and Rhizoctonia solani, the main soil-borne pathogens of Phaseolus vulgaris in Cuba. Plants 11:645. https://doi.org/10.3390/plants11050645

Ekwealor K, Echereme C, Ofobeze T, Okereke C (2019) Economic importance of weeds: a review. Asian Plant Res J 3:1–11. https://doi.org/10.9734/aprj/2019/v3i230063

El-Rahman A, Shaheen HA, El-Aziz A, Rabab M, Ibrahim DS (2019) Influence of hydrogen cyanide-producing rhizobacteria in controlling the crown gall and root-knot nematode, Meloidogyne incognita. Egypt J Biol Pest Control 29:41. https://doi.org/10.1186/s41938-019-0143-7

El-Saadony MT, Saad AM, Soliman SM, Salem HM, Ahmed AI, Mahmood M, El-Tahan AM, Ebrahim AA, El-Mageed A, Negm SH (2022) Plant growth-promoting microorganisms as biocontrol agents of plant diseases: mechanisms, challenges and future perspectives. Front Plant Sci 13:923880. https://doi.org/10.3389/fpls.2022.923880

Enebe MC, Babalola OO (2018) The influence of plant growth-promoting rhizobacteria in plant tolerance to abiotic stress: a survival strategy. Appl Microbiol Biotechnol 102:7821–7835. https://doi.org/10.1007/s00253-018-9214-z

Fatima S, Anjum T (2017) Identification of a potential ISR determinant from Pseudomonas aeruginosa PM12 against Fusarium wilt in tomato. Front Plant Sci 8:848. https://doi.org/10.3389/fpls.2017.00848

Fontana DC, De Paula S, Torres AG, De Souza VHM, Pascholati SF, Schmidt D, Dourado Neto D (2021) Endophytic fungi: biological control and induced resistance to phytopathogens and abiotic stresses. Pathogens 10:570. https://doi.org/10.3390/pathogens10050570

Garzo E, Moreno A, Plaza M, Fereres A (2020) Feeding behavior and virus-transmission ability of insect vectors exposed to systemic insecticides. Plants 9:895. https://doi.org/10.3390/plants9070895

Gull A, Lone AA, Wani NUI (2019) Biotic and abiotic stresses in plants. In: de Oliveira AB (ed) Abiotic and biotic stress in plants. IntechOpen, London, pp 1–19. https://doi.org/10.5772/intechopen.85832

Gupta A, Gupta R, Singh RL (2017) Microbes and environment. In: Singh RL (ed) Principles and applications of environmental biotechnology for a sustainable future. Springer, Singapore, pp 43–84. https://doi.org/10.1007/978-981-10-1866-4_3

Hua L, Zeng H, He L, Jiang Q, Ye P, Liu Y, Sun X, Zhang M (2021) Gliotoxin is an important secondary metabolite involved in suppression of Sclerotium rolfsii of Trichoderma virens T23. Phytopathology 111:1720–1725. https://doi.org/10.1094/PHYTO-09-20-0399-R

Iqbal Z, Iqbal MS, Hashem A, Abdallah EF, Ansari MI (2021) Plant defense responses to biotic stress and its interplay with fluctuating dark/light conditions. Front Plant Sci 12:631810. https://doi.org/10.3389/fpls.2021.631810

Jacoby R, Peukert M, Succurro A, Koprivova A, Kopriva S (2017) The role of soil microorganisms in plant mineral nutrition—current knowledge and future directions. Front Plant Sci 8:1617. https://doi.org/10.3389/fpls.2017.01617

Janakiev T, Dimkić I, Bojić S, Fira D, Stanković S, Berić T (2020) Bacterial communities of plum phyllosphere and characterization of indigenous antagonistic *Bacillus thuringiensis* R3/3 isolate. J Appl Microbiol 128:528–543. https://doi.org/10.1111/jam.14488

Kaur S, Samota MK, Choudhary M, Choudhary M, Pandey AK, Sharma A, Thakur J (2022) How do plants defend themselves against pathogens-biochemical mechanisms and genetic interventions. Physiol Mol Biol Plants 28:485–504. https://doi.org/10.1007/s12298-022-01146-y

Khalil MMR, Fierro-Coronado RA, Peñuelas-Rubio O, Villa-Lerma AG, Plascencia-Jatomea R, Félix-Gastélum R, Maldonado-Mendoza IE (2021) Rhizospheric bacteria as potential biocontrol agents against Fusarium wilt and crown and root rot diseases in tomato. Saudi J Biol Sci 28:7460–7471. https://doi.org/10.1016/j.sjbs.2021.08.043

Khati P, Parul, Bhatt P, Nisha Kumar R, Sharma A (2018) Effect of nanozeolite and plant growth promoting rhizobacteria on maize. 3 Biotech 8:141. https://doi.org/10.1007/s13205-018-1142-1

Khati P, Chaudhary P, Gangola S, Sharma P (2019) Influence of nanozeolite on plant growth promotory bacterial isolates recovered from nanocompound infested agriculture field. Environ Ecol 37:521–527

Kohl J, Kolnaar R, Ravensberg WJ (2019) Mode of action of microbial biological control agents against plant diseases: relevance beyond efficacy. Front Plant Sci 10:845. https://doi.org/10.3389/fpls.2019.00845

Kousar B, Bano A, Khan N (2020) PGPR modulation of secondary metabolites in tomato infested with *Spodoptera litura*. Agronomy 10:778. https://doi.org/10.3390/agronomy10060778

Kukreti B, Sharma A, Chaudhary P, Agri U, Maithani D (2020) Influence of nanosilicon dioxide along with bioinoculants on *Zea mays* and its rhizospheric soil. 3 Biotech 10:345. https://doi.org/10.1007/s13205-020-02329-8

Kumar A, Singh S, Gaurav AK, Srivastava S, Verma JP (2020) Plant growth-promoting bacteria: biological tools for the mitigation of salinity stress in plants. Front Microbiol 11:1216. https://doi.org/10.3389/fmicb.2020.01216

Kumari S, Sharma A, Chaudhary P, Khati P (2020) Management of plant vigor and soil health using two agriusable nanocompounds and plant growth promotory rhizobacteria in Fenugreek. 3 Biotech 10:1–11. https://doi.org/10.1007/s13205-020-02448-2

Kumari H, Khati P, Gangola S, Chaudhary P, Sharma A (2021) Performance of plant growth promotory rhizobacteria on Maize and soil characteristics under the influence of TiO_2 nanoparticles. Pantnagar J Res 19:28–39

Lahlali R, Ezrari S, Radouane N, Kenfaoui J, Esmaeel Q, El Hamss H, Belabess Z, Barka EA (2022) Biological control of plant pathogens: a global perspective. Microorganisms 10:596. https://doi.org/10.3390/microorganisms10030596

Lau ET, Tani A, Khew CY, Chua YQ, San Hwang S (2020) Plant growth-promoting bacteria as potential bio-inoculants and biocontrol agents to promote black pepper plant cultivation. Microbiol Res 240:126549. https://doi.org/10.1016/j.micres.2020.126549

Lawrance S, Varghese S, Varghese EM, Asok AK (2019) Quinoline derivatives producing Pseudomonas aeruginosa H6 as an efficient bioherbicide for weed management. Biocatal Agric Biotechnol 18:101096. https://doi.org/10.1016/j.bcab.2019.101096

Li H, Guan Y, Dong Y, Zhao L, Rong S, Chen W, Lv M, Xu H, Gao X, Chen R (2018) Isolation and evaluation of endophytic *Bacillus tequilensis* GYLH001 with potential application for biological control of *Magnaporthe oryzae*. PLoS One 13:e0203505. https://doi.org/10.1371/journal.pone.0203505

Liu Y, Yang Y, Wang B (2022) Entomopathogenic fungi *Beauveria bassiana* and *Metarhizium anisopliae* play roles of maize (*Zea mays*) growth promoter. Sci Rep 12:15706. https://doi.org/10.1038/s41598-022-19899-7

Madbouly AK, Abo Elyousr KAM, Ismail IM (2020) Biocontrol of *Monilinia fructigena*, causal agent of brown rot of apple fruit, by using endophytic yeasts. Biol Control 144:104239. https://doi.org/10.1016/j.biocontrol.2020.104239

Manzar N, Jayaswal D, Kishan G, Chauhan S, Joshi D, Singh J, Chaudhary P, Jamali H, Kushwaha P, Parveen H, Kashyap AS, Sahu PK, Singh UB (2022) Application of molecular ecology approaches in sustainable agriculture for a better understanding of plant–microbiome interactions. In: Singh UB, Sahu PK, Singh HV, Sharma PK, Sharma SK (eds) Rhizosphere microbes. Microorganisms for sustainability, vol 40. Springer, Singapore. https://doi.org/10.1007/978-981-19-5872-4_3

Marrez DA, Mohamad HS (2020) Biological activity and applications of pyocyanin produced by Pseudomonas aeruginosa. J Biomed Sci 2:OAJBS.ID.000133. https://doi.org/10.38125/OAJBS.000133

Masteling R, Lombard L, De Boer W, Raaijmakers JM, Dini-Andreote F (2019) Harnessing the microbiome to control plant parasitic weeds. Curr Opin Microbiol 49:26–33. https://doi.org/10.1016/j.mib.2019.09.006

Mishra S, Upadhyay RS, Nautiyal CS (2013) Unravelling the beneficial role of microbial contributors in reducing the allelopathic effects of weeds. Appl Microbiol Biotechnol 97:5659–5668. https://doi.org/10.1007/s00253-013-4885-y

Mishra R, Pandey S, Mishra M, Rathore U, Kumar K, Singh B (2020) Assessment of biocontrol potential of Trichoderma isolates against wilt in pulses. J Food Legumes 33:48–52

Moënne-Loccoz Y, Mavingui P, Combes C, Normand P, Steinberg C (2015) Microorganisms and biotic interactions. In: Bertrand J-C, Caumette P, Lebaron P, Matheron R, Normand P, Sime-Ngando T (eds) Environmental microbiology: fundamentals and applications: microbial ecology. Springer, Dordrecht, pp 395–444

Mon YY, Bidabadi SS, Oo KS, Zheng S-J (2021) The antagonistic mechanism of rhizosphere microbes and endophytes on the interaction between banana and *Fusarium oxysporum* f. sp. cubense. Physiol Mol Plant Pathol 116:101733. https://doi.org/10.1016/j.pmpp.2021.101733

Moustafa-Farag M, Almoneafy A, Mahmoud A, Elkelish A, Arnao MB, Li L, Ai S (2019) Melatonin and its protective role against biotic stress impacts on plants. Biomolecules 10:54. https://doi.org/10.3390/biom10010054

Nguyen SD, Trinh THT, Tran TD, Nguyen TV, Chuyen HV, Ngo VA, Nguyen AD (2020) Combined application of rhizosphere bacteria with endophytic bacteria suppresses root diseases and increases productivity of black pepper (*Piper nigrum* L.). Agriculture 11:15. https://doi.org/10.3390/agriculture11010015

Nusaibah SA, Musa H (2019) Review report on the mechanism of *Trichoderma* spp. as biological control agent of the Basal Stem Rot (BSR) disease of *Elaeis guineensis*. In: Shah MM, Sharif U, Buhari TR (eds) *Trichoderma*. IntechOpen, London, pp 79–90

Parveen H, Chaudhary A, Chaudhary P, Sultana R, Khati P, Kumar G, Khati P, Rana M, Bhatt P (2022) Omics approaches for the production of the microbial enzymes and applications. In: Bhatt P (ed) Industrial applications of microbial enzymes, 1st edn. CRC Press. https://doi.org/10.1201/9781003202998-16

Pathak P, Rai VK, Can H, Singh SK, Kumar D, Bhardwaj N, Roychowdhury R, de Azevedo LCB, Verma H, Kumar A (2022) Plant-endophyte interaction during biotic stress management. Plants 11:2203. https://doi.org/10.3390/plants11172203

Pellicciaro M, Padoan E, Lione G, Celi L, Gonthier P (2022) Pyoluteorin produced by the biocontrol agent pseudomonas protegens is involved in the inhibition of Heterobasidion species present in Europe. Pathogens 11:391. https://doi.org/10.3390/pathogens11040391

Ruiu L (2018) Microbial biopesticides in agroecosystems. Agronomy 8:235. https://doi.org/10.3390/agronomy8110235

Ruiu L (2020) Plant-growth-promoting bacteria (PGPB) against insects and other agricultural pests. Agronomy 10:861. https://doi.org/10.3390/agronomy10060861

Saeed Q, Xiukang W, Haider FU, Kučerik J, Mumtaz MZ, Holatko J, Naseem M, Kintl A, Ejaz M, Naveed M (2021) Rhizosphere bacteria in plant growth promotion, biocontrol, and bioremediation of contaminated sites: a comprehensive review of effects and mechanisms. Int J Mol Sci 22:10529. https://doi.org/10.3390/ijms221910529

Shahriar SA, Islam MN, Chun CNW, Kaur P, Rahim MA, Islam MM, Uddain J, Siddiquee S (2022) Microbial metabolomics interaction and ecological challenges of *Trichoderma* species as biocontrol inoculant in crop rhizosphere. Agronomy 12:900. https://doi.org/10.3390/agronomy12040900

Singh A, Bhardwaj R, Singh IK (2019) Biocontrol agents: potential of biopesticides for integrated pest management. In: Prasad R, Wu QS, Varma A (eds) Biofertilizers for sustainable agriculture and environment; Giri B. Springer, Cham, pp 413–433

Singh G, Tiwari A, Choudhir G, Kumar A, Kumar S, Hariprasad P, Sharma S (2022) Deciphering the role of *Trichoderma* sp. bioactives in combating the wilt causing cell wall degrading enzyme polygalacturonase produced by *Fusarium oxysporum*: an in-silico approach. Microb Pathog 168:105610. https://doi.org/10.1016/j.micpath.2022.105610

Smercina DN, Evans SE, Friesen ML, Tiemann LK (2019) To fix or not to fix: controls on free-living nitrogen fixation in the rhizosphere. Appl Environ Microbiol 85:2546–2618. https://doi.org/10.1128/aem.02546-18

Suresh P, Varathraju G, Shanmugaiah V, Almaary KS, Elbadawi YB, Mubarak A (2021) Partial purification and characterization of 2, 4-diacetylphloroglucinol producing *Pseudomonas fluorescens* VSMKU3054 against bacterial wilt disease of tomato. Saudi J Biol Sci 28:2155–2167. https://doi.org/10.1016/j.sjbs.2021.02.073

Tang X-X, Liu S-Z, Sun Y-Y, He F-M, Xu G-X, Fang M-J, Zhen W, Qiu Y-K (2021) New cyclopentenoneacrylic acid derivatives from a marine-derived fungus *Trichoderma atroviride* H548. Nat Prod Res 35:3772–3779. https://doi.org/10.1080/14786419.2020.1737053

Teshome DT, Zharare GE, Naidoo S (2020) The threat of the combined effect of biotic and abiotic stress factors in forestry under a changing climate. Front Plant Sci 1:1874. https://doi.org/10.3389/fpls.2020.601009

Thambugala KM, Daranagama DA, Phillips AJ, Kannangara SD, Promputtha I (2020) Fungi vs. fungi in biocontrol: an overview of fungal antagonists applied against fungal plant pathogens. Front Cell Infect Microbiol 10:604923. https://doi.org/10.3389/fcimb.2020.604923

Tyagi J, Chaudhary P, Jyotsana, Bhagwati U, Bhandari G, Chaudhary A (2022a) Chapter 17: Impact of endophytic fungi in biotic stress management. In: Soni R, Suyal D, Goel R (eds) Plant protection: from chemicals to biologicals. De Gruyter, Berlin, Boston, pp 447–462. https://doi.org/10.1515/9783110771558-017

Tyagi J, Chaudhary P, Mishra A, Khatwani M, Dey S, Varma A (2022b) Role of endophytes in abiotic stress tolerance: with special emphasis on *Serendipita indica*. Int J Environ Res 16:62. https://doi.org/10.1007/s41742-022-00439-0

Usta C (2013) Microorganisms in biological pest control—a review (bacterial toxin application and effect of environmental factors). Curr Prog Biol Res 13:287–317. https://doi.org/10.5772/55786

Verma D, Banjo T, Chawan M, Teli N (2020) Microbial control of pests and weeds. In: Egbuna C, Sawicka B (eds) Natural remedies for pest, disease and weed control. Academic, Cambridge, pp 119–126

Xu W, Wang F, Zhang M, Ou T, Wang R, Strobel G, Xiang Z, Zhou Z, Xie J (2019) Diversity of cultivable endophytic bacteria in mulberry and their potential for antimicrobial and plant growth-promoting activities. Microbiol Res 229:126328. https://doi.org/10.1016/j.micres.2019.126328

Xu Z, Wang M, Du J, Huang T, Liu J, Dong T, Chen Y (2020) Isolation of Burkholderia sp. HQB-1, a promising biocontrol bacteria to protect banana against Fusarium wilt through phenazine-1-carboxylic acid secretion. Front Microbiol 11:605152. https://doi.org/10.3389/fmicb.2020.605152

Yang M, Wei Q, Shi L, Wei Z, Lv Z, Asim N, Zhang K, Ge B (2021) Wuyiencin produced by *Streptomyces albulus* CK-15 displays biocontrol activities against cucumber powdery mildew. J Appl Microbiol 131:2957–2970. https://doi.org/10.1111/jam.15168

Yin C, Casa Vargas JM, Schlatter DC, Hagerty CH, Hulbert SH, Paulitz TC (2021) Rhizosphere community selection reveals bacteria associated with reduced root disease. Microbiome 9:1–18. https://doi.org/10.1186/s40168-020-00997-5

Yu W, Mijiti G, Huang Y, Fan H, Wang Y, Liu Z (2018) Functional analysis of eliciting plant response protein Epl1-Tas from *Trichoderma asperellum* ACCC30536. Sci Rep 8:1–13. https://doi.org/10.1038/s41598-018-26328-1

Yu Y, Zhang Y, Wang T, Huang T, Tang S, Jin Y, Mi D, Zheng Y, Niu D, Guo J-H (2022a) Kurstakin triggers multicellular behaviors in *Bacillus cereus* AR156 and enhances disease control efficacy against rice sheath blight. Plant Dis 107:1463–1470. https://doi.org/10.1094/PDIS-01-22-0078-RE

Yu Y, Gui Y, Li Z, Jiang C, Guo J, Niu D (2022b) Induced systemic resistance for improving plant immunity by beneficial microbes. Plants 11:386. https://doi.org/10.3390/plants11030386

Zoz F, Iaconelli C, Lang E, Iddir H, Guyot S, Grandvalet C, Gervais P, Beney L (2016) Control of relative air humidity as a potential means to improve hygiene on surfaces: a preliminary approach with *Listeria monocytogenes*. PLoS One 11:e0148418. https://doi.org/10.1371/journal.pone.0148418

Microbial Inoculants and Their Role in Abiotic Stress Management

Himanshu Rawal, Yukti Nautiyal, Barkha Sharma, and Shalini Tiwari

Abstract Along with exponentially burgeoning population, abiotic stresses like temperature, drought, heavy metal toxicity, and salinity are complementing the continuous pressure on agricultural productivity. Plants have developed various stress tolerance mechanisms, but there is still scarcity of food supply. To overcome this constraint, plant microbiome as plant growth-promoting microbiome (PGPM) has been explored lately, which shows enormous metabolic potential under abiotic stresses. PGPMs including rhizobacteria and mycorrhizae regulate stress alleviation by indirect and direct mechanisms, such as stimulating the production of phytohormone, siderophore, antioxidants, and phosphate solubilisation and reducing ethylene levels. They are employed in the field either solid, liquid, metabolite, or polymeric formulations, depending on the crop requirements. In this chapter, we explored the nature of abiotic stress-induced plant-microbiome interaction in terms of crop protection for translating higher productivity.

Keywords Abiotic stress · Direct and indirect mechanism · Plant microbiome · Microbial formulations · Crop productivity

1 Introduction

Abiotic stress resulting from unfavourable climatic conditions is one of the fundamental contributors preventing enhancement in agricultural output (Grayson 2013). Only 3.5% of the world's geographical area is said to be untroubled via environmental constraints, according to a 2007 FAO assessment.

H. Rawal · Y. Nautiyal
Department of Botany and Microbiology, H. N. B. Garhwal University, Srinagar, Uttarakhand, India

B. Sharma (✉) · S. Tiwari
Department of Microbiology, College of Basic Sciences & Humanities, G. B. Pant University of Agriculture & Technology, Pantnagar, Uttarakhand, India

P. Chaudhary, A. Chaudhary (eds.), *Microbial Inoculants*, Interdisciplinary Biotechnological Advances, https://doi.org/10.1007/978-981-97-0633-4_8

The most significant cause of the decline in agricultural yield of crops, especially in tropical regions, is increased rates of biotic and abiotic stress caused by both natural and human causes (Tyagi et al. 2022a). Submersion, low and high temperature, light intensity, soil pH, drought, anaerobiosis, nutrient deficiency, etc. are the most common abiotic stressors. The issues of soil degradation, erosion, and salinity impact 3.6 billion ha of the 5.2 billion hectares of dryland agriculture in the world (Riadh et al. 2010). The key factor contributing to reduced production in major crops is the increased prevalence of biotic and abiotic stressors. In many sections of the world, there is documentation of yield declines in a variety of crops due to escalating drought, a decreased precipitation rate, salt stress, and rising air temperatures. Abiotic stresses can affect an organism's physiology directly or indirectly by altering its metabolism, growth, and development (Tyagi et al. 2022b). Abiotic stressors are usually believed to be the primary cause of yield loss.

Abiotic stresses have varied effects on plants and lower agricultural output. Applying affordable solutions to manage stress situations is necessary to improve plant productivity (Tyagi et al. 2022b). When used effectively, soil microorganisms have demonstrated remarkable qualities that can aid agriculture in enhancing and maintaining crop output (Chaudhary et al. 2022a, c). While it is generally known that beneficial bacteria may boost productivity and drive development through processes including boosting nutrient availability, hormone synthesis, and disease management, it is also becoming more and more obvious that their benefits may be more extensive (Kumari et al. 2021; Chaudhary et al. 2022b).

Millions of living organisms make up the soil, which makes it a dynamic and living system. The intricate arrangement of soil particles and pore spaces, which are filled with air and water and contain millions of microorganisms, ranges in size from submicroscopic to macroscopic. In addition to aiding in soil development, these organisms are also the main force behind nutrient cycling, governing the dynamics of soil organic matter, raising the amount of nutrients that vegetation can absorb, conferring stress tolerance, exhibiting pathogen resistance, and enhancing plant health (Chaudhary et al. 2021b, d). The capacity for the production of enough food to achieve the requirement of world's population, which is expanding exponentially, is a great challenge for mankind in the twenty-first century. This problem is made more difficult by the loss of arable farmland brought on by human habitation, the degradation of the soil, and a range of environmental conditions, inclusive of floods, drought, alkalinity, temperature (high and low), agrochemicals, and heavy metal concentration (Chaudhary et al. 2022e). Microbial inoculants, such as bacteria, algae, and fungus, can promote plant development (Chaudhary et al. 2021e, 2022d). The major reason why microorganisms are regarded harmful is because they may cause diseases. Some, nonetheless, are useful in agricultural practices and are currently being employed in the growth of viable food crops. The useful microbes, which are employed as bio-inoculants to enhance interactions between plant and microbes, might come from two different sources:

2 Components of Microbiome

2.1 *Plant Microbiome*

Several different types of microbes live within plants. The term "plant microbiome" alludes to the whole genome of microorganisms, which are linked with the flora. The interactions of the plant microbiome have an impact on the crop production and health by giving the host plant a variety of advantages, including nutrient transformation, organic matter decomposition, disease suppression, nutrient absorption aid, and weed control (Manzar et al. 2022). In order for plants and the microbiome to operate and exhibit certain characteristics, they must work together. Plants fix carbon (C) through photosynthetic processes, which is then released into the environment and utilised as food by the soil biota.

2.2 *Rhizospheric Microbiome*

Compared to plants, the rhizospheric microflora that inhabits the area around their roots has a greater total genome (Mendes et al. 2013). The term "rhizospheric microbiome" allude to this. The most intricate interactions between microbes and plants take place in the rhizosphere, which is the region between soil and plant rhizomes. One gram of rhizospheric soil can contain numerous microorganisms with widespread characteristics. These microorganisms influence the plant microbiome in diverse ways to increase agricultural output (Chaudhary et al. 2023b, c). As their usage lessens reliance on inorganic fertilisers, the utilisation of advantageous microorganisms as inoculants for sustainable agricultural yield has come into emphasis in study. By taking advantage of the advantageous interactions of plant microbiome, crop output can be increased using microbial inoculants (Alori and Fawole 2017).

Naturally occurring populations of soil microorganisms (SMs) are crucial for soil fertility, plant development, and maintaining a healthy environment (Kumari et al. 2020; Agri et al. 2022). Numerous microorganisms, including bacteria, actinomycetes, cyanobacteria, and fungus, may be present in this microbial community. Some of them are thought to be effective because they can promote growth (Kukreti et al. 2020). Among these naturally existing populations, PGPR have encountered the most recognition owing to the beneficial impacts they have on plant development and environmental protection from numerous hazards (Chaudhary et al. 2023c, d). Free-living bacteria, called plant growth-promoting rhizobacteria (PGPR), encourage development of plant by colonising the roots. These are also known as nodule-promoting bacteria (NPB) or plant health-promoting bacteria (PHPB) (Hayat et al. 2010). Depending on how close to the host plant they are, they can be further classified as intracellular PGPR (iPGPR) or extracellular PGPR (ePGPR).

3 Stress under Changing Climatic Conditions

When a plant is exposed to a set of either biotic and abiotic factors, which are unsuitable for its growth during any of its various growth phases, from germination to fruiting, it may not only have a detrimental impact on its growth and productivity but also completely stop that growth and, as a result, its productivity (Tyagi et al. 2022a). In response, the plant may alter its physiological, molecular, morphological, or all of these characteristics. A plant's reaction to a variety of cellular and molecular processes that are provoked by the recognition of an adversity by the plant is referred to as a plant stress. The plant's capacity for a decrease or rise in plant metabolites that results in an increase in plant tolerance may be present in conjunction with these processes. These stresses can be biotic, which refers to the harm caused to a living entity by other living things like pathogens or herbivores, fungi, viruses, insects, parasites, bacteria, weeds, and cultivated or native plants, or they can be abiotic, such as drought (water scarcity or salinity), nutrient deficiency, excess water or flooding, drastic cold, frosting, heat, alkalinity, and heavy metal concentration (Tyagi et al. 2022b).

3.1 Abiotic Stress

Environmental stress known as abiotic stress restricts plant growth and metabolism (Pathak et al. 2021). Abiotic stressors are thought to have a more than 50% negative influence on the production and yields of important food and cash crops (Cai et al. 2020). Abiotic stresses may be divided into two categories: below- and aboveground abiotic stresses. Stresses that are caused by the atmosphere are known as atmospheric-induced abiotic stresses, whereas edaphic stresses take place in the soil (Reynaert et al. 2021).

In regions where precipitation patterns and climatic variability fluctuate with extended spells of drought interspersed with periods of copious rain, atmospheric-induced abiotic stresses are frequent (Rahman et al. 2021). Also, anthropogenic exercises like the utilisation of sewage and brackish water for watering, sludge for fertilisation, and inorganic pesticides for fumigation may result in abiotic pressures of edaphic origin (Ma et al. 2020). This issue is frequently made worse by bad cultural practises, the weathering of local rocks, and incorrect waste disposal techniques that have made vast tracts of land unusable for agricultural production (Mohammadi et al. 2021). It is well recognised that abiotic stresses set off a chain of molecular occurrences that change the biochemistry, physiology, and morphology of plants (Brito et al. 2019) (Fig. 1).

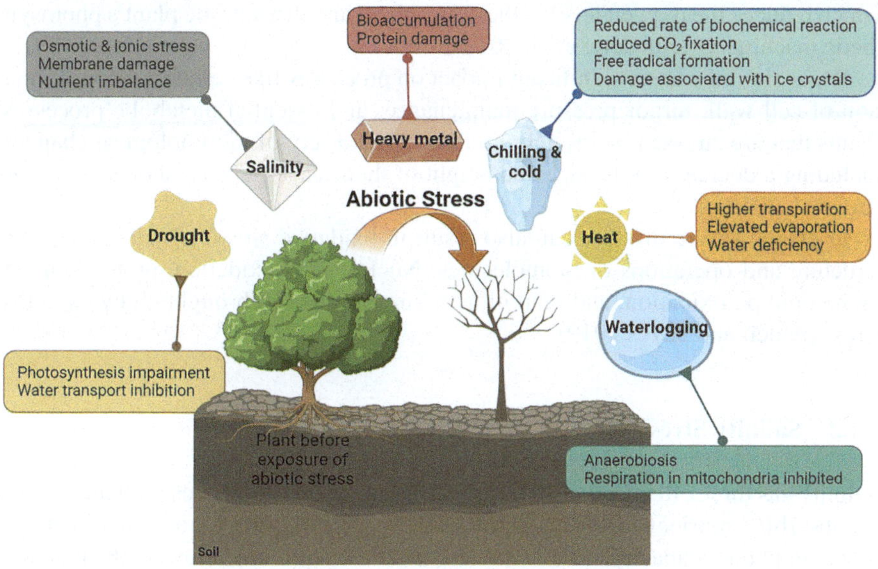

Fig. 1 Diverse abiotic stresses and their impact on the plant morphology, physiology, and biochemistry. Abiotic factors have negative effects on plants' growth, quantity, and quality; these effects can reduce plant productivity and permanently damages the plants when exposed for a longer period. Though the results of temperature, drought and salinity are different, the physiological responses are almost same

3.1.1 Drought Stress

A time without significant rainfall is referred to as a drought in meteorology, which influences 26% of productive lands and is regarded as a key constraining element amidst other abiotic pressures (Ross et al. 2020). In semiarid and arid regions, it limits agricultural productivity and yield stability (Bloomfield et al. 2019). It is a word used in meteorology to describe a stretch of time without much rain (Singh et al. 2020). According to statistics, one-third of the world's population is inhabited in dry regions (Ross et al. 2020). A region's soil's ability to hold moisture and evaporative needs both affect drought (Zhang et al. 2020). On top of reducing yield and agricultural production, drought stress limits plants' capability to devour water and nutrients from the soil, inhibits photosynthesis, and decreases endogenous levels of cytokinin while raising abscisic acid levels in the leaves, which causes the closure of stomata (Cernusak 2020). In response, plants minimise water loss by early closure of the stomata. However, this reaction has detrimental repercussions during gaseous exchange of oxygen (O_2), carbon dioxide (CO_2), and water vapour from the air as it also occurs through stomata (Coutinho et al. 2021). For example, CO_2 restriction results in limited photosynthetic fixation in plants as the drop in CO_2 has an impact on photosynthesis. Additionally, stomata closure slows down the water cycle by causing the production of peroxides (H_2O_2) and superoxide (O_2-), which reduces the oxygen level in the atmosphere and disrupts the plant's photosynthetic machinery (Ojuederie et al. 2019).

Water shortage has a significant impact on processes like cell expansion, formation of cell wall, turgor pressure maintenance, and essential metabolic processes. Plants that are stressed by drought go through a variety of morphological changes, including a decrease in the size and height of their leaves, stems, shoots, and roots (Ito et al. 2019).

In plants, severe drought can also result in oxidative stress, which changes the structure and operations of biomolecules. Nucleic acid oxidation, proteins, membrane lipid peroxidation, and inhibition of enzymes are all brought on by oxidative stress (Ameh and Sayes 2019).

3.1.2 Salinity Stress

Salinity has three effects on plants. Limiting water intake interferes with the process for absorbing nutrients and causes ion cytotoxicity. First, salt limits water intake by reducing plants' capacity to devour water from the soil, which lowers the potential water in the soil. In addition, it disrupts the plant's system for absorbing nutrients, leading to nutritional imbalance in plants (Tiwari et al. 2023).

Salt stress for long period elevates ions in plants, raising ion concentration—is extremely harmful to plants—and causing ion cytotoxicity (Ketehouli et al. 2019). High salt concentrations in plants cause membrane lipid peroxidation and hazardous Na + concentrations, which have a significant impact on the concentrations of both Cl- and Na+. Potassium (K+) ion absorption, which is crucial for physiological functions in plants such protein synthesis, osmosis, and activity of enzymes, can be inhibited by elevated Na + concentrations. Ionic disparity, deficit in protein functioning, and conformational alterations in metabolic activities in plants are all caused by K+ deficiency (Imran et al. 2021).

Together with yellow patches (necrosis) on the leaf margins, plants under high salt stress also exhibit reduced root and stem growth, stunted shoot, budding, as well as degraded fruit colour and flavour. These results are comparable to the one that drought stress has on plants, which slows growth and lowers yield (Cai et al. 2020).

3.1.3 Temperature Stress

Temperature stress is a significant barrier to agricultural productivity. Plants are subjected to numerous forms of freezing stress when temperatures drop below zero. Osmotic damage, desiccation, loss of stomatal control, pigment alteration reducing photosynthetic apparatus efficiency, fluorescence decrease, and damaged chloroplast apparatus are a few of these. Temperature changes can affect plants as well (Balliu et al. 2021). It has been demonstrated that plants experience heat stress when the environmental temperature suddenly rises by 5–7 °C. According to reports, these temperatures cause photosynthesis to be disrupted, plant water to be reduced, blooming and fruiting to be hampered, and pests and diseases to be attracted (Ljubej et al. 2021). In both temperate and tropical parts of the world, impacts of temperature on plants have been documented. For instance, plants undergo chilling stress, a nonfreezing kind of temperature stress, when the temperature decreases from 15 to 0 °C (Diamond et al. 2019). Susceptible plants are particularly vulnerable to the effects of this stress, which include a variety of physical limitations (membrane stability), osmotic stress, elevated respiration and ethylene synthesis, chlorophyll loss and reduced photosynthesis, irregular ripening, disease sensitivity, and water soaking (Sharma et al. 2023a).

3.1.4 Heavy Metal Stress

Heavy element pollution of soil is a type of mineral toxicity that has physiological significance for plants. Mining, parent rocks, and metal processing are all accountable for the escalated concentration of heavy metals in soils (Rai et al. 2021). Reduced photosynthetic, elevated heavy metal absorption, and the disturbance in root hydraulic conductivity via symplastic and apoplastic routes are examples of the often-observed impacts of heavy element toxicity in plants. Different heavy metals found in soil, including lead, mercury, copper, arsenic, zinc, chromium, cadmium, iron, and nickel, are vital to plant growth and development. The main heavy metal sources are releases from natural processes such volcanic activities, continental dust, and anthropogenic activities (Dumanovic et al. 2021). High levels of heavy elements have a detrimental effect on plant development, because they have toxic effects that prevent plants from absorbing nutrients, disrupt membrane integrity, and interfere with plant cell enzyme activity. Heavy metal poisoning can cause oxidative stress and reactive oxygen species, which disrupts activity of enzymes and block the functional groups of metabolites inside plants. Because iron (Fe2+) interacts strongly with oxygen to create ROS, a high quantity of this metal in plant cells is damaging (Vu et al. 2019). Overstimulating ROS causes oxidative stress, which damages plant organelles and ultimately causes the plant to die (Sharma et al. 2023b).

4 Chemicals in Agricultural Practices and Their Impact

The nutritional content of farm products as well as the health of farm workers and customers are negatively impacted by current agricultural techniques, which rely extensively on chemicals (such as fertilisers, pesticides, herbicides, etc.). A substantial impact on human health has been resulted by the excessive and irresponsive use of these chemicals, which has given rise to environmental damage, weed and disease resistance, and food contamination. The use of these chemicals encourages the buildup of hazardous substances in soils (Chaudhary et al. 2022e).

To preserve the high standard of agricultural goods and decrease yield losses, agrochemicals are frequently employed in agricultural practices to decrease or prevent diseases, for pest control, and elimination of weed. With this industrialised approach, farmers may produce food at lower prices and make more money from their farms, but substantial concerns have been raised about the health hazards related with occupational predisposition to chemicals and residues in food and drinking water (Alori et al. 2017b). According to Suyal et al. (2016), excessive chemical fertiliser usage can increase food production self-sufficiency but also has negative effects on the ecosystem and living things. The substance taints the food produced and further modifies the consumer's regular bodily functioning. Seventy-five percent of pesticide residues were found in conventionally cultivated food. Toxic pesticides, herbicides, and chemical fertilisers are employed, which contaminate water sources. Agricultural chemical inputs are presently posing significant public health issues due to alleviation in the population engaged in agriculture, living in agricultural regions, and purchasing contaminated farm goods (Alori et al. 2017a). A significant cause of surface-water pollution is agrochemical runoff (Wohlfahrt et al. 2010). Chemical fertiliser overuse and misuse lead to soil washing and water pollution in streams, lakes, and the ocean. There are three main ways that agricultural chemical inputs enter human body systems, first through oral intake, second by penetration via the skin, and last by inhalation (Roychowdhury et al. 2014). In the human body as well as in food such as vegetables, meat, and fruits, pesticides have demonstrated long-term resistance.

5 Microbes as an Alternative to Chemicals

Microbial inoculants, which function as biofertilisers, bioherbicides, biocontrol agents, and biopesticides, are a dependable substitute for the usage of chemical inputs (Kumar et al. 2021a, b, c, d). The most basic and alive system on earth is microbial life. They inherently become an essential component of the crop production from the time when seeds are peppered in the soil to begin its life cycle since they are an essential living component of the soils. Microorganisms are significant seed occupants as well, and when the seeds develop in the soil, they multiply to create symbiotic relationships outside or endophytic relationships within the plant. The key assistance that the plant microbiome gives to the plants is gaining micronutrients, fending off pathogens, and enduring abiotic challenges (Turner et al. 2013). Extreme environmental circumstances may be overcome by microorganisms, thanks to their inherent metabolic and genetic capacities (Sessitsch et al. 2012). The synergy of plants and soil microbes causes a variety of systemic reactions that enhance the plants' capacity to combat abiotic stresses through improved metabolism.

Microorganisms are capable of aiding plant development, controlling pests and diseases, and eliminating weeds (Pandit et al. 2022). Beneficial microorganisms are given to the plant or the soil as microbial inoculants to increase crop health and productivity (Chaudhary et al. 2021c, d, e). Microbial inoculants are commonly employed to manage pests, enhance the grade of the crop and soil quality, and ultimately improve human health. Microbial inoculants are made up of a variety of microbes that are accompanied by soil organisms to enhance soil health and fertility, which in turn benefits human health (Chaudhary et al. 2021g). The utilisation of microbial inoculants can decrease the detrimental effects of chemical input and, as a result, improve the quantity and quality of agricultural output. Plant nutrients are delivered to plants in a more feasible way via microbial inoculants, which are also favourable to the environment. Chemical fertiliser application can be decreased with the aid of microbial inoculants. Microbial inoculants may consist of fungus, bacteria, and algae. Numerous researches have demonstrated that mycorrhizae and rhizobacteria can aid in increase stress tolerance by promoting growth; promoting plant hormone production, siderophores, and phosphate solubilisation; decreasing levels of ethylene; and stimulating the expression of antioxidant and desiccation response genes (Tyagi et al. 2022a, b).

5.1 Microbes and the Plant Microbiome Interactions

It is quite complicated how plants interact with their environment. Microbes and plants have a close bond that makes it possible for them to coexist. A great variety of bacteria take part in the interactions in the plant microbiome, which frequently have synergistic consequences. The many interactions that contribute to the growth and development of agricultural plants are mediated and influenced by the soil, host plant, and the soil microorganisms (Jacoby et al. 2017). The way that plants and their environment interact is always changing. Plants keep an eye on their surroundings and respond to alterations in this process. Before, it was believed that a plant's root system solely served to support it and make it easier for it to absorb water and nutrients. Nevertheless, this root system is crucial in regulating how the plant microbiome interacts. Plants are able to communicate chemical messages from soil-dwelling microorganisms. Root exudates are secreted by plants in response to signals given off by microorganisms (Fig. 2). Even inside the same plant's roots, root exudates can vary in kind and content depending on the species of the plant. Sugars, amino acids, hormones, fatty acid, and proteins are among the wide variety of substances that plants exude as part of their root exudates (Canarini et al. 2019). Other substances, such as auxins, scopoletin, glycosides, inhibitors, glucosides and bacterial stimulants, etc., have also been found in plant roots in addition to the organic ones. The rhizosphere's environment is made specifically by these root exudates. These numerous substances (root exudates) start many symbiotic and other interactions in the rhizosphere. In accordance to the environmental and rhizospheric signals, plant age, soil type, and other environmental conditions, the content and concentration of root exudates alter. The substances that plant produce are used as growth substrates by soil microorganisms, and occasionally these root effluents can also serve as antimicrobials. Therefore, changes in the amount and makeup of root exudates have a beneficial, adverse, or neutral impact on the rhizospheric microorganisms (Upadhyay et al. 2022).

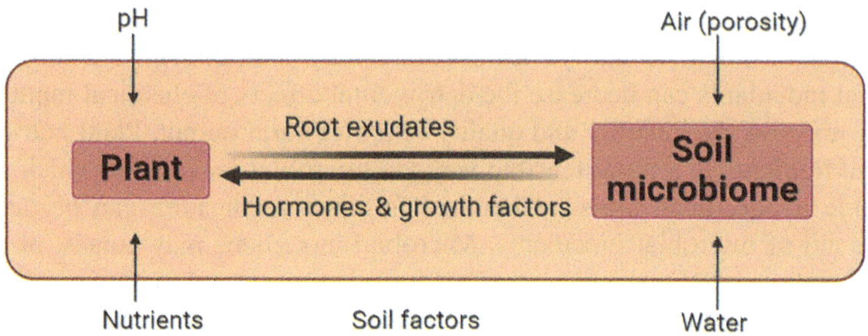

Fig. 2 Schematic illustration showing interaction between soil, soil microbiome, and host plant

5.2 *Microbial Inoculants*

These are formulations made of advantageous microorganisms, which are crucial for the soil ecosystems needed for healthy agriculture (Chaudhary et al. 2023d). Chemical fertilisers and pesticides might potentially be replaced with microbial inoculants since their sustainability towards the environment (Babalola and Glick 2012). They are made up of active microorganism strains that induce either direct or indirect promotion of microbial activity, improving the nutrient mobility from the soil. They could be microbial biocontrol agents, biofertilisers, or phytostimulants (Alori et al. 2017b; Khati et al. 2017). These act as defences against a variety of diseases and are efficient herbicides.

Microbial inoculants are introduced as part of microbial intervention for improved plant development, higher soil fertility, interactions between plant microbiomes, and biocontrol (Khati et al. 2019; Kukreti et al. 2020). Microbial inoculants are a remedy for several issues brought on the increased usage of chemical fertilisers. Popularly referred to as "bio-inoculants or biofertilisers," microbial inoculants are inexpensive, nonpolluting, nature friendly, and safe sustainable agricultural supplements, which use microorganisms with the capability to fix nitrogen, convert different inaccessible minerals into accessible forms, produce plant hormones, and promote the health of plants by chelating iron (Agri et al. 2021). Microbial inoculants are magnificent agricultural supplements for comprehending plant microbiome interactions and increasing crop production sustainably. The beneficial bacteria found in microbial inoculants are varied and exist naturally in soil. Their addition to the soil's biodiversity, soil quality, nutrient availability, development, production, and plant growth are all improved by their integration (Agri et al. 2022). The advantageous microorganisms that may be employed as microbial inoculants for crops include rhizobacteria, diazotrophs, bacteria that mobilise and solubilise phosphorus, plant diseases suppressing microbes, and microbes that can break down complex substances. These microbes are also capable of encouraging plant development (Chaudhary et al. 2021e).

According to studies, beneficial microorganisms counteract abiotic stress by promoting growth by the increased synthesis of plant hormones, siderophores, and phosphate solubilisers; reducing ethylene levels; and enhancing the gene expression related to the desiccation response and antioxidants. According to Kushwaha et al. (2020), bacteria that colonise roots create phytohormones that stimulate seedling development and reduce dormancy caused by salt. Moreover, Kumar et al. (2012) showed that the phosphate dissolution, siderophores, and indole acetic acid (IAA) secretion by the PGPB *Bacillus* and *Pseudomonas* sp. stimulated growth in plants, which are under stress. Moreover, several of these microbes that encourage plant development contain 1-aminocyclopropane-1-carboxylase (ACC) deaminase, an enzyme which breaks down the precursor to ethylene, ACC, preventing ethylene production in plants under stress. Root development and plant survival were enhanced by lower ethylene levels in plant stress (Stearns et al. 2012). Although it was evident that microbes which contain ACC-deaminase and plant hormones assisted in reducing a variety of plant stresses, it has been shown that other microbes use diverse techniques to provide plants with the ability to withstand stress.

5.3 Selection of Microbes for Inoculation

The nature and function of the microorganisms employed for inoculation determine the efficacy of microbial inoculants in field circumstances. Therefore, selecting the right microbial inoculants is an important step for improved outcomes. Complete understanding of the biology and functional activity of microorganisms is necessary for the selection of bacteria to be utilised as inoculant. When choosing microbial inoculants, aspects including climate adaptability, existence and permanence in soil, and effectiveness in field settings should be taken into account. A microbial strain has to have all the necessary traits or characteristics to function as an effective inoculant. For instance, a bacterium that is difficult to proliferate or maintain in a lab cannot be employed as an inoculant. Another approach that we may use to add desirable traits to microbial inoculants is genetic engineering (Sharma et al. 2021; Sharma and Kumawat 2022; Tiwari and Sharma 2023).

5.4 Types of Microbial Inoculants

In agriculture, soil microorganisms provide important functions, primarily through enhancing soil quality and plant nutrition and health. Fungi, bacteria, and blue-green algae are the principal microorganisms utilised as inoculants.

5.4.1 Bacteria in Abiotic Stress Tolerance

Bacteria are the most widespread type of microbe in soil (Carlson et al. 2020); as a result, some of them that are observed colonising plant roots are referred to as PGPR. It has been demonstrated that the enhancement of plant development under stressful circumstances is mediated by several bacterial families. It is demonstrated that PGPR, including *Achromobacter piechaudii, Burkholderia, Pseudomonas* sp., *Azospirillum brasilense, Micrococcus luteus, Paenibacillus polymyxa,* and *Bacillus*, improved the desiccation tolerance of drought-stressed plants, including *Arabidopsis thaliana* (Trivedi et al. 2020), pepper, tomato (Ali and Khan 2021), and wheat (Juric et al. 2020). Some of this bacterially induced resilience has been linked to an upregulated transcription of mRNA of the desiccation response gene ERD15 (early responsive to dehydration 15), 1-aminocycloropropane-1-carboxylic acid (ACC) deaminase production, greater synthesis of proline, and an enhancement in water content (Chukwuneme et al. 2021). By boosting the antioxidant enzymes (glutathione peroxidases (GPXs)) activities against ROS, found that two strains of PGPR, *Azospirillum brasilense* NO40 and *Bacillus amyloliquefaciens* 5113, mitigated effects of stress elicited by drought in *Triticum* plants (Juric et al. 2020). These results confirmed the PGPR's potential for counteracting the repercussions of drought stress and boosting crop output. The term "induced systemic tolerance" is used to describe the chemical and physical changes that PGPR produces in plants to protect them against abiotic stress (IST) (Samaddar et al. 2019).

It has also been shown that IST through PGPR may reduce salinity stress in a variety of plants. When exposed to NaCl, *Azospirillum* inoculation in lettuce seedlings increased germination and vegetative development (Samaddar et al. 2019). Decreased amounts of endogenous ethylene have been linked to bacterial resistance to salt stress, according to certain studies (Kerbab et al. 2021).

It has been shown that bacteria, such as *Bacillus* sp. and *Achromobacter piechaudii* ARV8, which contain ACC deaminase, alleviate salt resilience in tomato seedling development, groundnut growth, and pepper growth under salinity-induced osmotic stress (Juric et al. 2020). Additionally, *Pseudomonas fluorescens* has been found to increase tolerance by reducing the production of dehydrins and osmotins and the synthesis of osmolytes (Khan et al. 2021).

Extreme temperature impacts are likewise resisted by PGPR, as has long been demonstrated. When canola was exposed to 5 °C in both winter and spring, *Pseudomonas putida* GR 12–2 encouraged root and shoot elongation. *Bradyrhizobium japonicum, Serratia proteamaculans,* and *Serratia liquefaciens* all benefited the physiology and development of soybeans cultivated under suboptimal temperatures of the root area (Yang et al. 2020). Furthermore, it was shown that rhizobacteria such *Burkholderia phytofirmans* PsJN performed a protective effect in heat stress when potatoes were inoculated with them. The same bacterium, *Burkholderia phytofirmans* PsJN, reduced biomass loss and electrolyte leakage from membranes when infected with *Vitis vinifera*, a cold-stressed grapevine (Yang et al. 2018). The bacterial "IST" response to low temperature stress entails raising sugar levels, enhancing the production of proteins that contain antifreeze, and raising proline levels (Jinal et al. 2019).

The reduction of heavy metal stress has also been linked to PGPR. Plants that have been exposed to cadmium stress may benefit from the usage of the bacteria *Pseudomonas marginalis, Bacillus pumilus, Pseudomonas putida, Alcaligenes xylosoxidans, Alcaligenes* sp. ZN4, *Variovorax paradoxus, Pseudomonas tolaasii*

ACC23, *Pseudomonas oryzihabitans,* and *Rhodococcus* sp. The research showed that metal-accumulating plant *Brassica juncea* grew better after being exposed to these bacterial strains (Asad et al. 2019). The potential exists that bacteria that encourage plant development may help plants adapt to environmental challenges, regardless of the particular defence mechanisms employed by bacteria against these pressures. As a result, food production will rise concurrently, and this will aid in stabilising and replanting abiotically stressed abandoned areas (Chakraborty et al. 2023; Kumawat et al. 2023; Kumawat et al. 2024).

5.4.2 Fungi in Abiotic Stress Tolerance

Fungi are also linked to the adaptation of plants to different environments, including those that are impacted by abiotic stressors including salt, cold, drought, heat, toxic metals, and floods, in addition to bacteria's role in stress tolerance (Poveda et al. 2020). *Ectomycorrhiza* (EM) and *Arbuscular mycorrhiza* (AM) fungi are two different forms of fungus that have a part in stress tolerance.

5.4.2.1 Arbuscular Mycorrhizal Fungi in Mitigation of Abiotic Stress

Arbuscular mycorrhiza (AM) has been shown to elicit a range of stress tolerances while living harmlessly inside the host. Poveda et al. (2020) demonstrated that the plenitude of endophytic arbuscular mycorrhizal fungi, which coexist in mutually advantageous partnerships with around 80% of plants, might be used to generate abiotic stress tolerance. Tolerance-granting endophytes are further categorised into classes 1 and 2 (Sun et al. 2018). As a result of their restricted colonisation of shoots, stems, and rhizomes, class 1 endophytic fungi have a small host range. On the other hand, monocots and eudicots are given habitat-specific stress tolerance by class 2 endophytes (Zhang et al. 2019). They therefore have a wide spectrum of hosts on whom they may bestow the benefits of habitat adaptation. Both endophytes are said to increase resistance to abiotic stress (Abo 2019). *Fusarium culmorum, Paecilomyces formosus* LHL10, *Piriformospora indica, Neotyphodium lolii, Phoma glomerata* LWL2, *Penicillium* sp. LWL3, *Curvularia protuberata*, and *Trichoderma* are endophytic microbes that are found to impart resistance to abiotic stress (Okon et al. 2020). By altering the biochemical, physiological, and nutritional characteristics of the host plant, these fungi give stress tolerance. These include the siderophores overproduction, the inhibition of ethylene synthesis by ACC-deaminase, the increase in activity of antioxidants, the suppression of sodium absorption by increased electrolyte uptake such as K^+, the proline deposition, the improvement of root water assimilation, and the augmentation of antioxidant capacity (Diagne et al. 2020).

It is shown that AM can increase alkalinity tolerance through various mechanisms, including altering the ratio of Na^+/ K^+ in plant cells, moving salt ions into the vacuole, producing growth hormones, and improving soil and rhizospheric conditions as well as water use and photosynthetic capacity (Sofy et al. 2021). Furthermore, AM serves as osmoregulators by improving the consequences of salinity stress by raising concentration of electrolytes and sugars (Bahadur et al. 2019). In a different study, AM increased antioxidant capacities by turning on the glutathione-ascorbate cycle in plants to produce tolerance to salt stress. Additionally, it is shown that AM symbiosis alleviates the salt tolerance of plants' salt tolerance including clover, maize, cucumber, lettuce and tomato (Bahadur et al. 2019).

Tolerance induced by AM has been shown to be a prevalent aspect of abiotic stress tolerance in plants under drought stress (Ullah et al. 2019). Infection with *Neotyphodium* endophytic fungus is documented to increase biomass production and growth in various *Festuca* species in drought conditions (Okon et al. 2020). Furthermore, it has been documented that endophytic fungi have positive impacts on *Lolium perenne* ability to withstand drought and recover (Abo 2019). Endophytes altered physical, physiological, nutritional, and cellular processes in other plants to decrease the consequences of drought stress (Chandrasekaran et al. 2021). These include enhancing germination and water absorption (through aquaporins), decreasing transpiration (stomatal closure caused by excessive abscisic acid production), enhancing plants' drought resistance, accumulating proline, and enhancing antioxidant activities by minimising the harm resulting from free radicals formed in the course of drought (Chukwuneme et al. 2021). *Triticum aestivum*, *Coriandrum sativum* L., *Vitis vinifera*, and *Sorghum bicolor* are some examples of plants in which desiccation tolerance has been proven to be increased by inoculation of mycorrhiza (Ullah et al. 2019).

Further, AM has been shown to increase a plant's resistance to temperature changes. On tomato plants (*Lycopersicon esculentum*), AM *Glomus mosseae* is demonstrated to reduce the harm created by low temperature stress by lessening membrane lipid peroxidation, accumulation of osmolytes, elevation in photosynthetic pigments, and antioxidant enzyme activity (Sytar et al. 2019). AM provides stress tolerance on heat-stressed plants by raising phytohormone levels that delay untimely plant senescence and boost levels of metabolites like proline and anthocyanins (Kapoor and Hasanuzzaman 2020). AM has been reported to decrease peroxidation of membrane lipids and plasma membrane permeability in the presence of chilling stress, resulting in an alleviation in osmolyte accumulation, the activation of antioxidant enzymes, and an improvement in photosynthetic performance (Riaz et al. 2021).

AM also has a notable impact on heavy element tolerance in metal-toxic soils. For instance, inoculation of AM has been demonstrated to improve growth in tomatoes, maize, rice, and canola grown in soils contaminated with heavy metals by reducing toxic metal absorption; increasing the production of glycoproteins, glomalin, and chitin; and enhancing the absorption of essential soil nutrients (Sytar et al. 2019). Additionally, AM enhances the production of heavy element-stressed plants by ensuing buildup of soluble amino acids and boosting antioxidant enzyme activity. AM ecotypes, such as *Glomus claroideum, Glomus mosseae, Aspergillus niger, and Viola calaminaria,* are those isolated from heavy metal contaminated soils and are employed in the reduction of heavy metal stress (Betekhtina et al. 2021).

5.4.2.2 Ectomycorrhiza in the Alleviation of Abiotic Stress

Most of the researches are focused on how endomycorrhizal fungi AM reduce abiotic stress, but little is known about how ectomycorrhizal fungi are involved, and additionally, connected to the reduction of drought stress is ectomycorrhizal (Kapoor and Hasanuzzaman 2020). According to a hypothesis, ectomycorrhiza enhances osmotic adjustment, increases tissue flexibility, and controls gene expression to lessen the impacts of drought stress (Etesami 2021). Investigations have found that some species of ectomycorrhiza remain in dried soils and often impart resistance to stresses caused by drought on plants (Akram et al. 2021). *Pinus banksiana* (jack pine) and *Picea mariana* (black spruce) that had been infected with ectomycorrhizal fungi fared better under water stress. These plants have higher seedling performance as assessed by their shoot/root development and photosynthetic rate (Szuba et al. 2021). Mycorrhization, found in *Castanea sativa* (European chestnut), an economically significant forest tree, with ectomycorrhizal fungi like *Piloderma croceum, Laccaria laccata, Pisolithus tinctorius,* and *Amanita muscaria* enhanced resilience to water stress (Szuba et al. 2021).

Additionally, ectomycorrhiza has been demonstrated to influence the resistance to heavy metal stress. By boosting detoxification into vacuoles, ectomycorrhiza, *Paxillus involutus*, was shown by Ma and colleagues (2014) to improve *Populus canescens* nutritional and carbohydrate status (Domka et al. 2019). Heavy metals like zinc can build up in the hyphae of the ectomycorrhiza *Suillus bovinus* and can be accumulated in the fungi's cell wall or can sequester in their vacuoles (Zhang et al. 2022).

Plants exposed to cold temperatures have benefited from ectomycorrhiza's protective effects as well. Ectomycorrhizal inoculation of *Populus tremuloides* (aspen) and *Picea glauca* (white spruce) samplings at decreased soil temperatures of 4 °C and 8 °C enhanced resistance (Zhang et al. 2022). *Populus tinctorius* (cork oak tree) performed better when exposed to heat stress in temperate and boreal forests, thanks to ectomycorrhiza, which also boosted leaf area, nitrogen uptake, photosynthetic potential, and water usage efficiency. *Laccaria bicolor* and *Tuber melanosporum* are two ectomycorrhizal fungi that provide resistance from heat stress (Alsanius et al. 2019).

6 Microbes Induced Abiotic Stress Mitigation in Plants

The capability of microbes from a variety of genera, inclusive of *Paenibacillus, Azospirillum, Microbacterium, Achromobacter, Burkholderia, Paenibacillus, Pantoea, Methylobacterium, Pseudomonas, Bacillus, Enterobacter*, *Rhizobium,* etc., to offer host plants with resilience amid various abiotic stress environments has been recorded (Kazerooni et al. 2021). Rhizospheric microorganisms among them have several potential effects. Some PGPR strains produce antioxidants and cytokinin, which lead to the buildup of ABA and the breakdown of ROS. Gibberellins and

IAA, two growth-promoting chemicals produced by PGPR, enhance the area of the plant's total root system and its ability to absorb nutrients (Fig. 3, Table 1). PGPR carrying 1-aminocyclopropane-1-carboxylate (ACC) deaminase has a function in stress agriculture (Duan et al. 2021). Longer roots may be induced by inoculation of ACC deaminase-consisting bacteria during drought conditions, which may aid in the absorption of comparatively increased water from the soil layer. The effectiveness of the ACC-deaminase-producing bacteria that promote plant development is largely dependent on their capacity to reduce plant ethylene levels, which are frequently brought on by various stresses.

The bacteria functions as a sink for ACC due to the bacterial ACC deaminase. In addition, by reducing the levels of IAA-stimulated or endogenous ACC, the quantity of ethylene which may develop inside the plant is also decreased. As a result of

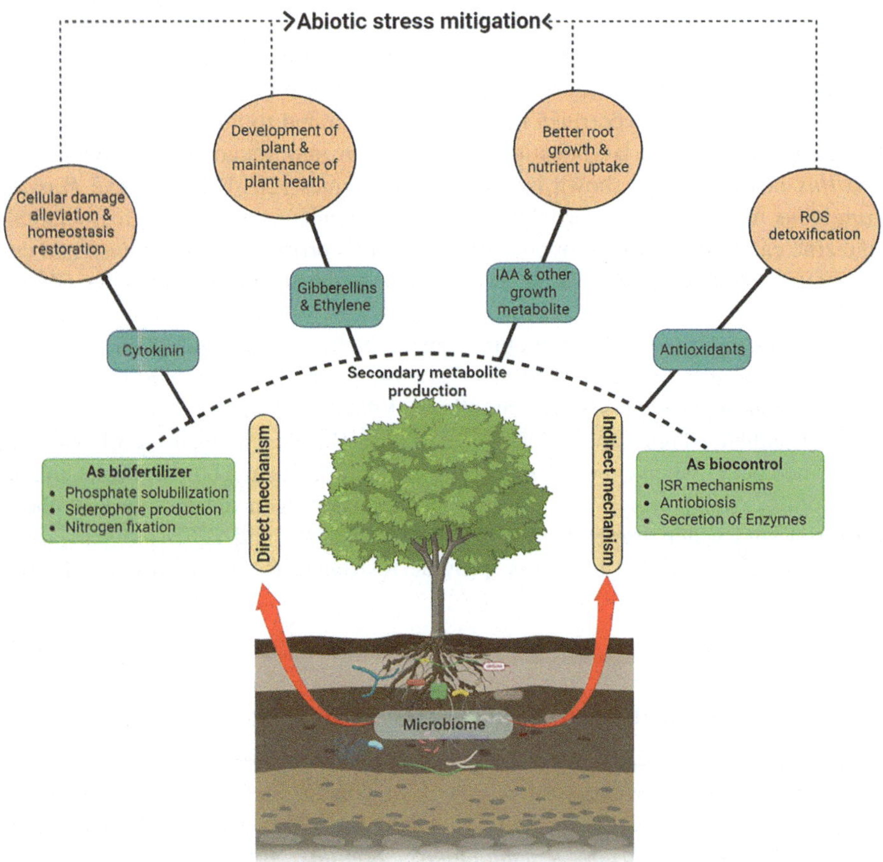

Fig. 3 Proposed model of PGPM against abiotic stress in plants: Microorganisms colonise roots by using different mechanisms such as triggering an ionic and osmotic response. Microbial inoculations elicit stress tolerance in plants by direct and indirect mechanism which help in yielding higher crop productivity

Table 1 Various tolerance strategies used to control abiotic stress in plants

Plants	Abiotic stress	Microbial inoculation	Tolerance method	References
Arabidopsis thaliana	Salt	*Bacillus subtilis* GB03	Tissue-specific regulation of sodium transporter HKT1	Guo et al. (2015)
Glycine max	Salt	*Pseudomonas simiae*	4-nitroguaiacol and quinolone promote soybean-seed germination	Yadav et al. (2014)
Oryza sativa	Salt	Root-associated growth-promoting rhizobacteria	Expression of salt-stress-related RAB18 plant gene	Ghosh et al. (2014)
Triticum aestivum	Heat	*Bacillus amyloliquefaciens, Azospirillum brasilense*	Reduced regeneration of reactive oxygen species, pre-activation of heat-shock transcription factors, and changes in metabolome	Kapoor et al. (2013)
Brassica juncea	Arsenic toxicity	*Staphylococcus arlettae*	Increased dehydrogenase, phosphatase, and available phosphorus in soil	Balestrini and Lumini (2018)
Miscanthus sinensis	Cd, AS, cu, Pb and Zn toxicity	*Pseudomonas koreensis* AGB-1	ACC deaminase and IAA production	Laxa et al. (2019)

reducing ethylene levels of plant, bacteria that promote plant development and contain ACC deaminase can lessen many effects of ethylene on plant growth after a various biotic and abiotic stresses (Gupta and Pandey 2019).

6.1 Role of Microbes to Mitigate Drought Stresses

A better utilisation of soil water because of the contribution of hyphae to water absorption and/or changes induced in root shape and structure of soil are some of the processes postulated in the mycorrhiza-mediated mitigation of drought stress (McNear Jr 2013). In fact, despite fungal modulation of numerous critical solute concentrations, the mycorrhizal preservation of root turgor during drought appears to be connected to altered apoplastic/symplastic water partitioning rather than osmotic adjustment (Boutasknit et al. 2020). While these local modifications aid in maintaining root development, soil bacteria also have an impact on the roots' capacity to absorb water in salty and drought conditions. Under drought and salt stress circumstances, it has been proposed that ABA might be the signal which differently modulates the aquaporins' behaviour in non-AMF and AMF plants.

Through producing antioxidants, cytokinins, and ACC deaminase-mediated degradation of the ethylene precursor ACC, PGPR can reduce drought through a mechanism known as "induced systemic tolerance." Abscisic acid (ABA) builds up in leaves due to the generation of cytokinins, which in turn induces the stomata to close. To combat drought, certain microorganisms create exopolysaccharides, trigger resistance genes, improve plant water circulation, and aid proline production (Morcillo and Manzanera 2021).

Zhang et al. (2020) showed that the drought-stressed tangerine plant's growth, biomass output, and osmotic adjustment were all enhanced by the *Arbuscular mycorrhizal* (AM) fungus (*Citrus tangerine*). By generating antioxidants that break down reactive oxygen species (ROS), some PGPR have also been reported to increase stress tolerance. In order to assimilate water and nutrients that non-AM plants are unable to assess, Arbuscular mycorrhizal fungal hyphae can penetrate pores of soil that are not accessible to plant rhizomes. This alters the plant-water connection, improving water usage efficiency and, ultimately, increasing crop output.

6.2 Role of Microbes to Alleviate Temperature Stress

An abrupt rise in temperature leads to the creation of a certain polypeptide classes known as HSP or heat-shock proteins in all living organisms. When exposed to high temperatures, a thermotolerant *P. aeruginosa* strain AMK-6 from a semiarid region demonstrated the increased expression of HSPs (Tian et al. 2021). Similar to humans, bacteria which are cold-tolerant produce cryoprotective proteins in response to a drop in temperature.

It is widely known that *Pseudomonas* spp. and other rhizobacteria may colonise the root cells of large agricultural crops and stimulate plant development by producing antagonistic chemicals, phytohormones, and enzymes (Agri et al. 2021). Abiotic stressors, such dryness, cold, salinity, metal toxicity, and high temperatures, may have been reduced by bacterial injection. Bacteria must be capable of quick colonisation of the root in order to be effective inoculants for boosting plant development (Alori and Babalola 2018). The fact that the introduced strain successfully colonised plant roots indicated a close plant-bacterial relationship which might be advantageous to plants, most likely due to the activation of specific responsive pathways by inoculation in response to stress, which allow the plants to withstand increased temperatures. By increasing the quantities of starch, proline, and phenols, the bacteria *Burkholderia phytofirmans* PSJN colonises grape vine residues and shields the plant from frost and heat (Fernandez et al. 2012).

6.3 Role of Microorganisms in Alleviation of Salinity Stress

To minimise the repercussions of plant salt stress on its development and agricultural productivity, it is crucial to maintain homeostasis of water and the functionality of photosynthetic organs. In salinised crops, there are reports that soil bacteria affect homeostasis of ions and enhance plant nutrition (Evelin et al. 2019). Additionally, variations in phytohormone concentrations may be used to control plant development in response to salinity stress. Different strategies are employed by microorganisms to reduce the salt stress on agricultural crops. Exopolysaccharides enhance plant growth in highly salinised conditions by limiting salt absorption (Morcillo and Manzanera 2021). Corn, beans, and clover are more resistant to salt when AM fungi produce osmoprotectants with osmoregulatory mechanisms. In some circumstances, free auxin is sufficient to have a preservative impact in defiance of salt stress. According to Siddikee et al. (2010), inoculation of 14 salt-tolerant bacterial strains alleviates salinity stress in canola crop by ACC deaminase activity-mediated decrease of ethylene production. Under saline circumstances, ACC deaminase containing *Pseudomonas fluorescens* greatly increased elongation of roots and fresh weight in maize plants. Through maintaining increased conductance of stomata and photosynthetic activity, the positive effect of PGPR under salt stress has also been linked to the reduction of osmotic stress. As a result, toxic ion buildup (Na + and Cl-) may be reduced, and the leaf K+:Na + ratio may be improved. This might postpone the harmful effects of ions through growth and/or energy maintenance of ion-exclusion mechanisms. Other indirect and new processes that promote plant salt tolerance may potentially be modulated by soil microorganisms (Egamberdieva et al. 2019).

7 Modern Strategies for Strengthening Bioinoculant Formulations

The bioinoculant formulation consists of a homogenous combination of a chosen beneficial strain and a substrate that can stabilise and protect the strain while it is being transported and stored. The carrier is the vehicle that houses latent living microorganisms and offers the microbial population shelter and a supportive environment. According to Sahu et al. (2018), a first-rate formulation should be efficient, pollution-free, and easily biodegradable, have high water holding potential, and have a long shelf life. Biofertilisers are more effective and have a longer shelf life thanks to the formulation process. Sometimes the formulation includes cell protectants with the targeted microorganisms to extend spore shelf life under severe environments (Bhattacharyya et al. 2020). Various formulation kinds are developed based on their effectiveness and survival rate. Some already commercialised microbe-based inoculants are mentioned in Table 2.

7.1 Solid Media

The desirable microbe is combined with a solid substrate in this formulation. It is utilised to move microbial strains from fields to laboratories. For this composition, peat, powder, and granules are employed mostly. It offers those microorganisms that

Table 2 Commercial PGPR-based plant bioinoculants (le Mire et al. 2016; Mishra and Arora 2016; Aamir et al. 2020)

Commercial products (manufacturer)	PGPR	Target crops	Function
FZB24®fl RhizoVital 42® (ABiTEP GmbH, Germany)	*Bacillus amyloliquefaciens* and *B. Amyloliquefaciens* sp. *plantarum*	Ornamentals, vegetable field crops	Phosphate availability and protection against pathogens
Inomix® biostimulant, Inomix® phosphore, and Inomix® Biofertilisant (IAB (Iabiotec), Spain)	*B. Subtilis* (IAB/BS/F1) and *B. Polymyxa* (IAB/BP/01); *Saccharomyces cerevisiae; B. Megaterium* and *P. Fluorescens;* and *rhizobium leguminosarum, Azotobacter vinelandii, B. Megaterium,* and *Saccharomyces cerevisiae*	Cereals	Plant growth promotion increases root and shoot weight, strong root system
Bio-gold (BioPower, Sri Lanka)	*Pseudomonas fluorescens* and *Azotobacter chroococcum*	All agricultural and horticultural crops	Growth promotion via nitrogen fixation, drought tolerance, control of root rot and wilt diseases, phosphorus solubilisation
Cedomon® (Lantmannen BioAgri AB, Sweden)	*P. Chlororaphis*	Barley and oats	Highly effective against various types of seed-borne diseases
Bioscrop BT16 (Motivos Campestres, Portugal)	*Bacillus thuringiensis* var. *kurstaki*	Deciduous fruit trees, horticultural brassicas, cotton, citrus, cauliflower, olives, pepper, banana, and tomato	Protection against pests (beetles)
Nitroguard®	*Azorhizobium caulinodans* NAB38, *Azospirillum brasilense* NAB317, *Azoarcus indigens* NAB04, and *bacillus* sp.	Cereals, rapeseed, and sugar	Growth promotion via nitrogen fixation
Symbion®-N, Symbion®-P, and Symbion®-K (T. Stanes & Company ltd., India)	*Rhizobium, Azotobacter, Azospirillum, Acetobacter; B. Megaterium* var. *phosphaticum;* and *Frateuria aurantia*	Beet, sugarcane, and vegetables	Promotion of plant growth, improved root and shoot weight, and a stronger root system

(continued)

Table 2 (continued)

Commercial products (manufacturer)	PGPR	Target crops	Function
Gmax® PGPR (Greenmax AgroTech, India)	*P. Fluorescens, Azotobacter,* and *phosphobacteria*	Field crops	Nitrogen and phosphatic nutrition, disease prevention and helps in plant growth promotion

establish microcolonies a safe and nutritious habitat. Peat should have a high-water holding capacity, be nontoxic, and be adaptable (Ceglie et al. 2015). Afterwards drying, the peat is mixed with the right strain and put through 250 μm sieves. Curing is the process of incubating peat at a specified temperature in order to promote bacterial multiplication. Peat is mostly utilised for ectomycorrhiza and AMF strains (Pathak et al. 2017). On a medium based upon glucose, ectomycorrhizal strains may grow swiftly and generate sporophore for use in field inoculations. Pure mycelial biomass is utilised in the form of formulation comprising 10–20% peat and vermiculite as a substrate with medium comprising salt and glucose to inhibit the growth of impurities and pathogens. Using this kind of formulation, more fulvic acid is produced, speeding up the chelating process. Applying inoculated peat to the seed surface prior to planting is carried on with the use of various equipment, such as big dough mixers, mechanical tumble machines, and cement mixers. Peat's diversity in composition and quality is its biggest disadvantage. Occasionally, harmful substances are also produced from peat during sterilisation, which may lower the rate of survival and growth of microbial strains (Malusa et al. 2012). Granules have recently replaced peat because they have certain benefits over peat. Granules are formed of silica grains, marble, and calcite and, like peat, are covered in the live cells of bacteria. In comparison to peat, the granules are significantly superior and offer several benefits. Granules are less dusty and simpler to handle, transport, and store. To facilitate the lateral-root interactions, inoculants are positioned in the furrow close to the seed surface during granule formation. A few experimental studies revealed that nitrogen fixation was not improved by using granular inoculants. The formulation of granules, in contrast, is better than peat and liquid inoculants in terms of nodule number, nitrogen buildup, nitrogen fixation, and total biomass, according to certain study (Zaidi et al. 2017). Granular inoculants have a huge benefit in stressful situations including moisture stress, wet soil, and acidic soil because of their strong nitrogen fixation and nodule forming abilities. In addition to peat, a carrier made of charcoal, or biochar, can be used to formulate bioinoculants. Biochar is environmentally safe and devoid of hazardous substances, and it increases the survival rate of bioinoculant (Osman et al. 2022). The fundamental advantage of employing charcoal is that, thanks to its low water content, it may be kept without having to be sterilised. It offers an effective assurance of plant health for crop fields and is devoid of waxy materials and blending issues.

7.2 Liquid Formulation

The liquid formulations are microbial preparations, which include those advantageous microorganisms that can solubilise, fix, or mobilise crucial plant nutrients through biological processes (Bahadur et al. 2016). The liquid formulation contains a variety of bacteria, including nitrogen-fixing microbes (NFM), potash mobilising microbes, and phosphate mobilising microbes (Surendra and Baby 2016). Researchers have recently focused on how microorganisms like manganese, sulphur, and zinc-solubilising bacteria mobilise and solubilise micronutrients. Due of its advantages over traditional carrier-based formulations, the liquid formulations are the newest and most propitious technique. The formulations based on carrier cannot be retained for the whole crop cycle and has a short shelf life of about 2–3 months. To solve this issue, a liquid formulation is employed, giving the facility the ability to extend the rate of survival of the chosen strains all over the crop cycle. The formulation's ability to withstand temperatures and stresses that carrier-based bioinoculants cannot is another benefit. In contrast to carrier-based formulations, where bulk sterilisation is not possible, this formulation allows for adequate sterilisation so that contamination may be managed. It increases the strain's viability and survival by maintaining a high moisture capacity for longer periods of time. It includes a desired strain and various preservatives to extend shelf life under stressful conditions by up to 19–25 months (Chandra et al. 2018). *Azospirillum* is a free-living, microaerophilic rhizobacterium that promotes plant development and is mostly utilised to create liquid formulations. Other phosphate-mobilising bacteria from the genera *Aspergillus*, *Penicillium*, *Bacillus*, and *Pseudomonas* are also utilised in addition to *Azospirillum*. According to studies, altering *Azospirillum* with trehalose, glycerol, and polyvinylpyrrolidone (PVP) in meat broth containing nitrogen-free bromothymol blue allows the preservation of 10^8 cells/ml for around 8 months. According to studies, PVP's exceptional capacity to hold water protects bacteria in hazardous and stressful conditions. *Azospirillum* formulations containing PVP exhibited a longer shelf life than those without. The addition of PVP revealed an increase in microbial population, or 10^8 CFU/ml. For this formulation, an *Azospirillum* population (1.66×10^8 CFU/ml) in trehalose (16 mM) and a phosphate-solubilising bacterial (PSB) strain (3.66×10^8 CFU/ml) in PVP (3%) are utilised (Surendra and Baby 2016). Thus, it was determined that trihalose (16 mM) and PVP (3%) are both excellent bio-inoculants that extend the shelf life of PSB and *Azospirillum* sp., respectively, up to 8–10 months with the greatest microbial population (10^8 CFU/ml).

7.3 Metabolite Formulation

The microbial metabolites in metabolite formulations assist in supplying bioregulators, vital nutrients, and defence against phytopathogens. Strains from the genera *Trichoderma*, *Pseudomonas*, *Rhizobium*, *Mesorhizobium*, and certain mycorrhiza are employed for metabolite formation. Metabolite bioformulation has recently been created to address the shortcomings of cell-based formulation (Tewari and Arora 2016). According to studies, under stressful conditions, the rhizobial strain's flavonoids biomolecules boost nitrogen fixation and nodulation. Biomolecules called lipochitooligosachharides are secreted by rhizobial organisms that are connected with leguminous host plants. In rhizobial-deficient areas, lipochitooligosachharide was said to aid in symbiosis. Through the use of flavonoids and a lipochitooligosachharide promoter, Novozymes crop output for both leguminous and non-leguminous crops have been greatly boosted. Mycorrhiza also releases Myc factors, which stimulate the signal transduction and aid in symbiosis, just like Nod factors do (Maillet et al. 2011). It has recently been shown that the exopolysaccharide (EPS) produced by bacteria that promote plant development, such as Rhizobia *and Pseudomonas*, increases nodulation, formation of biofilm, and root colonisation under stressful and toxic conditions. EPS produced by *Rhizobium* was essential for microbial defence. Strains can be shielded from toxic substances, desiccation, extremely high pH, radiation, predators, and osmotic shock with the use of EPS-containing bioformulation supplements. Tryptophan, for example, has been shown in studies to promote IAA secretion and increase grain yield, plant biomass, and the creation of root hairs. Growth was enhanced by modifying PGPR with l-methionine, which serves as the main metabolite and is a precursor to ethylene. Nutrients including molasses, amino acids, and starch effluent have reportedly been employed as helpful additions. These additions can lengthen the viability and increase the capacity of a useful strain to survive in adverse pedosphere (Timmusk et al. 2014). According to Arora and Mishra (2016), PSB (phosphate-solubilising bacteria) also secretes biosurfactants with antibacterial, emulsifying, antiviral, and insecticidal properties. Liquid bioinoculants with biosurfactants are mostly sprayed on the plant's aerial portions. Antimicrobials like as pyrrolnitrin, fanzines, omission A, and diacetylphloroglucinol are produced by strains such fluorescent *Pseudomonas*. Similarly, *Bacillus* spp. are also investigated for their potential to produce antibiotics. Additionally, it is revealed that secondary metabolites produced by *Bacillus* and *Pseudomonas* have a wide range of antiphytopathogenic properties (Perez-Garcia et al. 2011). This formulation's high cost and mass manufacture are its main disadvantages. Future metabolite formulations also combine the PGP strain with carriers and additives to increase microbial activity.

7.4 Polymeric Formulation

Alginate beads are typically utilised in polymeric formulations. It is a naturally occurring polymer consisting of l-glucuronic acid and d-mannuronic acid, both of which come from the fungus *Macrocystis pyrifera* (brown algae). Several strains of bacteria, including *Pseudomonas* and *Azotobacter*, are utilised in such kind of formulations. These formulations improve bacterial chemotaxis, plasmid, and mushroom growth in host cells. AMF is employed in the trapping method for alginate matrixes (Chaudhary et al. 2020). Recently, many methods for polymeric formulation exploiting latent cell encapsulation in gel matrixes for diverse purposes were developed. Gel matrix extends the viability of useful strains under stressed abiotic and biotic environments (Pereira et al. 2023). These days, nutritional additives that promote optimal growth both under anaerobic and aerobic conditions are also employed during the encapsulation process (Schoebitz and Belchi 2016). During formulation, this method was primarily applied to the *Aspergillus, A. brasilense*, and *Pseudomonas fluorescens,* strains (Singh et al. 2011). Adding beneficial nutrients like skimmed milk can increase the strain's shelf life in the presence of glycerol. Chitin-filled beads were found to be more porous than starch-filled encapsulated cells.

8 Drawbacks of Microbial Community Researches in Stressed Agro-Ecosystem

Because of the remarkable impacts of native microbes on the growth of plants and health as well as the possibility of modifying microbiomes to check plant characters and produce antimicrobial compounds, plant beneficial microbes are becoming more and more important for use in agriculture (Mueller and Sachs 2015; Gopal and Gupta 2016; Herrera Paredes et al. 2018; Helfrich et al. 2018). A significant scientific issue is to comprehend the molecular and ecological roles played by certain members of host-associated microbiomes. This is because microbial communities are very complex and genetically diverse at the species level, as well as because many microorganisms are difficult to cultivate and nucleic acid-based "omic" techniques can only be used for a limited number of microbes (Morales and Holben 2011). Moreover, the absence of model microbe for comprehending interactions of plant and microbe and systematic, extensive microbial culture collections for reconstruction studies hinders the advancement of this area.

8.1 Culture-Dependent Analysis of Community and Culture Collections

In comparison to culture-independent methodologies, several researches show the limits of culture dependent community analysis. When contrasting these two approaches, two major issues come up: culturability and the existence of uncommon species. Only a small portion of the bacterial community can be grown, and those organisms frequently exist in extremely low numbers (Lee et al. 2016; Eida et al. 2018). The technology used for sequencing—more particularly, the depth and quality of sequencing, the primer pairs and the amplicon size—is what determines whether rare species may be detected using culture-independent methods (Hiergeist et al. 2015). The distinction between extracellular DNA from dead or lysed cells and intracellular DNA from intact cells is frequently lost during marker gene sequencing and DNA extraction, as well. There are diverse causes for culturability challenges such as follows: (1) antagonistic or fast-growing microbes can restrict or inhibit growth of slow growing species; (2) some microbial strains are obligate endophytes and a host is required; (3) different species require different growth media and/or fastidious growth conditions; and (4) the presence of some strains is necessary for the growth of other strains (Yashiro et al. 2011; Sarhan et al. 2019).

8.2 Microbial Communities in Laboratory, Field, and Natural Conditions

Understanding community changes in agricultural or natural settings has additional limitations. First, there are several interdependent factors present in natural or field settings that are uncontrollable. When soil characteristics, biological elements, and climate come together, a complex ecosystem is created where a particular microbial community structure is developed and the function of the root microbiota may be impacted (Dastogeer et al. 2020). Any modification to one element might have an impact on all the others, creating spurious correlations as to which determining factor was responsible for community changes. For instance, it would be challenging to distinguish between pH and the other impacts on soil, since altering soil pH would also affect a number of other variables. Since other parameters, such as soil nutrients, may have a significant impact on the microbial population, it is frequently problematic to compare one component, such as soil pH, from many natural soils. The only way to acknowledge how every element of the soil biota plays a part in transforming the microbiomes is to conduct experiments on single components on microbial communities in a laboratory environment.

8.3 Technical Elements of Community Analysis

The sampling technique, like collecting soil samples at various depths, might also provide inconsistent results. In long-term fertilisation experiments, for instance, the bacterial and fungus populations change according to soil depth (Gu et al. 2017). Numerous reports have demonstrated that the variety of microbes often declines with depth of soil, most likely as a result of the reduced exposure to nutrients as soil depth increases from the topsoil to the subsoil (Feng et al. 2019). It is necessary to examine sampling procedures to account for this variability, since sampling of bulk soil might introduce significant variability. Similar to this, collecting plant tissue at various ages or stages of growth, such as leaves, may cause diversity. Depending on the bacterium, the quantity of per cell 16S rRNA copies and the efficiency of extraction of genomic DNA can change (Ketchum et al. 2018). As a result, another restriction is the inability to determine the precise richness of each bacterium without first identifying how many copies of the 16S rRNA are present in their genomes. Furthermore, sequences that are not typical of reality might result from sample contamination, whether it occurs during sampling or library preparation. Last but not least, while conducting community experiments, human aspects such as agriculture management techniques and land usage should also be taken into consideration (Reeve et al. 2010; Le Guillou et al. 2019).

9 A Blueprint for Effective Administrations of Plant-Associated Microbe-Based Inoculants

The development and use of tailored inoculants play an important part in advancing tenable agriculture by boosting plant output and health. When combined with initiatives to investigate and create model systems using these synthetic communities, microbiome researches and the use of SynComs would significantly increase our understanding of plant and microbe interactions. As was previously said, the composition, function, and structure of the microbial community are significantly affected by the plant host/genotype, soil characteristics, the native microbial population, and abiotic influences. Therefore, using microbial inoculants in actual, extensive agricultural field settings has several drawbacks and difficulties. A framework, which will allow farmers, scientists, and agricultural technology businesses to all work together, will enable us for successful administration of microbe-based inoculants.

The in-depth examination of the targeted field's surroundings and important crop is the first component of this framework. The inoculants must be adapted to the targeted field, crop, agricultural management techniques, and environmental conditions due to the existence of several elements that might alter microbial populations (Liu et al. 2022). Here, a thorough study of the soil's attributes, including temperature, microbial population, moisture content, nutrient status, and pH, as well as the local environment (including temperature, wind speeds, humidity levels, and yearly precipitation) is done. Next, the genotype of the crop plant and its suitability for the climate and soil as well as its economic viability are decided. Analysis of the local seed microbial population is also necessary since it can conflict with the inoculants that have been used.

The second element is a component of this architecture and is essential for the use of inoculants successfully across a broad geographical area. This stage necessitates additional culture-dependent isolations, lab-based experimentation, and scientific study of plant microbiomes. Environmental microorganisms can be difficult to cultivate, therefore increasing their viability will need intelligent fusion of omics (like metatranscriptomics) with cutting-edge culturomics approaches (like medium based on plant) (Bomar et al. 2011; Sarhan et al. 2019). More significantly, it is critical to characterise the single isolates' roles, capacities for survival, features that promote plant development, processes that promote growth and stress tolerance, and compatibility with the target crop. Apart from developing sequencing techniques, bioinformatics designs, and tools, an increase in pangenome sequencing, plant phenotyping, functional characterisation of isolates, and the use of meta-omics (metabolomics, proteomics, transcriptomics, and genomics) strategies are also required (Sergaki et al. 2018; Marco and Abram 2019). For instance, created a handy web application, which uses the numerous data sets of microbiomes to determine the core microbiome connected to various environments. Reverse genetics was employed by Cross et al. (2019) to extract and develop previously uncultured microorganisms. In order to circumvent the constraints of amplicon sequencing in differentiating individuals of a heterogeneous community while keeping desirable member traits, Carper et al. (2019) used sophisticated computational programmes.

In order to create an inoculant that is acceptable for the required purpose, agricultural technology businesses must integrate the field data from farmers with the present microbial resources from researches (Saad et al. 2020). This is the third component of the framework. First, depending on the required features and function, which are identified through field study, the required microbes are collected from the culture collections. The strains must subsequently be produced on a large scale before going through further procedures including lyophilisation for increased shelf life and transportation. SynComs must be assembled, and each member's survival and compatibility with the others must be tested, in order to create inoculants. Then, either as powdered formulation or by seed coating of the intended crop, the SynComs or single microbial inoculants may be given for field test.

The effectiveness of the inoculant is then assessed in a field environment that is comparable to the intended field or environment. The framework also depends on the ongoing communication between the three components. Additionally, a prediction of how SynComs would react, survive, and adapt in the real environment and crop plant may be possible, thanks to the growth in data on soil and plant microbiomes and the creation of models. For the achievement of reaching sustainability in food production in the future, rules for the application of these inoculants in various nations and increased financing for plant-microbiome research both need to be taken into account.

10 Conclusion

As environmental stresses continue to lower agricultural outputs, the demand for sustainability in food production is increasing. In order to bring stability in food production, we have to take such measures, which can protect the crops from these environmental stresses. In addition to lowering crop yield, these stressors have had a significant negative economic impact on the global agricultural industry. To combat these impacts, a variety of solutions have been devised, from whole-plant conventional breeding to molecular tactics, such as the transfer of resistant genes to enhance crop development. A straightforward, low-cost method of employing microbes to produce the much-needed food has been seen as a potential, all-purpose way to overcome the difficulties in these techniques. These bacteria have helped stressed plants grow better, but they have also provided protection from oxidative stress by employing secondary metabolites.

Microbes that are found in association with plants can significantly contribute to the development of abiotic stress tolerance. These organisms, like rhizobacteria and symbiotic fungi, work through a number of ways, including as inducing new genes in plants and generating an osmotic response. Microbial inoculation to relieve plant stresses may be a more economical, environmentally friendly solution that may be available in a shorter period of time than the establishment of stress resistant crop types through plant breeding and genetic engineering. It is well-known that beneficial soil microbes may stimulate growth and boost productivity through a variety of methods, including improving nutrient availability and absorption, producing phytohormones, ACC-deaminase, and biological control agents, among others. More research will be needed to optimise the mass production of these microbes and identify the best carriers for extending the shelf life of helpful microorganisms and, as a result, improving their storage capacity, as well as the best methods for their field application.

References

Aamir M, Rai KK, Zehra A, Dubey MK, Kumar S, Shukla V, Upadhyay RS (2020) Microbial bioformulation-based plant biostimulants: a plausible approach toward next generation of sustainable agriculture. In: Microbial endophytes. Woodhead Publishing, Cambridge, pp 195–225

Abo NF (2019) Endophytic fungi for sustainable agriculture. Microbial Biosyst 4:31–44. https://doi.org/10.21608/MB.2019.38886

Agri U, Chaudhary P, Sharma A (2021) In vitro compatibility evaluation of agriusable nanochitosan on beneficial plant growth-promoting rhizobacteria and maize plant. Natl Acad Sci Lett 44:555–559. https://doi.org/10.1007/s40009-021-01047-w

Agri U, Chaudhary P, Sharma A, Kukreti B (2022) Physiological response of maize plants and its rhizospheric microbiome under the influence of potential bioinoculants and nanochitosan. Plant Soil 474:451–468. https://doi.org/10.1007/s11104-022-05351-2

Akram W, Ahmad A, Yasin NA, Anjum T, Ali B, Fatima S, Ahmed S, Simirgiotis MJ, Li G (2021) Mechanical strengthening and metabolic re-modulations are involved in protection against *fusarium* wilt of tomato by *B. Subtilis* IAGS174. J Plant Interact 16:411–421. https://doi.org/10.1080/17429145.2021.1966107

Ali S, Khan N (2021) Delineation of mechanistic approaches employed by plant growth promoting microorganisms for improving drought stress tolerance in plants. Microbiol Res 249:126771. https://doi.org/10.1016/j.micres.2021.126771

Alori ET, Babalola OO (2018) Microbial inoculants for improving crop quality and human health in Africa. Front Microbiol 9:2213. https://doi.org/10.3389/fmicb.2018.02213

Alori ET, Fawole OB (2017) Microbial inoculants-assisted phytoremediation for sustainable soil management. In: Ansari AA, Gill SS, Lanza GR, Newman L (eds) Phytoremediation: management of environmental contaminants, Switzerland. Springer International Publishing, Berlin. https://doi.org/10.1007/978-3-319-52381-1_1

Alori ET, Dare MO, Babalola OO (2017a) Microbial inoculants for soil quality and plant fitness. In: Lichtfouse E (ed) Sustainable agriculture review. Springer, Berlin, pp 181–308. https://doi.org/10.1007/978-3-319-48006-0

Alori ET, Glick BR, Babalola OO (2017b) Microbial phosphorus solubilization and its potential for use in sustainable agriculture. Front Microbiol 8:971. https://doi.org/10.3389/fmicb.2017.00971

Alsanius B, Karlsson M, Rosberg A, Dorais M, Naznin M, Khalil S, Bergstrand KJ (2019) Light and microbial lifestyle: the impact of light quality on plant–microbe interactions in horticultural production systems—a review. Horticulturae 5:41. https://doi.org/10.3390/horticulturae5020041

Ameh T, Sayes CM (2019) The potential exposure and hazards of copper nanoparticles: a review. Environ Toxicol Pharmacol 71:103220. https://doi.org/10.1016/j.etap.2019.103220

Arora NK, Mishra J (2016) Prospecting the roles of metabolites and additives in future bioformulations for sustainable agriculture. Appl Soil Ecol 107:405–407. https://doi.org/10.1016/j.apsoil.2016.05.020

Asad SA, Farooq M, Afzal A, West H (2019) Integrated phytobial heavy metal remediation strategies for a sustainable clean environment—a review. Chemosphere 217:925–941. https://doi.org/10.1016/j.chemosphere.2018.11.021

Babalola OO, Glick BR (2012) The use of microbial inoculants in African agriculture: current practice and future prospects. J Food Agric Environ 10:540–549

Bahadur I, Maurya BR, Kumar A, Meena VS, Raghuwanshi R (2016) Towards the soil sustainability and potassium-solubilizing microorganisms. In: Meena VS, Maurya BR, Verma JP, Meena RS (eds) Potassium solubilizing microorganisms for sustainable agriculture. Springer, New Delhi, pp 255–266. https://doi.org/10.1007/978-81-322-2776-2_18

Bahadur A, Batool A, Nasir F, Jiang S, Mingsen Q, Zhang Q, Pan J, Liu Y, Feng H (2019) Mechanistic insights into arbuscular mycorrhizal fungi-mediated drought stress tolerance in plants. Int J Mol Sci 20:4199. https://doi.org/10.3390/ijms20174199

Balestrini R, Lumini E (2018) Focus on mycorrhizal symbioses. Appl Soil Ecol 123:299–304. https://doi.org/10.1016/j.apsoil.2017.09.001

Balliu A, Zheng Y, Sallaku G, Fernández JA, Gruda NS, Tuzel Y (2021) Environmental and cultivation factors affect the morphology, architecture and performance of root systems in soilless grown plants. Horticulturae 7:243. https://doi.org/10.3390/horticulturae7080243

Betekhtina AA, Ganem A, Nekrasova OA, Radchenko TA, Uchaev AP, Dergacheva MI, Veselkin DV (2021) Factors of carbon and nitrogen contents in the fine roots of plants in the middle Urals. Russ J Ecol 52:99–108. https://doi.org/10.1134/S106741362102003X

Bhattacharyya C, Roy R, Tribedi P, Ghosh A, Ghosh A (2020) Biofertilizers as substitute to commercial agrochemicals. In: Majeti N, Vara P (eds) Agrochemicals detection, treatment and remediation. Butterworth-Heinemann, Oxford, pp 263–290. https://doi.org/10.1016/B978-0-08-103017-2.00011-8

Bloomfield JP, Marchant BP, McKenzie AA (2019) Changes in groundwater drought associated with anthropogenic warming. Hydrol Earth Syst Sci 23:1393–1408. https://doi.org/10.5194/hess-23-1393-2019

Bomar L, Maltz M, Colston S, Graf J (2011) Directed culturing of microorganisms using metatranscriptomics. mBio 2:e00012–e00011. https://doi.org/10.1128/mBio.00012-11

Boutasknit A, Baslam M, Ait-El-Mokhtar M, Anli M, Ben-Laouane R, Douira A, El Modafar C, Mitsui T, Wahbi S, Meddich A (2020) Arbuscular mycorrhizal fungi mediate drought tolerance and recovery in two contrasting carob (*Ceratonia siliqua* L.) ecotypes by regulating stomatal, water relations, and (in)organic adjustments. Plants (Basel) 9:80. https://doi.org/10.3390/plants9010080

Brito C, Dinis LT, Moutinho P, Correia CM (2019) Drought stress effects and olive tree acclimation under a changing climate. Plan Theory 8:232. https://doi.org/10.3390/plants8070232

Cai Z, Wang X, Bhadra S, Gao Q (2020) Distinct factors drive the assembly of quinoa-associated microbiomes along elevation. Plant Soil 448:55–69. https://doi.org/10.1007/s11104-019-04387-1
Canarini A, Kaiser C, Merchant A, Richter A, Wanek W (2019) Root exudation of primary metabolites: mechanisms and their roles in plant responses to environmental stimuli. Front Plant Sci 10:157. https://doi.org/10.3389/fpls.2019.00157
Carlson R, Tugizimana F, Steenkamp PA, Dubery IA, Hassen AI, Labuschagne N (2020) Rhizobacteria-induced systemic tolerance against drought stress in *Sorghum bicolor* (L.) Moench. Microbiol Res 232:126388. https://doi.org/10.1016/j.micres.2019.126388
Carper DL, Lawrence TJ, Carrell AA, Pelletier DA, Weston DJ (2019) DISCo-microbe: design of an identifiable synthetic community of microbes. Peer J 7:e27898–e27891
Ceglie FG, Bustamante MA, Amara MB, Tittarelli F (2015) The challenge of peat substitution in organic seedling production: optimization of growing media formulation through mixture design and response surface analysis. PLoS One 10:e0128600. 10.1371/journal.pone.0128600
Cernusak LA (2020) Gas exchange and water use efficiency in plant canopies. Plant Biol 22:52–67. https://doi.org/10.1111/plb.12939
Chandra D, Barh A, Sharma IP (2018) Plant growth promoting bacteria: a gateway to sustainable agriculture. Microb Biotechnol Environ Monit Cleanup 2018:318–338. https://doi.org/10.4018/978-1-5225-3126-5.ch020
Chandrasekaran M, Boopathi T, Manivannan P (2021) Comprehensive assessment of ameliorative effects of AMF in alleviating abiotic stress in tomato plants. J Fungi 7:303. https://doi.org/10.3390/jof7040303
Chaudhary T, Dixit M, Gera R, Shukla AK, Prakash A, Gupta G, Shukla P (2020) Techniques for improving formulations of bioinoculants. 3 Biotech 10:199. https://doi.org/10.1007/s13205-020-02182-9
Chaudhary P, Chaudhary A, Parveen H, Rani A, Kumar G, Kumar A, Sharma A (2021a) Impact of nanophos in agriculture to improve functional bacterial community and crop productivity. BMC Plant Biol 21:519. https://doi.org/10.1186/s12870-021-03298-7
Chaudhary P, Khati P, Chaudhary A, Gangola S, Kumar R, Sharma A (2021b) Bioinoculation using indigenous *Bacillus* spp improves growth and yield of *Zea mays* under the influence of nanozeolite. 3 Biotech 11:11. https://doi.org/10.1007/s13205-020-02561-2
Chaudhary P, Khati P, Chaudhary A, Maithani D, Kumar G, Sharma A (2021c) Cultivable and metagenomic approach to study the combined impact of nanogypsum and *Pseudomonas taiwanensis* on maize plant health and its rhizospheric microbiome. PLoS One 16:e0250574. https://doi.org/10.1371/journal.pone.0250574
Chaudhary P, Khati P, Gangola S, Kumar A, Kumar R, Sharma A (2021d) Impact of nanochitosan and *Bacillus* spp on health, productivity and defence response in *Zea mays* under field condition. 3 Biotech 11:237. https://doi.org/10.1007/s13205-021-02790-z
Chaudhary A, Parveen H, Chaudhary P, Bhatt P, Khatoon H (2021e) Rhizospheric microbes and their mechanism. In: Bhatt P, Gangola S, Udayanga D, Kumar G (eds) Microbial technology for sustainable environment. Springer, Singapore. https://doi.org/10.1007/978-981-16-3840-4_6
Chaudhary P, Parveen H, Gangola S, Kumar G, Bhatt P, Chaudhary A (2021f) Plant growth promoting Rhizobacteria and their application in sustainable crop production. In: Bhatt P, Gangola S, Udayanga D, Kumar G (eds) Microbial technology for sustainable environment. Springer, Singapore. https://doi.org/10.1007/978-981-16-3840-4_13
Chaudhary P, Sharma A, Chaudhary A, Khati P, Gangola S, Maithani D (2021g) Illumina based high throughput analysis of microbial diversity of maize rhizosphere treated with nanocompounds and *bacillus* sp. Appl Soil Ecol 159:103836. https://doi.org/10.1016/j.apsoil.2020.103836
Chaudhary P, Agri U, Chaudhary A, Kumar A, Kumar G (2022a) Endophytes and their potential in biotic stress management and crop production. Front Microbiol 13:933017. https://doi.org/10.3389/fmicb.2022.933017
Chaudhary P, Chaudhary A, Agri U, Khatoon H, Singh A (2022b) Recent trends and advancements for agro-environmental sustainability at higher altitudes. In: Goel R, Soni R, Suyal DC, Khan

M (eds) Survival strategies in cold-adapted microorganisms. Springer, Singapore. https://doi.org/10.1007/978-981-16-2625-8_19
Chaudhary A, Chaudhary P, Upadhyay A, Kumar A, Singh A (2022c) Effect of gypsum on plant growth promoting Rhizobacteria. Environ Ecol 39:1248–1256
Chaudhary P, Singh S, Chaudhary A, Agri U, Bhandari G (2022d) Chapter 3 agrochemicals and their effects on soil microbial population. In: Soni R, Suyal D, Goel R (eds) Plant protection: from chemicals to biologicals. De Gruyter, Berlin, Boston, pp 45–58. https://doi.org/10.1515/9783110771558-003
Chaudhary P, Singh S, Chaudhary A, Sharma A, Kumar G (2022e) Overview of biofertilizers in crop production and stress management for sustainable agriculture. Front Plant Sci 13:930340. https://doi.org/10.3389/fpls.2022.930340
Chaudhary P, Ahamad L, Chaudhary A, Kumar G, Chen F-S, Chen S (2023a) Nanoparticle-mediated bioremediation as a powerful weapon in the removal of environmental pollutants. J Environ Chem Eng 11:109591. https://doi.org/10.1016/j.jece.2023.109591
Chaudhary P, Xu M, Ahamad L, Chaudhary A, Kumar G, Adeleke BS, Verma KK, Hu D-M, Širić I, Kumar P, Popescu SM, Abou Fayssal S (2023b) Application of synthetic consortia for improvement of soil fertility, pollution remediation, and agricultural productivity: a review. Agronomy 13:643. https://doi.org/10.3390/agronomy13030643
Chaudhary P, Chen S, Rajput VD, Jaiswal DK (2023c) Editorial: bioinoculants with nano-compounds to improve soil health: a step toward sustainable agriculture. Front Environ Sci 11:1270002. https://doi.org/10.3389/fenvs.2023.1270002
Chaudhary P, Chen S, Rajput VD, Jaiswal DK (eds) (2023d) Bioinoculants with nano-compounds to improve soil health: a step toward sustainable agriculture. Frontiers Media SA, Lausanne. https://doi.org/10.3389/978-2-8325-3287-4
Chakraborty M, Sharma B, Ghosh A, Sah D, Rai JPN (2023) Elicitation of E-waste (acrylonitrile-butadiene styrene) enriched soil bioremediation and detoxification using Priestia aryabhattai MGP1. Environ Res 238:117126
Chukwuneme CF, Ayangbenro AS, Babalola OO (2021) Impacts of land-use and management histories of maize fields on the structure, composition, and metabolic potentials of microbial communities. Curr Plant Biol 28:100228. https://doi.org/10.1016/j.cpb.2021.100228
Coutinho FS, Rodrigues JM, Lima LL, Mesquita RO, Carpinetti PA, Machado JPB, Vital CE, Vidigal PM, Ramos MES, Maximiano MR (2021) Remodeling of the cell wall as a drought-tolerance mechanism of a soybean genotype revealed by global gene expression analysis. aBIOTECH 2:14–31. https://doi.org/10.1007/s42994-021-00043-4
Cross KL, Campbell JH, Balachandran M (2019) Targeted isolation and cultivation of uncultivated bacteria by reverse genomics. Nat Biotechnol 37:1314–1321. https://doi.org/10.1038/s41587-019-0260-6
Dastogeer KMG, Tumpa FH, Sultana A, Akter MA, Chakraborty A (2020) Plant microbiome–an account of the factors that shape community composition and diversity. Curr Plant Biol 23:100161. https://doi.org/10.1016/j.cpb.2020.100161
Diagne N, Ngom M, Djighaly PI, Fall D, Hocher V, Svistoonoff S (2020) Roles of arbuscular mycorrhizal fungi on plant growth and performance: importance in biotic and abiotic stressed regulation. Diversity 12:370. https://doi.org/10.3390/d12100370
Diamond S, Andeer PF, Li Z, Crits-Christoph A, Burstein D, Anantharaman K, Lane KR, Thomas BC, Pan C, Northen TR (2019) Mediterranean grassland soil C–N compound turnover is dependent on rainfall and depth, and is mediated by genomically divergent microorganisms. Nat Microbiol 4:1356–1367. https://doi.org/10.1038/s41564-019-0449-y
Domka AM, Rozpaadek P, Turnau K (2019) Are fungal endophytes merely mycorrhizal copycats? The role of fungal endophytes in the adaptation of plants to metal toxicity. Front Microbiol 10:371. https://doi.org/10.3389/fmicb.2019.00371
Duan B, Li L, Chen G, Su-Zhou C, Li Y, Merkeryan H, Liu W, Liu X (2021) 1-Aminocyclopropane-1-carboxylate deaminase-producing plant growth-promoting Rhizobacteria improve drought

stress tolerance in grapevine (*Vitis vinifera* L.). Front Plant Sci 12:706990. https://doi.org/10.3389/fpls.2021.706990

Dumanovic J, Nepovimova E, Natic M, Kuca K, Jacevic V (2021) The significance of reactive oxygen species and antioxidant defense system in plants: a concise overview. Front Plant Sci 11:552969. https://doi.org/10.3389/fpls.2020.552969

Egamberdieva D, Wirth S, Bellingrath-Kimura SD, Mishra J, Arora NK (2019) Salt-tolerant plant growth promoting Rhizobacteria for enhancing crop productivity of saline soils. Front Microbiol 10:2791. https://doi.org/10.3389/fmicb.2019.02791

Eida AA, Ziegler M, Lafi FF, Michell CT, Voolstra CR, Hirt H, Saad MM (2018) Desert plant bacteria reveal host influence and beneficial plant growth properties. PLoS One 13:e0208223. https://doi.org/10.1371/journal.pone.0208223

Etesami H (2021) Potential advantage of rhizosheath microbiome, in contrast to rhizosphere microbiome, to improve drought tolerance in crops. Rhizosphere 20:100439. https://doi.org/10.1016/j.rhisph.2021.100439

Evelin H, Devi TS, Gupta S, Kapoor R (2019) Mitigation of salinity stress in plants by arbuscular mycorrhizal Symbiosis: current understanding and new challenges. Front Plant Sci 10:470. https://doi.org/10.3389/fpls.2019.00470

Feng H, Guo J, Wang W, Song X, Yu S (2019) Soil depth determines the composition and diversity of bacterial and archaeal communities in a poplar plantation. Forests 10:550. https://doi.org/10.3390/f10070550

Fernandez O, Theocharis A, Bordiec S, Feil R, Jacquens L, Clément C, Fontaine F, Barka EA (2012) *Burkholderia phytofirmans* PsJN acclimates grapevine to cold by modulating carbohydrate metabolism. Mol Plant Microbe Interact 25:496–504. https://doi.org/10.1094/MPMI-09-11-0245

Ghosh U-D, Saha C, Maiti M, Lahiri S, Ghosh S, Seal A, MitraGhosh M (2014) Root associated iron oxidizing bacteria increase phosphate nutrition and influence root to shoot partitioning of iron in tolerant plant typha angustifolia. Plant Soil 381:279–295. https://doi.org/10.1007/s11104-014-2085-x

Gopal M, Gupta A (2016) Microbiome selection could spur next-generation plant breeding strategies. Front Microbiol 7:1971. https://doi.org/10.3389/fmicb.2016.01971

Grayson M (2013) Agriculture and drought. Nature 501:S1. https://doi.org/10.1038/501S1a

Gu Y, Wang Y, Lu S, Xiang Q, Yu X, Zhao K, Zou L, Chen Q, Tu S, Zhang X (2017) Long-term fertilization structures bacterial and archaeal communities along soil depth gradient in a paddy soil. Front Microbiol 8:1516. https://doi.org/10.3389/fmicb.2017.01516

Guo J-K, Ding Y-Z, Feng R-W, Wang R-G, Xu Y-M, Chen C, Wei X-L, Chen W-M (2015) *Burkholderia metalliresistens* sp. Nov. a multiple metal-resistant and phosphate-solubilising species isolated from heavy metal-polluted soil in Southeast China. Antonie Van Leeuwenhoek 107:1591–1598. https://doi.org/10.1007/s10482-015-0453-z

Gupta S, Pandey S (2019) ACC deaminase producing bacteria with multifarious plant growth promoting traits alleviates salinity stress in French bean (*Phaseolus vulgaris*) plants. Front Microbiol 10:1506. https://doi.org/10.3389/fmicb.2019.01506

Hayat R, Ali S, Amara U, Khalid R, Ahmed I (2010) Soil beneficial bacteria and their role in plant growth promotion: a review. Ann Microbiol 60:579–598. https://doi.org/10.1007/s13213-010-0117-1

Helfrich EJN, Vogel CM, Ueoka R, Schafer M, Ryffel F, Muller DB, Probst S, Kreuzer M, Piel J, Vorholt JA (2018) Bipartite interactions, antibiotic production and biosynthetic potential of the Arabidopsis leaf microbiome. Nat Microbiol 3:909–919. https://doi.org/10.1038/s41564-018-0200-0

Herrera Paredes S, Gao T, Law TF (2018) Design of synthetic bacterial communities for predictable plant phenotypes. PLoS Biol 16:e2003962. https://doi.org/10.1371/journal.pbio.2003962

Hiergeist A, Glasner J, Reischl U, Gessner A (2015) Analyses of intestinal microbiota: culture versus sequencing. ILAR J 56:228–240. https://doi.org/10.1093/ilar/ilv017

Imran M, Shazad R, Bilal S, Imran QM, Khan M, Kang SM, Khan AL, Yun BW, Lee IJ (2021) Exogenous melatonin mediates the regulation of endogenous nitric oxide in *Glycine max* L. to reduce effects of drought stress. Environ Exp Bot 188:104511. https://doi.org/10.1016/j.envexpbot.2021.104511

Ito F, Sono Y, Ito T (2019) Measurement and clinical significance of lipid peroxidation as a biomarker of oxidative stress: oxidative stress in diabetes, atherosclerosis, and chronic inflammation. Antioxidants 8:72. https://doi.org/10.3390/antiox8030072

Jacoby R, Peukert M, Succurro A, Koprivova A, Kopriva S (2017) The role of soil microorganisms in plant mineral nutrition—current knowledge and future directions. Front Plant Sci 8:1617. https://doi.org/10.3389/fpls.2017.01617

Jinal HN, Gopi K, Prittesh P, Kartik VP, Amaresan N (2019) Phytoextraction of iron from contaminated soils by inoculation of iron-tolerant plant growth-promoting bacteria in *Brassica juncea* L. Czern. Environ Sci Pollut Res 26:32815–32823. https://doi.org/10.1007/s11356-019-06394-2

Juric S, Sopko Stracenski K, Król-Kilinska Z, Žutic I, Uher SF, Đermic E, Topolovec-Pintaric S, Vincekovic M (2020) The enhancement of plant secondary metabolites content in *Lactuca sativa* L. by encapsulated bioactive agents. Sci Rep 10:3737. https://doi.org/10.1038/s41598-020-60690-3

Kapoor RT, Hasanuzzaman M (2020) Exogenous kinetin and putrescine synergistically mitigate salt stress in *Luffa acutangula* by modulating physiology and antioxidant defense. Physiol Mol Biol Plants 26:2125–2137. https://doi.org/10.1007/s12298-020-00894-z

Kapoor R, Evelin H, Mathur P, Giri B (2013) Plant acclimation to environmental stress. Springer, Berlin, pp 359–401

Kazerooni EA, Maharachchikumbura SSN, Adhikari A, Al-Sadi AM, Kang SM, Kim LR, Lee IJ (2021) Rhizospheric *bacillus amyloliquefaciens* protects *Capsicum annuum* cv. Geumsugangsan from multiple abiotic stresses via multifarious plant growth-promoting attributes. Frontiers. Plant Sci 12:669693. https://doi.org/10.3389/fpls.2021.669693

Kerbab S, Silini A, Chenari Bouket A, Cherif-Silini H, Eshelli M, El Houda RN, Belbahri L (2021) Mitigation of NaCl stress in wheat by rhizosphere engineering using salt habitat adapted PGPR halotolerant bacteria. Appl Sci 11:1034. https://doi.org/10.3390/app11031034

Ketchum RN, Smith EG, Vaughan GO, Phippen BL, McParland D, Al-Mansoori N, Carrier TJ, Burt JA, Reitzel AM (2018) DNA extraction method plays a significant role when defining bacterial community composition in the marine invertebrate *Echinometra mathaei*. Front Mar Sci 5:255. https://doi.org/10.3389/fmars.2018.00255

Ketehouli T, Idrice Carther KF, Noman M, Wang FW, Li XW, Li HY (2019) Adaptation of plants to salt stress: characterization of Na+ and K+ transporters and role of CBL gene family in regulating salt stress response. Agronomy 9:687. https://doi.org/10.3390/agronomy9110687

Khan MA, Hamayun M, Asaf S, Khan M, Yun BW, Kang SM, Lee IJ (2021) Rhizospheric *bacillus* spp. rescues plant growth under salinity stress via regulating gene expression, endogenous hormones, and antioxidant system of *Oryza sativa* L. Front Plant Sci 12:665590. https://doi.org/10.3389/fpls.2021.665590

Khati P, Chaudhary P, Gangola S, Bhatt P, Sharma A (2017) Nanochitosan supports growth of *Zea mays* and also maintains soil health following growth. 3 Biotech 7:81. https://doi.org/10.1007/s13205-017-0668-y

Khati P, Chaudhary P, Gangola S, Sharma P (2019) Influence of nanozeolite on plant growth promotory bacterial isolates recovered from nanocompound infested agriculture field. Environ Ecol 37:521–527

Kukreti B, Sharma A, Chaudhary P, Agri U, Maithani D (2020) Influence of nanosilicon dioxide along with bioinoculants on *Zea mays* and its rhizospheric soil. 3Biotech 10:345. https://doi.org/10.1007/s13205-020-02329-8

Kumar A, Kumar A, Devi S, Patil S, Payal C, Negi S (2012) Isolation, screening and characterization of bacteria from Rhizospheric soils for different plant growth promotion (PGP) activities: an in vitro study. Recent Res Sci Technol 4:1–5

Kumar G, Lal S, Maurya SK, Bhattacherjee AK, Chaudhary P, Gangola S, Rajan S (2021a) Exploration of *Klebsiella pneumoniae* M6 for paclobutrazol degradation, plant growth attributes, and biocontrol action under subtropical ecosystem. PLoS One 16:e0261338. https://doi.org/10.1371/journal.pone.0261338

Kumar A, Sharma A, Chaudhary P, Gangola S (2021b) Chlorpyrifos degradation using binary fungal strains isolated from industrial waste soil. Biologia 10:3071–3081. https://doi.org/10.1007/s11756-021-00816-8

Kumar G, Suman A, Lal S, Ram RA, Bhatt P, Pandey G, Chaudhary P, Rajan S (2021c) Bacterial structure and dynamics in mango (*Mangifera indica*) orchards after long term organic and conventional treatments under subtropical ecosystem. Sci Rep 11:20554. https://doi.org/10.1038/s41598-021-00112-0

Kumar A, Tripti MM, Bruno LB, Rajkumar M (2021d) Synergistic effect of ACC deaminase producing *pseudomonas* sp. TR15a and siderophore producing *Bacillus aerophilus* TR15c for enhanced growth and copper accumulation in *Helianthus annuus* L. Chemosphere 276:130038. https://doi.org/10.1016/j.chemosphere.2021.130038

Kumari S, Sharma A, Chaudhary P, Khati P (2020) Management of plant vigor and soil health using two agriusable nanocompounds and plant growth promotory rhizobacteria in Fenugreek. 3 Biotech 10:1–11. https://doi.org/10.1007/s13205-020-02448-2

Kumari H, Khati P, Gangola S, Chaudhary P, Sharma A (2021) Performance of plant growth promotory rhizobacteria on maize and soil characteristics under the influence of TiO_2 nanoparticles. Pantnagar J Res 19:28–39

Kumawat KC, Sharma B, Nagpal S, Kumar A, Tiwari S, Nair RM (2023) Plant growth-promoting rhizobacteria: Salt stress alleviators to improve crop productivity for sustainable agriculture development Front. Plant Sci 13:1101862

Kumawat KC, Sharma P, Sirari A, Sharma B, Kumawat G, Nair RM, Bindumadhava H (2024) Co-existence of halo-tolerant Pseudomonas fluorescens and Enterococcus hirae with multifunctional growth promoting traits to ameliorate salinity stress in Vigna radiata. Chemosphere 349:140953

Kushwaha P, Kashyap PL, Bhardwaj AK, Kuppusamy P, Srivastava AK, Tiwari RK (2020) Bacterial endophyte mediated plant tolerance to salinity: growth responses and mechanisms of action. World J Microbiol Biotechnol 36:1–16. https://doi.org/10.1007/s11274-020-2804-9

Laxa M, Liebthal M, Telman W, Chibani K, Dietz K-J (2019) The role of the plant antioxidant system in drought tolerance. Antioxidants 8:94. https://doi.org/10.3390/antiox8040094

Le Guillou C, Chemidlin Prevost-Boure N, Karimi B (2019) Tillage intensity and pasture in rotation effectively shape soil microbial communities at a landscape scale. Microbiol Open 8:e00676. https://doi.org/10.1002/mbo3.676

Le Mire G, Nguyen M, Fassotte B, du Jardin P, Verheggen F, Delaplace P, Jijakli H (2016) Implementing biostimulants and biocontrol strategies in the agroecological management of cultivated ecosystems. BASE 20:1–15. https://doi.org/10.25518/1780-4507.12717

Lee SA, Park J, Chu B, Kim JM, Joa JH, Sang MK, Song J, Weon HY (2016) Comparative analysis of bacterial diversity in the rhizosphere of tomato by culture-dependent and -independent approaches. J Microbiol 54:823–831. https://doi.org/10.1007/s12275-016-6410-3

Liu X, Roux XL, Salles JF (2022) The legacy of microbial inoculants in agroecosystems and potential for tackling climate change challenges. iScience 25:103821. https://doi.org/10.1016/j.isci.2022.103821

Ljubej V, Karalija E, Salopek-Sondi B, Šamec D (2021) Effects of short-term exposure to low temperatures on proline, pigments, and phytochemicals level in kale (*Brassica oleracea* var. *acephala*). Horticulturae 7:341. https://doi.org/10.3390/horticulturae7100341

Ma X, Su Z, Ma H (2020) Molecular genetic analyses of abiotic stress responses during plant reproductive development. J Exp Bot 71:2870–2885. https://doi.org/10.1093/jxb/eraa089

Maillet F, Poinsot V, André O (2011) Fungal lipochitooligosaccharide symbiotic signals in arbuscular mycorrhiza. Nature 469:58. https://doi.org/10.1038/nature09622

Malusa E, Sas-Paszt L, Ciesielska J (2012) Technologies for beneficial microorganisms inocula used as biofertilizers. Sci World J 2012:1. https://doi.org/10.1100/2012/491206

Manzar N, Jayaswal D, Kishan G, Chauhan S, Joshi D, Singh J, Chaudhary P, Jamali H, Kushwaha P, Parveen H, Kashyap AS, Sahu PK, Singh UB (2022) Application of molecular ecology approaches in sustainable agriculture for a better understanding of plant–microbiome interactions. In: Singh UB, Sahu PK, Singh HV, Sharma PK, Sharma SK (eds) Rhizosphere microbes. Microorganisms for sustainability, vol 40. Springer, Singapore. https://doi.org/10.1007/978-981-19-5872-4_3

Marco DE, Abram F (2019) Editorial: using genomics, metagenomics and other 'omics' to assess valuable microbial ecosystem services and novel biotechnological applications. Front Microbiol 10:151. https://doi.org/10.3389/fmicb.2019.00151

McNear DH Jr (2013) The rhizosphere—roots, soil and everything in between. Nat Educ Knowl 4:1

Mendes R, Garbeva P, Raaijmakers JM (2013) The rhizosphere microbiome: significance of plant beneficial, plant pathogenic, and human pathogenic microorganisms. FEMS Microbiol Rev 37:634–663. https://doi.org/10.1111/1574-6976.12028

Mishra J, Arora N-K (2016) Bioformulations for plant growth promotion and combating phytopathogens: a sustainable approach. In: Bioformulations: for sustainable agriculture. Springer, New Delhi, pp 3–33

Mohammadi MA, Cheng Y, Aslam M, Jakada BH, Wai MH, Ye K, He X, Luo T, Ye L, Dong C (2021) ROS and oxidative response systems in plants under biotic and abiotic stresses: revisiting the crucial role of phosphite triggered plants defense response. Front Microbiol 12:631318. https://doi.org/10.3389/fmicb.2021.631318

Morales SE, Holben WE (2011) Linking bacterial identities and ecosystem processes: can 'omic' analyses be more than the sum of their parts? FEMS Microbiol Ecol 75:2–16. https://doi.org/10.1111/j.1574-6941.2010.00938.x

Morcillo RJL, Manzanera M (2021) The effects of plant-associated bacterial exopolysaccharides on plant abiotic stress tolerance. Meta 11:337. https://doi.org/10.3390/metabo11060337

Mueller UG, Sachs JL (2015) Engineering microbiomes to improve plant and animal health. Trends Microbiol 23:606–617. https://doi.org/10.1016/j.tim.2015.07.009

Ojuederie O, Olanrewaju O, Babalola O (2019) Plant growth promoting Rhizobacterial mitigation of drought stress in crop plants: implications for sustainable agriculture. Agronomy 9:712. https://doi.org/10.3390/agronomy9110712

Okon O, Matrood A, Okon J (2020) Na+ exclusion and Na+/K+ ratio adjustment by mycorrhiza enhances macro/micro nutrients uptake in two members of Cucurbitaceae family under salt stress. Int J Bot Stud 5:108–114

Osman AI, Fawzy S, Farghali M, El-Azazy M, Elgarahy AM, Fahim RA, Maksoud AIAA, Ajlan AA, Yousry M, Saleem Y, Rooney DW (2022) Biochar for agronomy, animal farming, anaerobic digestion, composting, water treatment, soil remediation, construction, energy storage, and carbon sequestration: a review. Environ Chem Lett 20:2385–2485. https://doi.org/10.1007/s10311-022-01424-x

Pandit MA, Kumar J, Gulati S, Bhandari N, Mehta P, Katyal R, Rawat CD, Mishra V, Kaur J (2022) Major biological control strategies for plant pathogens. Pathogens 11:273. https://doi.org/10.3390/pathogens11020273

Pathak D, Lone R, Koul KK (2017) Arbuscular mycorrhizal fungi (amf) and plant growth-promoting rhizobacteria (pgpr) association in potato (*Solanum tuberosum* L.): a brief review. In: Kumar V, Kumar M, Sharma S, Prasad R (eds) Probiotics and plant health. Springer, Singapore, pp 401–420. https://doi.org/10.1007/978-981-10-3473-2_18

Pathak H, Kumar M, Molla KA, Chakraborty K (2021) Abiotic stresses in rice production: impacts and management. Oryza 58:103–125. https://doi.org/10.35709/ory.2021.58.spl.4

Pereira JF, Oliveira ALM, Sartori D, Yamashita F, Mali S (2023) Perspectives on the use of biopolymeric matrices as carriers for plant-growth promoting bacteria in agricultural systems. Microorganisms 11:467. https://doi.org/10.3390/microorganisms11020467

Perez-Garcia A, Romero D, De Vicente A (2011) Plant protection and growth stimulation by microorganisms: biotechnological applications of *bacilli* in agriculture. Curr Opin Biotechnology 22:187–193. https://doi.org/10.1021/bp025532t

Poveda J, Abril-Urias P, Escobar C (2020) Biological control of plant-parasitic nematodes by filamentous fungi inducers of resistance: *Trichoderma*, mycorrhizal and endophytic fungi. Front Microbiol 11:992. https://doi.org/10.3389/fmicb.2020.00992

Rahman NSN, Abdul Hamid NW, Nadarajah K (2021) Effects of abiotic stress on soil microbiome. Int J Mol Sci 22:9036. https://doi.org/10.3390/ijms22169036

Rai GK, Bhat BA, Mushtaq M, Tariq L, Rai PK, Basu U, Dar AA, Islam ST, Dar TUH, Bhat JA (2021) Insights into decontamination of soils by phytoremediation: a detailed account on heavy metal toxicity and mitigation strategies. Physiol Plant 173:287–304. https://doi.org/10.1111/ppl.13433

Reeve J, Schadt C, Carpenter-Boggs L, Kang S, Zhou J, Reganold JP (2010) Effects of soil type and farm management on soil ecological functional genes and microbial activities. ISME J 4:1099–1107. https://doi.org/10.1038/ismej.2010.42

Reynaert S, De Boeck HJ, Verbruggen E, Verlinden M, Flowers N, Nijs I (2021) Risk of short-term biodiversity loss under more persistent precipitation regimes. Glob Chang Biol 27:1614–1626. https://doi.org/10.1111/gcb.15501

Riadh K, Wided M, Hans-Werner K, Chedly A (2010) Responses of halophytes to environmental stresses with special emphasis to salinity. Adv Bot Res 53:117–145. https://doi.org/10.1016/S0065-2296(10)53004-0

Riaz M, Kamran M, Fang Y, Wang Q, Cao H, Yang G, Deng L, Wang Y, Zhou Y, Anastopoulos I (2021) Arbuscular mycorrhizal fungi-induced mitigation of heavy metal phytotoxicity in metal contaminated soils: a critical review. J Hazard Mater 402:123919. https://doi.org/10.1016/j.jhazmat.2020.123919

Ross F, Di Matteo J, Cerrudo A (2020) Maize prolificacy: a source of reproductive plasticity that contributes to yield stability when plant population varies in drought-prone environments. Field Crop Res 247:107699. https://doi.org/10.1016/j.fcr.2019.107699

Roychowdhury D, Paul M, Banerjee SK (2014) Review on the effects of biofertilizers and biopesticides on rice and tea cultivation and productivity. Int J Eng Sci Technol 2:96–108

Saad MM, Eida AA, Hirt H (2020) Tailoring plant-associated microbial inoculants in agriculture: a roadmap for successful application. J Exp Bot 71:3878–3901. https://doi.org/10.1093/jxb/eraa111

Sahu PK, Gupta A, Singh M (2018) Bioformulation and fluid bed drying. A new approach towards an improved biofertilizer formulation. In: Sengar RS, Singh A (eds) Eco-friendly agrobiological techniques for enhancing crop productivity. Springer, Singapore, pp 47–62. https://doi.org/10.1007/978-981-10-6934-5_3

Samaddar S, Chatterjee P, Roy Choudhury A, Ahmed S, Sa T (2019) Interactions between *pseudomonas* spp. and their role in improving the red pepper plant growth under salinity stress. Microbiol Res 219:66–73. https://doi.org/10.1016/j.micres.2018.11.005

Sarhan MS, Hamza MA, Youssef HH (2019) Culturomics of the plant prokaryotic microbiome and the dawn of plant-based culture media—a review. J Adv Res 19:15–27. https://doi.org/10.1016/j.jare.2019.04.002

Schoebitz M, Belchi MDL (2016) Encapsulation techniques for plant growth-promoting rhizobacteria. In: Arora NK, Mehnaz S, Balestrini R (eds) Bioformulations for sustainable agriculture. Springer, New Delhi, pp 251–265. https://doi.org/10.1007/978-81-322-2779-3_14

Sergaki C, Lagunas B, Lidbury I, Gifford ML, Schäfer P (2018) Challenges and approaches in microbiome research: from fundamental to applied. Front Plant Sci 9:1205. https://doi.org/10.3389/fpls.2018.01205

Sessitsch A, Hardoim P, Döring J, Weilharter A, Krause A, Woyke T (2012) Functional characteristics of an endophyte community colonizing roots as revealed by metagenomic analysis. Mol Plant-Microbe Interact 25:28–36. https://doi.org/10.1094/MPMI-08-110204

Siddikee MA, Chauhan PS, Anandham R, Han GH, Sa TM (2010) Isolation, characterization, and use for plant growth promotion under salt stress, of ACC deaminase-producing halotolerant bacteria derived from coastal soil. J Microbiol Biotechnol 20:1577–1584. https://doi.org/10.4014/jmb.1007.07011

Singh JS, Pandey VC, Singh DP (2011) Efficient soil microorganisms: a new dimension for sustainable agriculture and environmental development. Agric Ecosyst Environ 140:339–353. https://doi.org/10.1016/j.agee.2011.01.017

Singh C, Wang-Erlandsson L, Fetzer I, Rockström J, Van Der Ent R (2020) Rootzone storage capacity reveals drought coping strategies along rainforest-savanna transitions. Environ Res Lett 15:124021. https://doi.org/10.1088/1748-9326/abc377

Sofy MR, Aboseidah AA, Heneidak SA, Ahmed HR (2021) ACC deaminase containing endophytic bacteria ameliorate salt stress in *Pisum sativum* through reduced oxidative damage and induction of antioxidative defense systems. Environ Sci Pollut Res 28:40971–40991. https://doi.org/10.1007/s11356-021-13585-3

Sharma B, Tiwari S, Bisht N, Tewari L (2021) Eco-friendly bioprocess using agar plug immobilized Penicillium crustosum PWWS-6 biomass for treatment of wastewater contaminated with toxic Congo red dye for use in agriculture. Ind Crop Prod 170:113755

Sharma B, Kumawat KC (2022) Functional diversity of microbes in rhizosphere: A key player for soil health conservation under changing climatic conditions In Plant Stress Mitigators: Action and Application. Springer Nature Singapore, Singapore, pp 469–493

Sharma B, Tiwari S, Kumawat KC, Cardinale M (2023a) Nano-biofertilizers as bio-emerging strategies for sustainable agriculture development: Potentiality and their limitations. Sci Total Environ 860:160476

Sharma B, Tiwari S, Bisht S, Bhrdwaj A, Nayarisseri A, Tewari L (2023b) Coupling effect of ionophore and oxidoreductases produced by halotolerant novel fungal strain Trametes flavida WTFP2 on dye wastewater treatment: An optimized green bioprocess. J Environ Chem Eng 11(3):109629

Stearns JC, Woody OZ, McConkey BJ, Glick BR (2012) Effects of bacterial ACC deaminase on *Brassica napus* gene expression. Mol Plant-Microbe Interact 25:668–676. https://doi.org/10.1094/MPMI-08-11-0213

Sun Z, Song J, Xin XA, Xie X, Zhao B (2018) Arbuscular mycorrhizal fungal 14-3-3 proteins are involved in arbuscule formation and responses to abiotic stresses during AM symbiosis. Front Microbiol 9:91. https://doi.org/10.3389/fmicb.2018.00091

Surendra KA, Baby A (2016) Enhanced shelf life of *Azospirillum* and PSB (phosphate solubilizing bacteria) through addition of chemical additives in liquid formulations. Int J Environ Sci Technol 5:2023–2029

Suyal DC, Soni R, Sai S, Goel R (2016) Microbial inoculants as biofertilizer. In: Singh DP (ed) Microbial inoculants in sustainable agricultural productivity. Springer, New Delhi, pp 311–318. https://doi.org/10.1007/978-81-322-2647-5_18

Sytar O, Kumari P, Yadav S, Brestic M, Rastogi A (2019) Phytohormone priming: regulator for heavy metal stress in plants. J Plant Growth Regul 38:739–752. https://doi.org/10.1007/s00344-018-9886-8

Szuba A, Marczak Ł, Kozłowski R (2021) Pb stress and Ectomycorrhizas: strong protective proteomic responses in poplar roots inoculated with *Paxillus involutus* isolate and characterized by low root colonization intensity. Int J Mol Sci 22:4300. https://doi.org/10.3390/ijms22094300

Tewari S, Arora NK (2016) Exopolysaccharides based bioformulation from *Pseudomonas aeruginosa* combating saline stress. Recent Trends PGPR Res Sustain Crop Product 2016:93

Tiwari S, Sharma B, Bisht N, Tewari L (2023) Role of beneficial microbial gene pool in mitigating salt/nutrient stress of plants in saline soils through underground phytostimulating signalling molecules. Pedosphere 33(1):153–171

Tiwari S, Sharma B (2023) 4 Green Synthesis of Plant-Based Nanoparticles Advances in Nanotechnology for Smart Agriculture. Techniques and Applications 71

Tian F, Hu X-L, Yao T, Yang X, Chen J-G, Lu M-Z, Zhang J (2021) Recent advances in the roles of HSFs and HSPs in heat stress response in Woody plants. Front Plant Sci 12:704905. https://doi.org/10.3389/fpls.2021.704905

Timmusk S, El-Daim IAA, Copolovici L (2014) Drought-tolerance of wheat improved by rhizosphere bacteria from harsh environments: enhanced biomass production and reduced emissions of stress volatiles. PLoS One 9:e96086. https://doi.org/10.1371/journal.pone.0096086

Trivedi P, Leach JE, Tringe SG, Sa T, Singh BK (2020) Plant–microbiome interactions: from community assembly to plant health. Nat Rev Microbiol 18:607–621. https://doi.org/10.1038/s41579-020-0412-1

Turner TR, James EK, Poole PS (2013) The plant microbiome. Genome Biol 14:209. https://doi.org/10.1186/gb-2013-14-6-209

Tyagi J, Chaudhary P, Jyotsana BU, Bhandari G, Chaudhary A (2022a) Chapter 17 impact of endophytic fungi in biotic stress management. In: Soni R, Suyal D, Goel R (eds) Plant protection: from chemicals to biologicals. De Gruyter, Berlin, Boston, pp 447–462. https://doi.org/10.1515/9783110771558-017

Tyagi J, Chaudhary P, Mishra A, Khatwani M, Dey S, Varma A (2022b) Role of endophytes in abiotic stress tolerance: with special emphasis on *Serendipita indica*. Int J Environ Res 16:62. https://doi.org/10.1007/s41742-022-00439-0

Ullah A, Nisar M, Ali H, Hazrat A, Hayat K, Keerio AA, Ihsan M, Laiq M, Ullah S, Fahad S (2019) Drought tolerance improvement in plants: an endophytic bacterial approach. Appl Microbiol Biotechnol 103:7385–7397. https://doi.org/10.1007/s00253-019-10045-4

Upadhyay SK, Srivastava AK, Rajput VD, Chauhan PK, Bhojiya AA, Jain D, Chaubey G, Dwivedi P, Sharma B, Minkina T (2022) Root exudates: mechanistic insight of plant growth promoting Rhizobacteria for sustainable crop production. Front Microbiol 13:916488. https://doi.org/10.3389/fmicb.2022.916488

Vu LD, Gevaert K, De Smet I (2019) Feeling the heat: searching for plant thermosensors. Trends Plant Sci 24:210–219. https://doi.org/10.1016/j.tplants.2018.11.004

Wohlfahrt J, Colin F, Assaghir Z, Bockstaller C (2010) Assessing the impact of the spatial arrangement of agricultural practices on pesticide runoff in small catchments: combining hydrological modeling and supervised learning. Ecol Indic 10:826–839. https://doi.org/10.1016/j.ecolind.2010.01.004

Yadav J, Verma J-P, Jaiswal D-K, Kumar A (2014) Evaluation of pgpr and different concentration of phosphorus level on plant growth, yield and nutrient content of rice (*Oryza sativa*). Ecol Eng 62:123–128

Yang L, Wen KS, Ruan X, Zhao YX, Wei F, Wang Q (2018) Response of plant secondary metabolites to environmental factors. Molecules 23:762. https://doi.org/10.3390/molecules23040762

Yang A, Akhtar SS, Fu Q, Naveed M, Iqbal S, Roitsch T, Jacobsen SE (2020) *Burkholderia phytofirmans* PsJN stimulate growth and yield of quinoa under salinity stress. Plan Theory 9:672. https://doi.org/10.3390/plants9060672

Yashiro E, Spear RN, McManus PS (2011) Culture-dependent and culture-independent assessment of bacteria in the apple phyllosphere. J Appl Microbiol 110:1284–1296. https://doi.org/10.1111/j.1365-2672.2011.04975.x

Zaidi A, Khan MS, Saif S (2017) Role of nitrogen-fixing plant growth-promoting rhizobacteria in sustainable production of vegetables. Current perspective. In: Zaidi A, Khan MS (eds) Microbial strategies for vegetable production. Springer, Cham, pp 49–79. https://doi.org/10.1007/978-3-319-54401-4

Zhang Y, Yu X, Zhang W, Lang D, Zhang X, Cui G, Zhang X (2019) Interactions between endophytes and plants: beneficial effect of endophytes to ameliorate biotic and abiotic stresses in plants. J Plant Biol 62:1–13. https://doi.org/10.1007/s12374-018-0274-5

Zhang Y, Ding J, Wang H, Su L, Zhao C (2020) Biochar addition alleviate the negative effects of drought and salinity stress on soybean productivity and water use efficiency. BMC Plant Biol 20:288. https://doi.org/10.1186/s12870-020-02493-2

Zhang F, Labourel A, Haon M, Kemppainen M, Da Silva ME, Brouilly N, Veneault-Fourrey C, Kohler A, Rosso MN, Pardo A (2022) The ectomycorrhizal basidiomycete *Laccaria bicolor* releases a GH28 polygalacturonase that plays a key role in symbiosis establishment. New Phytol 233:2534–2547. https://doi.org/10.1111/nph.17940

Importance of Microbial Inoculants in Maintaining Plant Growth and Productivity

Shivani Bhutani, Anshi Mehra, Anita Mishra, Deep Chandra Suyal, Shruti Bhasin, Nisha Dinkar, Nishtha Srivastava, Sami Abou Fayssal, Parul Chaudhary, and Manali Singh

Abstract Advanced plant rearing operations and reasonable developments are required to boost crop production and support better innovative techniques. To achieve these goals, it is necessary to acquire deep understanding onto soil microbial communities, plant-microbiome collaboration, and their steadiness during the natural transformation. Individual microorganisms are known as the vital controllers of microbial communities. Microbial networks, having beneficial characteris-

S. Bhutani
Department of Biotechnology, Invertis Village, Bareilly, Uttar Pradesh, India

A. Mehra
Department of Chemical Engineering, B.R Ambedkar National Institute of Technology, Jalandhar, Punjab, India

A. Mishra · D. C. Suyal
Vidyadayini Institute of Science, Management and Technology, Bhopal, Madhya Pradesh, India

S. Bhasin
Department of Biotechnology, Banasthali Vidyapith, Jaipur, Rajasthan, India

N. Dinkar
Department of Biotechnology, Khandelwal College of Management Science and Technology, Bareilly, Uttar Pradesh, India

N. Srivastava
Department of Applied Science and Humanities, Invertis Village, Bareilly, Uttar Pradesh, India

S. Abou Fayssal
Department of Agronomy, Faculty of Agronomy, University of Forestry, Sofia, Bulgaria

Department of Plant Production, Faculty of Agriculture, Lebanese University, Beirut, Lebanon

P. Chaudhary
School of Agriculture, Graphic Era Hill University, Dehradun, Uttarakhand, India

M. Singh (✉)
Department of Life Sciences, Parul Institute of Applied Sciences, Parul University, Waghodia, Vadodara, Gujarat, India

P. Chaudhary, A. Chaudhary (eds.), *Microbial Inoculants*, Interdisciplinary Biotechnological Advances, https://doi.org/10.1007/978-981-97-0633-4_9

tics go about, are a robust tool to host the supportable agronomics by diminishing fertilizers use and fighting plant disorders for crop production improvement. The capability of microflora is to inspire plant development, mitigate stresses, and further develop plant robustness. Different soil inoculants are generated to make field progress a reality; this is possible via the planning of an effective microbial consortium or the design of microflora with useful attributes. Purposely, the present chapter summarizes the effect of microorganisms on crop yields, disease resistance, and nutrient cycle. The investigation of new approaches for microflora designing to attain sustainable cultivation is also denoted in this chapter.

Keywords Crop production · Microbial inoculants · Microbiota · Sustainable agriculture

1 Introduction

India's economy is primarily based on agriculture; the need for food to feed the population's rapid growth has led to extensive agricultural output intensification. India needs to increase the productivity factor to feed its alarmingly growing population (Gulati and Juneja 2022). FAO forecasts that by 2030, the agricultural product demand will increase by 60%. A greater portion of this demand will be attributed to the rest of the globe. To meet such demand, global harvesting mostly needs to depend on vertical yield development. The grossly excessive application of agrochemicals including fertilizers, herbicides, and insecticides is a key component of the traditional agricultural paradigm (Powlson et al. 2014; Chaudhary et al. 2022a). However, there is a number of sustainability issues that overcome agricultural practices, and increasing the agricultural production on finite arable areas is relatively difficult. For instance, improper and excessive use of agricultural chemicals reduces soil health, contaminates water, and produces a chemical imbalance in various biodiversity categories (Bindraban et al. 2015).

The adoption of sustainable agriculture methods, e.g., microbiological tools, is increasingly advised to solve such persistent economic and ecological difficulties (Sammauria et al. 2020; Gómez-Godínez et al. 2021). The latter specifically targets the active nourishing system, i.e., soil that is the contributor in plant production system. Leveraging the botanical microbiome is thought to be a certainly attainable strategy for the upcoming eco-innovation in an ecological conservative manner. It is intricately tied to plant health, growth, and production. In order to increase agricultural productivity and sustainability, a number of scientific and logistical issues need to be resolved. The primary forces behind raising farm output for centuries have been the advances in farming techniques and crop varieties (De Buck et al. 2016; Altieri 2019). Furthermore, global warming already showed and will also have additionally a considerable negative influence on crop production; thus, placing enormous strain on farming systems that depend heavily on seasonal precipitation (Malhi et al. 2021). Similar problems, i.e., the vigorous use of chemicals, pesticides, and fertilizers, are endangering natural habitats and polluting water, soil, and air. Majorly, because of the structural reduction in farm fertility, additional inorganic analog inputs are unable to boost livestock production in various broad yield locations (Trivedi et al. 2017).

Microbes continue to play a major role in the functionality and health of soil. Such biota has a wide range of effects on soil quality and, subsequently, plants development and productivity (Khati et al. 2018; Chaudhary et al. 2021a, b). The use of significantly important microbes in various production systems can be seen as a good chance for achieving global food security via environmentally benign methods while also keeping up with the growing population. These are anticipated to help providing food and nutrition security for an expanding global population; it will also serve as a crucial tool for agricultural sustainability (Chaudhary et al. 2021c, d). The National Academies of Sciences, Engineering, and Medicine (2018) noted that embracing microbiome features and works in the perspective of livestock fabrication undoubtedly promise for addressing some of the most pressing problems currently facing, i.e., crop yield, food security, and land degradation. Indeed, bacteria have a significant role in the production of agricultural crops and soil processes (Singh and Trivedi 2017; Kukreti et al. 2020; Kumari et al. 2020; Chaudhary et al. 2021e).

Nowadays, the overall market mainly emphasizes the fabrication and development of soil inoculants for its uses in livestock, based upon the eventual advantages indicated by excessive (primarily lab) examination (Kumari et al. 2023). Many soil inoculants (most commonly fungi and bacteria rather than other microbes) are outlined as particular purpose organisms, i.e., its role in biological control or plant growth promotion (Kaminsky et al. 2019; Kumari et al. 2021; Tyagi et al. 2022a). Biopesticides that are an alternative to synthetic pesticides and biofertilizers that increasingly replace conventional inorganic fertilizers are more and more available on the global market (Ayilara et al. 2023). Between 2022 and 2027, a compound yearly expansion of 12.1% is predicted for the biofertilizer industry, which had a market value of USD 1.57 billion in 2018 (Mordor Intelligence 2023). Presently, the sector is very disintegrated, for numerous small scales and a handful of large-scale industries operating over various zonal areas. However, there is still a market for biofertilizers. Their commercial domination is due to a number of variables, i.e., their well-established field efficacy, the development of standardized production and administration methods, and, even possibly, the extremely particular character of the symbiotic microbe-plant relationship (Manzar et al. 2022).

Under various circumstances, soil inoculants have varied strategies for promoting plant development (Kumar et al. 2021a). For instance, in a nutrient-limited situation, inoculating peppermint with triple superphosphate and PSB (*Pseudomonas putida* and *Pantoea agglomerans*) increased P-use efficiency and crop output in comparison with non-inoculated soils (Sahandi et al. 2019). The effective soil solubilization of phosphate is done using PSB in conjunction with P-fertilization; this showed a remarkable ability to increase soil P nutrients along with the reduction of chemical fertilizers (Adnan et al. 2017). Additionally, soil inoculants can increase plant tolerance to various abiotic and biotic stresses (Tyagi et al. 2022b). For instance, the inoculation of 1-aminocyclopropane-1-carboxylate (ACC) deaminase and the preservation of ion homoeostasis might promote the growth of plant seedlings under salt and drought stresses (Xiong et al. 2019; Agri et al. 2021). Soil inoculants have the ability to produce antimicrobial substances under biotic stress, which arrest the development and infection of possible hazardous microorganisms (He et al. 2020). The induced systemic resistance (ISR) of plants can be also activated by advantageous microorganisms to ward off infections and parasites (Tyagi et al. 2022b).

2 Advancement in Plant Microbiome Studies

The use of determined variables having impacts on microflora local area's organization is an advanced method to generate gainful components. A few instances of these persuasive elements are natural conditions, legitimate agricultural practices, and bioinoculants use (Maitra et al. 2021). Afterward, microbes accompanied by basic microbiome functions are extracted and designed and overtake another natural creation. The primary discoveries of this analysis demonstrate the positive impact of microbiome on different plant cultures, along with the reduction of both abiotic and biotic stresses, which impacts their shape (synthesis/variety) (Munir et al. 2022). Several tools like sequencing, PCR, TRFLP, microscopy, RT-polymerase chain reaction, biochemical assay, automatic ribosomal intergenic spacer analysis, high performance liquid chromatography-mass spectrometry, spectrophotometry, PhyloChip, and nuclear magnetic resonance practically utilized chain sequences as their focal strategy (Panigrahi et al. 2019) (Fig. 1). This depicts the significance of advanced sequencing methods for the investigation of microbes' community. Such types of unrestrained development methods permit us to store a more prominent variety of microbes than usual development strategies. Nonetheless, these tools can be also used altogether to fundamentally separate various communities of microbes and investigate their removal from local areas. This will further lead to the extraction and disintegration of these microbes (Santos and Olivares 2021).

2.1 Molecular Methods

There is a sensational development in data due to culture-independent technique, i.e., single-cell genomics or by the gathering of genome obtained through metagenome investigation (Parveen et al. 2022). Although the genomic information of soil microorganisms is gathered, their usefulness still needs to be figured out. Subsequently, sorting microorganisms in natural taxa by considering their genomic data is hazardous (Peršoh 2015). Besides, the moderate creating of microorganisms is another challenge. Thus, approaches, which utilize genomic data to capture the information around perfect confinement and development of uncultured microorganisms, will verifiably be valuable in forward course (Liu et al. 2022).

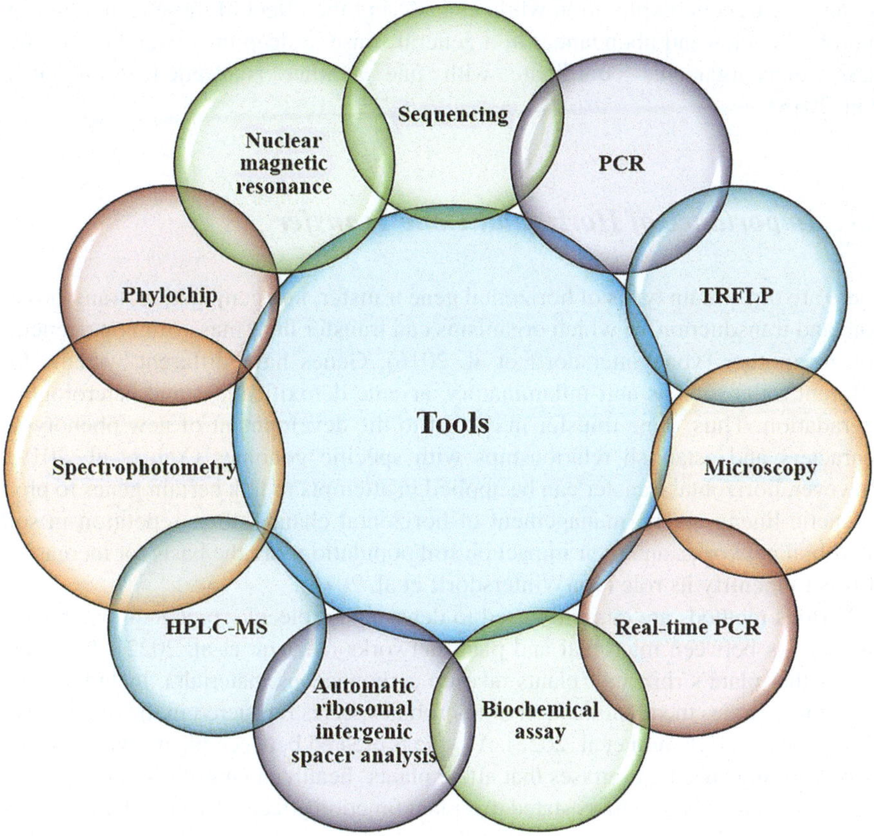

Fig. 1 Plant microbiome study tools

2.2 *Role of Viruses in Soil Microbiome Aspect*

Around 107–109 disease particles are followed down in 1 g of soil, and this number is not precisely the number of viruses found within the oceanic environment. Clearly, these soil diseases are significantly distinctive and remain dismissed (Williamson et al. 2017). Marine examinations have revealed that phages play chief portion in supplement components, as they keep up with the supplement elements by killing 20–40% microbial populations display within water. Diverse investigations have illustrated the soil to be phage-supplied, which unequivocally targets organisms like *Xanthomonas* spp. and *Rhizobium* spp. (Kizheva et al. 2021). However, the effect of these infections on soil composition and microorganisms' mode of action remains unacknowledged (Buttimer et al. 2017). It was estimated that more than 90% of soil contaminations get basically expended by soil itself. Moreover, it is dubious that these diseases may be adequately capable to contaminate the plant microbiome (Andika et al. 2016). Due to headway in viral metagenomic field and progression in unused techniques, the extension of viral populations incites researchers' exploration while considering the effect of these techniques on microbial cycles and abundance. In a general sense, a deep understanding of how these microorganisms coordinate with one another is needed (Tshikantwa et al. 2018).

2.3 Importance of Horizontal Gene Transfer

There are three main types of horizontal gene transfer, i.e., conjugation, transformation, and transduction, in which organisms can transfer their mass from one generation to another (vonWintersdorff et al. 2016). Genes have different systems for different traits such as anti-inflammatory, arsenic detoxification, and heterologous degradation. Thus, gene transfer may lead to the development of new phenotypic characters and establish relationships with specific genomes (Yan et al. 2019). However, horizontal transfer can be applied in attempts to link certain genes to phylogenetic lineages. The management of horizontal change, their repetition in soil microbial networks, and their impact on soil populations are the basis for increasing efforts to identify its role (vonWintersdorff et al. 2016).

Various methods are currently used to determine molecular preconditions for the interactions between microbial and plant networks (Manzar et al. 2022). The main idea is that plant's rhizome (plants take up carbonaceous materials), including root secretions, guides the identification of microbial species registered in the rhizosphere of target plants (Dlamini et al. 2022). A cycle is created between plants and microbes by performing specific exercises that affect plants' health and nutrition, starting from the root system. Many authors stated that plant function affected by microbial network activity can be detected by advanced plant phenotyping, while the effect of plant genotype on microbial network value cost and diversity can be differentiated by molecular ecology tools, metagroup studies, and so on (Liu et al. 2023).

3 Soil Inoculants as Biofertilizers and Biocontrol Agents and Their Mode of Action

Bacteria act as biofertilizers and biocontrol agents. Biofertilizers are defined as micronutrients that help plants to grow and maintain a healthy environment (Söğüt and Çiğ 2019; Chaudhary et al. 2022). Microorganisms are bacteria and fungi that help produce, add, store, and convert nutrients (Chaudhary et al. 2021). Thus, biodiversity is contributed by adding beneficial bacteria and fungi to the soil. Biological agents include bacteria such as arbuscular mycorrhizal fungi (AMF), nitrogen-fixing bacteria, phosphate-soluble bacteria, and soil organic fertilizers (Chaudhary et al. 2022b). All these biological agents release chemicals that promote plant growth and development and secrete antibiotics to fight pathogenic bacteria

(Chaudhary et al. 2022c). It participates in the nutrient cycle and is environmentally friendly and sustainable; therefore, biofertilizers form the backbone of nutritional management (Agri et al. 2022; Chaudhary et al. 2023a, b, c). Because plant survival and crop production depend on both biotic and abiotic ecosystems, biotechnology is being tested to improve stress response and mitigation (Chaudhary et al. 2022d, e). Table 1 summarizes the role of some soil inoculants in the improvement of plant production and development and biocontrol of non-desirable diseases/parasites.

4 Bacteria as Biofertilizers

Bacteria are a group of organisms that do not belong to any type of biofertilizers but, at the same time, can be considered as biofertilizers (Mitter et al. 2021). These bacteria support plant growth by facilitating plant-microbe interaction with their three-phase process. For instance, rhizosphere actinomycetes isolated from fire and actinorhizal nitrogen stimulate nodule formation, which ultimately helps nitrogen fixation within the plant. Such third linkages are not very well understood but have been demonstrated to improve the biofertilizers efficiency (AbdElgawad et al. 2020).

4.1 "Bio" Concept

Organic fertilizers are plant or animal products such as animal feed and green feed. Biofertilizers are micronutrients that help increase soil fertility and crop yield. Biocides are pesticides that protect plants by destroying harmful insects (Mazid and Khan 2015). Bio-binding agents are microbial organisms that increase plant growth by secreting hormones. The application of biofertilizers increases nutrients, minerals, and salts in soil (Kumar et al. 2022). Specific and selected bacterial strains are grown on a large scale in labs and are becoming among the best materials for biofertilizers preparation. Biofertilizers are also used altogether with bacteria and nutrients in liquid form as cell protectors. New biofertilizers contain bacteria that are beneficial to plants and with some advantageous characteristics, e.g., environmentally friendly and organic food safety (Allouzi et al. 2022).

Biofertilizing microorganisms promote plant growth hormones, i.e., auxin, cytokinin, gibberellic acid, and ethylene, directly involved in whole plant growth, production, and yield (Orozco-Mosqueda et al. 2023). Rhizobacteria with plant growth-promoting activity (PGPR) have received a lot of attention in this regard (Gamez et al. 2019; Söğüt and Çiğ 2019). The latter are also generally mixed with nitrogenous materials and provide oxides, phosphate, potassium, sulfur, and iron (Širić et al. 2022).

Table 1 Soil inoculants as biofertilizer/biocontrol agents

Soil inoculant	Role	References
Azotobacter, *Rhizobium*, *anabaena*, *Pseudomonas* sp., AM fungi	Helped plants accessing nutrients	Umesha et al. (2018)
Auxin-producing bacteria and cellulolytic microorganisms	Increased the availability of easily accessible nutrients to plants	Umesha et al. (2018)
PGPR	Reduced ethylene production of plants by the help of ACC deaminase	Olanrewaju et al. (2017)
Pseudomonas spp., *Azotobacter* spp., *Burkholderia* spp., *Enterobacter* spp., *Erwinia* spp., *Kushneria* spp., *Paenibacillus* spp., *Ralstonia* spp., *Rhizobium* spp.	Produced inorganic acids and chelating substances to solubilize mineral phosphate	Alori et al. (2017)
PGPR	Sustained intrinsic resistance of plant toward pathogens	Enebe and Babalola (2018)
PGPR	HCN synthesized by PGPR enhanced the efficacy of plants antifungal activity	Olanrewaju et al. (2017)
Bradyrhizobium and *Sinorhizobium*	Increased grain yield and grain N yield via increased N fixation	Thilakarathna and Raizada (2017)
Azospirillum	Increased phytohormone production, stimulated root growth, and improved N fixation of plants	Cassán et al. (2020)
Pseudomonas simiae PICF7	Increased the systemic defenses of olive plants	Montes-Osuna et al. (2021)
Streptomyces strains CBQ-EA-2	Increased the extracellular enzyme activities of bean plants	Díaz-Díaz et al. (2022)
Pseudomonas spp.	Increased the production of bioactive metabolites, (antibiotics and cyclic peptides)	Bonaterra et al. (2022)
B. velezensis	Controlled pepper gray mold via hydrolase activity	Jiang et al. (2018)
S. lydicus M01	Reduced ROS accumulation and enhanced resistance of cucumbers against *Alternaria alternata*	Wang et al. (2020)
Mycorrhizal fungi	Improved biocontrol activities against *Fusarium*, *Pythium*, and *Phytophthora*	Benami et al. (2020)

Biofertilizers aid plants growth by releasing sulforaphane, antibiotics, enzymes, and/or fungicides to eliminate bacteria (Chakraborty and Akhtar 2021). Plants growth improvement by microbial biofertilizers occurs via the promotion of plant life, insurance of nutrients, and secretion of plant hormones called bio stimulants

(Htwe et al. 2019; Kumari et al. 2021). Such hormones are added to increase the effectiveness of biofertilizers. A good carrier should have a small size, be light in weight, be easily sterilized, operate easily, and be nonstick, nontoxic, microbial load-bearer, and waterproof.

5 Plant Growth-Promoting Rhizobacteria (PGPR) Used in Biological Control and Biofertilization

Most biofertilizers and organic control specialists have nowadays a place to PGPR (Pii et al. 2015; Paterson et al. 2017). PGPRs dwell within the rhizosphere of numerous plant species, where it has advantageous impacts, i.e., expanding plant efficiency and decreasing plant pathogens, such as nematodes, parasites, and microscopic organisms (Fig. 2). Most common PGPRs belong to the following classes: *Alcaligenes, Azotobacter, Bacillus, Burkholderia. Enterobacter, Pseudomonas, Rhizobium*, and *Serratia* (Saeed et al. 2021). There are several benefits from PGPR, i.e., expanded abdicate, root thickness, surrender, leaf range, chlorophyll content, protein content, pressure driven effectiveness, resilience to abiotic push, root stretching, and deferred senescence (Neshat et al. 2022). *Azotobacter chroococcum* and *A. vinelandii* are used as nitrogen biofertilizers globally. *Bacillus megaterium, B. amyloliquefaciens* IT45, and *P. fluorescens* are used as phosphorus biofertilizers (Sumbul et al. 2020). For zinc, sulfur, and silicate mineralization, species like *Thiobacillus thiooxidans, Delftia acidovorans,* and *Bacillus* spp. are basically used in Asia (Mumtaz et al. 2017). Some *Pseudomonas* strains create a wide assortment

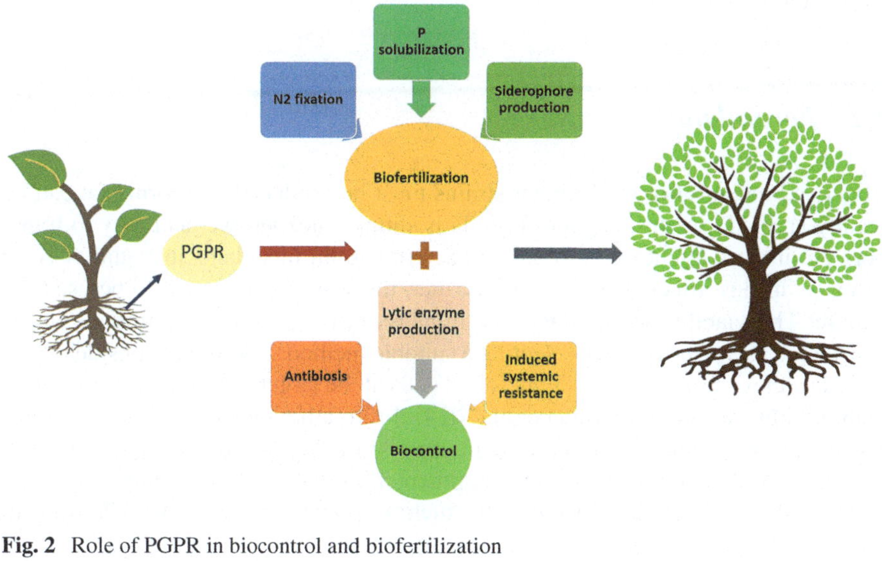

Fig. 2 Role of PGPR in biocontrol and biofertilization

of antimicrobials, counting pyoluteorin (PLT), pyrrolnitrin (PRN), 2,4-diacetylploroglucinol, and hydrogen cyanide, whereas *P. fluorescens* ZX secretes lytic chemicals (Sharma et al. 2017). A few PGPRs are accessible for natural control of *Bacillus* spp., *Pseudomonas* spp., and *Streptomyces griseoviridis* (Pii et al. 2015). On the other hand, *P. chlororaphis* is viable against unadulterated grain (*Drechslera teres*), grain spot (*Drechslera graminae*), and *Fusarium* pathogens (Huang et al. 2018).

5.1 *Rhizobia*

Rhizobium is one of the oldest agricultural tools dating back to the end of the nineteenth century. Nitrogen is fixed by α-amebocytes of *Agrobacterium*, *Heterozobium*, *Azotobacter*, and *Bradyrhizobium*. In addition to specific mycorrhizal plant interactions, the outcome of mycorrhizal inoculation also depends on the environment, i.e., soil fertility, bio-composite, salinity, and pH. To better predict mycorrhizal disease for a 10-year period, the MycoDB database for plant responses to mycorrhizal treatments has been widely used (Chaudhary et al. 2016). Despite the benefits of mycorrhizal inoculum, many difficulties remain related to its manufacture capability. Reinforced mycorrhizal fungi are often difficult to grow, because some symbionts cannot proliferate in cast iron conditions. The mycorrhizal inoculum may not be clean but fungal species exist; thus, a molecular analysis is required to evaluate the composition of the inoculum. However, there are many biofertilizers and mycorrhizal fungal biocontrol products, mainly based on *Glomus iranicum* species, that improve plant water and nutrient uptake while preventing nematodes. Another mycorrhizal vaccine containing spores of *Rhizophagus irregularis*, *Funneliformis mosseae*, and *Claroideoglomus etunicatum* helped increase fruit yield (Aamir et al. 2020).

5.2 *Formulation*

Effective biocontrol or biofertilizer strains must be produced in a form that can be readily applied to crops. A good formula is simple, inexpensive, and easy to transport. An important choice is to obtain bacterial strain inoculants in liquid or solid forms as it may affect shelf life and, further, the way it is intended to be used for culture. The viability of the microbial biomass in the process is important and is considered as a priority when selecting a suitable method. Additional considerations include adhesion and coating of microbial cells at the target site as well as microbial viability after injection (Bashan et al. 2014). From a biological perspective, gram-positive spore-forming bacteria turned out to be the best strains owing to the high resistance of their spores to various treatments. Similarly, spore-forming fungi are often well-suited for dry formulations such as powders or granules (Woo et al.

2014). However, some biofertilizers strains or biocontrol agents are gram-negative bacteria and are thus more sensitive to various environmental conditions, e.g., drought or heat, which complicate the biological formulations (Kamilova et al. 2015). Therefore, a formulation can be a bottleneck in the development of biocontrol products or biofertilizers. It is important to avoid the contamination of products as unwanted microorganisms may be completely inactivated or the strain properties may be altered. Although a simple and inexpensive manufacturing process may be required, generating strains under sterile conditions significantly increases the manufacturing costs (Arora and Mishra 2016). Arbuscular mycorrhiza may have positive consequences in terms of increased photosynthetic potential and antioxidant properties. Different microorganisms have different mechanisms to promote plant growth. For instance, *Azospirillum* spp. promote plant growth by releasing auxins, gibberellins, and ethylene. They also produce antibiotics and vitamin B complex that help fighting plant diseases (Yadav et al. 2020). Some bacteria induce plant hormone synthesis, i.e., *Paenibacillus polymyxa* that induces IAA production in Scots pine roots. The addition of *Bacillus subtilis* N11 to banana plant compost makes *Fusarium* infection controllable (Meena et al. 2020). *Azospirillum* produces a compound called cadaverine that helps rice resist to osmotic pressure (Pirttilä et al. 2021).

5.3 Beneficial Microbes for Plant Growth and Productivity

According to reports, microbes boost photosynthesis; generate biologically active substances, such as hormones and enzymes that regulate soil illnesses; and speed up the breakdown of soil lignin, thus promoting crop development and productivity (Chaudhary et al. 2021e, f). A natural microbial consortium (or may contain genetically-modified native microbial strains) has the necessary microbial variety for good agricultural outputs. Plant-associated microbes have some initial significance as effective microbes (EM) candidates (Yap and Al-Mutairi 2023). EO compositions often contain phytonutrients, lactobacilli (*Lactobacillus plantarum, Lactobacillus casei,* and *Streptococcus lactis*), photosynthetic bacteria (*Rhodopseudomonas palustris,* and *Rhodobacter sphaeroides*), Actinomycetes (*Streptomyces albus* and *Streptomyces griseus*), fermentative fungi (*Aspergillus oryzae* and *Mucor hiemalis*), and yeasts. In general, plant growth-enhancing microorganisms are divided into three subgroups: plant growth-promoting rhizobacteria (PGPR), plant growth-promoting bacteria (PGPB), and vesicular mycorrhizal fungi (PGPM) (Diagne et al. 2020). According to Moreira et al. (2016), some PGPRs, like *Pseudomonas reactans* and *Chryseobacterium humi*, and vesicular-arbuscular mycorrhiza (VAM), like *Rhizophagus*, stimulate plant growth and productivity. Bacteria like *Aspergillus*, *Penicillium*, *Bacillus*, and *Pseudomonas* have been thought to be effective solvents for phosphate. However, various performances can be improved under field conditions through genetic engineering or co-inoculation

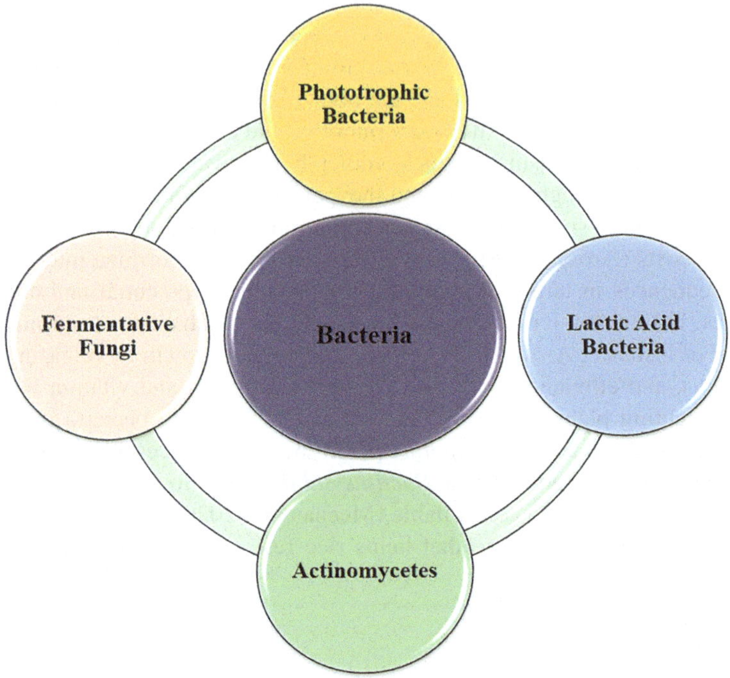

Fig. 3 Bacteria used for the enhancement of plant growth and productivity

techniques (Sharma et al. 2013). The discussion can be continued further by only focusing on the easily approachable group (bacteria) among the aforementioned ones. It includes following varieties that are also summarized in Fig. 3.

5.3.1 Phototrophic Bacteria

These organisms, known as autotrophs, create bioactive compounds such as carbohydrates, amino acids, and nucleic acids using sunlight as their energy source. In result, they are regarded as autonomous and self-sufficient. Among many solar radiation wavelengths, the infrared band between 700 and 1800 nm is beneficial for both organisms and crops (Olle and Williams 2013). These microorganisms are frequently used in agriculture to improve crop quality and plant growth.

Rhodopseudomonas palustris is a helpful strain that aids plants in defending themselves against viral infections. It generates phytohormones like 5-aminolevulinic acid and IAA, which promote tobacco germination and growth. The inoculation of this strain demonstrates an enhanced immunological reaction of plants to possible TMV infection (Su et al. 2017). It is discovered that phototrophs are helpful in promoting plant development when combined with other microorganisms. VAM, a unique example of biological agents with the capacity for both crop improvement and disease management, symbiotically promotes plant growth. Due to their facilitated access to nitrogenous substances (amino acids) produced by phototrophs (*Rhodopseudomonas* sp.), the activity of phototrophic bacteria is increased (Su et al. 2017).

N_2-fixing cyanobacteria are promising biofertilizers that can enhance soil's physicochemical properties. Moreover, crop development is greatly improved due to the increased water holding capacity and mineral uptake (Chittapun et al. 2018). More often, these bioinoculants are used to enhance soil fertility and environmental standards (Singh et al. 2016). *Anabaena, Nostoc, Calothrix, Hapalosiphon, Scytonema, Tolypothrix, Aulosira,* and *Westiellopsis* fix molecular nitrogen (N_2) in NH_3 form from plants and release amino acids, such as glutamic acid, vitamin B12, and auxin-like substances (Issa et al. 2014). Vegetable crops benefit from the use of *Anabaena vaginicola* and *Nostoc calcicola* as biofertilizers. The positive effects of cyanobacteria on crops, such as soybean, tomato, sugarcane, maize, radish, cotton, lettuce, and chili, make them very promising (Kollmen and Strieth 2022).

5.3.2 Lactic Acid Bacteria (LAB)

Salmonella, E. coli, and other microbial diseases have been well avoided in food preparation by using lactic acid bacteria (LAB). These ubiquitous microorganisms are useful in agricultural and livestock cultivation as they produce lactic acid, a potent disinfectant (Mgomi et al. 2023). The latter has the power to stop the spread of *Fusarium* wilt. During continuous cropping, *Fusarium* sp. both spreads illness and promotes the growth of dangerous nematodes (Olle and Williams 2013). Nitric oxide (NO) is produced by *Lactobacillus* sp., which also inhibits nematodes (Yarullina et al. 2014). *Ralstonia solanacearum*-induced tomato wilt is controlled by LAB (Konappa et al. 2016). *Ralstonia solanacearum* was found antagonistic to *Lactobacillus* sp. KLF01 (Limanska et al. 2013). In an ecological environment with poor nutrition, *L. plantarum* promotes plant growth (Shrestha et al. 2014). Cucumber growth was aided by a combination of *L. plantarum, Sphaeroides, Rhodobacter*, and *Saccharomyces cerevisiae* (Kang et al. 2015).

5.3.3 Actinomycetes

By acting as an antibacterial agent, actinomycetes cohabit alongside photosynthetic bacteria and enhance soil quality (Lamont et al. 2017). These antimicrobials prevent the growth of harmful bacteria and fungi. Amino acids generated by photosynthetic bacteria are used to create antimicrobial chemicals by Actinomycetes and act as the missing link between bacteria and fungi. By destroying chitin in the fungal cell wall and other cellular constituents including lipids and proteins, lytic enzymes from *Streptomyces griseus* defend pepper plants from *Phytophthora capsici* (Nguyen et al. 2012). *Streptomyces* species, specifically *S. griseus, S. albus, S. antibioticus*, and *S. champavati*, protect *Solanum lycopersicon* from *Rhizoctonia solani* (Singh et al. 2016). Actinomycetes also aid in encouraging crop growth besides their involvement in biocontrol. *Streptomyces* boosted the number of nodules, shoot weight, and yield in chickpea plants as well as soil overall N, accessible P, and organic C contents (Sreevidya et al. 2016). When cultivating saline soils, Actinobacteria is a preferred inoculant (Qin et al. 2017).

5.3.4 Fermenting Fungi

Despite having a bad reputation as major plant pathogens, some fermenting fungi restrict fungal pathogens, which promotes plant growth. The antibiotics of *Penicillium*, *Trichoderma*, and *Aspergillus* are abundant in "disease suppressive soils." Organic molecules are quickly broken down by *Aspergillus* and *Penicillium* into esters, alcohol, and antibacterial compounds. This lessens odors and stop the entry of hazardous insects and maggots (Sequeira 2016). *Trichoderma* sp. produces a variety of enzymes, including cellulases, hemicellulases, proteases, and 1,3-glucanase, which are widely used as antagonists against many pests.

VAM connected to plant roots may include *Rhizobium* and *Azotobacter* that fix N_2. In addition to their participation in the detritus food chain and the manufacturing of antimicrobials and other beneficial compounds to plant growth, unicellular yeasts also undertake a number of ecological tasks that ensure plants welfare. In the root system, bioactive substances that promote cell division include yeast hormones and enzymes. Through the use of phytase and polyamine, *Candida tropicalis* was able to release phosphate from non-dissolvable tricalcium phosphate and boost rice seedlings growth (Amprayn et al. 2012).

6 Methods of Application of Soil Inoculants

EMs are PGPMs that improve soil quality, siderophores production, ACC deaminase activity, phytohormone production, phosphorus and mineral solubilization, and crop output (Duan et al. 2021). In addition, microorganisms can serve as a biopesticide/biocontrol agent for plants. According to reports, EMs boost photosynthesis, create bioactive compounds like hormones and enzymes that regulate soil illnesses, and speed up the breakdown of soil lignin to promote crop development and productivity (Yap and Al-Mutairi 2023). There are numerous ways for microorganisms to interact with crops and encourage growth, whether directly or indirectly (Fig. 4). Indirect actions comprise the generation of biocontrol agents that inactivate or kill plant diseases in order to provide a healthy crop

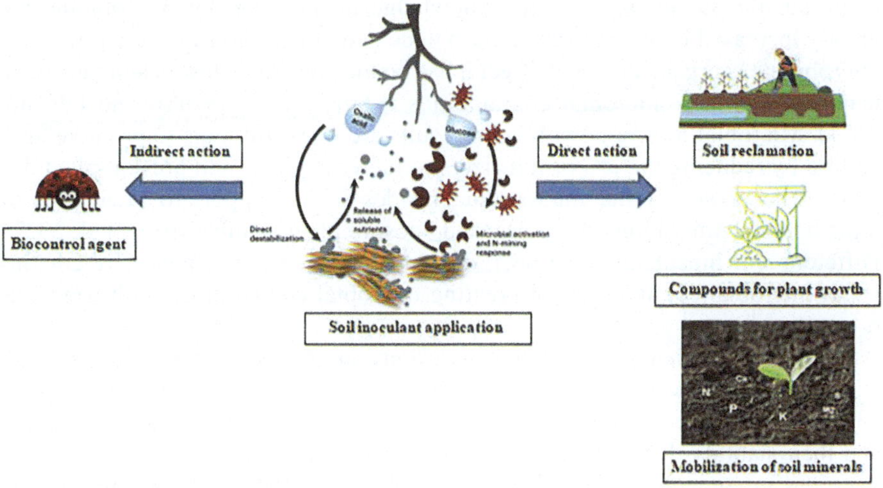

Fig. 4 Soil inoculants application methods

environment. Direct actions include soil reclamation, production of compounds for plant growth, and improvement of soil fertility through the mobilization of soil minerals (Kumar et al. 2021b).

7 Future Perspectives

Global food demand will increase by 70% by 2050; farmers will face more climate change impacts, soil pollution, food insecurity, and water scarcity (Islam and Karim 2020). In such cases, using natural resources such as the plant microbiome is the best option. Data collected on the microbiome in the scientific literature, expanding molecular methods (including testing, extraction, and remediation of DNA), reduced sequencing costs, and developments in bioinformatics, along with the traditional techniques, are important for a more comprehensive understanding of how the plant microbiome behaves and functions (Parveen et al. 2022). Once the functions, structures, and features that govern the microbiome structure are identified, they can be incorporated into agricultural systems for plant improvement (Santos and Olivares 2021). Microbiome engineering consists on the selection and introduction of beneficial bacteria. Some factors alter the structure of the microbiome, such as the selection of agricultural practices that benefit the microbial community, planting crops that benefit the microbiome, and the use of chemicals and biological agents that work synergistically with the microbiome (Moreira et al. 2023). For plants from independent cultures, it is possible to identify and exclude individuals from the microbiome. New hybrid biological inoculants can be harvested from these organisms to develop their potential in production facilities. Although knowledge about the plant microbiome has greatly increased in recent years, a deep understanding of how the structure of the microbiome is organized, what effect it has on the plant, and how beneficial chemicals are linked to microbial communities is still needed (Santos and Olivares 2021). Microbial networks with good results are powerful tools to promote agriculture by reducing the use of fertilizers to increase yields and prevent plant diseases. The power of the microbiome resides in its support to plant growth, reduction and mitigation of stresses, and boost of plant health (Afridi et al. 2022). Different soil inoculants have been developed for job performance by creating smart microbial communities or creating microbial communities with beneficial properties.

New growing methods and efficient designs are needed to increase crop yields and support the development of new technologies. Achieving these goals requires an understanding of soil microbial communities, plant-microbiome interactions, and their stability during natural changes (Michl et al. 2023). Because individual microbes are important regulators in microbial communities, a comprehensive analysis of these organisms as well as soil microbial communities can help us broaden the horizon. The effects of microbes on crop development, disease resistance, and cycling push for the exploration of new strategies for microbiome design in order to increase agricultural productivity (Afridi et al. 2022).

8 Challenges Facing Conventional Soil Inoculants Application in Agriculture

Soil fertility and productivity can be restored from organic fertilization due to nitrogen-fixing soil inoculants (Zhao et al. 2022). *Rhizobium* bacteria, found in bean root nodules, have long been used as biological fertilizers. In addition, nitrogen-fixing bacteria are also used to promote growth (Jaiswal et al. 2021). *Azospirillum* and blue-green algae are bacteria that can be used as biological fertilizers (Fukami et al. 2018). Organic compost production improves soil minerals by decomposing organic residues and agricultural by-products leading to increased yields (Goldan et al. 2023). The first famous commercial use of fertilizers began in 1895 with the introduction of "Nitragin" by Nobbe and Hilthner. Many rhizobia are used together with *Azotobacter* as nitrogen-fixing bacteria in biofertilizers. Some nitrogen-fixing cyanobacteria are also powerful biofertilizers (Aasfar et al. 2021). Many organisms are rich in phosphorus, e.g., *Bacillus megaterium*. Fungi such as *Aspergillus fumigatus* can also bind phosphorus. PGPRs have been also able to dissolve phosphorus when used with VAM fungi in biofertilizers (Etesami et al. 2021). Biofertilizers can maintain environmental sustainability by reducing air, water, and soil pollution promoted by traditional fertilizers. It increases the biodiversity of organisms, increases soil fertility, helps plants survival against various stresses, and delays the occurrence of plant diseases (Nosheen et al. 2021). However, some limitations of biofertilizers have prevented their widespread acceptance and demand in the market. More precisely, lower nutrient content, lower storage capacities in uncontrolled conditions (e.g., light, temperature, and humidity) compared to chemical fertilizers, lack of quality insurance, unskilled and incompetent workers, low productivity and poor-quality systems, and change in production and trade are important influencing factors that constitute a barrier for its commercialization on an industrial level (Mazid and Khan 2015; Gao et al. 2023).

9 Conclusion

Microbial inoculants constitute one of many solutions for the uprising traditional fertilizers issue. They can act as biofertilizers and biocontrol agents, leading to the improvement of soil texture, plant growth and development, and resultant yields. Although microbial inoculants are more and more involved in the agricultural sector, there is still a lot to know about their mode of action, mechanisms, and reactions facing several environmental conditions. It is clear that a research gap chain is lacking some of its rings in this context. More precisely, research is still more focused on the action of single or dual implications of microbial inoculants as biocontrol agents with no clear outline of how they co-interact if a multiple implication takes place. However, there is an acute need to engineer and test combinations of microbes (microbial consortia) not only on a lab scale but also on large field scale for the same purposes. Such step may not only contribute to the development of the agricultural sector but also to a higher accuracy of biological control in its large title. Training of workers on such new biotechnology for better quality preservation (facilitated by improved storage facilities) is a must if developing countries aim to continue meeting their society's needs in the upcoming decade.

References

Aamir M, Rai KK, Zehra A, Dubey MK, Kumar S, Shukla V, Upadhyay RS (2020) Microbial bioformulation-based plant biostimulants: a plausible approach toward next generation of sustainable agriculture. In: Kumar A, Radhakrishnan EK (eds) Microbial endophytes. Woodhead Publishing, Sawston, pp 195–225. https://doi.org/10.1016/B978-0-12-819654-0.00008-9

Aasfar A, Bargaz A, Yaakoubi K, Hilali A, Bennis I, Zeroual Y, Meftah Kadmiri I (2021) Nitrogen fixing Azotobacter species as potential soil biological enhancers for crop nutrition and yield stability. Front Microbiol 12:628379. https://doi.org/10.3389/fmicb.2021.628379

AbdElgawad H, Abuelsoud W, Madany MMY, Selim S, Zinta G, Mousa ASM, Hozzein WN (2020) Actinomycetes enrich soil rhizosphere and improve seed quality as well as productivity of legumes by boosting nitrogen availability and metabolism. Biomol Ther 10(12):1675. https://doi.org/10.3390/biom10121675

Adnan M, Shah Z, Fahad S, Arif M, Alam M, Khan IA, Mian IA, Basir A, Ullah H, Arshad M, Rahman IU, Saud S, Ihsan MZ, Jamal Y, Amanullah Hammad HM, Nasim W (2017) Phosphate-solubilizing bacteria nullify the antagonistic effect of soil calcification on bioavailability of phosphorus in alkaline soils. Sci Rep 7:16131. https://doi.org/10.1038/s41598-017-16537-5

Afridi MS, Javed MA, Ali S, De Medeiros FHV, Ali B, Salam A, Sumaira MRA, Alkhalifah DHM, Selim S, Santoyo G (2022) New opportunities in plant microbiome engineering for increasing agricultural sustainability under stressful conditions. Front Plant Sci 13:899464. https://doi.org/10.3389/fpls.2022.899464

Agri U, Chaudhary P, Sharma A (2021) In vitro compatibility evaluation of agriusable nanochitosan on beneficial plant growth-promoting rhizobacteria and maize plant. Natl Acad Sci Lett 44:555. https://doi.org/10.1007/s40009-021-01047-w

Agri U, Chaudhary P, Sharma A, Kukreti B (2022) Physiological response of maize plants and its rhizospheric microbiome under the influence of potential bioinoculants and nanochitosan. Plant Soil 474:451–468. https://doi.org/10.1007/s11104-022-05351-2

Allouzi MMA, Allouzi SMA, Keng ZX, Supramaniam CV, Singh A, Chong S (2022) Liquid biofertilizers as a sustainable solution for agriculture. Heliyon 8(12):e12609. https://doi.org/10.1016/j.heliyon.2022.e12609

Alori ET, Glick BR, Babalola OO (2017) Microbial phosphorus solubilization and its potential for use in sustainable agriculture. Front Microbiol 8:971. https://doi.org/10.3389/fmicb.2017.00971

Altieri MA (2019) Agroecology: the science of sustainable agriculture. CRC Press, Boca Raton. https://doi.org/10.1201/9780429495465

Amprayn KO, Rose MT, Kecskés M, Pereg L, Nguyen HT, Kennedy IR (2012) Plant growth promoting characteristics of soil yeast (*Candida tropicalis* HY) and its effectiveness for promoting rice growth. Appl Soil Ecol 61:295–299. https://doi.org/10.1016/j.apsoil.2011.11.009

Andika IB, Kondo H, Sun L (2016) Interplays between soil-borne plant viruses and RNA silencing-mediated antiviral defense in roots. Front Microbiol 7:1458. https://doi.org/10.3389/fmicb.2016.01458

Arora NK, Mishra J (2016) Prospecting the roles of metabolites and additives in future bioformulations for sustainable agriculture. Appl Soil Ecol 107:405–407. https://doi.org/10.1016/j.apsoil.2016.05.020

Ayilara MS, Adeleke BS, Akinola SA, Fayose CA, Adeyemi UT, Gbadegesin LA, Omole RK, Johnson RM, Uthman QO, Babalola OO (2023) Biopesticides as a promising alternative to synthetic pesticides: a case for microbial pesticides, phytopesticides, and nanobiopesticides. Front Microbiol 14:1040901. https://doi.org/10.3389/fmicb.2023.1040901

Bashan Y, de Bashan LE, Prabhu SR, Hernandez JP (2014) Advances in plant growth-promoting bacterial inoculant technology: formulations and practical perspectives (1998–2013). Plant Soil 378:1–33. https://doi.org/10.1007/s11104-013-1956-x

Benami M, Isack Y, Grotsky D, Levy D, Kofman Y (2020) The economic potential of arbuscular mycorrhizal fungi in agriculture. Extremophiles Biotechnol 2020:239–279. https://doi.org/10.1007/978-3-030-29541-7_9

Bindraban PS, Dimkpa C, Nagarajan L, Roy A, Rabbinge R (2015) Revisiting fertilizers and fertilization strategies for improved nutrient uptake by plants. Biol Fertil Soils 51:897–911. https://doi.org/10.1007/s00374-015-1039-7

Bonaterra A, Badosa E, Daranas N, Francés J, Roselló G, Montesinos E (2022) Bacteria as biological control agents of plant diseases. Microorganisms 10:1759. https://doi.org/10.3390/microorganisms10091759

Buttimer C, McAuliffe O, Ross RP, Hill C, O'Mahony J, Coffey A (2017) Bacteriophages and bacterial plant diseases. Front Microbiol 20:34

Cassán F, Coniglio A, López G, Molina R, Nievas S, de Carlan CLN, Donadio F, Torres D, Rosas S, Pedrosa FO, de Souza E, Díaz Zorita M, de Bashan L, Mora V (2020) Everything you must know about Azospirillum and its impact on agriculture and beyond. Biol Fertil Soils 56:461–479. https://doi.org/10.1007/s00374-020-01463-y

Chakraborty T, Akhtar N (2021) Biofertilizers: prospects and challenges for future. In: MIA I, Boddula R, Rezakazemi M (eds) Biofertilizers: study and impact. Scrivener Publishing LLC, Beverly, MA, pp 575–590. https://doi.org/10.1002/9781119724995.ch20

Chaudhary VB, Rúa MA, Antoninka A, Bever JD, Cannon J, Craig A, Duchicela J, Frame A, Gardes M, Gehring C et al (2016) MycoDB, a global database of plant response to mycorrhizal fungi. Sci Data 3:1–10. https://doi.org/10.1038/sdata.2016.28

Chaudhary A, Parveen H, Chaudhary P, Bhatt P, Khatoon H (2021) Rhizospheric microbes and their mechanism. In: Bhatt P, Gangola S, Udayanga D, Kumar G (eds) Microbial technology for sustainable environment. Springer, Singapore. https://doi.org/10.1007/978-981-16-3840-4_6

Chaudhary P, Chaudhary A, Parveen H, Rani A, Kumar G, Kumar A, Sharma A (2021a) Impact of nanophos in agriculture to improve functional bacterial community and crop productivity. BMC Plant Biol 21:519. https://doi.org/10.1186/s12870-021-03298-7

Chaudhary P, Khati P, Chaudhary A, Gangola S, Kumar R, Sharma A (2021b) Bioinoculation using indigenous *Bacillus* spp improves growth and yield of *Zea mays* under the influence of nanozeolite. 3 Biotech 11:11. https://doi.org/10.1007/s13205-020-02561-2

Chaudhary P, Khati P, Chaudhary A, Maithani D, Kumar G, Sharma A (2021c) Cultivable and metagenomic approach to study the combined impact of nanogypsum and Pseudomonas taiwanensis on maize plant health and its rhizospheric microbiome. PLoS One 16:e0250574. https://doi.org/10.1371/journal.pone.0250574

Chaudhary P, Khati P, Gangola S, Kumar A, Kumar R, Sharma A (2021d) Impact of nanochitosan and Bacillus spp on health, productivity and defence response in *Zea mays* under field condition. 3 Biotech 11:237. https://doi.org/10.1007/s13205-021-02790-z

Chaudhary P, Parveen H, Gangola S, Kumar G, Bhatt P, Chaudhary A (2021e) Plant growth promoting rhizobacteria and their application in sustainable crop production. In: Bhatt P, Gangola S, Udayanga D, Kumar G (eds) Microbial technology for sustainable environment. Springer, Singapore. https://doi.org/10.1007/978-981-16-3840-4_13

Chaudhary P, Sharma A, Chaudhary A, Khati P, Gangola S, Maithani D (2021f) Illumina based high throughput analysis of microbial diversity of maize rhizosphere treated with nanocompounds and *Bacillus* sp. Appl Soil Ecol 159:103836. https://doi.org/10.1016/j.apsoil.2020.103836

Chaudhary A, Chaudhary P, Upadhyay A, Kumar A, Singh A (2022) Effect of gypsum on plant growth promoting rhizobacteria. Environ Ecol 39(4A):1248–1256

Chaudhary P, Singh S, Chaudhary A, Agri U, Bhandari G (2022a) Chapter 3 agrochemicals and their effects on soil microbial population. In: Soni R, Suyal D, Goel R (eds) Plant protection: from chemicals to biologicals. De Gruyter, Berlin, Boston, pp 45–58. https://doi.org/10.1515/9783110771558-003

Chaudhary P, Singh S, Chaudhary A, Sharma A, Kumar G (2022b) Overview of biofertilizers in crop production and stress management for sustainable agriculture. Front Plant Sci 13:930340. https://doi.org/10.3389/fpls.2022.930340

Chaudhary P, Agri U, Chaudhary A, Kumar A, Kumar G (2022c) Endophytes and their potential in biotic stress management and crop production. Front Microbiol 13:933017. https://doi.org/10.3389/fmicb.2022.933017

Chaudhary P, Chaudhary A, Agri U, Khatoon H, Singh A (2022d) Recent trends and advancements for agro-environmental sustainability at higher altitudes. In: Goel R, Soni R, Suyal DC, Khan M (eds) Survival strategies in cold-adapted microorganisms. Springer, Singapore. https://doi.org/10.1007/978-981-16-2625-8_19

Chaudhary P, Chaudhary A, Bhatt P, Kumar G, Khatoon H, Rani A, Kumar S, Sharma A (2022e) Assessment of soil health indicators under the influence of nanocompounds and Bacillus spp. in field condition. Front Environ Sci 9:769871. https://doi.org/10.3389/fenvs.2021.769871

Chaudhary P, Chen S, Rajput VD, Jaiswal DK (2023a) Editorial: bioinoculants with nanocompounds to improve soil health: a step toward sustainable agriculture. Front Environ Sci 11:1270002. https://doi.org/10.3389/fenvs.2023.1270002

Chaudhary P, Xu M, Ahamad L, Chaudhary A, Kumar G, Adeleke BS, Verma KK, Hu D-M, Širić I, Kumar P, Popescu SM, Abou Fayssal S (2023b) Application of synthetic consortia for improvement of soil fertility, pollution remediation, and agricultural productivity: a review. Agronomy 13:643. https://doi.org/10.3390/agronomy13030643

Chaudhary P, Chen S, Rajput VD, Jaiswal DK (2023c) Bioinoculants with nano-compounds to improve soil health: a step toward sustainable agriculture. Frontiers Media SA, Lausanne. https://doi.org/10.3389/978-2-8325-3287-4

Chittapun S, Limbipichai S, Amnuaysin N, Boonkerd R, Charoensook M (2018) Effects of using cyanobacteria and fertilizer on growth and yield of rice, Pathum Thani I: a pot experiment. J Appl Psychol 30:79–85. https://doi.org/10.1007/s10811-017-1138-y

De Buck S, De Oliveira D, Van Montagu M (2016) Key innovations in plant biotechnology and applications in agriculture, industrial processes, and healthcare. International Plant Biotechnology Outreach, Ghent, pp 13–33

Diagne N, Ndour M, Djighaly PI, Ngom D, Ngom MCN, Ndong G, Svistoonoff S, Cherif-Silini H (2020) Effect of plant growth promoting rhizobacteria (PGPR) and arbuscular mycorrhizal fungi (AMF) on salt stress tolerance of Casuarina obesa (Miq.). Front Sustain Food Syst 4:601004. https://doi.org/10.3389/fsufs.2020.601004

Díaz-Díaz M, Bernal-Cabrera A, Trapero A, Medina-Marrero R, Sifontes-Rodríguez S, Cupull-Santana RD, García-Bernal M, Agustí-Brisach C (2022) Characterization of actinobacterial strains as potential biocontrol agents against Macrophomina phaseolina and Rhizoctonia solani, the main soil-borne pathogens of Phaseolus vulgaris in Cuba. Plan Theory 11:645. https://doi.org/10.3390/plants11050645

Dlamini SP, Akanmu AO, Babalola OO (2022) Rhizospheric microorganisms: the gateway to a sustainable plant health. Front Sustain Food Syst 6:925802. https://doi.org/10.3389/fsufs.2022.925802

Duan B, Li L, Chen G, Su-Zhou C, Li Y, Merkeryan H, Liu W, Liu X (2021) 1-Aminocyclopropane-1-carboxylate deaminase-producing plant growth-promoting rhizobacteria improve drought stress tolerance in grapevine (Vitis vinifera L.). Front Plant Sci 12:706990. https://doi.org/10.3389/fpls.2021.706990

Enebe MC, Babalola OO (2018) The influence of plant growth-promoting rhizobacteria in plant tolerance to abiotic stress: a survival strategy. Appl Microbiol Biotechnol 102:7821–7835. https://doi.org/10.1007/s00253-018-9214-z

Etesami H, Jeong BR, Glick BR (2021) Contribution of Arbuscular Mycorrhizal fungi, phosphate–solubilizing bacteria, and silicon to P uptake by plant. Front Plant Sci 12:699618. https://doi.org/10.3389/fpls.2021.699618

Fukami J, Cerezini P, Hungria M (2018) Azospirillum: benefits that go far beyond biological nitrogen fixation. AMB Express 8:73. https://doi.org/10.1186/s13568-018-0608-1

Gamez R, Cardinale M, Montes M, Ramirez S, Schnell S, Rodriguez F (2019) Screening, plant growth promotion and root colonization pattern of two rhizobacteria (*Pseudomonas fluorescens* Ps006 and *Bacillus amyloliquefaciens* Bs006) on banana cv. Williams (*Musa acuminata* Colla). Microbiol Res 220:12–20. https://doi.org/10.1016/j.micres.2018.11.006

Gao F, Li H, Mu X, Gao H, Zhang Y, Li R, Cao K, Ye L (2023) Effects of organic fertilizer application on tomato yield and quality: a meta-analysis. Appl Sci 13(4):2184. https://doi.org/10.3390/app13042184

Goldan E, Nedeff V, Barsan N, Culea M, Panainte-Lehadus M, Mosnegutu E, Tomozei C, Chitimus D, Irimia O (2023) Assessment of manure compost used as soil amendment—a review. PRO 11(4):1167. https://doi.org/10.3390/pr11041167

Gómez-Godínez LJ, Martínez-Romero E, Banuelos J, Arteaga-Garibay RI (2021) Tools and challenges to exploit microbial communities in agriculture. Curr Res Microb Sci 2:100062. https://doi.org/10.1016/j.crmicr.2021.100062

Gulati A, Juneja R (2022) Transforming Indian agriculture. In: Chand R, Joshi P, Khadka S (eds) Indian agriculture towards 2030. India studies in business and economics. Springer, Singapore. https://doi.org/10.1007/978-981-19-0763-0_2

He H, Hao X, Zhou W, Shi N, Feng J, Han L (2020) Identification of antimicrobial metabolites produced by a potential biocontrol Actinomycete strain A217. J Appl Microbiol 128:1143–1152. https://doi.org/10.1111/jam.14548

Htwe AZ, Moh SM, Soe KM, Moe K, Yamakawa T (2019) Effects of biofertilizer produced from *Bradyrhizobium* and *Streptomyces griseoflavus* on plant growth, nodulation, nitrogen fixation, nutrient uptake, and seed yield of mung bean, cowpea, and soybean. Agronomy 9:77. https://doi.org/10.3390/agronomy9020077

Huang R, Feng Z, Chi X, Sun X, Lu Y, Zhang B, Lu R, Luo W, Wang Y, Miao J, Ge Y (2018) Pyrrolnitrin is more essential than phenazines for pseudomonas chlororaphis G05 in its suppression of fusarium graminearum. Microbiol Res 215:55–64. https://doi.org/10.1016/j.micres.2018.06.008

Islam MFS, Karim Z (2020) World's demand for food and water: the consequences of climate change. IntechOpen. https://doi.org/10.5772/intechopen.85919

Issa AA, Abd-Alla MH, Ohyama T (2014) Nitrogen fixing cyanobacteria: future prospect. In: Ohyama T (ed) Advances in biology and ecology of nitrogen fixation. InTech, Janeza, Croatia, pp 23–46

Jaiswal SK, Mohammed M, Ibny FYI, Dakora FD (2021) Rhizobia as a source of plant growth-promoting molecules: potential applications and possible operational mechanisms. Front Sustain Food Syst 4:619676. https://doi.org/10.3389/fsufs.2020.619676

Jiang C-H, Liao M-J, Wang H-K, Zheng M-Z, Xu J-J, Guo J-H (2018) Bacillus velezensis a potential and efficient biocontrol agent in control of pepper gray mold caused by botrytis cinerea. Biol Control 126:147–157. https://doi.org/10.1016/j.biocontrol.2018.07.017

Kamilova F, Okon Y, de Weert S, Hora K (2015) Commercialization of microbes: manufacturing, inoculation, best practice for objective field testing, and registration. In: Lugtenberg B (ed) Principles of plant-microbe interactions. Springer, Cham, pp 319–327. https://doi.org/10.1007/978-3-319-08575-3_33

Kaminsky LM, Trexler RV, Malik RJ, Hockett KL, Bell TH (2019) The inherent conflicts in developing soil microbial inoculants. Trends Biotechnol 37:140–151. https://doi.org/10.1016/j.tibtech.2018.11.011

Kang SM, Radhakrishnan R, You YH, Khan AL, Park JM, Lee SM, Lee IJ (2015) Cucumber performance is improved by inoculation with plant growth promoting microorganisms. Acta agriculturae Scandinavica section B soil and plant. Science 65:36–44. https://doi.org/10.1080/09064710.2014.960889

Khati P, Parul, Bhatt P, Nisha Kumar R, Sharma A (2018) Effect of nanozeolite and plant growth promoting rhizobacteria on maize. 3 Biotech 8:141. https://doi.org/10.1007/s13205-018-1142-1

Kizheva Y, Eftimova M, Rangelov R, Micheva N, Urshev Z, Rasheva I, Hristova P (2021) Broad host range bacteriophages found in rhizosphere soil of a healthy tomato plant in Bulgaria. Heliyon 7(5):e07084. https://doi.org/10.1016/j.heliyon.2021.e07084

Kollmen J, Strieth D (2022) The beneficial effects of cyanobacterial co-culture on plant growth. Life (Basel) 12(2):223. https://doi.org/10.3390/life12020223

Konappa NM, Maria M, Uzma F, Krishnamurthy S, Nayaka SC, Niranjana SR, Chowdappa S (2016) Lactic acid bacteria mediated induction of defense enzymes to enhance the resistance in tomato against *Ralstonia solanacearum* causing bacterial wilt. Sci Hortic 207:183–192. https://doi.org/10.1016/j.scienta.2016.05.029

Kukreti B, Sharma A, Chaudhary P, Agri U, Maithani D (2020) Influence of nanosilicon dioxide along with bioinoculants on *Zea mays* and its rhizospheric soil. 3 Biotech 10:345. https://doi.org/10.1007/s13205-020-02329-8

Kumar G, Lal S, Maurya SK, Bhattacherjee AK, Chaudhary P, Gangola S, Rajan S (2021a) Exploration of *Klebsiella pneumoniae* M6 for paclobutrazol degradation, plant growth attributes, and biocontrol action under subtropical ecosystem. PLoS One 16:e0261338. https://doi.org/10.1371/journal.pone.0261338

Kumar G, Suman A, Lal S, Ram RA, Bhatt P, Pandey G, Chaudhary P, Rajan S (2021b) Bacterial structure and dynamics in mango (*Mangifera indica*) orchards after long term organic and conventional treatments under subtropical ecosystem. Sci Rep 11:20554. https://doi.org/10.1038/s41598-021-00112-0

Kumar S, Diksha SS, Sindhu KR (2022) Biofertilizers: an ecofriendly technology for nutrient recycling and environmental sustainability. Curr Res Microb Sci 3:100094. https://doi.org/10.1016/j.crmicr.2021.100094

Kumari S, Sharma A, Chaudhary P, Khati P (2020) Management of plant vigor and soil health using two agriusable nanocompounds and plant growth promotory rhizobacteria in fenugreek. 3 Biotech 10:1–11. https://doi.org/10.1007/s13205-020-02448-2

Kumari H, Khati P, Gangola S, Chaudhary P, Sharma A (2021) Performance of plant growth promotory rhizobacteria on maize and soil characteristics under the influence of TiO2 nanoparticles. Pantnagar J Res 19:28–39

Kumari M, Swarupa P, Kesari KK, Kumar A (2023) Microbial inoculants as plant biostimulants: a review on risk status. Life 13(1):12. https://doi.org/10.3390/life13010012

Lamont JR, Wilkins O, Bywater-Ekegard M, Smith DL (2017) From yogurt to yield: potential applications of lactic acid bacteria in plant production. Soil Biol Biochem 111:1–9. https://doi.org/10.1016/j.soilbio.2017.03.015

Limanska N, Ivanytsia T, Basiul O, Krylova K, Biscola V, Chobert JM, Haertle T (2013) Effect of *Lactobacillus plantarum* on germination and growth of tomato seedlings. Acta Physiol Plant 35:1587–1595. https://doi.org/10.1007/s11738-012-1200-y

Liu S, Moon CD, Zheng N, Huws S, Zhao S, Wang J (2022) Opportunities and challenges of using metagenomic data to bring uncultured microbes into cultivation. Microbiome 10:76. https://doi.org/10.1186/s40168-022-01272-5

Liu Q, Cheng L, Nian H, Jin J, Lian T (2023) Linking plant functional genes to rhizosphere microbes: a review. Plant Biotechnol J 21(5):902–917. https://doi.org/10.1111/pbi.13950

Maitra S, Brestic M, Bhadra P, Shankar T, Praharaj S, Palai JB, Shah MMR, Barek V, Ondrisik P, Skalický M, Hossain A (2021) Bioinoculants-natural biological resources for sustainable plant production. Microorganisms 10(1):51. https://doi.org/10.3390/microorganisms10010051

Malhi GS, Kaur M, Kaushik P (2021) Impact of climate change on agriculture and its mitigation strategies: a review. Sustain For 13(3):1318. https://doi.org/10.3390/su13031318

Manzar N, Jayaswal D, Kishan G, Chauhan S, Joshi D, Singh J, Chaudhary P, Jamali H, Kushwaha P, Parveen H, Kashyap AS, Sahu PK, Singh UB (2022) Application of molecular ecology approaches in sustainable agriculture for a better understanding of plant–microbiome interactions. In: Singh UB, Sahu PK, Singh HV, Sharma PK, Sharma SK (eds) Rhizosphere microbes. Microorganisms for sustainability, vol 40. Springer, Singapore. https://doi.org/10.1007/978-981-19-5872-4_3

Mazid M, Khan TA (2015) Future of bio-fertilizers in Indian agriculture: an overview. Int J Agric Food Res 3:10–23

Meena RS, Kumar S, Datta R, Lal R, Vijayakumar V, Brtnicky M, Marfo TD (2020) Impact of agrochemicals on soil microbiota and management: a review. Land 9:34. https://doi.org/10.3390/land9020034

Mgomi FC, Yang YR, Cheng G, Yang ZQ (2023) Lactic acid bacteria biofilms and their antimicrobial potential against pathogenic microorganisms. Biofilms 5:100118. https://doi.org/10.1016/j.bioflm.2023.100118

Michl K, Berg G, Cernava T (2023) The microbiome of cereal plants: the current state of knowledge and the potential for future applications. Environ Microbiome 18:28. https://doi.org/10.1186/s40793-023-00484-y

Mitter EK, Tosi M, Obregón D, Dunfield KE, Germida JJ (2021) Rethinking crop nutrition in times of modern microbiology: innovative biofertilizer technologies. Front Sustain Food Syst 5:606815. https://doi.org/10.3389/fsufs.2021.606815

Montes-Osuna N, Gómez-Lama Cabanás C, Valverde-Corredor A, Berendsen RL, Prieto P, Mercado-Blanco J (2021) Assessing the involvement of selected phenotypes of *Pseudomonas simiae* PICF7 in olive root colonization and biological control of Verticillium dahliae. Plan Theory 10:412. https://doi.org/10.3390/plants10020412

Mordor Intelligence (2023) Biofertilizers market size & share analysis—Growth trends & forecasts (2023–2028). https://www.mordorintelligence.com/industry-reports/global-biofertilizers-market-industry/. Accessed 28 May 2023

Moreira H, Pereira SI, Marques AP, Rangel AO, Castro PM (2016) Mine land valorization through energy maize production enhanced by the application of plant growth-promoting rhizobacteria and arbuscular mycorrhizal fungi. Environ Sci Pollut Res 23:6940–6950. https://doi.org/10.1007/s11356-015-5914-4

Moreira ZPM, Chen MY, Ortuno DLY, Haney CH (2023) Engineering plant microbiomes by integrating eco-evolutionary principles into current strategies. Curr Opin Plant Biol 71:102316. https://doi.org/10.1016/j.pbi.2022.102316

Mumtaz MZ, Ahmad M, Jamil M, Hussain T (2017) Zinc solubilizing bacillus spp. potential candidates for biofortification in maize. Microbiol Res 202:51–60. https://doi.org/10.1016/j.micres.2017.06.001

Munir N, Hanif M, Abideen Z, Sohail M, El-Keblawy A, Radicetti E, Mancinelli R, Haider G (2022) Mechanisms and strategies of plant microbiome interactions to mitigate abiotic stresses. Agronomy 12(9):2069. https://doi.org/10.3390/agronomy12092069

National Academies of Sciences, Engineers and Medicine (2018) Science breakthroughs to advance food and agricultural research by 2030. National Academies Press, Washington, DC

Neshat M, Abbasi A, Hosseinzadeh A, Sarikhani MR, Dadashi Chavan D, Rasoulnia A (2022) Plant growth promoting bacteria (PGPR) induce antioxidant tolerance against salinity stress through biochemical and physiological mechanisms. Physiol Mol Biol Plants 28(2):347–361. https://doi.org/10.1007/s12298-022-01128-0

Nguyen XH, Naing KW, Lee YS, Tindwa H, Lee GH, Jeong BK, Kim KY (2012) Biocontrol potential of *Streptomyces griseus* H7602 against root rot disease (*Phytophthora capsici*) in pepper. Plant Pathol J 28:282–289. https://doi.org/10.5423/PPJ.OA.03.2012.0040

Nosheen S, Ajmal I, Song Y (2021) Microbes as biofertilizers, a potential approach for sustainable crop production. Sustain For 13(4):1868. https://doi.org/10.3390/su13041868

Olanrewaju OO, Glick BR, Babalola OO (2017) Mechanisms of action of plant growth promoting bacteria. World J Microbiol Biotechnol 33:197. https://doi.org/10.1007/s11274-017-2364-9

Olle M, Williams IH (2013) Effective microorganisms and their influence on vegetable production–a review. J Hortic Sci Biotechnol 88:380–386. https://doi.org/10.1080/14620316.2013.11512979

Orozco-Mosqueda MC, Santoyo G, Glick BR (2023) Recent advances in the bacterial phytohormone modulation of plant growth. Plan Theory 12(3):606. https://doi.org/10.3390/plants12030606

Panigrahi S, Velraj P, Rao TS (2019) Functional microbial diversity in contaminated environment and application in bioremediation. In: Microbial diversity in the genomic era. Academic Press, pp 359–385. https://doi.org/10.1016/B978-0-12-814849-5.00021-6

Parveen H, Chaudhary A, Chaudhary P, Sultana R, Khati P, Kumar G, Khati P, Rana M, Bhatt P (2022) Omics approaches for the production of the microbial enzymes and applications. In: Bhatt P (ed) Industrial applications of microbial enzymes, 1st edn. CRC Press. https://doi.org/10.1201/9781003202998-16

Paterson J, Jahanshah G, Li Y, Wang Q, Mehnaz S, Gross H (2017) The contribution of genome mining strategies to the understanding of active principles of PGPR strains. FEMS Microbiol Ecol 93:fiw249. https://doi.org/10.1093/femsec/fiw249

Peršoh D (2015) Plant-associated fungal communities in the light of meta' omics. Fungal Divers 75:1–25

Pii Y, Mimmo T, Tomasi N, Terzano R, Cesco S, Crecchio C (2015) Microbial interactions in the rhizosphere: beneficial influences of plant growth-promoting rhizobacteria on nutrient acquisition process. A review. Biol Fertil Soils 51:403–415. https://doi.org/10.1007/s00374-015-0996-1

Pirttilä AM, Mohammad Parast Tabas H, Baruah N, Koskimäki JJ (2021) Biofertilizers and biocontrol agents for agriculture: how to identify and develop new potent microbial strains and traits. Microorganisms 9:817. https://doi.org/10.3390/microorganisms9040817

Powlson DS, Stirling CM, Jat ML, Gerard BG, Palm CA, Sanchez PA, Cassman KG (2014) Limited potential of no-till agriculture for climate change mitigation. Nat Clim Chang 4:678–683. https://doi.org/10.1038/nclimate2292

Qin S, Feng WW, Wang TT, Ding P, Xing K, Jiang JH (2017) Plant growth promoting effect and genomic analysis of the beneficial endophyte *Streptomyces* sp. KLBMP 5084 isolated from halophyte *Limonium sinense*. Plant Soil 416:117–132. https://doi.org/10.1007/s11104-017-3192-2
Saeed Q, Xiukang W, Haider FU, Kučerik J, Mumtaz MZ, Holatko J, Naseem M, Kintl A, Ejaz M, Naveed M, Brtnicky M, Mustafa A (2021) Rhizosphere bacteria in plant growth promotion, biocontrol, and bioremediation of contaminated sites: a comprehensive review of effects and mechanisms. Int J Mol Sci 22(19):10529. https://doi.org/10.3390/ijms221910529
Sahandi MS, Mehrafarin A, Badi HN, Khalighi-Sigaroodi F, Sharifi M (2019) Improving growth, phytochemical, and antioxidant characteristics of peppermint by phosphate-solubilizing bacteria along with reducing phosphorus fertilizer use. Ind Crops Product 141:111777. https://doi.org/10.1016/j.indcrop.2019.111777
Sammauria R, Kumawat S, Kumawat P, Singh J, Jatwa TK (2020) Microbial inoculants: potential tool for sustainability of agricultural production systems. Arch Microbiol 202:677–693. https://doi.org/10.1007/s00203-019-01795-w
Santos LF, Olivares FL (2021) Plant microbiome structure and benefits for sustainable agriculture. Curr Plant Biol 26:100198. https://doi.org/10.1016/j.cpb.2021.100198
Sequeira SO (2016) Fungal biodeterioration of paper: development of safer and accessible conservation treatments. PhD thesis, New University of Lisbon, Lisbon
Sharma SB, Sayyed RZ, Trivedi MH, Gobi TA (2013) Phosphate solubilizing microbes: sustainable approach for managing phosphorus deficiency in agricultural soils. Springerplus 2:2–14. https://doi.org/10.1186/2193-1801-2-587
Sharma P, Bora LC, Puzari KC, Baruah AM, Baruah R, Talukdar K, Kataky L, Phukan A (2017) Review on bacterial blight of rice caused by *Xanthomonas oryzae* pv. *Oryzae*: different management approaches and role of *Pseudomonas fluorescens* as a potential biocontrol agent. Int J Curr Microbiol App Sci 6:982–1005
Shrestha A, Kim BS, Park DH (2014) Biological control of bacterial spot disease and plant growth-promoting effects of lactic acid bacteria on pepper. Biocontrol Sci Tech 24:763–779. https://doi.org/10.1080/09583157.2014.894495
Singh BK, Trivedi P (2017) Microbiome and the future for food and nutrient security. Microb Biotechnol 10:50–53. https://doi.org/10.1111/1751-7915.12592
Singh JS, Kumar A, Rai AN, Singh DP (2016) Cyanobacteria: a precious bioresource in agriculture, ecosystem, and environmental sustainability. Front Microbiol 7:1–19. https://doi.org/10.3389/fmicb.2016.00529
Širić I, Eid EM, Taher MA, El-Morsy MHE, Osman HEM, Kumar P, Adelodun B, Abou Fayssal S, Mioč B, Andabaka Ž, Goala M, Kumari S, Bachheti A, Choi KS, Kumar V (2022) Combined use of spent mushroom substrate biochar and PGPR improves growth, yield, and biochemical response of cauliflower (*Brassica oleracea* var. *botrytis*): a preliminary study on greenhouse cultivation. Horticulturae 8:830. https://doi.org/10.3390/horticulturae8090830
Söğüt S, Çığ F (2019) Determination of the effect of plant growth promoting bacteria on wheat (*Triticum aestivum* L.) development under salinity stress conditions. Appl Ecol Environ Res 17:1129–1141. https://doi.org/10.15666/aeer/1701_11291141
Sreevidya M, Gopalakrishnan S, Kudapa H, Varshney RK (2016) Exploring plant growth-promotion actinomycetes from vermicompost and rhizosphere soil for yield enhancement in chickpea. Braz J Microbiol 47:85–95. https://doi.org/10.1016/j.bjm.2015.11.030
Su P, Tan X, Li C, Zhang D, Cheng JE, Zhang S, Liu Y (2017) Photosynthetic bacterium *Rhodopseudomonas palustris* GJ-22 induces systemic resistance against viruses. Microb Biotechnol 10:612–624. https://doi.org/10.1111/1751-7915.12704
Sumbul A, Ansari RA, Rizvi R, Mahmood I (2020) *Azotobacter*: a potential bio-fertilizer for soil and plant health management. Saudi J Biol Sci 27(12):3634–3640. https://doi.org/10.1016/j.sjbs.2020.08.004
Thilakarathna MS, Raizada MN (2017) A meta-analysis of the effectiveness of diverse rhizobia inoculants on soybean traits under field conditions. Soil Biol Biochem 105:177–196. https://doi.org/10.1016/j.soilbio.2016.11.022

Trivedi P, Schenk PM, Wallenstein MD, Singh BK (2017) Tiny microbes, big yields: enhancing food crop production with biological solutions. Microb Biotechnol 10:999–1003. https://doi.org/10.1111/1751-7915.12804

Tshikantwa TS, Ullah MW, He F, Yang G (2018) Current trends and potential applications of microbial interactions for human welfare. Front Microbiol 9:1156. https://doi.org/10.3389/fmicb.2018.01156

Tyagi J, Chaudhary P, Jyotsana BU, Bhandari G, Chaudhary A (2022a) Chapter 17 impact of endophytic fungi in biotic stress management. In: Soni R, Suyal D, Goel R (eds) Plant protection: from chemicals to biologicals. De Gruyter, Berlin, Boston, pp 447–462. https://doi.org/10.1515/9783110771558-017

Tyagi J, Chaudhary P, Mishra A, Khatwani M, Dey S, Varma A (2022b) Role of endophytes in abiotic stress tolerance: with special emphasis on Serendipita indica. Int J Environ Res 16:62. https://doi.org/10.1007/s41742-022-00439-0

Umesha S, Singh PK, Singh, RP (2018) Microbial biotechnology and sustainable agriculture. Biotechnology for sustainable agriculture: emerging approaches and strategies. 185–205. https://doi.org/10.1016/b978-0-12-812160-3.00006-4

VonWintersdorff CJ, Penders J, vanNiekerk JM, Mills ND, Majumder S, vanAlphen LB (2016) Dissemination of antimicrobial resistance in microbial ecosystems through horizontal gene transfer. Front Microbiol 7:173. https://doi.org/10.3389/fmicb.2016.00173

Wang M, Xue J, Ma J, Feng X, Ying H, Xu H (2020) *Streptomyces lydicus* M01 regulates soil microbial community and alleviates foliar disease caused by *Alternaria alternata* on cucumbers. Front Microbiol 11:942. https://doi.org/10.3389/fmicb.2020.00942

Williamson KE, Fuhrmann JJ, Wommack KE, Radosevich M (2017) Viruses in soil ecosystems: an unknown quantity within an unexplored territory. Ann Rev Virol 4:201–219. https://doi.org/10.1146/annurev-virology-101416-041639

Woo SL, Ruocco M, Vinale F, Nigro M, Marra R, Lombardi N, Pascale A, Lanzuise S, Manganiello G, Lorito M (2014) *Trichoderma*-based products and their widespread use in agriculture. Open Mycol J 8:71–126. https://doi.org/10.2174/1874437001408010071

Xiong Y-W, Gong Y, Li X-W, Chen P, Ju X-Y, Zhang C-M, Yuan B, Lv Z-P, Xing K, Qin S (2019) Enhancement of growth and salt tolerance of tomato seedlings by a natural halotolerant actinobacterium *Glutamicibacter halophytocola* KLBMP 5180 isolated from a coastal halophyte. Plant Soil 445:307–322. https://doi.org/10.1007/s11104-019-04310-8

Yadav AN, Singh J, Rastegari AA, Yadav N (2020) Plant microbiomes for sustainable agriculture: current research and future challenges. Springer, Cham, pp 475–482. https://doi.org/10.1007/978-3-030-38453-1_16

Yan G, Chen X, Du S, Deng Z, Wang L, Chen S (2019) Genetic mechanisms of arsenic detoxification and metabolism in bacteria. Curr Genet 65(2):329–338. https://doi.org/10.1007/s00294-018-0894-9

Yap CK, Al-Mutairi KA (2023) Effective microorganisms as halal-based sources for biofertilizer production and some socio-economic insights: a review. Foods 12(8):1702. https://doi.org/10.3390/foods12081702

Yarullina DR, Asafova EV, Kartunova JE, Ziyatdinova GK, Ilinskaya ON (2014) Probiotics for plants: NO-producing lactobacilli protect plants from drought. Appl Biochem Microbiol 50:166–168. https://doi.org/10.1134/S0003683814020197

Zhao Y, Lu G, Jin X, Wang Y, Ma K, Zhang H, Yan H, Zhou X (2022) Effects of microbial fertilizer on soil fertility and alfalfa rhizosphere microbiota in alpine grassland. Agronomy 12(7):1722. https://doi.org/10.3390/agronomy12071722

Commercialized Microbial Inoculants for Crop Improvement

Bhavya Trivedi, Anuj Chaudhary, Sami Abou Fayssal, Viabhav Kumar Upadhayay, and Govind Kumar

Abstract Since the development of a practical use of particular natural or genetically altered microorganisms, there has been a high interest in understanding the persistence of inoculation fungal and bacterial strains in field and the potential effects of their release on the native microbial communities. Structural changes of the native microbial communities may result from soil inoculation or seed bacterization, which is crucial for a secure bacterial introduction into the environment. It was acknowledged that applying microbial inoculants can change temporarily the vicinity's microbial communities. However, the key issue resides in how affected taxonomic groups can alter the functional capacities of soil microbial communities. These alterations may be caused directly by trophic competitions and interactions with local microbial communities or indirectly by increased root development and exudation. Blending inoculants might not always result in additive or synergistic impact and lead instead to a competitive process. Future studies should concentrate on determining the influence extent of the inoculants on crops and the ability of plant-soil-microbiota to act as a protective barrier.

Keywords Biofertilizers · Crop Production · Micronutrients · Soil fertility

---

B. Trivedi
Department of Microbiology, Maya Group of Colleges, Dehradun, Uttarakhand, India

A. Chaudhary
School of Agriculture and Environmental Sciences, Shobhit University, Gangoh, Uttar Pradesh, India

S. Abou Fayssal
Department of Agronomy, Faculty of Agronomy, University of Forestry, Sofia, Bulgaria

Department of Plant Production, Faculty of Agriculture, Lebanese University, Beirut, Lebanon

V. K. Upadhayay
Department of Microbiology, College of Basic Sciences and Humanities, Dr. Rajendra Prasad Central Agricultural University, Pusa, Samastipur, Bihar, India

G. Kumar (✉)
Crop Production Division, ICAR-Central Institute for Subtropical Horticulture, Lucknow, Uttar Pradesh, India

P. Chaudhary, A. Chaudhary (eds.), *Microbial Inoculants*, Interdisciplinary Biotechnological Advances, https://doi.org/10.1007/978-981-97-0633-4_10

1 Introduction

Biofertilizers are natural fertilizers based on living microorganisms that colonize rhizosphere and enhance plant growth through boosting the intake of essential nutrients by the host plant (Kumar et al. 2022a, b). The function of natural fertilizer in agriculture is becoming increasingly important, particularly due to the rising rate of chemical fertilizers as well as their detrimental consequences on soil health (Chaudhary et al. 2022a). Unregulated application of fertilizers has run to the contamination of water and soil ecosystems. Soil is depleted of vital nutrients and organic matter for plants leading to the unavailability of beneficial microbes, lower soil fertility, and higher crop vulnerability to diseases (Akanmu et al. 2021). Figure 1 outlines the beneficial inoculation of biofertilizers in agricultural practices.

Environmental risks and challenges to sustainable agriculture are causing widespread concern. Given the foregoing facts, the long-term usage of organic fertilizers has proven to be more eco-friendly, productive, cost-effective, and available to marginal and small farms as compared to synthetic fertilizers (Muhie 2022). Thus, the requirement for biofertilizer arises for a pair of reasons. The use of biofertilizers leads to higher agricultural yields while limiting soil texture deterioration and the raise of other environmental concerns (Fasusi et al. 2021). Microbes that enhance plant development via controlling damaging microbes are considered as biopesticides. Likewise, bacteria that improve plant growth by secreting phytohormones are considered bioenhancers (Chaudhary et al. 2021f, 2022e). For instance, some PGPR can improve plant development by functioning as both biofertilizer and bioenhancer

Fig. 1 Some advantages of biofertilizers inoculation in agriculture

or biopesticide. This natural pest management method not only saves the environment, but it also protects human health by lowering exposure to hazardous chemicals (Souza et al. 2015). Biofertilizers improved the soil structure while also encouraging microbial diversity resulting healthier soils (Agri et al. 2021, 2022; Khati et al. 2017, 2018).

2 Organic Fertilizers: Timely Need in Agriculture

The term "biofertilizer," which is also known as "microbial inoculant," repurposes the term "biological fertilizer" (Arora et al. 2010). A biofertilizer can be characterized as a constituent made up of living microbes that colonize the rhizosphere surrounding the underside of a plant and enhance its growth by rendering minerals and nutrients more easily available and absorbed (Malusa and Vassilev 2014; Kukreti et al. 2020; Manzar et al. 2022). Microbial inoculants or microbial cultures, bioformulations, bioinoculants, and bacterial fertilizers are other terms that are interchangeable with "biofertilizers." *Rhizobium* was the first biofertilizer identified from a variety of cultivated legumes. Table 1 lists the key achievements in biofertilizer development, production, and promotion in India.

Table 1 Commercialized microbial inoculants

Inoculant	Formulation	Product	Crop
Glomus mosseae, G. etunicatum, G. intraradices	In-furrow powder and liquid forms, granule at seedling, liquid seed treatment	MycoApply	Multiple
Bacillus amyloliquefaciens + Trichoderma	In-furrow wettable powder and granule	Quick roots	Wheat, soy, pulses, gram
Penicillium fluorescens	Liquid seed coat	Rhizofos	Multiple
Azorhizobium caulinodans, Azospirillum brasilense, Azoarcus indigens	Freeze-dried powder	TwinN	Multiple
Azospirillum	In-furrow liquid, granule, seed inoculants	MicroAZ-IF, MicroAZ-ST	Corn, turf, wheat
Bradyrhizobium japonicum + penicillium bilaiae	Granular in-furrow inoculants	TagTeam soybean	Soybean
Penicillium bilaiae	In-furrow liquid as a wettable powder or in-furrow granule	Jumpstart	Multiple
Gluconacetobacter diazotrophicus	In-furrow or foliar spray	Envita	Multiple
Sinorhizobium meliloti, Rhizobium leguminosarum	Inoculant for seed coating	Nitragin Gold	Alfalfa, sweet clover, clover

In order to create biofertilizers, a number of microorganisms are used in conjunction with agricultural plants. Various types of biofertilizers used in agriculture, along with their contribution in crop improvement, are listed in Table 2.

Table 2 Various types of biofertilizers

Biofertilizer type	Biofertilizer name	Impact	References
Nitrogen-fixing	*Actinobacteria* *Azotobacter* *Azospirillum* *Rhizobium*	Transformation of nitrogen into organic compounds Phytohormone production and N2 fixation	Aasfar et al. (2021), Gedamu et al. (2021), Aloo et al. (2022), Ketema and Tefera (2022)
Phosphorus-solubilizing	*Bacillus* *Pseudomonas* *Aspergillus*	– Improvement of P availability in soil – Rendering of insoluble P able to be utilized by plants	Kalayu (2019), Soumare et al. (2020), Mayadunna et al. (2023), Wang et al. (2023)
Zinc-solubilizing	*Pseudomonas protegens* *Bacillus megaterium* *Bacillus altitudinis*	– Improvement of Zn uptake – Improvement of Zn availability in inoculated soils	Bhatt and Maheshwari (2020), Kushwaha et al. (2021), Yasmin et al. (2021)
Siderophore-producing (Iron-sequestering)	*Bacillus subtilis*, *Bacillus licheniformis*, *Bacillus coagulans*, *Bacillus cereus*, *Pseudomonas fluorescens*, *Pseudomonas koreensis*	– Assimilation of Fe^{2+}	Sinha et al. (2019), Subramanium and Sundaram (2020), Ghazy and El-Nahrawy (2021)
Potassium-solubilizing	*Bacillus* *Burkholderia* *Pseudomonas* *Paenibacillus*	– Improvement of germination – Increased nutrient uptake and K availability in soil	Laxita and Shruti (2020), Imran et al. (2020), Raji and Thangavelu (2021)
Phytostimulators	*Azospirillum* *Bacillus* spp.	– Modulation of plant growth – Improvement of plant tolerance to abiotic stress – Plant nutrient solubilization – Nitrogen fixation – Phytohormone production	Kalimuthu et al. (2019), Coniglio et al. (2019), Kang et al. (2019), Bandopadhyay (2020), Haerani et al. (2021), Santhi et al. (2021)

3 Commercialized Microbial Inoculants for Crop Improvement

The following types of microorganisms have been commercialized as biofertilizers for crop improvement.

3.1 Nitrogen-Fixing Biofertilizers

Bacteria that perform N_2 fixation and increase soil N supply serve as nitrogen-fixing biofertilizers. Nitrogen can be fixed via both symbiotic or nonsymbiotic fixation methods. Legume nitrogen-fixing rhizobia have been explored for decades. Martinus Beijerinck discovered *Rhizobium leguminosarum* from legumes and demonstrated that this species increased the amount of fixed nitrogen in inoculated plants compared to non-inoculated ones (Sanyal et al. 2020; Ketema and Tefera 2022). The type of legume-rhizobia symbiosis, which is determined by a number of parameters, e.g., genotype, geographical distribution, and cultivar type, greatly affects the variability in fixed nitrogen amounts (Ramoneda et al. 2020; Riah et al. 2021).

Nonsymbiotic N2-fixing bacteria serve as a primary natural supply of nitrogen in natural and agricultural habitats without symbiotic N2 fixation (Tahir et al. 2020). *Clostridium, Azotobacter, Herbaspirillum, Burkholderia, Methanosarcina, Azospirillum, Gluconacetobacter, Paenibacillus*, and *Beijerinckia* have demonstrated notable efficacy in cereal crops (improvement of growth and grain yield) among nonsymbiotic N_2-fixing bacteria (Ladha et al. 2016; Yongbin et al. 2019). Soil carbon and sulfur contents are enhanced in response to *Azotobacter* species inoculation as a result of faster mineralization of soil organic wastes, which subsequently decreases roots' absorption of heavy metals (Velmourougane et al. 2019).

Commercial biofertilizers have been developed effectively using several potent N-fixing bacteria, including *Rhizobium* and *Azotobacter* species (Adeleke et al. 2019; Din et al. 2019). By inoculating these biofertilizers into farm and crop, it is possible to both meet plant N requirements and significantly reduce the need for artificial fertilizers (Aloo et al. 2022). Various nitrogen-fixing biofertilizers are found in the market with such following trade names: *Rhizo-Enrich, Bioazoto, Biospirillum, Bio-N & Spironik*, among others (Matoso et al. 2021).

3.2 Phosphate-Solubilizing Biofertilizers

Along with nitrogen fixation, biofertilizers can also provide a better bioavailability of soil nutrients and crop absorption (El-Deen et al. 2020; Abdelmoteleb and Gonzalez-Mendoza 2020). Plants require phosphorus to carry out a variety of physiological processes, e.g., cell division, photosynthesis, development of a strong root system, and carbohydrates utilization (Mahantesh and Patil 2011). However, the lack of soil soluble phosphate for agricultural productivity is frequently brought up in many countries (Zulkifli and Tarmizi 2010). This is because soils typically contain around 95% P in insoluble form. Plants can only uptake phosphorus in soluble form; however, soil P solubility depends on its origin (organic or inorganic). Phosphate-solubilizing microbial inoculants can improve soil P availability and offer biocontrol properties against agricultural pests; such biofertilizers are expected to be widely used in the future (Soumare et al. 2020). Phosphatic biofertilizers can significantly improve crop yields due to the limited economically mineable P deposits. Using P solubilizing as bioinoculants can create new possibilities for maintaining soil P levels and, in a large extent, for sustainable crop production (Rizvi et al. 2022). These biofertilizers are found in the market with such following names: Biophos, Get-Phos, and Kisan P.S.B. Culture.

Phosphate-solubilizing bacteria can increase root growth of roots by providing them with a significant amount of auxin (Kudoyarova et al. 2017; Haidar et al. 2018). The potential of *Enterobacter cancerogenus* to improve P accessibility in several organic manures such as poultry manure and cow dung was reported by Walpola and Hettiarachchi (2020). KSB are also equally crucial to crop production since they are limited in agricultural soils. KSB play a vital role in boosting crop growth and productivity (Raji and Thangavelu 2021). The aforementioned findings illustrate that, in both uncontrolled and controlled environments, these bacteria can immensely increase crop yields, nutrient intake, and growth. Studies reveled that this unique strategy may greatly increase K availability in croplands, even though K solubilization may not completely satisfy plant K requirements compared to chemical fertilizers (Laxita and Shruti 2020). Bio-Phos, Get-Phos, and PSB culture are some examples of KSB found on the market. The microbes that take part in soil P mobilization are known as phosphate-mobilizing microorganisms, e.g., *Bacillus megaterium* and mycorrhizal fungi (Tyagi et al. 2022a, b). Phosphate-solubilizing bacteria make up the vast bulk of phosphate-mobilizing microbes. They cause the release of fixed and insoluble forms of phosphorous in the soil, making plants able to sustainably absorb phosphorus (Chaudhary et al. 2021a). Microbes that mobilize phosphate play a role in the alteration of soil P. P-mobilizing biofertilizers are available in the market with MycoRhiz, Reap P, and Phosphonive as trade names (Khoshru et al. 2023).

3.3 *Biofertilizers for Micronutrients*

Even while micronutrients are only minimally necessary, their lack results in low production. Zinc (Zn) is regarded as the most important micronutrient among them. It lacks causes widespread abiotic stress in Asian countries like India, Bangladesh, Turkey, Iraq, and Syria, as well as in Africa and the USA. Zn is required for several enzymatic processes, including protein synthesis, auxin synthesis, glucose metabolism, and the maintenance of cellular membranes integrity (Gontia-Mishra et al.

2016). Due to global deficiency of Zn in soil, Zn-solubilizing biofertilizers (ZSB) are needed to improve crop productivity. Most arable areas suffer from this shortage, brought on by nutrient mining from crop harvesting. Crop yields and nutritional quality are both impacted by Zn deficiency, and crop products are often compromised, leading to nutritional and human health issues, particularly in children (Zhang et al. 2017). Zn deficit can hinder plant growth by inhibiting the production of phytohormones, lowering glucose levels, and diminishing the growth of flowers and fruits. These signs of Zn deficiency can lead to lower crop yields and lower nutritional quality (Mumtaz et al. 2017). Crops cannot absorb the insoluble Zn, which causes a Zn deficit. Zn fertilizers are useful to crops in order to address Zn shortage. Still, this method is pricey and might be hazardous for the environment and people's health. Consequently, eco-friendly and economical agrotechnology's are required to boost agricultural output and diminish Zn deficit (Zaheer et al. 2019; Khanghahi et al. 2021). Recent research on rhizobacterial Zn solubilization implied that ZSB improved plant growth and yields by increasing Zn uptake by plants (Perumal et al. 2019). *Azotobacter, Rhizobium, Pseudomonas*, and *Azospirillum* spp. were demonstrated to dramatically increase Zn uptake in wheat assessed with various rhizospheric ZSBs (Dinesh et al. 2018; Geries and Elsadany Abdelgawad 2021). Treated plants had higher Zn concentrations than untreated ones. According to Goteti et al. (2013), Zn-solubilizing bacteria, i.e., *Pseudomonas* and *Bacillus* spp., led to improved maize plant development after 60 days of inoculation. Bhakat et al. (2021) isolated rhizospheric bacteria from rice fields and demonstrated arsenic tolerance ability and Zn-solubilizing capacity (6 isolates out of 20). The most prominent bacteria were *Burkholderia vietnamiensis* and *Burkholderia seminalis*, and *Burkholderia* spp. were found to be a contender for improving soil-based Zn solubilization, which could lower the rate of use of inorganic Zn during soil fertilization (Bhakat et al. 2021).

PGPR enclose regulators and transporters that encourage the uptake of specialized nutrients and supply plants with synthesized substances (e.g., phytohormones) and improve plant growth (Kumari et al. 2020, 2021). Plant growth is positively impacted by PGPR either directly or indirectly by diminishing the negative effects of phytopathogenic organisms. This may be accomplished by generating antagonistic chemicals or by fostering pathogen resistance. PGPR are generally known for being optimistic inputs for sustainable agriculture output (Gupta et al. 2015; Backer et al. 2018; Chaudhary et al. 2022b, c, d).

Numerous bacterial species, including *Bacillus*, *Rhizobium*, and *Pseudomonas*, have been identified as primary phosphorus solubilizers (Chaudhary and Sharma 2019; Khati et al. 2019a). Numerous bacterial strains, including *Herbaspirillum, Azospirillum, Enterobacter, Azotobacter, Acetobacter*, and *Pseudomonas*, accelerate plant's developmental processes by either enhancing PGPR linked or not to grains (Gupta et al. 2021). Various studies have clearly demonstrated the beneficial effects of PGPR on the general growth and development of different plant species, the bioavailability of both micro- and macronutrients (e.g., iron (Fe), Zn, and P), the production of phytohormones, or fixing atmospheric nitrogen (Arya et al. 2020; Nawaz et al. 2020). According to Reis et al. (2020), PGPR's multi-trait mechanisms are very helpful in controlling plant pathogens and diseases. Moreover, environmental contaminants, such as pesticides, chemicals, and heavy metals, can be detoxified by PGPR (Reis et al. 2020; Kumar et al. 2021, 2022a, b; Chaudhary et al. 2023a, b).

3.4 Siderophore-Producing Biofertilizers

The presence of microbes, such as bacteria and fungi, play an important role in the entire process of plant growth and development (Bukhat et al. 2020). PGPR can boost plant growth via the improvement of heavy metal resistance, siderophore synthesis, stress responses, adaptation to drought and halotolerance, among others (Bennis et al. 2022; Ghazy and El-Nahrawy 2021; Wang et al. 2022; Chaudhary et al. 2023c). Using distinct methods, some microorganisms can sequester Fe via siderophores, which are metabolites with high Fe affinity in low Fe settings. The previous is reduced to Fe^{2+} and then released into the cell via an input mechanism after the creation of Fe^{3+} microbial siderophores complexes. Using this process, plants directly access and adjust Fe^{2+} from siderophores through ligand exchange processes (Wang et al. 2022). The main siderophore-producing microbes are *Pseudomonas fluorescens* and *Bacillus* spp. (Table 3). Plant microorganisms in the rhizosphere release one or many siderophores for protection against plant diseases

Table 3 Commercialized biofertilizers available in market

Country	Manufactures	Product name	Crop	Microorganisms	References
Australia	Nutri-Tech Solutions	Bio-N, Twin N, and Myco-Tea	Tea	*Azotobacter* spp., *Azospirillum* sp. *Bacillus polymyxa*	Adeleke et al. (2019)
China	China Bio- Fertilizer AG	CBF	Various crops	*Bacillus subtilis*	Celador-Lera et al. (2018)
Germany	AbiTEP GmbH	FZB 24 fl, BactoFil A 10	Cereals, Vegetables	*Pseudomonas fluorescens, B. amyloliquefaciens*	Odoh et al. (2019)
India	Biomax	GreenMax AgroTech Life Biomix	Various cereals	*Azotobacter* sp., *P. fluorescens*	Odoh et al. (2019)
Hungary	AgroBio	BactoFil A10	Cereals	*A. vinelandii, A. brasilense, B. megaterium*	Mustafa et al. (2019)
Kenya	MEA Fertilizer Ltd	Biofix	–	Rhizobia spp.	Adeleke et al. (2019)
South Africa	Amka Products Ltd	Organico	–	*Rhizobium*, *Enterobacter* spp., *Bacillus* spp., *Pseudomonas* spp.	Adeleke et al. (2019)

(continued)

Table 3 (continued)

Country	Manufactures	Product name	Crop	Microorganisms	References
	Biocontrol Products Ltd	Azo-N, Azo-N-Plus	–	*A. brasiliense, A. lipoferum*	Raimi et al. (2021)
USA	Flozyme Corporation	Inogro	Rice	30 bacterial species	Celador-Lera et al. (2018)
	Becker Underwood	Chickpea Nodulator	Chickpea	*Mesorhizobium cicero*	Adeleke et al. (2019)
	Plant Health Care Inc.	PHC BioPak	NA	*B, subtilis, B. megaterium, B. licheniformis*	Adeleke et al. (2019)
United Kingdom	Ammonite A 100	Cleveland Biotech	Cucumber, tomato, pepper	*Azotobacter* sp., *Rhizobium* sp., *Bacillus* sp., *Pseudomonas* sp.	Odoh et al. (2019)
	Legume Fix	Legume Technology	Common bean, soyabean	*Rhizobium* sp., *B. japonicum*	Adeleke et al. (2019)
Brazil	Embrafros Ltd.	Bioativo	Bean, sugarcane, cereals, maize	PGPR consortia	Odoh et al. (2019)
Argentina	Rhizobacteria	Rizo-Liq	Chickpea, soyabean, common beans, green gram, groundnut	*Rhizobium* spp., *Bradyrhizobium* sp., *Mesorhizobium cicero*	Adeleke et al. (2019)
Nigeria	IJTA	NoduMax	–	*Bradyrhizobia*	Tairo and Ndakidemi (2014)
Canada	Lallen and plant care BASF	Rhizocell GC Nodulator	Beans, carrot, cotton, rice, maize	*B. amyloliquefaciens* IT 45	Odoh et al. (2019)

(Zhou et al. 2019). Optimal conditions increase siderophores chelation by defeating the fungi that cause stem and root rots, which could be beneficial in a financially sensible and environmentally responsible way (Kalyan et al. 2022).

4 Commercialization Issues for Bacterial and Fungal Inoculants under Field Conditions

Figure 2 shows some commercialization issues associated with the use of bioinoculants in agriculture. Soil microbes are largely variable and play a vital role in the key biogeochemical cycles of carbon, nutrients, and water (Jacoby et al. 2017; Khati

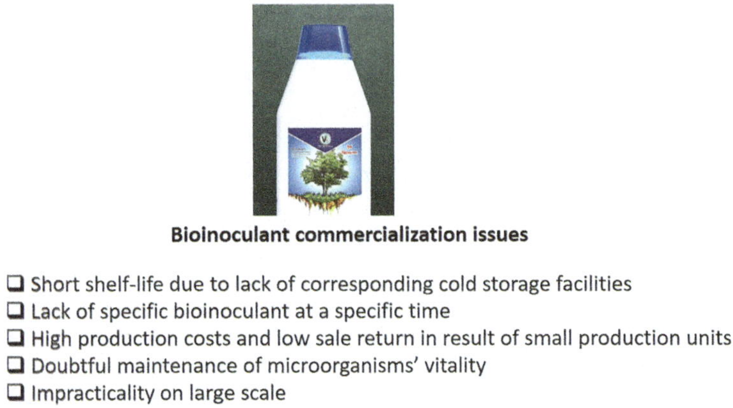

Fig. 2 Bioinoculant commercialization issues

et al. 2019b; Chaudhary et al. 2021b, c, d, e). Diverse growth habitats and microbial populations can enhance stress tolerance and nutrient delivery to plants. As bacterial cultivation is to some extent limited, the description of a microbial community, in which the bacterial members are isolated, presents a challenge. Consequently, the primary problem with the application of microbial inoculants for crop enhancement emerges from the different growing habitats and community structure of plant roots (Timmusk et al. 2017). The application of microbial inoculants in research studies as dry or liquid formulations (sprays, root dips, or drenches) arises as an addition problem as the majority of these methods are impractical on a broad scale. This is due to the fact that a significant amount of inoculant is necessary to perform appropriately. Instead of being a broad-spectrum product like chemical inputs, biofertilizers are very selective and targeted. This leads to inconsistent quality and efficacy under field circumstances, where multiple species act synergistically. The variability in biofertilizer's shelf life is another obstacle to the widespread use of microbial inoculants in agriculture (Alori and Babalola 2018). Maintaining the vitality of the microorganisms included in microbial inoculant formulations is a significant challenge. The treatment method and storage temperature have a substantial impact on the viability of microorganisms in inoculated seeds. It is advised that microbial inoculants shall have an extended shelf life under ambient storage conditions being widely used in agriculture. Additionally, preserving the viability of seeds and microorganisms throughout storage comes at a very significant cost (Callaghan 2016).

5 Limitations of Microbial Inoculants

Microbial inoculants are seen as advantageous for both the environment and crops. They serve as nitrogen fixers, phosphate solubilizers, siderophores producers, and solubilizers of micronutrients, among others. In spite of the fact that biofertilizers

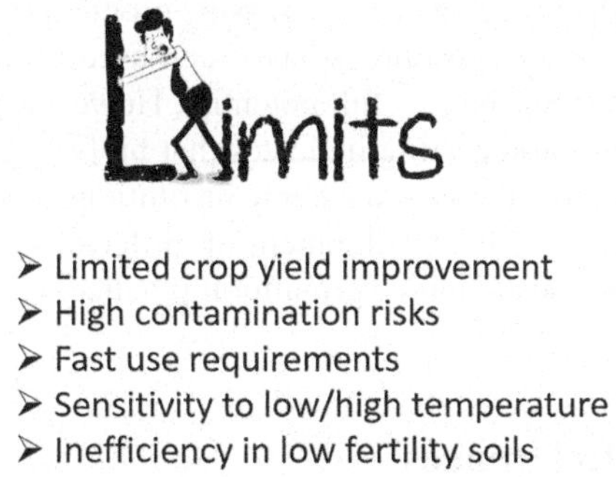

Fig. 3 Microbial inoculants' limitations

are highly effective, there is still a lot of research to be done in this area, because they are still only partial substitutes of chemical fertilizers. This assertion is supported by the fact that microbial inoculants continue to have serious downsides (Fig. 3), which include:

- The crop yield is only increased by 25–35% when using microbial fertilizers; they do not significantly boost productivity in the same way as chemical fertilizers do.
- Due to contamination, which is a major problem during microbial mass production, microbial fertilizers require rigorous aseptic conditions at the time of manufacturing.
- When kept at room temperature, microbial fertilizers must be utilized within 6 months of manufacturing, while when kept at a cooling temperature, it must be used within 2 years.
- Because microbes are temperature-sensitive, they become inactive or die when exposed to sunlight for an extended period of time.
- The effectiveness of microbial fertilizers is influenced by soil characteristics, including moisture content, pH, temperature, organic matter, and bacterial species. Microbial fertilizers may not be efficient at boosting soil fertility when these conditions exist.

6 Conclusion

The development and adoption of sustainable farming practices represents one of the major tasks. Dealing with such a challenge can only be possible when new and advanced technologies like microbial inoculants as biofertilizers are adopted. By fixing N_2 from the air and mobilizing fixed macro- and micronutrients in the soil into plant-assimilable forms, biofertilizers—essential elements of organic farming—can play an important role in preserving soil fertility on a long-term period. Ten million tons of plant nutrients are currently lost through crop clearance compared to delivery through chemical fertilizers. In this situation, farmers would have the best chance of increasing productivity per unit area using biofertilizers. Although these bioresources have been used in many world regions, their efficiency has not yet reached its full potential. However, practice will undoubtedly expand with time and forecasting indicates that biofertilizers will eventually have a sizable commercial potential. Academic institutions, researchers, and agricultural organizations can hasten the development of biofertilizers and promote their application to/adaption for sustainable agricultural practices.

References

Aasfar A, Bargaz A, Yaakoubi K, Hilali A, Bennis I, Zeroual Y, Kadmiri MI (2021) Nitrogen fixing azotobacter species as potential soil biological enhancers for crop nutrition and yield stability. Front Microbiol 12:628379. https://doi.org/10.3389/fmicb.2021.628379

Abdelmoteleb A, Gonzalez-Mendoza D (2020) Isolation and identification of phosphate solubilizing *Bacillus* spp. from *Tamarix ramosissima* rhizosphere and their effect on growth of *Phaseolus vulgaris* under salinity stress. Geomicrobiol J 37:901–908. https://doi.org/10.1080/01490451.2020.1795321

Adeleke RA, Raimi AR, Roopnarain A, Mokubedi SM (2019) In: Giri B, Prasad R, Wu QS, Varma A (eds) Status and prospects of bacterial inoculants for sustainable management of agroecosystems in biofertilizers for sustainable agriculture and environment. Springer International Publishing, Cham, pp 137–172

Agri U, Chaudhary P, Sharma A (2021) In vitro compatibility evaluation of agriusable nanochitosan on beneficial plant growth-promoting rhizobacteria and maize plant. Natl Acad Sci Lett 44:555–559. https://doi.org/10.1007/s40009-021-01047-w

Agri U, Chaudhary P, Sharma A, Kukreti B (2022) Physiological response of maize plants and its rhizospheric microbiome under the influence of potential bioinoculants and nanochitosan. Plant Soil 474:451–468. https://doi.org/10.1007/s11104-022-05351-2

Akanmu AO, Babalola OO, Venturi V, Ayilara MS, Adeleke BS, Amoo AE, Sobowale AA, Fadiji AE, Glick BR (2021) Plant disease management: leveraging on the plant-microbe-soil interface in the biorational use of organic amendments. Front Plant Sci 12:700507. https://doi.org/10.3389/fpls.2021.700507

Aloo BN, Mbega ER, Makumba BA, Tumuhairwe JB (2022) Effects of carrier materials and storage temperatures on the viability and stability of three biofertilizer inoculants obtained from potato (Solanum tuberosum L.) rhizosphere. Agriculture 12:140. https://doi.org/10.3390/agriculture12020140

Alori ET, Babalola OO (2018) Microbial inoculants for improving crop quality and human health in Africa. Front Microbiol 9:2213. https://doi.org/10.3389/fmicb.2018.02213

Arora NK, Khare E, Maheshwari DK (2010) Plant growth promoting rhizobacteria: constraints in bioformulation, commercialization, and future strategies. In: Maheshwari D (ed) Plant growth and health promoting bacteria. Microbiology monographs, vol 18. Springer, Berlin, Heidelberg. https://doi.org/10.1007/978-3-642-13612-2_5

Arya R, Pandey C, Dheeman S, Aeron A, Dubey RC, Maheshwari DK, Chen L, Ahmad P, Bajpai VK (2020) Fertilizer adaptive bacteria *Acidovorax valerianellae* and *Sinorhizobium fredii* in integrated nutrient management of pigeon pea (*Cajanus cajan* L.). S Afr J Bot 134:84–90. https://doi.org/10.1016/j.sajb.2020.03.018

Backer R, Rokem JS, Ilangumaran G, Lamont J, Praslickova D, Ricci E, Subramanian S, Smith DL (2018) Plant growth-promoting rhizobacteria: context, mechanisms of action, and roadmap to commercialization of biostimulants for sustainable agriculture. Front Plant Sci 9:1473. https://doi.org/10.3389/fpls.2018.01473

Bandopadhyay S (2020) Application of plant growth promoting *Bacillus thuringiensis* as biofertilizer on *Abelmoschus esculentus* plants under field condition. J Pure Appl Microbiol 14:1287–1294. https://doi.org/10.22207/JPAM.14.2.24

Bennis M, Perez-Tapia V, Alami S, Bouhnik O, Lamin H, Abdelmoumen H, Bedmar EJ, Missbah El Idrissi M (2022) Characterization of plant growth-promoting bacteria isolated from the rhizosphere of *Robinia pseudoacacia* growing in metal-contaminated mine tailings in eastern Morocco. J Environ Manag 304:114321. https://doi.org/10.1016/j.jenvman.2021.114321

Bhakat K, Chakraborty A, Islam E (2021) Characterization of zinc solubilization potential of arsenic tolerant *Burkholderia* spp. isolated from rice rhizospheric soil. World J Microbiol Biotechnol 37:39. https://doi.org/10.1007/s11274-021-03003-8

Bhatt K, Maheshwari DK (2020) Zinc solubilizing bacteria (*Bacillus megaterium*) with multifarious plant growth promoting activities alleviates growth in *Capsicum annuum* L. 3 Biotech 10:36. https://doi.org/10.1007/s13205-019-2033-9

Bukhat S, Imran A, Javaid S, Shahid M, Majeed A, Naqqash T (2020) Communication of plants with microbial world: exploring the regulatory networks for PGPR mediated defense signaling. Microbiol Res 238:126486. https://doi.org/10.1016/j.micres.2020.126486

Callaghan MO (2016) Microbial inoculation of seed for improved crop performance: issues and opportunities. Appl Microbiol Biotechnol 100:5729–5746. https://doi.org/10.1007/s00253-016-7590-9

Celador-Lera L, Jiménez-Gómez A, Menéndez E, Rivas R (2018) Biofertilizers based on bacterial endophytes isolated from cereals: potential solution to enhance these crops. In: Meena VS (ed) Stress management and agricultural sustainability. Springer, Singapore, pp 175–203. https://doi.org/10.3389/fpls.2022.1002448

Chaudhary A, Chaudhary P, Upadhyay A, Kumar A, Singh A (2022e) Effect of gypsum on plant growth promoting rhizobacteria. Environ Ecol 39(4A):1248–1256

Chaudhary A, Parveen H, Chaudhary P, Bhatt P, Khatoon H (2021f) Rhizospheric microbes and their mechanism. In: Bhatt P, Gangola S, Udayanga D, Kumar G (eds) Microbial Technology for Sustainable Environment. Springer, Singapore. https://doi.org/10.1007/978-981-16-3840-4_6

Chaudhary P, Agri U, Chaudhary A, Kumar A, Kumar G (2022c) Endophytes and their potential in biotic stress management and crop production. Front Microbiol 13:933017. https://doi.org/10.3389/fmicb.2022.933017

Chaudhary P, Ahamad L, Chaudhary A, Kumar G, Chen F-S, Chen S (2023b) Nanoparticle-mediated bioremediation as a powerful weapon in the removal of environmental pollutants. J Environ Chem Eng 11:109591. https://doi.org/10.1016/j.jece.2023.109591

Chaudhary P, Chaudhary A, Agri U, Khatoon H, Singh A (2022d) Recent trends and advancements for agro-environmental sustainability at higher altitudes. In: Goel R, Soni R, Suyal DC, Khan M (eds) Survival strategies in cold-adapted microorganisms. Springer, Singapore. https://doi.org/10.1007/978-981-16-2625-8_19

Chaudhary P, Chaudhary A, Parveen H, Rani A, Kumar G, Kumar A, Sharma A (2021a) Impact of nanophos in agriculture to improve functional bacterial community and crop productivity. BMC Plant Biol 21:519. https://doi.org/10.1186/s12870-021-03298-7

Chaudhary P, Chen S, Rajput VD, Jaiswal DK (2023c) Editorial: bioinoculants with nanocompounds to improve soil health: a step toward sustainable agriculture. Front Environ Sci 11:1270002. https://doi.org/10.3389/fenvs.2023.1270002

Chaudhary P, Khati P, Chaudhary A, Gangola S, Kumar R, Sharma A (2021b) Bioinoculation using indigenous Bacillus spp. improves growth and yield of *Zea mays* under the influence of nanozeolite. 3 Biotech 11:11. https://doi.org/10.1007/s13205-020-02561-2

Chaudhary P, Khati P, Chaudhary A, Maithani D, Kumar G, Sharma A (2021c) Cultivable and metagenomic approach to study the combined impact of nanogypsum and *Pseudomonas taiwanensis* on maize plant health and its rhizospheric microbiome. PLoS One 16:e0250574. https://doi.org/10.1371/journal.pone.0250574

Chaudhary P, Khati P, Gangola S, Kumar A, Kumar R, Sharma A (2021d) Impact of nanochitosan and Bacillus spp. on health, productivity and defence response in Zea mays under field condition. 3 Biotech 11:237. https://doi.org/10.1007/s13205-021-02790-z

Chaudhary P, Parveen H, Gangola S, Kumar G, Bhatt P, Chaudhary A (2021e) Plant growth promoting rhizobacteria and their application in sustainable crop production. In: Bhatt P, Gangola S, Udayanga D, Kumar G (eds) Microbial Technology for Sustainable Environment. Springer, Singapore. https://doi.org/10.1007/978-981-16-3840-4_13

Chaudhary P, Sharma A (2019) Response of nanogypsum on the performance of plant growth promotory bacteria recovered from nanocompound infested agriculture field. Environ Ecol 37:363–372

Chaudhary P, Singh S, Chaudhary A, Agri U, Bhandari G (2022a) Chapter 3 agrochemicals and their effects on soil microbial population. In: Soni R, Suyal D, Goel R (eds) Plant protection: from chemicals to biologicals. De Gruyter, Berlin, Boston, pp 45–58. https://doi.org/10.1515/9783110771558-003

Chaudhary P, Singh S, Chaudhary A, Sharma A, Kumar G (2022b) Overview of biofertilizers in crop production and stress management for sustainable agriculture. Front Plant Sci 13:930340. https://doi.org/10.3389/fpls.2022.930340

Chaudhary P, Xu M, Ahamad L, Chaudhary A, Kumar G, Adeleke BS, Verma KK, Hu D-M, Širić I, Kumar P, Popescu SM, Abou Fayssal S (2023a) Application of synthetic consortia for improvement of soil fertility, pollution remediation, and agricultural productivity: a review. Agronomy 13:643. https://doi.org/10.3390/agronomy13030643

Coniglio A, Mora V, Puente M, Cassán F (2019) *Azospirillum* as biofertilizer for sustainable agriculture: *Azospirillum brasilense* AZ39 as a model of PGPR and field traceability. In: Zúñiga-Dávila D, González-Andrés F, Ormeño-Orrillo E (eds) Microbial probiotics for agricultural systems. Sustainability in plant and crop protection. Springer, Cham, pp 45–70

Din M, Nelofer R, Salman M, Abdullah KFH, Khan A, Ahmad M, Jalil F, Din JU, Khan M (2019) Production of nitrogen fixing *Azotobacter* (SR-4) and phosphorus solubilizing *Aspergillus niger* and their evaluation on *Lagenaria siceraria* and *Abelmoschus esculentus*. Biotechnology Reports 22:e00323. https://doi.org/10.1016/j.btre.2019.e00323

Dinesh R, Srinivasan V, Hamza S, Sarathambal C, Anke Gowda SJ, Ganeshamurthy AN, Subila KP, Lijina A, Divya VC (2018) Isolation and characterization of potential Zn solubilizing bacteria from soil and its effects on soil Zn release rates, soil available Zn and plant Zn content. Geoderma 321:173–186. https://doi.org/10.1016/j.geoderma.2018.02.013

El-Deen SRO, El-Azeem A, Samy A, Abd Elwahab AF, Mabrouk SS (2020) Effects of phosphate solubilizing microorganisms on wheat yield and phosphatase activity. Egypt J Med Microbiol 55:71–86. https://doi.org/10.21608/ejm.2020.20675.1137

Fasusi OA, Cruz C, Babalola OO (2021) Agricultural sustainability: microbial biofertilizers in rhizosphere management. Agriculture 11(2):163. https://doi.org/10.3390/agriculture11020163

Gedamu SA, Tsegaye EA, Beyene TF (2021) Effect of rhizobial inoculants on yield and yield components of faba bean (*Vicia fabae* L.) on vertisol of Wereillu District, South Wollo, Ethiopia. CABI Agric Biosci 2:1–10. https://doi.org/10.1186/s43170-021-00025-y

Geries LSM, Elsadany Abdelgawad Y (2021) Maximizing growth and productivity of onion (*Allium cepa* L.) by *Spirulina platensis* extract and nitrogen-fixing endophyte *pseudomonas* stutzeri. Arch Microbiol 203(1):169–181. https://doi.org/10.1007/s00203-020-01991-z

Ghazy N, El-Nahrawy S (2021) Siderophore production by *Bacillus subtilis* MF497446 and *Pseudomonas koreensis* MG209738 and their efficacy in controlling *Cephalosporium maydis* in maize plant. Arch Microbiol 203:1195–1209. https://doi.org/10.1007/s00203-020-02113-5

Gontia-Mishra I, Sapre S, Sharma A, Tiwari S (2016) Amelioration of drought tolerance in wheat by the interaction of plant growth-promoting rhizobacteria. Plant Biol 18:992–1000. https://doi.org/10.1111/plb.12505

Goteti PK, Emmanuel LDA, Desai S, Shaik MHA (2013) Prospective zinc solubilising bacteria for enhanced nutrient uptake and growth promotion in maize (*Zea mays* L.). Int J Microbiol 2013:869697. https://doi.org/10.1155/2013/869697

Gupta A, Bano A, Rai S, Kumar M, Ali J, Sharma S, Pathak N (2021) ACC deaminase producing plant growth promoting rhizobacteria enhance salinity stress tolerance in *Pisum sativum*. 3 Biotech 11:514. https://doi.org/10.1007/s13205-021-03047-5

Gupta G, Parihar SS, Ahirwar NK, Snehi SK, Singh V (2015) Plant growth promoting rhizobacteria (PGPR): current and future prospects for development of sustainable agriculture. J Microbiol Biochem Technol 7:1013–1020. https://doi.org/10.4172/1948-5948.1000188

Haerani N, SyamUn E, Rasyid B, Haring F (2021) Isolation and characterization of N-fixing and IAA producing rhizobacteria from two rice field agro-ecosystems in South Sulawesi. Biodivers J Biol Divers 22:2497–2503. https://doi.org/10.13057/biodiv/d220506

Haidar B, Ferdous M, Fatema B, Ferdous AS, Islam MR, Khan H (2018) Population diversity of bacterial endophytes from jute (*Corchorus olitorius*) and evaluation of their potential role as bioinoculants. Microbiol Res 208:43–53. https://doi.org/10.1016/j.micres.2018.01.008

Imran M, Shahzad SM, Arif MS, Yasmeen T, Ali B, Tanveer A (2020) Inoculation of potassium solubilizing bacteria with different potassium fertilization sources mediates maize growth and productivity. Pak J Agric Sci 57:1045–1055

Jacoby R, Peukert M, Succurro A, Koprivova A, Kopriva S (2017) The role of soil microorganisms in plant mineral nutrition—current knowledge and future directions. Front Plant Sci 8:1617. https://doi.org/10.3389/fpls.2017.01617

Kalayu G (2019) Phosphate solubilizing microorganisms: promising approach as biofertilizers. Int J Agron 7:4917256. https://doi.org/10.1155/2019/4917256

Kalimuthu R, Suresh P, Varatharaju G, Balasubramanian N, Rajasekaran KM, Shanmugaiah V (2019) Isolation and characterization of indole acetic acid [IAA] producing tomato rhizobacterium *pseudomonas* sp VSMKU4050 and its potential for plant growth promotion. Int J Curr Microbiol Appl Sci 8:443–455. https://doi.org/10.20546/ijcmas.2019.806.050

Kalyan VSRK, Meena S, Karthikeyan S, Jawahar D (2022) Isolation, screening, characterization, and optimization of bacteria isolated from calcareous soils for siderophore production. Arch Microbiol 204:21. https://doi.org/10.1007/s00203-022-03322-w

Kang S-M, Khan AL, Waqas M, Asaf S, Lee K-E, Park Y-G (2019) Integrated phytohormone production by the plant growth-promoting rhizobacterium *bacillus tequilensis* SSB07 induced thermotolerance in soybean. J Plant Interaction 14:416–423. https://doi.org/10.1080/17429145.2019.1640294

Ketema P, Tefera T (2022) Effectiveness of *Rhizobium* strains on faba bean (*Vicia fabae* L.) at Gumer district, highland area of Southern Ethiopia. Ukr J Ecol 12:13–18. https://doi.org/10.15421/2022_329

Khanghahi MY, Strafella S, Allegretta I, Crecchio C (2021) Isolation of bacteria with potential plant-promoting traits and optimization of their growth conditions. Curr Microbiol 78:464–478. https://doi.org/10.1007/s00284-020-02303-w

Khati P, Chaudhary P, Gangola S, Bhatt P, Sharma A (2017) Nanochitosan supports growth of Zea mays and also maintains soil health following growth. 3 Biotech 7:81. https://doi.org/10.1007/s13205-017-0668-y

Khati P, Chaudhary P, Gangola S, Sharma P (2019a) Influence of Nanozeolite on plant growth Promotory bacterial isolates recovered from nanocompound infested agriculture field. Environ Ecol 37:521–527

Khati P, Parul BP, Nisha Kumar R, Sharma A (2018) Effect of nanozeolite and plant growth promoting rhizobacteria on maize. 3 Biotech 8:141. https://doi.org/10.1007/s13205-018-1142-1

Khati P, Sharma A, Chaudhary P, Singh AK, Gangola S, Kumar R (2019b) High- throughput sequencing approach to access the impact of nanozeolite treatment on species richness and evens of soil metagenome. Biocatal Agric Biotechnol 20:101249. https://doi.org/10.1016/j.bcab.2019.101249

Khoshru B, Nosratabad AF, Mitra D, Chaithra M, Danesh YR, Boyno G, Chattaraj S, Priyadarshini A, Anđelković S, Pellegrini M (2023) Rock phosphate solubilizing potential of soil microorganisms: advances in sustainable crop production. Bacteria 2(2):98–115. https://doi.org/10.3390/bacteria2020008

Kudoyarova GR, Vysotskaya LB, Arkhipova TN, Kuzmina LY, Galimsyanova NF, Sidorova LV, Gabbasova IM, Melentiev AI, Veselov SY (2017) Effect of auxin producing and phosphate solubilizing bacteria on mobility of soil phosphorus, growth rate, and P acquisition by wheat plants. Acta Physiol Plant 39:1–8. https://doi.org/10.1007/s11738-017-2556-9

Kukreti B, Sharma A, Chaudhary P, Agri U, Maithani D (2020) Influence of nanosilicon dioxide along with bioinoculants on Zea mays and its rhizospheric soil. 3 Biotech 10:345. https://doi.org/10.1007/s13205-020-02329-8

Kumar G, Lal S, Maurya SK, Bhattacherjee AK, Chaudhary P, Gangola S, Rajan S (2021) Exploration of Klebsiella pneumoniae M6 for paclobutrazol degradation, plant growth attributes, and biocontrol action under subtropical ecosystem. PLoS One 16:e0261338. https://doi.org/10.1371/journal.pone.0261338

Kumar G, Lal S, Soni SK, Maurya SK, Shukla PK, Chaudhary P, Bhattacherjee AK, Garg N (2022a) Mechanism and kinetics of chlorpyrifos co-metabolism by using environment restoring microbes isolated from rhizosphere of horticultural crops under subtropics. Front Microbiol 13:891870. https://doi.org/10.3389/fmicb.2022.891870

Kumar S, Diksha, Sindhu SS, Kumar R (2022b) Biofertilizers: an ecofriendly technology for nutrient recycling and environmental sustainability. Curr Res Microb Sci 3:100094. https://doi.org/10.1016/j.crmicr.2021.100094

Kumari H, Khati P, Gangola S, Chaudhary P, Sharma A (2021) Performance of plant growth promotory rhizobacteria on maize and soil characteristics under the influence of TiO2 nanoparticles. Pantnagar J Res 19:28–39

Kumari S, Sharma A, Chaudhary P, Khati P (2020) Management of plant vigor and soil health using two agriusable nanocompounds and plant growth promotory rhizobacteria in fenugreek. 3 Biotech 10:1–11. https://doi.org/10.1007/s13205-020-02448-2

Kushwaha P, Srivastava R, Pandiyan K, Singh A, Chakdar H, Kashyap PL, Bhardwaj AK, Murugan K, Karthikeyan N, Bagul SY, Srivastava AK, Saxena AK (2021) Enhancement in plant growth and zinc biofortification of chickpea (*Cicer arietinum* L.) by *Bacillus altitudinis*. J Soil Sci Plant Nutr 21:922–935. https://doi.org/10.1007/s42729-021-00411-5

Ladha JK, Tirol-Padre A, Reddy CK, Cassman KG, Verma S, Powlson DS (2016) Global nitrogen budgets in cereals: a 50-year assessment for maize, rice and wheat production systems. Sci Rep 6:19355. https://doi.org/10.1038/srep19355

Laxita L, Shruti S (2020) Isolation and characterization of potassium solubilizing microorganisms from South Gujarat region and their effects on wheat plant. Mukta Shabad 9:7483–7496

Mahantesh P, Patil CS (2011) Isolation and biochemical characterization of phosphate solubilizing microbes. Int J Microbiol Res 3:67–70

Malusa E, Vassilev NA (2014) Contribution to set a legal framework for biofertilisers. Appl Microbiol Biotechnol 98(15):6599–6607. https://doi.org/10.1007/s00253-014-5828-y

Manzar N, Jayaswal D, Kishan G, Chauhan S, Joshi D, Singh J, Chaudhary P, Jamali H, Kushwaha P, Parveen H, Kashyap AS, Sahu PK, Singh UB (2022) Application of molecular ecology approaches in sustainable agriculture for a better understanding of plant–microbiome interactions. In: Singh UB, Sahu PK, Singh HV, Sharma PK, Sharma SK (eds) Rhizosphere microbes. Microorganisms for sustainability, vol 40. Springer, Singapore. https://doi.org/10.1007/978-981-19-5872-4_3

Matoso ES, Reis VM, Giacomini SJ, Silva MTD, Avancini AR, Silva SDDA (2021) Diazotrophic bacteria and substrates in the growth and nitrogen accumulation of sugarcane seedlings. Sci Agric 78(1):184. https://doi.org/10.1590/1678-992x-2019-0035

Mayadunna N, Karunarathna SC, Asad S, Stephenson SL, Elgorban AM, Al-Rejaie S, Kumla J, Yapa N, Suwannarach N (2023) Isolation of phosphate-solubilizing microorganisms and the formulation of biofertilizer for sustainable processing of phosphate rock. Life 13:782. https://doi.org/10.3390/life13030782

Muhie SH (2022) Novel approaches and practices to sustainable agriculture. J Agric Food Res 10:100446. https://doi.org/10.1016/j.jafr.2022.100446

Mumtaz MZ, Ahmad M, Jamil M, Hussain T (2017) Zinc solubilizing *Bacillus* spp. potential candidates for biofortification in maize. Microbiol Res 202:51–60. https://doi.org/10.1016/j.micres.2017.06.001

Mustafa S, Kabir S, Shabbir U, Batool R (2019) Plant growth promoting rhizobacteria in sustainable agriculture: from theoretical to pragmatic approach. Symbiosis 78:115–123. https://doi.org/10.1007/s13199-019-00602-w

Nawaz N, Ali S, Shabir G, Rizwan M, Shakoor MB, Shahid MJ, Afzal M, Arslan M, Hashem A, Abd-Allah EF, Alyemeni MN, Ahmad P (2020) Bacterial augmented floating treatment wetlands for efficient treatment of synthetic textile dye wastewater. Sustain For 12(9):3731. https://doi.org/10.3390/su12093731

Odoh CK, Eze CN, Akpi UK, Unah VU (2019) Plant growth promoting rhizobacteria (PGPR): a novel agent for sustainable food production. Am J Agric Biol Sci 14:35–54. https://doi.org/10.3844/ajabssp.2019.35.54

Perumal MD, Selvi D, Chitdeshwari T, Balachandar D (2019) Enhanced zinc nutrient and enzyme activity of rice crop by zinc solubilizing bacteria with Zn sources in Zn deficient Rice soil. Madras Agric J 106:71–177. https://doi.org/10.29321/MAJ2019.000242

Raimi A, Roopnarain A, Adeleke R (2021) Biofertilizer production in Africa: current status, factors impeding adoption and strategies for success. Sci Afr 11:e00694. https://doi.org/10.1016/j.sciaf.2021.e00694

Raji M, Thangavelu M (2021) Isolation and screening of potassium solubilizing bacteria from saxicolous habitat and their impact on tomato growth in different soil types. Arch Microbiol 203:3147–3161. https://doi.org/10.1007/s00203-021-02284-9

Ramoneda J, Roux JJL, Frossard E, Frey B, Gamper HA (2020) Geographical patterns of root nodule bacterial diversity in cultivated and wild populations of a woody legume crop. FEMS Microbiol Ecol 96:45. https://doi.org/10.1093/femsec/fiaa145

Reis VM, Alves BJR, Hartmann A, James EK, Zilli JE (2020) Beneficial microorganisms in agriculture: the future of plant growth-promoting rhizobacteria. Plant Soil 451:1–3. https://doi.org/10.1007/s11104-020-04482-8

Riah N, de Lajudie P, Béna G, Heulin K, Djekoun A (2021) Variability in symbiotic efficiency with respect to the growth of pea and lentil inoculated with various rhizobial genotypes originating from sub-humid and semi-arid regions of eastern Algeria. Symbiosis 85:371–384. https://doi.org/10.1007/s13199-021-00821-0

Rizvi A, Ahmed B, Khan MS, El-Beltagi HS, Umar S, Lee J (2022) Bioprospecting plant growth promoting rhizobacteria for enhancing the biological properties and phytochemical composition of medicinally important crops. Molecules 27:1407. https://doi.org/10.3390/molecules27041407

Santhi C, Rajesh M, Ramesh S, Muralikrishna K, Gangaraj K, Payal G (2021) Genome-wide exploration of auxin response factors (ARFs) and their expression dynamics in response to abiotic stresses and growth regulators in coconut (*Cocos nucifera* L.). *Plant.* Gene 28:100344. https://doi.org/10.1016/j.plgene.2021.100344

Sanyal D, Osorno JM, Chatterjee A (2020) Influence of *Rhizobium* inoculation on dry bean yield and symbiotic nitrogen fixation potential. J Plant Nutr 43:798–810. https://doi.org/10.1080/01904167.2020.1711946

Sinha AK, Parli VB, Tripathy SC, Sarkar A, Prabhakaran S (2019) Effects of growth conditions on siderophore producing bacteria and siderophore production from Indian Ocean sector of Southern Ocean. J Basic Microbiol 59:412–424. https://doi.org/10.1002/jobm.201800537

Soumare A, Boubekri K, Lyamlouli K, Hafidi M, Ouhdouch Y, Kouisni L (2020) From isolation of phosphate solubilizing microbes to their formulation and use as biofertilizers: status and needs. Front Bioeng Biotechnol 7:425. https://doi.org/10.3389/fbioe.2019.00425

Souza R, Ambrosini A, Passaglia LM (2015) Plant growth-promoting bacteria as inoculants in agricultural soils. Genet Mol Biol 38(4):401–419. https://doi.org/10.1590/S1415-475738420150053

Subramanium N, Sundaram L (2020) Siderophore producing Pseudomonas spp. isolated from rhizospheric soil and enhancing iron content in *Arachis hypogaea* L. plant. Int J Agric Technol 16:429–442

Tahir N, Asma I, Sohail H, Muhammad S, Afshan M, Javed I, Kashif HM, Shaghef E, Abdullah MK (2020) First report of diazotrophic *Brevundimonas* spp. as growth enhancer and root colonizer of potato. Sci Rep 10(1):12893. https://doi.org/10.1038/s41598-020-69782-6

Tairo EV, Ndakidemi PA (2014) Macronutrients uptake in soybean as affected by *Bradyrhizobium* japonicum inoculation and phosphorus (P) supplements. Am J Plant Sci 5:488–496. https://doi.org/10.4236/ajps.2014.54063

Timmusk S, Behers L, Muthoni J, Muraya A, Aronsson A-C (2017) Perspectives and challenges of microbial application for crop improvement. Front Plant Sci 8:49. https://doi.org/10.3389/fpls.2017.00049

Tyagi J, Chaudhary P, Jyotsana BU, Bhandari G, Chaudhary A (2022a) Chapter 17 impact of endophytic fungi in biotic stress management. In: Soni R, Suyal D, Goel R (eds) Plant protection: from chemicals to biologicals. De Gruyter, Berlin, Boston, pp 447–462. https://doi.org/10.1515/9783110771558-017

Tyagi J, Chaudhary P, Mishra A, Khatwani M, Dey S, Varma A (2022b) Role of endophytes in abiotic stress tolerance: with special emphasis on Serendipita indica. Int J Environ Res 16:62. https://doi.org/10.1007/s41742-022-00439-0

Velmourougane K, Prasanna R, Chawla G, Nain L, Kumar A, Saxena AK (2019) Trichoderma–Azotobacter biofilm inoculation improves soil nutrient availability and plant growth in wheat and cotton. J Basic Microbiol 59:632–644. https://doi.org/10.1002/jobm.201900009

Walpola BC, Hettiarachchi RHAN (2020) Organic manure amended with phosphate solubilizing bacteria on soil phosphorous availability. The Journal of Agricultural Science- Sri Lanka 15:142–153. https://doi.org/10.4038/jas.v15i2.8796

Wang C, Pan G, Lu X, Qi W (2023) Phosphorus solubilizing microorganisms: potential promoters of agricultural and environmental engineering. Front Bioeng Biotechnol 11:1181078. https://doi.org/10.3389/fbioe.2023.1181078

Wang Y, Zhang G, Huang Y, Guo M, Song J, Zhang T, Long Y, Wang B, Liu H (2022) A potential biofertilizer-Siderophilic bacteria isolated from the Rhizosphere of *Paris polyphylla var. yunnanensis*. Front Microbiol 13:870413. https://doi.org/10.3389/fmicb.2022.870413

Yasmin R, Hussain S, Rasool MH, Siddique MH, Muzammil S (2021) Isolation, characterization of Zn solubilizing bacterium (*Pseudomonas protegens* RY2) and its contribution in growth of chickpea (*Cicer arietinum* L) as deciphered by improved growth parameters and Zn content. Dose-Response 19(3):1–12. https://doi.org/10.1177/15593258211036791

Yongbin L, Yunlong L, Haowei Z, Minyang W, Sanfeng C (2019) Diazotrophic *Paenibacillus* beijingensis BJ-18 provides nitrogen for plant and promotes plant growth, nitrogen uptake and metabolism. Front Microbiol 29(10):1119. https://doi.org/10.3389/fmicb.2019.01119

Zaheer A, Malik A, Sher A, Qaisrani MM, Mehmood A, Khan SU, Ashraf M, Mirza Z, Karim S, Rasool M (2019) Isolation, characterization, and effect of phosphate-zinc-solubilizing bacterial strains on chickpea (*Cicer arietinum* L.) growth. Saudi J Biol Sci 26:1061–1067. https://doi.org/10.1016/j.sjbs.2019.04.004

Zhang Y, Wang Y, Ding Z, Wang H, Song L, Jia S, Ma D (2017) Zinc stress affects ionome and metabolome in tea plants. Plant Physiol Biochem 111:318–328. https://doi.org/10.1016/j.plaphy.2016.12.014

Zhou HY, Han Y, Shi Q, Chen CF (2019) Directional transportation of a helic[6]arene along a nonsymmetric molecular axle. J Org Chem 84:5872–5876. https://doi.org/10.1021/acs.joc.9b00229

Zulkifli H, Tarmizi AM (2010) Phosphorus fractions in soil amended with empty fruit bunches and phosphate fertilizer—an incubation study. J Oil Palm Res 22:823–834. https://doi.org/10.1021/acs.joc.9b00229

Plant Growth-Promoting Psychrotrophic Microbes: Current Research and Future Challenges

Divya Joshi, Deep Chandra Suyal, Jyoti Singh, Damini Maithani, Jyoti Rajwar, Preeti, Manali Singh, and Dipti Singh

Abstract Food production should be raised to keep up with the growing world population, yet not at the expense of the environment by reducing the use of agrochemicals. This key insight has encouraged the development of microorganisms as cost- and sustainably-efficient substitutes for these dangerous compounds. Ecosystems with a preference for low temperatures are home to psychrophiles and psychrotrophs. Also, they serve as study models for the metabolites and biomolecules connected to low-temperature survival that are beneficial to several sectors. Biological nitrogen fixation, or siderophore, HCN, and ammonia synthesis are only a few of the secondary metabolites that plant growth-promoting bacteria employ directly or indirectly to benefit the plant. The low temperatures and less fertile soil

D. Joshi (✉)
Patanjali Herbal Research Division, Patanjali Research Institute, Haridwar, Uttarakhand, India

Department of Microbiology, Govind Ballabh Pant University of Agriculture & Technology, Pantnagar, Uttarakhand, India

D. C. Suyal
Vidyadayini Institute of Science, Management and Technology, Bhopal, Madhya Pradesh, India

J. Singh · J. Rajwar
Department of Microbiology, Govind Ballabh Pant University of Agriculture & Technology, Pantnagar, Uttarakhand, India

D. Maithani
Department of Microbiology, Lovely Professional University, Phagwara, Punjab, India

Preeti
Department of Veterinary Public Health, IVRI, Izzatnagar, Bareilly, Uttar Pradesh, India

M. Singh
Department of Life Sciences, Parul Institute of Applied Sciences, Parul University, Waghodia, Vadodara, Gujarat, India

D. Singh
Department of Biochemistry, Govind Ballabh Pant University of Agriculture & Technology, Pantnagar, Uttarakhand, India

P. Chaudhary, A. Chaudhary (eds.), *Microbial Inoculants*, Interdisciplinary Biotechnological Advances, https://doi.org/10.1007/978-981-97-0633-4_11

provide various difficulties for agriculture in high-altitude soils. During the past few decades, a variety of strategies have been used to address this issue, including the employment of psychrophilic or psychrotrophic organisms. These microbial inoculants can therefore successfully help mountain agro-systems to fight the severity of cold settings and encourage plant development. A wide variety of psychrotrophic microbiomes have been created as a result of exploration of cold environments over time. Despite facing many difficulties, the resulting bacterial diversity from the cold environment has the potential to act as a large reservoir for bio-inoculants that have the ability to promote plant growth. This could be used to enhance the growth and yield of crops grown at high altitudes, where low temperatures are the norm.

Keywords Agrochemicals · Biological nitrogen fixation · Low temperature · Microbiomes · Psychrotrophs

1 Introduction

Global food security is negatively impacted by the growing environmental deterioration, which might lead to even more serious issues if not promptly handled. Enhancing and improving agricultural output is crucial, but not at the expense of the environment (Islam and Kieu 2020). This can only be accomplished by putting all of the emphasis on sustainable and ecologically friendly agriculture techniques (Parveen et al. 2022). Thus, the focus of our agriculture should be on the utilization of transgenic plants and bacteria that promote plant development (Wang et al. 2016a, b; Khati et al. 2018; Kumari et al. 2020; Agri et al. 2021; Chaudhary et al. 2023).

By means of both indirect and direct methods, as well as through its effect on organic molecules (Fig. 1), temperature has a substantial impact on a particular type of creature so that it can live (Maayer et al. 2014). Moreover, cold weather has a range of effects on agriculture. These chilly conditions are home to a specific species of bacteria known as "psychrophiles," which are acclimated to the cold. The Indian Himalayan Area is home to a huge variety of microorganisms, including fungus, cyanobacteria, bacteria, and archaebacteria. In frigid habitats, these creatures persist and sustain vitality (Joshi et al. 2017). The ecology of the Indian Himalayas is a storehouse of advantageous bacteria that promotes the development and growth of plants through a number of mechanisms, including increased plant biomass, which eventually increases yield (Sati et al. 2023). These specialized microbe groups may find use in a variety of industries, including agriculture, medicine, and others. This variety acts as a stockpile of microbial inoculants with plant growth-promoting (PGP) properties that may still produce a good yield under such difficult circumstances. *Exiguobacterium, Pseudomonas,* and *Serratia* are three unique cold-adaptive potential bacterial species that are abundant in these regions

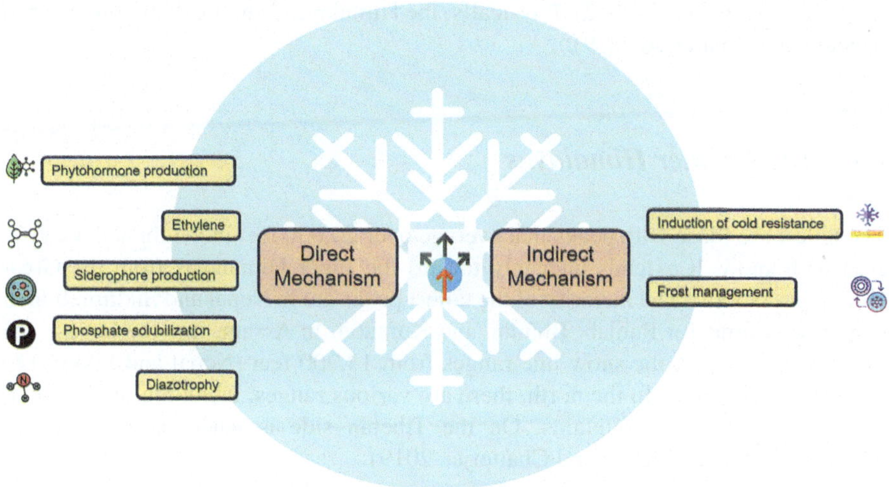

Fig. 1 The direct and indirect mechanisms adapted by psychrotrophs to benefit plants

(Yadav et al. 2019). Research on bioinoculants is widely used and of the highest relevance as a result of growing knowledge of and reports of adverse effects from the widespread usage of agrochemicals (Chaudhary et al. 2022a, b). In light of recent concerns about environmental degradation and conservation, biodiversity preservation, mitigating climate change, and food security, the relevance of the microbial community and its potential for alteration to boost impacts on crop yield have come to light (Raza et al. 2019; Bouabdelli et al. 2022; Chaudhary et al. 2022c). Also, because mountain soils typically lack nutrients, maintaining and producing crops is much more difficult. Hence, hill agroecosystems are currently actively promoting and using bioinoculants-based ways to increase crop production or achieve long-term development and growth (Joshi et al. 2017).

2 Mountain Ecosystem

From Burma (east) to Afghanistan, the Himalayas are continuous ranges of tertiary mountains that span more than 3000 kilometers (west). The complicated succession of parallel, convergent, and diverging ranges that they give rise to rise from the Indian plains and extend to the high plateau of Tibet (north). There is a specific distribution of height, temperature, pressure, slope, winds, and radiation since the breadth varies from 80 km to more than 300 km. The winter westerlies and the summer monsoon are naturally barred from air circulation by these mountains. These climatic factors cause yearly cycles of freezing and thawing, which creates the ideal environment for local microbes that are particularly suited to low-temperature settings (Shrestha et al. 2012). Physically, the Himalayas is divided into three zones (Bishop and Chatterjee 2019).

2.1 The Greater Himalayas

The tallest are those with peaks that average over 20,000 feet in height and are covered with snow. It extends unevenly toward the Inner Himalayas from the Great Himalayas (south). The majority of the tallest peaks are in Nepal and Sikkim. It then contains sections for Punjab, Bhutan, and Kumaon. In Assam, it is the lowest. On the southern slopes, the snow line ranges from 14,700 feet (Nepal and Sikkim) to 17,000 feet (Punjab). In the north, there are various ranges, including the Zanskar, Ladakh, and Kailas mountains. On the Tibetan side is where you'll find the Karakoram Range (Bishop and Chatterjee 2019).

2.2 The Inner Himalayas

This region's typical elevation ranges from 1500 to 4500 m above mean sea level. This area shows a steady ascent toward the Pir Panjal and Dhauladhar Mountains. Tehsil Pachhad (upper), District Sirmaur, Karsog and Chachiote tehsils of the Mandi district, higher Kangra and Palampur tehsil of the Kangra district, upper hills of the District Shimla, and upper portions of tehsil Churah of the Chamba district are regions that come within this zone. Churdhar, a mountain with a height of 3647 m, is located south of Shimla. With one exception—the Kullu valley, which runs anti-clockwise to the main alignment—the longitudinal valleys in Himachal Pradesh separate the chain of parallel ranges (Bishop and Chatterjee 2019).

2.3 The Outer Himalayas

Between 350 and 1500 m above the average sea level is the typical height. The "Shivalik Hills" are a popular name for this section of the outer Himalayas. Shivalik means "tresses of the Lord Shiva" in the literal sense. It includes the lower hills of the districts of Mandi, Solan, and Sirmaur, as well as those of Kangra, Hamirpur, Una, and Bilaspur (Bishop and Chatterjee 2019).

2.4 *Climatic Conditions*

The Great Himalaya Range, due to its geographic location and immense altitude, prevents cold continental air from the north from entering India in the winter and compels the south-westerly monsoon winds to lose the majority of their moisture before reaching the range northward (Joshi et al. 2017). As a result, the Indian side of the border sees substantial precipitation in the form of both rain and snow, whereas Tibet suffers dry conditions. In the south slopes, the average annual rainfall ranges from 1530 mm at Shimla and Mussoorie to 3050 mm at Darjeeling. Just 75 to 150 mm of precipitation per year falls in places like Skardu, Gilgit, and Leh in the Kashmir region of the Indus valley (north of the Great Mountains). The two seasons of precipitation are the lighter one in summer that is followed by south-westerly monsoon winds and the heavier one in winter that is brought on by storms. During the winter, the Himalayas experience heavy snowfall as a result of low-pressure weather systems. Condensation takes place at higher air levels in the locations where the western disturbances are felt, which causes precipitation to be heavier over high mountains. During that period, snow builds up around the Himalayan high peaks, causing more precipitation to fall in the west than the east (Joshi et al. 2017).

2.5 *Challenges for Diversity of Microbes*

The unregulated rate of climate change is putting a lot of pressure on natural resources and the ecosystem. The availability and accessibility of natural resources would surely be significantly impacted. These changes have altered the extent and frequency of meteorological phenomena. There is considerable agreement that extreme fluctuations are directly connected to enhanced and irregular monsoon circulations in particular (Singh and Singh 2012). The quick portrayal of major challenges of these ecosystems has been depicted in Fig. 2. The Himalayan region's delicate landscapes are particularly vulnerable to a number of natural calamities. A greater percentage of total precipitation appears to be falling as rain than before in several regions. The shorter winter and early snowmelt have an impact on river cycles, natural disasters, water supply, local residents' way of life, and infrastructure. Also impacted are the well-being of high-altitude wetlands, green-water flows from terrestrial ecosystems, reservoirs, and water flow and sediment movement along rivers and in lakes. Flash floods might also be caused by high meltwaters mixed with liquid precipitation. More than one-sixth of the world's population totally depends on glaciers and seasonal snow for the water supply needed for their home, agricultural, and industrial requirements, while mountain regions account for more than half of the world's river flow (Biemans et al. 2019). Total and peak river runoff are directly impacted by changes in precipitation type, volume, intensity, and distribution throughout time and location. The variance in downstream runoff, however, will someday grow suddenly and finally represent direct rainfall-runoff, which

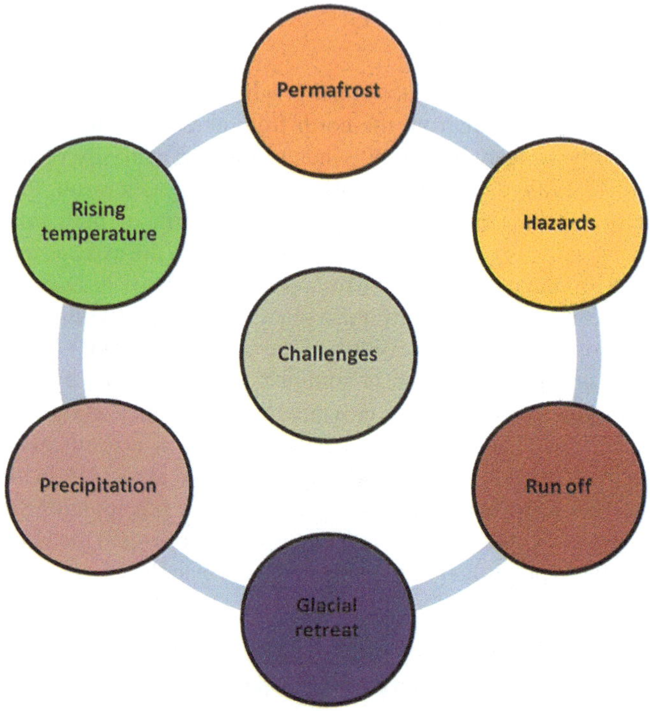

Fig. 2 Major challenges of mountain ecosystems

then in return will mimic precipitation and evapotranspiration rates as the reservoirs of ice and snow at high altitude continue to diminish. Additionally, permafrost lies under several regions in the high mountains. These locations are vulnerable to deterioration as a result of climate change. Additionally, the percentage of seasonal thawing has accelerated over sizable permafrost areas, causing increased erosion and ground instability. In the meanwhile, certain places' desertification will be accelerated by the melting of permafrost (Biemans et al. 2019).

2.6 *Ecological Diversity of Psychrotrophic Plant Growth-Promoting Microorganisms*

The most unusual living forms can be found in extreme environmental environments. Around 80% of the terrestrial and aquatic ecosystems on earth are continuously or seasonally exposed to frigid temperatures. Regions including permafrost soils, snow coverings, cloud glaciers, glacial lakes, and mountainous terrain cover 20% of the earth's surface (Yadav et al. 2017a). Temperature has a significant impact on the functioning and production of an ecosystem, since it is one of the key factors in determining the survival of living things. In low-temperature locations, cold stress is a key limiting factor for the development and production of commercially significant crops, because it causes a number of changes at the molecular and physiological levels. Low temperatures are a persistent challenge to agroecosystems in cold climates, and producing enough food requires significant work (Altieri et al. 2015). Each ecosystem's foundation is its microbial diversity. These natural microbial populations are unique in their ability to withstand low temperatures and adapt physiologically and metabolically. All of these cold-adapted microbes generate a wide range of enzymes, anti-freezing substances, and other biomolecules that have potential uses in the agricultural, industrial, and pharmaceutical fields (Hamid et al. 2022). Differential gene expression, modifications to membrane lipids, changes in protein, and total soluble protein inside the cell are all components of cold stress adaptation (Barrero-Sicilia et al. 2017). These potential microorganisms have recently captured the attention of the applied research community. These areas' soil microbiome is made up of a wide range of psychotropic microorganisms, including archaea, bacteria, fungus, protozoa, and mycorrhiza, which creates a diverse ecology (Yadav et al. 2017b). Several of these microbes have been shown to have biocontrol abilities and features that support plant development. They have been thoroughly researched for new plant growth-promoting microorganisms (PGPM) utilizing culture-dependent and independent methods (Bhandari et al. 2020).

The diversity of microorganisms that support plant development and are indigenous to low-temperature environments are outlined in this section. In cold or hilly situations, these microorganisms and their derivatives can be employed to improve plant health. Also, these microorganisms can be utilized as a database for research into new microorganisms with characteristics that promote plant development (Verma et al. 2016). In the past several years, the variety of psychrotolerant and psychotropic bacteria and fungi in cold environments has drawn a lot of interest for their potential use as bioinoculants for plants subjected to cold stress, particularly in mountainous areas (Vega-Celedón et al. 2021).

The two types of cold-adapted microorganisms are psychrophiles and pscychrotrophs. The former category consists of microorganisms with temperature optimums of 15 °C or below and environments that are extremely cold (Santiago et al. 2016). They include several Gram-positive (*Bacillus* and *Micrococcus* sp.), Gram-negative (*Polaribacter, Pseudoalteromonas, Moraxella,* etc.), and Gram-neutral (other species of Archaea, yeast, and fungus). Later ones, however, are those that can survive in a broad range of temperatures, from 4 to 42 °C, but have temperature optimum ranges of more than 15 °C. Various psychrotrophs are reported from bacterial genera *Pseudomonas, Halomonas, Leuconostoc, Acinetobacter, Exiguobacterium, Serratia, Rhodococcus, Rahnella, and Pantoea* (Subramanian et al. 2011). Many of these microorganisms have been found to produce a variety of beneficial traits, such as nitrogen fixation, phytohormone production, ACC deaminase and other hydrolytic enzymes, antimicrobial compounds, siderophores, and plant stress tolerance (Gupta and Pandey 2019; Vega-Celedón et al. 2021). These microorganisms are commonly found in the rhizosphere or phyllosphere of cold region plants (Tiryaki et al. 2019).

Various bacterial genera native to low-temperature have been reported with plant growth-promoting attributes, including *Bacillus*, *Pseudomonas*, *Serratia*, *Exiguobacterium*, *Enterobacter*, *Sinorhizobium*, *Burkholderia*, *Flavobacterium*, *Clostridium*, *Providencia*, *Microbacterium*, *Janthinobacterium*, *Paenibacillus*, *Kocuria*, *Lysinibacillus*, *Methylobacterium*, *Arthrobacter*, *Brevundimonas Serratia*, *Citricoccus*, *Azotobacter*, *Hydrogenophaga*, *Burkholderia*, *Enterobacter*, *Azospirillum, and Bradyrhizobium* (Bhandari et al. 2020). Both direct and indirect processes employed by these bacteria enhance plant development. For instance, the apple rhizosphere-isolated *Exiguobacterium acetylicum* strain 1 P was able to prevent the growth of several phytopathogenic fungi, such as *F. oxysporum*, *Pythium*, *R. solani*, and *S. rolfsii* (Selvakumar et al. 2010). This strain, which was found in the high-altitude Himalayas, was well adapted to the effects of cold stress and was shown to produce P-solubilization, IAA, and HCN. It has been found that *Pantoea dispersa* 1A produces phytohormones at low temperatures and has antagonistic action against *S. rolfsii*. Several psychrotrophic bacterial strains from the proteobacteria (α, β, γ), firmicutes, actinobacteria, and Bacteroidetes have been described by Yadav and colleagues. These strains have multiple functions for supporting plant development (Yadav et al. 2015). Eight strains were identified in the same investigation as producing HCN. A substantial amount of ammonia was being created by 21% of the strains, 4% of them were able to inhibit the growth of phytopathogens, and 8% of them produced gibberellic acid. Whereas siderophore, phosphate, and ACC deaminase manufacturers made up 15%, 16%, and 8% of the total, respectively. Nine psychrotolerant bacterial strains that can solubilize potassium from Northern India were described by Verma et al. (2015). In the same investigation, it was discovered that 25 bacterial isolates produced ACC deaminase. *Mycobacterium phlei* and *Mycoplana bullata* were shown to be able to enhance the development of winter wheat in nutrient-deficient soil. Several tactics are used by microorganisms to shield plants from cold damage (Rilling et al. 2018). Several of them create substances that prevent freezing, such as ice-nucleating proteins. *Pseudomonas*, *Pantoea*, *Xanthomonas*, and *Erwinia* are among the microbial genera that have been shown to generate ice-nucleating proteins. The term "ice plus strains" refers to bacteria that generate ice-nucleating proteins (Bhandari et al. 2020). Other psychrotrophic PGPBs that are offered for sale include *Raoultella terrigena*, *Azospirillum brasiliense*, and *Bacillus subtilis*. According to reports, these strains either increase the synthesis of antioxidants or antifreeze substances in crops as barley and wheat, which helps to reduce the negative impacts of cold temperatures (Turan et al. 2013).

Many psychrotrophic microbes from a variety of domains, including Actinobacteria, Euryarchaeota, Proteobacteria, Basidiomycota, Bacteroidetes, and Firmicutes, have been identified via investigations of culturable and unculturable diversity studies (Yadav et al. 2017a). Bacterial diversity in the Himalayan cold deserts varied with altitude, according to a metagenomic research. *Proteobacteria*, *Acidobacteria*, and *Actinobacteria* were found to be prevalent in soils of higher altitude locations, while Bacteroidetes and Firmicutes were found to be prevalent in soils of lower altitude regions. Several other psychrotrophic plant beneficiary bacterial species with multifunctional growth promotion traits reported till date include

Arthrobacter nicotianae, Brevundimonas terrae, Paenibacillus tylopili Pseudomonas cedrina, Alishewanella sp., *Aurantimonas altamirensis, Bacillus baekryungensis, B. marisflavi, Desemzia incerta, Paenibacillus xylanexedens, Pontibacillus* sp., Providencia sp., *P. frederiksbergensis, Sinobaca beijingensis, and Vibrio metschnikovii* (Yadav et al. 2015, 2016). *Low-temperature-adapted Bacillus and Bacillus-derived genera have been reported with economically important multifunctional traits, including Bacillus psychrosaccharolyticus, Paenibacillus pabuli, Bacillus amyloliquefaciens, Paenibacillus tylopili, B. altitudinis, B. Muralis, Paenibacillus tylopili, and P. lautus* (Yadav et al. 2016, 2017a, b).

Moreover, it has been reported that a number of psychrotolerant plant growth-promoting fungi exist. *Trichoderma, T. harzianum, T. koningii,* and *T. viride* spp. from soils in low-temperature areas of the Indian Himalayas were identified. Each of the three species could sporulate at low temperatures and produce volatile antifungal compounds to inhibit the growth of a variety of phytopathogenic fungus. *Trichoderma polysporum*, a different fungus species, was discovered in ginseng (*Panax ginseng* Meyer) grown in Korea's hilly areas. A range of ginseng pathogens were able to be inhibited by the isolate (Park et al. 2017). There have been reports of *Trichoderma gamsii* in root tissues. In the chilly Indian Himalayas, the plant *Lens esculenta* may hydrolyze chitin, solubilize phosphorus, and create ammonia and salicylic acid (Rinu et al. 2014). Six out of seven phytopathogenic fungi could not develop due to the fungus strain. *Paecilomyces hepiali*, a psychrotolerant phosphate-solubilizing fungus, has been discovered in low-temperature Himalayan locations (Rinu and Pandey 2011). Moreover, *Penicillium* species have been linked to the solubilization of inorganic phosphate in cold climates. Yet, compared to bacteria and filamentous fungus, yeast is less common in plants and soil, and their contribution to agricultural systems is likewise modest. In contrast to bacteria, their role in promoting plant development is also less well understood, particularly in agroecosystems that are suited to the cold. Nevertheless, some yeast strains, such as *Aureobasidium pullulans*, *Holtermaniella takashimae*, and *Candida maritime*, have been shown to produce auxin, P-solubilize, and siderophores, among other features that may promote plant development. Also, these strains were able to inhibit the development of phytopathogens such as *Pythium aphanidermatum* and *Verticillium dahliae* (Mestre et al. 2016). The phyla Ascomycota and Basidiomycota of yeast were found to predominate. There have also been reports of the psychrotolerant yeast strains *Rhodotorula* sp., *Naganishia* sp., and *Mrakia* sp. that promote plant development (Tapia-Vázquez et al. 2020). These yeast strains were able to make indole acetic acid as well as cellulose, xylanase, and chitinase, among other hydrolytic enzymes. Agroecosystems from cold locations are reservoirs of potential multifunctional diversity, which need to be mined in order to be used as biofertilizers or biopesticides in the future to promote the growth of crops grown in these places (Kumar et al. 2022). Also, a clear plan is needed to preserve microbial diversity, use it wisely, and turn it into economically viable bioproducts for agricultural systems in high altitudes and cold climates.

2.7 *Adaptations and Mechanisms of Psychrotrophs*

Low temperatures are a common occurrence on the majority of the agricultural surface, which is one of the main causes of abiotic stress and lower agricultural output (Kopecká et al. 2023). Under such harsh circumstances, psychrotrophic bacteria are essential for promoting plant development. Utilizing scarce resources, they perform numerous essential tasks like producing 1-aminocyclopropane-1-carboxylate (ACC), aiding in the solubilization of phosphorus, potassium, and zinc essential micronutrients, promoting biological nitrogen fixation, and producing various bioactive substances like ammonia, hydrogen cyanide, indole-3-acetic acid, and Fe-chelating compounds to support plant growth (Rizvi et al. 2021). A few psychrotrophic microorganisms also function as biocontrol agents, preventing plant pathogenic microbes, particularly those with ACC deaminase activity, from damaging plants (Vega-Celedón et al. 2021). Recent discoveries of these psychrotrophic microorganisms have sparked extensive studies in this area. In the last decades, their isolation and research from frigid habitats throughout the world have been greatly aided by the draught genome sequencing. Acc"rdin' to reports, these bacteria's genes for DNA repair, carotenoid/terpenoids biosynthesis, cold adaptations and cold shock proteins, as well as a collection of chaperone proteins, have all been examined (Yadav et al. 2019; Wang et al. 2023).

2.8 *Mechanisms of Adaptation of Microbes at Low Temperature*

The majority of research on adaption mechanisms has historically focused on changes in enzyme structure, preservation of membrane fluidity, expression of cold shock proteins, and the predominance of compatible solutes (Keto-Timonen et al. 2016). New research has also shown nonclassical characteristics connected to energy production, compounds, and stress resistance and cold adaption processes. The chapter summarizes some of the investigations that have been conducted to comprehend the processes and manifestations of stress tolerance in psychotropic bacteria under cold environment. A diagrammatic representation of mechanism adapted by cold-adapted microbes is given in Fig. 3.

2.8.1 Adjustments in Reactive Oxygen Species (ROS) Producing Pathways

Low temperatures cause a rise in oxygen solubility, which raises ROS levels and causes oxidative stress. In order to combat this condition, oxidative metabolisms such as glycolysis, the pentose phosphate pathway (PPP), the tricarboxylic acid cycle (TCA), and the electron transport chain are often suppressed (Piette et al. 2012). For instance, the glycolysis genes are absent in the Siberian permafrost

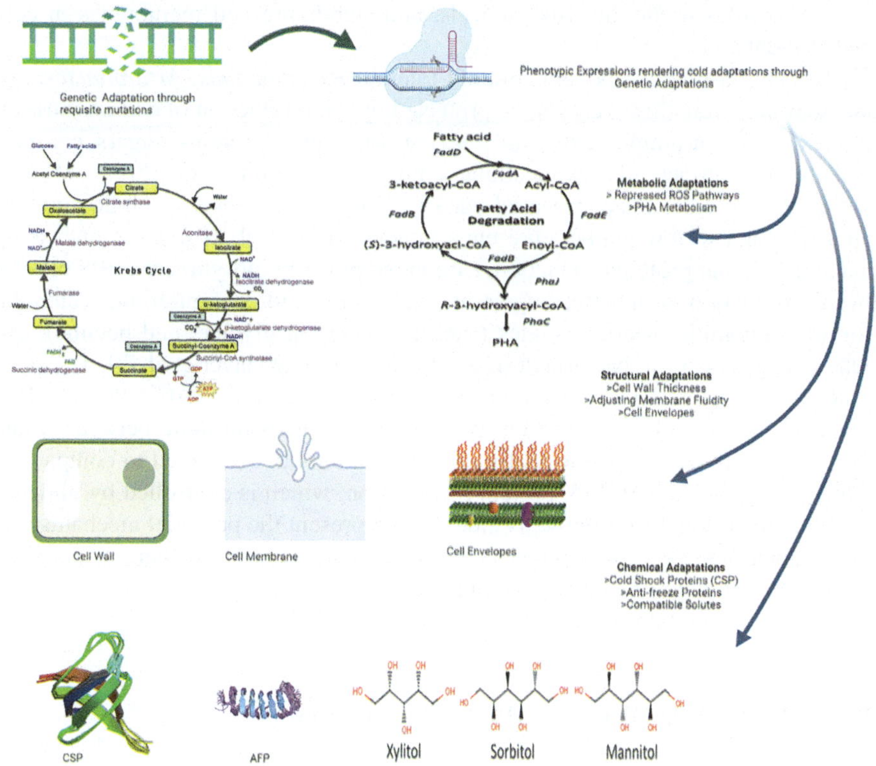

Fig. 3 Diagrammatic representation of cold-adapted mechanism in psychrotrophs

bacteria *Psychrobacter arcticus*, which can survive at −10 °C. This bacterium's energy metabolism is dependent on acetate, which is easily dispersed inside of cells and is widely accessible in the environment. Another species of the bacterium PAMC 21119, which was discovered in Antarctic permafrost, exhibits upregulation of the acetyl-CoA pathway while downregulation of the proteins involved in energy synthesis and conversion (Koh et al. 2017). It was discovered that glycolysis and the TCA cycle are suppressed in many other bacterial species that are growing between −15 and 10 °C, with other routes of energy production taking the lead (Tribelli et al. 2015).

2.8.2 Compatible Solutes and Related Compounds

Several cold-adapted bacteria have compatible solutes that are essential for osmoprotection, cryo-protection, energy generation, and a supply of carbon and nitrogen (Sajjad et al. 2020). Glycine, glycerol, trehalose, betaine, sucrose, mannitol, and sorbitol have all been explored as compatible solutes for managing cold stress. According to Tribelli and López (2018), the key functions of these substances include lowering the freezing point of cytoplasm, preventing the aggregation of macromolecules, removing free radicals, and stabilizing cell membranes in cold environments.

The cytoplasmic components of cold-tolerant *Pseudomonas*, *Arthrobacter*, and *Sporosarcina* were discovered to be protected by the production of certain intracellular/extracellular amino acids, anti-freezing compounds/proteins, sugars, cryoprotectants, and intracellular/extracellular amino acids (Kim et al. 2017). Modern "omic" technologies have increased our knowledge of how psychrophilic microorganisms adapt to low-temperature environments (Liu et al. 2019). Ice-nucleating and anti-freezing protein production, carotenoid pigment biosynthesis, EPS biosynthesis, production of anti-freezing compounds, fatty acid configuration, accumulation of compatible osmotic solutes (such as mannitol, glycine, and betaine), and other changes occur in bacteria during cold adaptation (Maayer et al. 2014).

Recent investigations of Antarctic *Pseudoalteromonas* sp. TAC 125 and TB41 global phenotypic characterization and genome comparison done between 4 and 15°C indicated diverse pathways for cold adaptation but connected to comparable solute processes. In TAC125, protein S-thiolation, which is controlled by glutathione and glutathionylspermidine, is thought to represent the potential mechanism of cold adaptations. The research revealed that the catabolism of substances, including spermidine, glutathione, and ornithine that may promote cryo-tolerance, was boosted via the glutathione mechanism (Mocali et al. 2017).

2.8.3 Polyhydroxyalkanoates (PHAs) Metabolism

Many bacteria are able to produce PHAs, which are unique polymers with crucial physiological functions. PHAs are shown to accumulate more frequently than other nutrients like nitrogen or phosphorus under adverse development situations like excess carbon sources (Vicente et al. 2023). Bacteria are given the capacity to endure a range of environmental challenges, thanks to PHAs. According to López et al. (2015), higher abundance of PHA synthesis-related enzymes in *S. alaskensis*' proteome suggests that PHA production also takes place under conditions of cold stress. The paper also mentions compensating for slower rates of nutrition transport and enzyme activity as an adaptation strategy to maintain functioning in cold weather. *P. extremaustralis*, a different bacterium, exhibits the function of PHAs for adaption in low temperatures. This bacterium generates medium-chain PHAs, short-chain PHAs, and polyhydroxybutyrate (PHB). Particularly, PHBs are essential for development and survival in cold climates. The PHBs are seen to boost the motility and improve the survival of planktonic cells in biofilms formed in cold environments in the bacteria. The PGB genes in *P. extremaustralis* are placed on a genomic island, which further emphasizes the significance of lateral gene transfer events for adaptability, despite the fact that PHB synthesis is uncommon in *Pseudomonas* sp. (Catone et al. 2014). Moreover, it has been noted that PHB monomer, 3-hydroxybutrate, has a cryoprotective role since, even when frozen, PHB granules retain a high degree of flexibility, indicating that they may provide defence against both intracellular and extracellular ice. While PHAs are known to increase anti-oxidative enzymes including catalase, superoxide dismutase, and glutathione peroxidase under the stated setting, some studies have also demonstrated that PHAs may also give protection from oxidative stress under cold temperatures (Margesin and Miteva 2011). Although studies have revealed a relationship between the existence of highly unsaturated mclPHA in *Pseudomonas* sp. UMAB-40 and survival in extremely low temperatures, there is no conclusive evidence linking the creation of PHAs to the development of cold tolerance (Catone et al. 2014). Moreover, it has been shown that Antarctic isolates of alpha, beta, and gamma protobacteria are most frequently found to be capable of synthesizing PHA (Ciesielski et al. 2014). Comparable research from the Baltic Sea and Greenland Sea ice have also revealed that the bacterial community contains PHA synthases gene encodings (Parnanen et al. 2015). The effect of PHAs in either enhancing or impairing the cold endurance in psychrotrophs has to be further investigated, though.

2.8.4 Envelopes and Cold Adaptation

Envelopes serve as a protective covering and a dynamic contact between the cell and the environment in a range of unicellular species. A variety of cold-adapted bacteria have modified membrane fluidity to counteract cold rigidity. This is accomplished by altering the lipid composition of cell membranes, lowering lipid saturation, and preferring shorter chain lipids (Hassan et al. 2020). Gram-negative bacteria, which have an envelope made up of an inner and outer membrane divided by a periplasmic gap that includes a thin peptidoglycan layer, are the most prevalent example of enveloped protection. Proteins, phospholipids, and lipopolysaccharides (LPS) make up the outer membrane (Maayer et al. 2014). According to studies, *P. haloplanktis* has a larger percentage of cell envelope genes in its genome, which is a unique adaptation characteristic for cold tolerance (Tribelli and López 2018). Similar genomic analyses of *P. arcticus* indicated that the genes responsible for lipid production in membranes and cell walls were important for growth in cold climates. Low temperatures are what cause changes in the composition and fluidity of LPS in the Antarctic bacterium *Pseudomonas syringae* Lz4w. This was seen because more hydroxyl-fatty acids were found to increase polymyxin B sensitivity (Jagannadham and Chowdhury 2012). The majority of studies on *Pseudomonas* sp. point out the significance of LPS as a necessary trait for active development in cold climates.

At −2.5 °C, *E. sibiricum* showed an increase in cell wall biosynthesis genes, which is typical of Gram-positive bacteria found in colder climates (Tribelli and López 2018). This supports the finding that cells' cell walls have thickened to defend them from the extremely cold environment. Cells may provide protection by preventing the production of ice or by maintaining osmotic pressure to withstand changes brought on by extremely cold temperatures. The production of cell wall, membrane, and envelope components was shown to be essential for supporting active growth in freezing temperatures in *Planococcus halocryophilus*, which made a similar finding. In the cell membrane of *P. halocryophilus*, which thrives in extremely cold temperatures, increased hydrophobicity and the formation of unique extracellular encrustations of 20% CaCo3, 29% choline, and 50% peptidoglycan were also noticed (Mykytczuk et al. 2013).

Exopolysaccharides (EPS), which constitute a significant portion of the extra-cellular polymers encircling the bacterial cells, are another crucial characteristic of bacteria that can survive in cold climates (Ibrahim et al. 2022). It has been shown that cold-adapted bacteria share EPS synthesis that has unique properties. More evidence for their cryoprotective effect comes from the observation that they are associated with an increase in the freeze-thaw survival frequency in bacteria from the *Winogradskyella*, *Colwellia*, and *Shewanella* genera recovered from the Antarctic sponge (Caruso et al. 2018).

2.8.5 Cold Shock Proteins

Due to a number of factors, most notably the possible use of cold-active enzymes and genetic expressions, there is an extraordinary amount of interest in the scientific community about the cold survival and adaptability of bacteria. Moreover, research on low-temperature adaptation to freezing conditions may reveal obscure areas of life science, such as how life functions under extremely low temperatures (Bhatia et al. 2021). The function of cold shock and antifreeze proteins, the function of cryoprotectants, the preservation of membrane fluidity, and the function of hydrolytic enzymes are all closely related to adaptation at low temperatures (Yadav et al. 2019). A collection of naturally occurring proteins known as "cold shock proteins" (CSPs) is thought to provide their producer organisms with the capacity to withstand severe cold and to develop there. There are 236 shock proteins occur in psychrophilic/psychrotrophic bacteria, such as *Arthrobacter sulfureus, Bacillus licheniformis, Exiguobacterium undae, Janthinobacterium lividum, Pseudomonas stutzeri, Psychrobacter marincola*, and *Sporosarcina pasteurii* (Yadav et al. 2015). The cold acclimation proteins (Caps) are a different family of cold stress proteins that have been described in the psychrotrophic bacteria*Arthrobacter, Exiguobacterium, Janthinobacterium,* and *Pseudomonas* (Yadav et al. 2019). Numerous Antarctic bacteria were found to contain cspA gene homologs. Among 11 bacterial isolates collected from several lakes in Antarctica, anti-freezing proteins (AFPs), a different family of proteins, have been identified as being in charge of bacterial cold adaptation. Anti-freezing substances, such as sugars, organic acids, cryoprotectants, amino acids, and anti-freezing proteins, are present, indicating their involvement in preserving organisms at subfreezing temperatures (Eskandari et al. 2020; Gharib et al. 2022).

2.9 Psychrotrophs: Revolutionizing the Hill Agriculture

A hill agricultural system is made up of a number of climatically distinct zones that are close together and at different elevations, showcasing a variety of microhabitats and a large amount of biodiversity (Aneva et al. 2020). Their unusual biogeographical and climatic characteristics make them vulnerable and changeable. Moreover, hill agricultural systems are under strain from changes in land use patterns, infrastructural expansion, unsustainable tourism, overexploitation of natural resources, habitat fragmentation, and climate change. However, several crops are on the verge of extinction, and the area planted with conventional crops has decreased dramatically (>65%), mostly in the recent five decades (Negi et al. 2012). Many bacteria are capable of surviving and even growing in such harsh settings, in addition to thriving at low temperatures. The ideal temperature range for psychrotrophic microbe development is between 15 and 25°C; however, they may also thrive at 0°C. Given their prevalence and dominance, psychrotrophic bacteria are widely recognized as the most successful colonists of our planet. An understanding of these microbial enzymes, which have a huge potential for utilization in hill agriculture, has led to a significant amount of scientific attention being given to psychrotrophic microorganisms during the past few decades (Mishra et al. 2010). Due to the dependence of the cropping cycle in hilly regions on periodic cold periods that are detrimental to crop growth and output, psychrotrophic microorganisms are important in hill agriculture. Most microbial activity either slow down or halt entirely under these circumstances, which negatively affects crop output. The temperature of the root system also has a key role in the growth and functioning of many rhizospheric communities since most physiological responses that control plant growth effectively stop at inadequate temperatures. This effect is particularly pronounced in nutritional transformations, where bacteria play a significant role.

In a situation like this, when time and temperature play key roles in determining both plant and microbial development, psychrotrophic bacteria that can continue to function in low temperatures are essential. Many psychotropic substances can significantly aid the growth of several crucial agricultural crops (Singh and Singh 2017). Through a variety of direct and indirect mechanisms, including the manufacture of multiple plant hormones, the solubility and decomposition of organic matter, and an improvement in the bioavailability of numerous macro and micronutrients, like zinc, iron, and phosphorus, various types of psychrotolerant in soil suppress many plant pathogens and promote crop growth (Araya et al. 2020). For biocontrol, bio-fertilization, and bio-stimulation, these psychrotrophs are frequently utilized as bio-inoculants. These bio-inoculants boost crop development under various abiotic stress situations, together with other microorganisms. These bioinoculants are affordable and convenient options for reducing cold stress (Anderson et al. 2015).

2.9.1 Direct Mechanism

2.9.1.1 Phytohormone Production

By promoting exogenous hormones and metabolites, soil microorganisms modify the state of phytohormones, which may improve cold tolerance (Raza et al. 2023a, b). The stimulatory phytohormones that are generated by psychrotrophic bacteria in the root zone promote the length and quantity of root hairs and are a key factor in plant development and production (Rizvi et al. 2021). The capacity of the plant to absorb minerals and water from the soil system is improved by the increase in root surface area. Tryptophan consumption and subsequent conversion to indole-3-acetic acid is one of the several mechanisms through which auxin is produced by rhizobacteria (IAA). An auxin signaling pathway is triggered in conjunction with the plant's endogenous IAA pool, causing plant growth and proliferation (Singh et al. 2017). *Pantoea dispersa* strain 1A and *Serratia marcescens* strain SRM, two cold-tolerant PGPB strains, were found in the Northwestern Indian Himalayas (Sangiorgio et al. 2020). These bio-inoculants continue to produce IAA between 4 and 15 °C. These bio-inoculants considerably increase the mineral absorption and biomass of wheat seedlings cultivated at low temperatures. They also characterized the cold tolerance and IAA generation by *Pseudomonas* sp. strain PGERs17 and NARs9 at cold temperature. It is possible to find new cold-tolerant PGPB species in the future given the metabolic plasticity of pseudomonads (Sangiorgio et al. 2020).

2.9.1.2 Ethylene

In reaction to cold stress, plants may produce more ethylene, which may speed up senescence or increase plant tolerance. In stressful circumstances, ethylene preserves plant adaptability at the expense of growth and development. Stress causes ethylene levels to rise, which suppresses transcription of auxin stimulation factors and limits plant growth (Raza et al. 2023a, b). Plant ethylene production is constrained by psychrotolerant bacteria that release the enzyme 1-aminocyclopropane-1-carboxylase (ACC) deaminase. The ethylene precursor ACC is converted to ammonia and ketobutyrate by the enzyme ACC deaminase. Many studies noted that ACC deaminase-producing bacteria enhanced plant growth promotion and stress tolerance (Singh and Singh 2017). According to earlier research, bacterial ACC deaminase is essential for promoting plant development in cold-stressed environments. Moreover, the synthesis of the ACC deaminase enzyme by *Enterobacter* sp. and *Pseudomonas fluorescens* is claimed to reduce the effects of cold stress. It is clear from the significance of ethylene in stress physiology that much more work is needed to understand how bacterial strains that produce ACC deaminase promote plant development at low temperatures (Singh and Singh 2017; Shahid et al. 2023).

2.9.1.3 Siderophore Production

Nearly all kinds of life require iron, the fourth most common metal in the crust of the planet. Its availability to the organism is severely constrained, though, as a result of the quick transition from the ferrous (Fe++) to the ferric (Fe+++) state. Iron is present in its oxidized form ($Fe3+$, insoluble at neutral pH) rather than its reduced form ($Fe2+$, soluble at neutral pH), which is taken up by plants, due to the aerobic environment conditions. Several fungi, bacteria, and certain plants have an uncommon adaption to make siderophores, a class of low-molecular-weight molecules

(10 KD) that have a strong affinity for Fe3+ ions, in order to sequester the iron (Ahmed and Holmström 2014). Iron molecules are transported to plant cells through siderophores, which are substances released by PGPM and subsequently absorbed by plants (de Andrade et al. 2023). It's possible that plant roots can absorb siderophore and utilize them as iron sources. As a result, microbial siderophore can promote plant development by increasing iron intake and preventing plant pathogens through competitive symbiosis, which in turn results in iron biofortification in plants and their grains (Singh et al. 2020; Lurthy et al. 2021). *Arthrobacter, Azotobacter, Azospirillum, Bacillus, Burkholderia, Enterobacter, Pseudomonas, Rhizobium, Rhodospirillum,* and *Serratia*, as well as fungal species *Aspergillus, Syncephalastrum, Penicillium,* and *Rhizopus*, have all been linked to siderophore synthesis by several studies (Misslinger et al. 2021).

2.9.1.4 Phosphate Solubilization

The bioresource in cold regions is generally "untapped," and because of the flourishing microorganisms there, there is a lot of opportunity to isolate new and unusual microorganisms. The enhancement of plant health and implications for low-input sustainable agriculture for both food and non-food crops can arise from the exploitation of cold-tolerant PSB-plant communication. The essential element phosphorus (P), which is regarded as a macronutrient since crops require a significant quantity of it, is essential. P availability for crops that is sufficient encourages early plant development and speeds up maturity (Rajwar et al. 2018; Chaudhary et al. 2021a, b, 2022d). Phosphorous, while being crucial for plant development, is not readily available to them. When the temperature is low, plants can use the phosphate that is available in the soluble form that has been solubilized by psychrotolerant microbes. Many studies on phosphate-solubilizing microorganisms and their function in the mineralization of organic P form to inorganic form, which is readily accessible to plants, have been conducted worldwide (Suarez et al. 2019; Chaudhary et al. 2021a, b). The bacterial genera *Pseudomonas* and *Bacillus*, as well as the fungal species *Penicillium* and *Aspergillus*, include the most effective phosphorus-solubilizing microorganisms. Microorganisms that transform organic phosphate into inorganic phosphate present in the soil aid in the mineralization process. The major enzyme responsible for the direct oxidation of glucose to gluconic acid during bacterial mineral phosphate solubilization has been identified as glucose dehydrogenase (Rasul et al. 2019). After that, 2,5-diketogluconic acid and 2-ketogluconic acid are produced by enzymes from gluconic acid. In terms of solubilizing phosphate, 2-ketogluconic acid is more effective than gluconic acid. As a result, the primary source of plant nourishment is phosphate-solubilizing microorganism-enriched soil (Silva et al. 2023). There are nearly 200 different mineral forms of phosphorus, and apatite, an insoluble form of phosphorus, is the most often found form in the soil system. Around 95–99% of the total phosphates in the soil are insoluble phosphates that are not available to the crops. A significant portion of the soluble inorganic phosphate that was added to agricultural soil as chemical fertilizer quickly became immobilized after the addition and eventually became inaccessible to crops (Mishra et al. 2010; Elhaissoufi et al. 2022).

2.9.1.5 Diazotrophy Under Low Temperature

Low-temperature environments are home to microorganisms that have evolved to survive there. One of the main ways that life on this planet is maintained is through the fixation of nitrogen by symbiotic and asymbiotic microbial groups. Yet cold temperature stress has a significant impact on this process. Rhizobial activities are affected by low temperatures in a way that lowers nodule functionality and competitiveness (Mendoza-Suárez et al. 2021). By reducing the temperature, *Rhizobium leguminosarum* bv. *trifolii* produces less Nod metabolites, which may have an impact on the nodulation and yield of the host legumes. Native legumes in the high arctic can nodulate and fix nitrogen at rates comparable to those found for legumes in temperate climes, despite the freezing temperatures having an influence on nodule formation and nitrogen fixation (Mishra et al. 2012). In agriculture and horticulture, microbial inoculants that promote crop development in low-temperature environments are receiving a lot of interest (Joshi et al. 2019).

2.9.2 Indirect Mechanism

2.9.2.1 Induction of Resistance in Response to Cold Stress

The metabolic activity of plants is impacted by cold stress in a variety of ways, which significantly reduces agricultural output (Ihtisham et al. 2023). Furthermore, by encouraging the growth of saprophytic fungi and disrupting the normal cycle of soil nutrients, which lowers soil fertility, these circumstances can cause crop loss. The disruption of the plant's cellular homeostasis caused by exposure to low temperatures and the generation of reactive oxygen species (ROS) are two of the primary outcomes of stress-induced cellular changes. Hydrogen peroxide (H_2O_2), singlet oxygen (O_2), and HO are ROS that harm biomolecules, including DNA, lipids, proteins, and carbohydrates, ultimately causing complete plant cell death (Subramanian et al. 2016). Psychrotolerant microorganisms have a variety of methods at their disposal to help them deal with cold stress, but root exudates and the preservation of membrane fluidity are two of the most important. It has been determined that the main components of plant root exudates are oxalate and malate, and it has been discovered that bacteria that may colonize root surfaces use these substances as carbon sources. Malate may be used by *Arthrobacter*, *Flavimonas*, *Flavobacterium*, *Massilia*, and *Pseudomonas*. Consequently, these bioinoculants can be used as helpful bacteria that colonize plants to enhance plant development or provide protection from disease assault (de Andrade et al. 2023).

Also, a number of psychrotolerant bacteria are routinely involved in the creation of antioxidant enzymes, the activation of ROS scavenging compounds such as

Table 1 List of cold regulated gene and their putative functions

S. no.	Genes	Annotated species	Description	Reference
1.	*sod*A, *trx*A, *kat*A, and *per*	*Bacillus* strains CJCL2 and RJGP41	Antioxidant enzyme production	Zubair et al. (2019)
2.	*ohr*R and *Opu*AC	*Bacillus* strains CJCL2 and RJGP41	Osmotic balance maintenance	Zubair et al. (2019)
3.	*Coma*	*Bacillus subtilis*	Quorum sensing regulator	Wolf et al. (2016)
4.	*gcr*1 and *gpa*1	*Pseudomonas* spp.	Root elongation, lateral root growth	Zhao et al. (2016)
5.	*dreB1*A	*Pseudomonas putida*	Dehydration responsive element binding	Backer et al. (2018)
6.	*cut*A	*Yersinia pestis*	Divalent-cation tolerance	
7.	*ots*A and *ots*B	*E. coli*	Trehalose synthesis	Mishra et al. (2012)
8.	*hrs* and *gyr* A	*Pseudomonas* spp.	cold-shock induction	Mishra et al. (2012)
9.	*pqq*BCDE	*Hartmannibacter diazotrophicus*	PQQ biosynthetic	Suarez et al. (2019)
10.	*tor*D	*Shewanella amazonensis*	Chaperone synthesis	Mishra et al. (2010)
11.	*sep*F	*Mycobacterium smegmatis*	Cell division	Mishra et al. (2010)
12.	*cip5*	***B. cereus***	Cellular processes and signaling	Brillard et al. (2010)
13.	*csp*A	*B. weihenstephanensis*	Cold shock gene	Beno et al. (2019)
14.	*ds-26-16*	*E. coli, Haematococcus pluvialis*	ROS alleviation	Wang et al. (2016a, b)
15.	*dwhy*1	*E. coli*	Water hypersensitivity against cold and freeze stress	Anderson et al. (2015)

proline, and the mitigation of damage to plant cell membranes (Hasanuzzaman et al. 2021). These reactive oxygen species (ROS), which have been well-described as being connected to temperature-induced stress in plants, are highly reactive and harmful metabolic products. These molecules can also regulate the expression of a number of genes that affect functions, including development, hypersensitivity response that results in programmed cell death, growth, and defence. When expressed in tobacco, the mitogen-activated protein kinase kinase (MAPKK) gene ZmMKK1 was found to improve antioxidant enzyme activity, significantly enhance the expression of ROS-related genes, and influence chilling tolerance (Mishra et al. 2012). Some other cold regulated genes with putative functions are listed in Table 1. Other techniques, such as sugar and melatonin buildup via the ROS scavenging mechanism, which is strengthened in psychrotrophic inoculation plants, also confers freezing resistance. As a result, increased ROS scavenging activity in the plant can be at least partially blamed for the increase in freezing tolerance (Mishra et al. 2010).

2.9.2.2 Frost Management

The main reason that freezing damage to traditional crops is so difficult is that different plant elements, such as buds, flowers, leaves, roots, and stems, don't all react in the same way. Crop ice nucleation is frequently not endogenous but is instead brought on by microbial parasites that can be found on leaves, fruits, or stems and have catalytic sites. Frost forms on plants during cold weather conditions and can seriously harm them. Crops destroyed by frost result in significant losses for the agriculture sector every year. Ice-nucleating activities (INAs) imitate the surface structure of an ice crystal to restrict supercooling and promote freezing at high subzero temperatures. They reduce the energy required for the commencement of ice formation by imposing an ice-like collection on the water molecule in contact with their surface. The "ice plus" microorganisms have the Ina protein, which is located on the surface of bacteria and serves as the nucleation site for ice crystals. This promotes the production of ice at extremely low temperatures, whereas "ice-minus" bacteria do not have Ina proteins and lower the freezing point of ice (Hill et al. 2014). Scientists believe that spraying ice-free microbes on a variety of crops might help prevent the yearly damage brought on by icy temperatures. In addition to being a disease of these crops, *P. syringae* ice-nucleating strains enhance the frost sensitivity of tomato and soybean when sprayed on leaves prior to low-temperature stress (Fadiji et al. 2022). The "ice-minus" mutant synthesis was created after the gene responsible for ice nucleation in *P. syringae* was discovered. This mutant was discovered to be ineffective at encouraging ice nucleation in plant leaves (Karimi et al. 2019). An efficient and ecologically safe strategy of preventing freezing damage in plants is to reduce the population of ice-nucleating microorganisms, which is regarded as a typical example of the displacement of a bacterial pathogen by a biocontrol agent.

2.10 Current research, Challenges, and Possible Solutions

The persistent low temperature (below 5 °C) that is colonized by a variety of extremophiles, including bacteria, algae, fungus, yeasts, and filamentous fungi, covers more than 85% of the earth's biosphere (Joshi et al. 2017). In contrast to psychrotolerant microorganisms, which have growth at an optimal temperature of 25 °C, psychrophiles are described as having cardinal growth at lowest, optimum, and maximum temperatures, such as at or below 0, 15, and 20 °C, respectively (Sandle and Skinner 2013). Many studies on the genetics, acclimation processes, and other adaption mechanisms of cold-loving microorganisms are motivated by their diverse appealing features (Yadav et al. 2019).

2.11 Psychrophiles: An Overview

Plant diversity and the soil ecosystem are able to interact positively thanks to microorganisms living in the soil ecosystem (Agri et al. 2022; Manzar et al. 2022). The Himalaya's diverse plant life supports a great diversity of microorganisms, including fungus, bacteria, archaebacteria, and unicellular algae (Dhakar and Pandey 2020). Reduced nutrient bioavailability, low enzymatic activity, altered soil pH, water activity, and increased soil salinity are all effects of these low-temperature conditions. This group of microbes has various mechanisms for coping with cold stress and thriving in frigid environments. Additionally, they have a wide range of plant growth-promoting qualities that make them an environmentally friendly substitute for dangerous chemicals in hill agriculture systems. From Himalayan soil, a number of bacteria that fix nitrogen under cold stress were isolated and studied for their proteomes. For example, the protein profile under low-temperature diazotrophy was examined in *Pseudomonas migulae* S10724 (Suyal et al. 2014), *Rhodococcus qingshengii* S10107 (Suyal et al. 2019), and *Pseudomonas palleroniana* N26 (Soni et al. 2015). Some of the reported species of bacterial species are summed up in Table 2.

The genomic sequences and proteomic methods of biotechnology are used to investigate the metabolic and structural changes at the cellular level. Based on proteomic analysis of three Siberian permafrost bacteria (upregulation of protein synthesis-transcription and translation, RNA and protein folding, regulation of particular metabolic activities in low temperature, membrane homeostasis, and antioxidant activities), three Siberian permafrost bacteria are identified (Barria et al. 2013). Psychrophiles are heat labile, and low and moderate temperatures make their enzymes 10 times more active.

Table 2 Some bacterial species isolated from cold habitats

S. no.	Bacterial species	Reference
1.	*Cryobacterium roopkundense*	Reddy et al. (2010)
2.	*Glaciimonas frigoris*	Margesin et al. (2016)
3.	*Lacinutrix jangbogonensis*	Lee et al. (2014)
4.	*Massilia eurypsychrophila*	Shen et al. (2015)
5.	*Pedobacter arcticus*	Zhou et al. (2012)
6.	*Sphingobacterium psychroaquaticum*	Albert et al. (2013)
7.	*Sphingomonas glacialis*	Zhang et al. (2011)
8.	*Arthrobacter nicotianae* *Brevundimonas terrae*	Yadav et al. (2015)
9.	*Arthrobacter methylotrophus* *Pseudomonas rhodesiae*	Verma et al. (2015)
10.	*Rhodococcus*	De Mandal et al. (2015)
11.	*Burkholderia*	Soni et al. (2016)
12.	*Yersinia*	Sangwan et al. (2015)
13.	*Flavobacterium*	Gangwar et al. (2011)

These traits are responsible for the importance of psychrophiles in biotechnology:

1. Requires low amount of catalyst for activity and reduces the cost of enzyme preparation.
2. Does not require heat to activate the enzymatic activity as they remain ambient at tap water temperature.
3. Where they aren't required, they can be easily eliminated by heating.

It has been discovered that cold-active enzymes differently stabilize the whole molecule or at active regions, maintaining the flexibility, dynamics, and temperature of the active areas. Because proline residues are less common, the thermal stability is primarily determined by electrostatic interactions, including ionic pairing, h-bonding, aromatic interactions, weakening the hydrophobic effect, carefully placing glycine residues, improving surface interactions, destabilizing factors, and adaptive destabilization of enzymes (Margesin and Miteva 2011).

Various areas of interest related to low temperature are:

1. **Enzymes**: In the field of food technology, reagents in molecular biology or medicinal/pharmaceutical research.
2. **Biomolecules**: Importantly, polyunsaturated fatty acids and antifreeze proteins are used as food additives and nutritional supplements in cattle, aquaculture, and human diets.
3. **Pharmaceutical and medicinal use**: Screening of novel vaccines, antibiotics, cancer treatments, and many other medications, as well as pharmaceuticals used in nutraceuticals and cosmetics.
4. **Bioremediation**: For the degradation of contaminants or harmful environmental residues in cold-adapted and other areas, biostimulation or bioaugmentation is used.

2.12 Biotechnological Approaches

Applications of psychrophiles in a variety of industries, including waste processing, food production, bioremediation, mining, agriculture, pharmaceuticals, and molecular diagnostics, highlight their biotechnological potential for growing in extremely cold environments and their capacity for adaptation (Banerjee et al. 2016; Araya et al. 2020). High catalytic efficiency at low temperatures, special substrate requirements, and decreased thermal stability are the key drivers of the inventive biotechnological exploitation (Cavicchioli et al. 2011).

2.13 Bioremediation

Due to human activity and population increase, the demand for fossil fuels is at an all-time high, and environmental threats are posed by petroleum product leaks. *Rhodococcus* (Xu et al. 2018), *Pseudoalteromonas* (Gutierrez et al. 2018), and *Pseudomonas* (Ojewumi et al. 2018) are psychrophiles that can reduce these impacts, because they have specific catabolic pathways to break down petroleum hydrocarbons, such as n-alkanes and polycyclic aromatic hydrocarbons (Giudice et al. 2010). Organic pollutants (petroleum hydrocarbons) are degraded by the local psychrophilic microbial communities by mineralizing (using hydrocarbons as the only source of carbon and energy) them into less harmful, non-hazardous compounds so that they can be naturally incorporated into the biogeochemical cycles (Margesin and Feller 2010). Polar and alpine soils and sediments, subsoils, seawater, and groundwater with temperatures below 10°C are among the cold areas that are polluted. Twenty years ago, it was reported that altered psychrophiles based on the transfer of the TOL plasmid from the mesophile *Pseudomonas putida* to the psychrophile *Pseudomonas putida* via conjugation had been discovered. By transferring the gene for a monooxygenase from the mesophilic *Pseudomonas stutzeri* into the *Pseudomonas haloplanktis*, a novel microbe for the breakdown of aromatic hydrocarbons has just been discovered, although its effectiveness has not yet been shown. The most effective microbial starter for wastewater treatment at 10 °C comes from *Arthrobacter psychrolactophilus*, which increases the biodegradability of organic components in the turbid water by hydrolyzing protein, starch, and lipids. *Pseudomonas* (a cold tolerant bacteria) involved in degradation in phenol (99%) at 20 °C within 24 h (Sepehr et al. 2019).

2.14 Food Industry

The tendency to store food at lower temperatures to preserve flavor, nutritional content, texture, and energy is luring cold-loving microorganisms that are exploited in the dairy industry to produce lactose-free milk-derived meals for those who are lactose intolerant (Hamdan 2018). Moreover, it may be utilized to create a natural sweetener with a high added value and low calorie and glycemic index that transforms lactose into D-tagatose (Struvay and Feller 2012). In the wine industry, cold-active pectinases from yeasts and fungi are utilized to retain volatile components for the wine's aromatic character (Singh et al. 2012). With increased loaf volume, the mechanical dough qualities and final bread quality are enhanced by the application of xylanases from the *Pseudoalteromonas haloplanktis* (Struvay and Feller 2012).

2.15 Cosmetic and Pharmaceutical Application

When more and more potential psychrophilic strains are discovered, they are employed to make priceless antibacterial chemicals that are beneficial to the pharmaceutical and cosmetic sectors (Tomova et al. 2015). *Nocardioides* sp. strain A-1, a halophilic actinomycete, produces a substance that has antibacterial properties against the rice bacterial blight-causing *Xanthomonas oryzae* (Gesheva and Vasileva-Tonkova 2012). Scar therapy and cosmetic regeneration creams benefit from the antifreeze glycoprotein "Antarcticine-NF3" generated by *Pseudoalteromonas* found in the Antarctic (Hamdan 2018).

2.16 Detergent and Fabric Industries

The detergent industry, which relies on enzymes for 30–40% of all enzyme production globally, needs novel enzymes that can function at low temperatures in order to reduce energy consumption. Important enzymes including lipases, proteases, and amylases increase detergent effectiveness and safeguard the texture and color of fabrics to prevent tearing (Kuddus et al. 2011, Maiangwa et al. 2015; Liu et al. 2023a, b). Cold-active detergents are now combined with *Bacillus* sp. from the Antarctic to store the alkaline stability and activity needed for washing (Struvay and Feller 2012). The denim tissues are smoothed and made softer by the cold-adapted cellulase, which also decolorizes textile waste and serves as a textile bleach (Kasana and Gulati 2011).

2.17 Use of Cold-Adapted Enzymes for Cleaning

The ability of the enzymes to hydrolyze substrates is helpful in the treatment of water (Pourmohammadi and Abedi 2021). The increased regulatory demands and commercial requirements are the reason behind the use of the enzymes for improved efficacy and environmental sustainability for or more than 20°C washing the proteases, amylases, lipases, and cellulases (alcalase, natalase, and lipolase) are used (Liu et al. 2023a, b). To increase the efficacy of cleaning, amylase from metagenomic analysis of glacier water and proteases from *Serratia rubidaea* and *Stenotrophomonas maltophilia* are utilized (Sharma et al. 2010).

3 Conclusion and Future Prospectives

In modern agricultural practice, the use of plant growth promotory microorganisms (PGPM) is essential. In a nation like India, where steep terrain areas with chilly temperatures make up a sizeable component of the land area under cultivation, microbial diversity studies for mining psychrotrophic plant benefit microorganisms are crucial. In a number of nations, these microorganisms are already being employed successfully, and this trend is expected to continue. Yet, as the usage of plant growth promotory microorganisms' spreads, attention will be drawn to a number of crucial challenges. For instance, new strategies and perspectives for the growth, storage, shipping, formulation, and application of these potential bioinoculants with respect to hill agriculture will be necessary as the journey from laboratory experiments to greenhouse experiments to full-fledged field trials and eventually to commercial field. Second, it will be necessary to extensively educate the community and farmers about the application and advantages of PGPB. Before releasing these eco-friendly, cost-effective alternatives to agrochemicals on a broad scale, it is essential to address the common public's notion that "bacteria" are solely detrimental agents. Third, even though the initial PGPM are likely to be wild-type bacterial strains that are strictly selected for their positive traits, it is possible that in the future, as detailed knowledge of the mechanisms for bacterial stimulation of plant growth is attained, scientists will be able to engineer genetically more efficient strains while guaranteeing low to none hazards or risks. The ability of psychoactive soil bacteria to survive at low temperatures is the foundation for the creation of efficient biofertilizers for encouraging agriculture in colder climates as well as the manufacture of useful food sources such as single-cell proteins in extremely cold climates. examining how these specialized microorganisms may be utilized to imitate human adaptations in daily life. In all lower-temperature-based enterprises, for instance, a large number of low-temperature microorganisms play a key role. Given the aforementioned limitations and potential remedies, there are excellent odds that a sustainable agriculture practice will be adopted that is solely focused on the effectiveness of these little miracles. Thus, this technology has a very bright and hopeful future for the hill agricultural practices.

References

Agri U, Chaudhary P, Sharma A (2021) In vitro compatibility evaluation of agriusable nanochitosan on beneficial plant growth-promoting rhizobacteria and maize plant. Natl Acad Sci Lett. https://doi.org/10.1007/s40009-021-01047-w

Agri U, Chaudhary P, Sharma A, Kukreti B (2022) Physiological response of maize plants and its rhizospheric microbiome under the influence of potential bioinoculants and nanochitosan. Plant Soil 474:451–468. https://doi.org/10.1007/s11104-022-05351-2

Ahmed E, Holmström SJ (2014) Siderophores in environmental research: roles and applications. Microb Biotechnol 7(3):196–208. https://doi.org/10.1111/1751-7915.12117

Albert RA, Waas NE, Pavlons SC, Pearson JL, Ketelboeter L, Rossello-Mora R, Busse HJ (2013) *Sphingobacterium psychroaquaticum* sp. nov., a psychrophilic bacterium isolated from Lake Michigan water. Int J Syst Evol Microbiol 63(Pt_3):952–958. https://doi.org/10.1099/ijs.0.043844-0

Altieri MA, Nicholls CI, Henao A, Lana MA (2015) Agroecology and the design of climate change-resilient farming systems. Agron Sustain Dev 35:869–890. https://doi.org/10.1007/s13593-015-0285-2

Anderson D, Ferreras E, Trindade M, Cowan D (2015) A novel bacterial water hypersensitivity-like protein shows in vivo protection against cold and freeze damage. FEMS Microbiol Lett 362(15):110. https://doi.org/10.1093/femsle/fnv110. Epub 2015 Jul 17

Aneva I, Zhelev P, Lukanov S, Peneva M, Vassilev K, Zheljazkov VD (2020) Influence of the land use type on the wild plant diversity. Plants (Basel) 9(5):602. https://doi.org/10.3390/plants9050602

Araya MA, Valenzuela T, Inostroza NG, Maruyama F, Jorquera MA, Acuña JJ (2020) Isolation and characterization of cold-tolerant hyper-ACC-degrading bacteria from the rhizosphere, endosphere, and phyllosphere of antarctic vascular plants. Microorganisms 8(11):1788. https://doi.org/10.3390/microorganisms8111788

Backer R, Rokem JS, Ilangumaran G, Lamont J, Praslickova D, Ricci E, Subramanian S, Smith DL (2018) Plant growth-promoting rhizobacteria: context, mechanisms of action, and roadmap to commercialization of biostimulants for sustainable agriculture. Front Plant Sci 9:1473. https://doi.org/10.3389/fpls.2018.01473

Banerjee R, Halder A, Natta A (2016) Psychrophilic microorganisms: habitats and exploitation potentials. Eur J Biotechnol Biosci 4:16–24

Barrero-Sicilia C, Silvestre S, Haslam RP, Michaelson LV (2017) Lipid remodelling: unravelling the response to cold stress in Arabidopsis and its extremophile relative *Eutrema salsugineum*. Plant Sci 263:194–200. https://doi.org/10.1016/j.plantsci.2017.07.017

Barria C, Malecki M, Arraiano CM (2013) Bacterial adaptation to cold. Microbiology 159(Pt_12):2437–2443. https://doi.org/10.1099/mic.0.052209-0

Beno SM, Orsi RH, Cheng RA, Kent DJ, Kovac J, Duncan DR, Martin NH, Wiedmann M (2019) Genes associated with psychrotolerant *Bacillus cereus* group isolates. Front Microbiol 10:662. https://doi.org/10.3389/fmicb.2019.00662

Bhandari G, Nautiyal N, Sharma M (2020) Assessment of genes and enzymes of microorganisms of high altitudes and their application in agriculture. In: Microbiological advancements for higher altitude agro-ecosystems & sustainability. Springer, Singapore, pp 307–326. https://doi.org/10.1007/978-981-15-1902-4_16

Bhatia RK, Ullah S, Hoque MZ, Ahmad I, Yang Y-H, Bhatt AK, Bhatia SK (2021) Psychrophiles: a source of cold-adapted enzymes for energy efficient biotechnological industrial processes. J Environ Chem Eng 9:104607. https://doi.org/10.1016/j.jece.2020.104607

Biemans H, Siderius C, Lutz A, Nepal S, Ahmad B, Hassan T, Von Bloh W, Wijngaard R, Wester P, Shrestha A, Immerzeel W (2019) Importance of snow and glacier meltwater for agriculture on the Indo-Gangetic Plain. Nat Sustain 2:594–601. https://doi.org/10.1038/s41893-019-0305-3

Bishop BC, Chatterjee SP (2019) Himalayas. Encyclopaedia Britannica. Encyclopaedia Britannica, inc. December 30, 2019 (https://www.britannica.com/place/Himalayas)

Bouabdelli S, Zeroual A, Meddi M, Assani A (2022) Impact of temperature on agricultural drought occurrence under the effects of climate change. Theor Appl Climatol 148(1-2):191–209. https://doi.org/10.1007/s00704-022-03935-7

Brillard J, Jehanno I, Dargaignaratz C, Barbosa I, Ginies C, Carlin F, Fedhila S, Broussolle V, Sanchis V (2010) Identification of *Bacillus cereus* genes specifically expressed during growth at low temperatures. Appl Environ Microbiol 76(8):2562–2573. https://doi.org/10.1128/AEM.02348-09

Caruso C, Rizzo C, Mangano S, Poli A, Di Donato P, Finore I, Giudice AL (2018) Production and biotechnological potential of extracellular polymeric substances from sponge-associated Antarctic bacteria. Appl Environ Microbiol 84:4. https://doi.org/10.1128/AEM.01624-17

Catone MV, Ruiz JA, Castellanos M, Segura D, Espin G, López NI (2014) High polyhydroxybutyrate production in *Pseudomonas extremaustralis* is associated with differential expression of horizontally acquired and core genome polyhydroxyalkanoate synthase genes. PLoS One 9(6):e98873. https://doi.org/10.1371/journal.pone.0098873

Cavicchioli R, Charlton T, Ertan H, Mohd Omar S, Siddiqui K, Williams T (2011) Biotechnological uses of enzymes from psychrophiles. Microb Biotechnol 4:449–460. https://doi.org/10.1111/j.1751-7915.2011.00258.x

Chaudhary A, Parveen H, Chaudhary P, Bhatt P, Khatoon H (2021a) Rhizospheric microbes and their mechanism. In: Bhatt P, Gangola S, Udayanga D, Kumar G (eds) Microbial technology for sustainable environment. Springer, Singapore. https://doi.org/10.1007/978-981-16-3840-4_6

Chaudhary P, Agri U, Chaudhary A, Kumar A, Kumar G (2022a) Endophytes and their potential in biotic stress management and crop production. Front Microbiol 13:933017. https://doi.org/10.3389/fmicb.2022.933017

Chaudhary P, Chaudhary A, Agri U, Khatoon H, Singh A (2022b) Recent trends and advancements for agro-environmental sustainability at higher altitudes. In: Goel R, Soni R, Suyal DC, Khan M (eds) Survival strategies in cold-adapted microorganisms. Springer, Singapore. https://doi.org/10.1007/978-981-16-2625-8_19

Chaudhary P, Parveen H, Gangola S, Kumar G, Bhatt P, Chaudhary A (2021b) Plant growth promoting rhizobacteria and their application in sustainable crop production. In: Bhatt P, Gangola S, Udayanga D, Kumar G (eds) Microbial technology for sustainable environment. Springer, Singapore. https://doi.org/10.1007/978-981-16-3840-4_13

Chaudhary P, Singh S, Chaudhary A, Agri U, Bhandari G (2022c) Chapter 3 Agrochemicals and their effects on soil microbial population. In: Soni R, Suyal D, Goel R (eds) Plant protection: from chemicals to biologicals. De Gruyter, Berlin, Boston, pp 45–58. https://doi.org/10.1515/9783110771558-003

Chaudhary P, Singh S, Chaudhary A, Sharma A, Kumar G (2022d) Overview of biofertilizers in crop production and stress management for sustainable agriculture. Front Plant Sci 13:930340. https://doi.org/10.3389/fpls.2022.930340

Chaudhary P, Xu M, Ahamad L, Chaudhary A, Kumar G, Adeleke BS, Verma KK, Hu D-M, Širić I, Kumar P, Popescu SM, Abou Fayssal S (2023) Application of synthetic consortia for improvement of soil fertility, pollution remediation, and agricultural productivity: a review. Agronomy 13:643. https://doi.org/10.3390/agronomy13030643

Ciesielski S, Górniak D, Mozejko J, Swiatecki A, Grzesiak J, Zdanowski M (2014) The diversity of bacteria isolated from Antarctic freshwater reservoirs possessing the ability to produce polyhydroxyalkanoates. Curr Microbiol 69:594–603. https://doi.org/10.1007/s00284-014-0629-1

de Andrade LA, Santos CHB, Frezarin ET, Sales LR, Rigobelo EC (2023) Plant growth-promoting rhizobacteria for sustainable agricultural production. Microorganisms 11(4):1088

De Mandal S, Sanga Z, Kumar NS (2015) Metagenome sequencing reveals *Rhodococcus* dominance in Farpuk Cave, Mizoram, India, an Eastern Himalayan biodiversity hot spot region. Genome Announc 3:1–2. https://doi.org/10.1128/genomeA.00610-15

Dhakar K, Pandey A (2020) Microbial ecology from the Himalayan cryosphere perspective. Microorganisms 8(2):257. https://doi.org/10.3390/microorganisms8020257

Elhaissoufi W, Ghoulam C, Barakat A, Zeroual Y, Bargaz A (2022) Phosphate bacterial solubilization: a key rhizosphere driving force enabling higher P use efficiency and crop productivity. J Adv Res 38:13–28. https://doi.org/10.1016/j.jare.2021.08.014

Eskandari A, Leow TC, Rahman MBA, Oslan SN (2020) Antifreeze proteins and their practical utilization in industry, medicine, and agriculture. Biomolecules 10(12):1649. https://doi.org/10.3390/biom10121649

Fadiji AE, Babalola OO, Santoyo G, Perazzolli M (2022) The potential role of microbial biostimulants in the amelioration of climate change-associated abiotic stresses on crops. Front Microbiol 12:829099. https://doi.org/10.3389/fmicb.2021.829099

Gangwar P, Alam SI, Singh L (2011) Metabolic characterization of cold active *Pseudomonas*, *Arthrobacter*, *Bacillus*, and *Flavobacterium* spp. from Western Himalayas. Indian J Microbiol 51:70–75. https://doi.org/10.1007/s12088-011-0092-7

Gesheva V, Vasileva-Tonkova E (2012) Production of enzymes and antimicrobial compounds by halophilic Antarctic *Nocardioides* sp. grown on different carbon sources. World J Microbiol Biotechnol 28:2069–2076. https://doi.org/10.1007/s11274-012-1009-2

Gharib G, Saeidiharzand S, Sadaghiani AK, Koşar A (2022) Antifreeze proteins: a tale of evolution from origin to energy applications. Front Bioeng Biotechnol 9:770588. https://doi.org/10.3389/fbioe.2021.770588

Giudice AL, Bruni V, De Domenico M, Michaud L (2010) Psychrophiles-cold-adapted hydrocarbon-degrading microorganisms. In: Handbook of hydrocarbon and lipid microbiology. Springer, Berlin, Heidelberg. https://doi.org/10.1007/978-3-540-77587-4_139

Gupta S, Pandey S (2019) ACC deaminase producing bacteria with multifarious plant growth promoting traits alleviates salinity stress in French Bean (*Phaseolus vulgaris*) plants. Front Microbiol 10:1–17. https://doi.org/10.3389/fmicb.2019.01506

Gutierrez T, Morris G, Ellis D, Bowler B, Jones M, Salek K, Teske A (2018) Hydrocarbon-degradation and MOS-formation capabilities of the dominant bacteria enriched in sea surface oil slicks during the Deepwater Horizon oil spill. Mar Pollut Bull 135:205–215. https://doi.org/10.1016/j.marpolbul.2018.07.027

Hamdan A (2018) Psychrophiles: ecological significance and potential industrial application. S Afr J Sci 114(5-6):1–6. https://doi.org/10.17159/sajs.2018/20170254

Hamid B, Bashir Z, Yatoo AM, Mohiddin F, Majeed N, Bansal M, Poczai P, Almalki WH, Sayyed RZ, Shati AA, Alfaifi MY (2022) Cold-active enzymes and their potential industrial applications-a review. Molecules 27(18):5885. https://doi.org/10.3390/molecules27185885

Hasanuzzaman M, Parvin K, Bardhan K, Nahar K, Anee TI, Masud AAC, Fotopoulos V (2021) Biostimulants for the regulation of reactive oxygen species metabolism in plants under abiotic stress. Cells 10(10):2537. https://doi.org/10.3390/cells10102537

Hassan N, Anesio AM, Rafiq M, Holtvoeth J, Bull I, Haleem A, Shah AA, Hasan F (2020) Temperature driven membrane lipid adaptation in glacial psychrophilic bacteria. Front Microbiol 11:824. https://doi.org/10.3389/fmicb.2020.00824

Hill TC, Moffett BF, Demott PJ, Georgakopoulos DG, Stump WL, Franc GD (2014) Measurement of ice nucleation-active bacteria on plants and in precipitation by quantitative PCR. Appl Environ Microbiol 80(4):1256–1267. https://doi.org/10.1128/AEM.02967-13

Ibrahim HAH, Abou Elhassayeb HE, El-Sayed WMM (2022) Potential functions and applications of diverse microbial exopolysaccharides in marine environments. J Genet Eng Biotechnol 20:151. https://doi.org/10.1186/s43141-022-00432-2

Ihtisham M, Hasanuzzaman M, El-Sappah AH, Zaman F, Khan N, Raza A, Sarraf M, Khan S, Abbas M, Hassan MJ, Li J, Zhao X, Zhao X (2023) Primary plant nutrients modulate the reactive oxygen species metabolism and mitigate the impact of cold stress in overseeded perennial ryegrass. Front Plant Sci 14:1149832. https://doi.org/10.3389/fpls.2023.1149832

Islam MS, Kieu E (2020) Tackling regional climate change impacts and food security issues: a critical analysis across ASEAN, PIF and SAARC. Sustainability 12(3):883. https://doi.org/10.3390/su12030883

Jagannadham MV, Chowdhury C (2012) Differential expression of membrane proteins helps Antarctic *Pseudomonas syringae* to acclimatize upon temperature variations. J Proteomics 75(8):2488–2499. https://doi.org/10.1016/j.jprot.2012.02.033

Joshi D, Chandra R, Suyal DC, Kumar S, Goel R (2019) Impacts of bioinoculants *Pseudomonas jesenii* MP1 and *Rhodococcus qingshengii* S10107 on chickpea (*Cicer arietinum* L.) yield and soil nitrogen status. Pedosphere 29(3):388–399. https://doi.org/10.1016/S1002-0160(19)60807-6

Joshi D, Kumar S, Suyal DC, Goel R (2017) The microbiome of the Himalayan ecosystem. In: Kalia VC et al (eds) Mining of microbial wealth and metagenomics. Springer, Singapore. https://doi.org/10.1007/978-981-10-5708-3_6

Karimi B, Nosrati R, Fazly Bazzaz BS, Mirsasan M, Malboobi M, Owlia P (2019) A comparative evaluation of freezing criteria and molecular characterization of epiphytic ice-nucleating (Ice+) and non-ice-nucleating (Ice−) *Pseudomonas syringae* and *Pseudomonas fluorescens*. J Plant Pathol 102:169–178. https://doi.org/10.1007/s42161-019-00402-7

Kasana RC, Gulati A (2011) Cellulases from psychrophilic microorganisms: a review. J Basic Microbiol 51(6):572–579. https://doi.org/10.1002/jobm.201000385

Keto-Timonen R, Hietala N, Palonen E, Hakakorpi A, Lindström M, Korkeala H (2016) Cold shock proteins: a minireview with special emphasis on Csp-family of *Enteropathogenic Yersinia*. Front Microbiol 7:1151. https://doi.org/10.3389/fmicb.2016.01151

Khati P, Parul BP, Nisha Kumar R, Sharma A (2018) Effect of nanozeolite and plant growth promoting rhizobacteria on maize. 3 Biotech 8:141. https://doi.org/10.1007/s13205-018-1142-1

Kim HJ, Lee JH, Hur YB, Lee CW, Park SH, Koo BW (2017) Marine antifreeze proteins: structure, function, and application to cryopreservation as a potential cryoprotectant. Marina Drugs 15(2):27. https://doi.org/10.3390/md15020027

Koh HY, Park H, Lee JH, Han SJ, Sohn YC, Lee SG (2017) Proteomic and transcriptomic investigations on cold-responsive properties of the psychrophilic Antarctic bacterium *Psychrobacter sp.* PAMC 21119 at sub-zero temperatures. Environ Microbiol 19(2):628–644. https://doi.org/10.1111/1462-2920.13578

Kopecká R, Kameniarová M, Černý M, Brzobohatý B, Novák J (2023) Abiotic stress in crop production. Int J Mol Sci 24(7):6603. https://doi.org/10.3390/ijms24076603

Kuddus M, Roohi AJ, Jamal MA, Ramteke PW (2011) An overview of cold-active microbial α-amylase: adaptation strategies and biotechnological potentials. Biotechnology 10:246–258. https://doi.org/10.3923/biotech.2011.246.258

Kumar S, Diksha, Sindhu SS, Kumar R (2022) Biofertilizers: an ecofriendly technology for nutrient recycling and environmental sustainability. Curr Res Microb Sci 3:100094. https://doi.org/10.1016/j.crmicr.2021.100094

Kumari S, Sharma A, Chaudhary P, Khati P (2020) Management of plant vigor and soil health using two agriusable nanocompounds and plant growth promotory rhizobacteria in Fenugreek. 3 Biotech 10:1–11. https://doi.org/10.1007/s13205-020-02448-2

Lee YM, Hwang CY, Lee I, Jung YJ, Cho Y, Baek K (2014) *Lacinutrix jangbogonensis* sp. nov. A psychrophilic bacterium isolated from Antarctic marine sediment and emended description of the genus *Lacinutrix*. Antonie Van Leeuwenhoek 106:527–533. https://doi.org/10.1007/s10482-014-0221-5

Liu Q, Liu HC, Zhou YG, Xin YH (2019) Microevolution and adaptive strategy of psychrophilic species *Flavobacterium bomense* sp. nov. isolated from glaciers. Front Microbiol 10:1069. https://doi.org/10.3389/fmicb.2019.01069

Liu Y, Jia K, Chen H, Wang Z, Zhao W, Zhu L (2023a) Cold-adapted enzymes: mechanisms, engineering and biotechnological application. Bioprocess Biosyst Eng. https://doi.org/10.1007/s00449-023-02904-2

Liu Y, Zhang N, Ma J, Zhou Y, Wei Q, Tian C, Fang Y, Zhong R, Chen G, Zhang S (2023b) Advances in cold-adapted enzymes derived from microorganisms. Front Microbiol 14:1152847. https://doi.org/10.3389/fmicb.2023.1152847

López NI, Pettinari MJ, Nikel PI, Méndez BS (2015) Polyhydroxyalkanoates: much more than biodegradable plastics. In: Advances in applied microbiology, vol 93. Academic Press, pp 73–106. https://doi.org/10.1016/bs.aambs.2015.06.001

Lurthy T, Pivato B, Lemanceau P, Mazurier S (2021) Importance of the rhizosphere microbiota in iron biofortification of plants. Front Plant Sci 12:744445. https://doi.org/10.3389/fpls.2021.744445

Maayer PD, Anderson D, Cary C, Cowan DA (2014) Some like it cold: understanding the survival strategies of psychrophiles. EMBO Rep 15:508–517. https://doi.org/10.1002/embr.201338170

Maiangwa J, Ali M, Salleh A, Abd Rahman R, Shariff F, Leow T (2015) Adaptational properties and applications of cold-active lipases from psychrophilic bacteria. Extremophiles 19:235–247. https://doi.org/10.1007/s00792-014-0710-5

Manzar N, Jayaswal D, Kishan G, Chauhan S, Joshi D, Singh J, Chaudhary P, Jamali H, Kushwaha P, Parveen H, Kashyap AS, Sahu PK, Singh UB (2022) Application of molecular ecology approaches in sustainable agriculture for a better understanding of plant–microbiome interactions. In: Singh UB, Sahu PK, Singh HV, Sharma PK, Sharma SK (eds) Rhizosphere microbes. microorganisms for sustainability, vol 40. Springer, Singapore. https://doi.org/10.1007/978-981-19-5872-4_3

Margesin R, Feller G (2010) Biotechnological applications of psychrophiles. Environ Technol 31(8-9):835–844. https://doi.org/10.1080/09593331003663328

Margesin R, Miteva V (2011) Diversity and ecology of psychrophilic microorganisms. Res Microbiol 162(3):346–361. https://doi.org/10.1016/j.resmic.2010.12.004

Margesin R, Zhang DC, Frasson D, Brouchkov A (2016) *Glaciimonas frigoris* sp. nov., a psychrophilic bacterium isolated from ancient Siberian permafrost sediment, and emended description of the genus *Glaciimonas*. Int J Syst Evol Microbiol 66:744–748. https://doi.org/10.1099/ijsem.0.000783

Mendoza-Suárez M, Andersen SU, Poole PS, Sánchez-Cañizares C (2021) Competition, nodule occupancy, and persistence of inoculant strains: key factors in the rhizobium-legume symbioses. Front Plant Sci 12:690567. https://doi.org/10.3389/fpls.2021.690567

Mestre MC, Fontenla S, Bruzone MC, Fernández NV, Dames J (2016) Detection of plant growth enhancing features in psychrotolerant yeasts from *Patagonia* (Argentina). J Basic Microbiol 56:1098–1106. https://doi.org/10.1002/jobm.201500728

Mishra PK, Bisht SC, Bisht JK, Bhatt JC (2012) Cold-tolerant PGPRs as bioinoculants for stress management. In: Bacteria in agrobiology: stress management. Springer, Berlin, Heidelberg, pp 95–118. https://doi.org/10.1007/978-3-642-23465-1_6

Mishra PK, Joshi P, Bisht SC, Bisht JK, Selvakumar G (2010) Cold-tolerant agriculturally important microorganisms. In: Plant growth and health promoting bacteria. Springer, Berlin, Heidelberg, pp 273–296. https://doi.org/10.1007/978-3-642-13612-2_12

Misslinger M, Hortschansky P, Brakhage AA, Haas H (2021) Fungal iron homeostasis with a focus on *Aspergillus fumigatus*. Biochim Biophys Acta Mol Cell Res 1868:118885. https://doi.org/10.1016/j.bbamcr.2020.118885

Mocali S, Chiellini C, Fabiani A, Decuzzi S, de Pascale D, Parrilli E, Giudice AL (2017) Ecology of cold environments: new insights of bacterial metabolic adaptation through an integrated genomic-phenomic approach. Sci Rep 7(1):1–13. https://doi.org/10.1038/s41598-017-00876-4

Mykytczuk NC, Foote SJ, Omelon CR, Southam G, Greer CW, Whyte LG (2013) Bacterial growth at– 15 C; molecular insights from the permafrost bacterium *Planococcus halocryophilus Or1*. ISME J 7(6):1211–1226. https://doi.org/10.1038/ismej.2013.8

Negi GCS, Samal PK, Kuniyal JC, Kothyari BP, Sharma RK, Dhyani PP (2012) Impact of climate change on the western Himalayan mountain ecosystems: an overview. Trop Ecol 53(3):345–356

Ojewumi ME, Okeniyi JO, Ikotun JO, Okeniyi ET, Ejemen VA, Popoola API (2018) Bioremediation: data on *Pseudomonas aeruginosa* effects on the bioremediation of crude oil polluted soil. Data Brief 19:101–113. https://doi.org/10.1016/j.dib.2018.04.102

Park YH, Kim Y, Mishra RC, Bae H (2017) Fungal endophytes inhabiting mountain-cultivated ginseng (*Panax ginseng* Meyer): diversity and biocontrol activity against ginseng pathogens. Sci Rep 7:1–10. https://doi.org/10.1038/s41598-017-16181-z

Parnanen K, Karkman A, Virta M, Eronen-Rasimus E, Kaartokallio H (2015) Discovery of bacterial polyhydroxyalkanoate synthase (PhaC)-encoding genes from seasonal Baltic Sea ice and cold estuarine waters. Extremophiles 19:197–206. https://doi.org/10.1007/s00792-014-0699-9

Parveen H, Chaudhary A, Chaudhary P, Sultana R, Khati P, Kumar G, Khati P, Rana M, Bhatt P (2022) Omics approaches for the production of the microbial enzymes and applications. In: Bhatt P (ed) Industrial applications of microbial enzymes. Ist Edition. CRC Press. https://doi.org/10.1201/9781003202998-16

Piette F, Leprince P, Feller G (2012) Is there a cold shock response in the Antarctic psychrophile *Pseudoalteromonas haloplanktis*? Extremophiles 16(4):681–683. https://doi.org/10.1007/s00792-012-0456-x

Pourmohammadi K, Abedi E (2021) Hydrolytic enzymes and their directly and indirectly effects on gluten and dough properties: an extensive review. Food Sci Nutr 9(7):3988–4006. https://doi.org/10.1002/fsn3.2344

Rajwar J, Chandra R, Suyal DC, Tomer S, Kumar S, Goel R (2018) Comparative phosphate solubilizing efficiency of psychrotolerant *Pseudomonas jesenii* MP1 and *Acinetobacter* sp. ST02 against chickpea for sustainable hill agriculture. Biologia 73(8):793–802. https://doi.org/10.2478/s11756-018-0089-3

Rasul M, Yasmin S, Suleman M, Zaheer A, Reitz T, Tarkka MT, Islam E, Mirza MS (2019) Glucose dehydrogenase gene containing phosphobacteria for biofortification of Phosphorus with growth promotion of rice. Microbiol Res 223:1–12. https://doi.org/10.1016/j.micres.2019.03.004

Raza A, Charagh S, Abbas S, Hassan MU, Saeed F, Haider S, Varshney RK (2023a) Assessment of proline function in higher plants under extreme temperatures. Plant Biol 25(3):379–395. https://doi.org/10.1111/plb.13510

Raza A, Charagh S, Najafi-Kakavand S, Abbas S, Shoaib Y, Anwar S, Sharifi S, Lu G, Siddique KHM (2023b) Role of phytohormones in regulating cold stress tolerance: physiological and molecular approaches for developing cold-smart crop plants. Plant Stress 8:100152. https://doi.org/10.1016/j.stress.2023.100152

Raza A, Razzaq A, Mehmood SS, Zou X, Zhang X, Lv Y, Xu J (2019) Impact of climate change on crops adaptation and strategies to tackle its outcome: a review. Plants 8:34. https://doi.org/10.3390/plants8020034

Reddy G, Pradhan S, Manorama R, Shivaji S (2010) *Cryobacterium roopkundense* sp. nov., a psychrophilic bacterium isolated from glacial soil. Int J Syst Evol Microbiol 60:866–870. https://doi.org/10.1099/ijs.0.011775-0

Rilling JI, Acuña JJ, Sadowsky MJ, Jorquera MA (2018) Putative nitrogen-fixing bacteria associated with the rhizosphere and root endosphere of wheat plants grown in an andisol from Southern Chile. Front Microbiol 9:2710. https://doi.org/10.3389/fmicb.2018.02710

Rinu K, Pandey A (2011) Slow and steady phosphate solubilization by a psychrotolerant strain of *Paecilomyces hepiali* (MTCC 9621). World J Microbiol Biotechnol 27:1055–1062. https://doi.org/10.1007/s11274-010-0550-0

Rinu K, Sati P, Pandey A (2014) *Trichoderma gamsii* (NFCCI 2177): a newly isolated endophytic, psychrotolerant, plant growth promoting, and antagonistic fungal strain. J Basic Microbiol 54:408–417. https://doi.org/10.1002/jobm.201200579

Rizvi A, Ahmed B, Khan MS, Umar S, Lee J (2021) Psychrophilic bacterial phosphate-biofertilizers: a novel extremophile for sustainable crop production under cold environment. Microorganisms 9(12):2451. https://doi.org/10.3390/microorganisms9122451

Sajjad W, Din G, Rafiq M, Iqbal A, Khan S, Zada S, Ali B, Kang S (2020) Pigment production by cold-adapted bacteria and fungi: colorful tale of cryosphere with wide range applications. Extremophiles 24(4):447–473. https://doi.org/10.1007/s00792-020-01180-2

Sandle T, Skinner K (2013) Study of psychrophilic and psychrotolerant micro-organisms isolated in cold rooms used for pharmaceutical processing. J Appl Microbiol 114(4):1166–1174. https://doi.org/10.1111/jam.12101

Sangiorgio D, Cellini A, Donati I, Pastore C, Onofrietti C, Spinelli F (2020) Facing climate change: application of microbial biostimulants to mitigate stress in horticultural crops. Agronomy 10(6):794. https://doi.org/10.3390/agronomy10060794

Sangwan N, Lambert C, Sharma A, Gupta V, Khurana P, Khurana JP, Sockett RE, Gilbert JE, Lal R (2015) Arsenic rich Himalayan hot spring metagenomics reveal genetically novel predator-prey genotypes. Environ Microbiol Rep 7:812–823. https://doi.org/10.1111/1758-2229.12297

Santiago M, Ramírez-Sarmiento CA, Zamora RA, Parra LP (2016) Discovery, molecular mechanisms, and industrial applications of cold-active enzymes. Front Microbiol 7:1408. https://doi.org/10.3389/fmicb.2016.01408

Sati D, Pande V, Samant M (2023) Plant-beneficial Bacillus, Pseudomonas, and Staphylococcus spp. from Kumaon Himalayas and their drought tolerance response. Front Sustain Food Syst 7:1085223. https://doi.org/10.3389/fsufs.2023.1085223

Selvakumar G, Kundu S, Joshi P, Nazim S, Gupta AD, Gupta HS (2010) Growth promotion of wheat seedlings by Exiguobacterium acetylicum 1P (MTCC 8707) a cold tolerant bacterial strain from the Uttarakhand Himalayas. Indian J Microbiol 50:50–56. https://doi.org/10.1007/s12088-009-0024-y

Sepehr S, Shahnavaz B, Asoodeh A, Karrabi M (2019) Biodegradation of phenol by cold-tolerant bacteria isolated from alpine soils of Binaloud Mountains in Iran. J Environ Sci Health A Tox Hazard Subst Environ Eng 54(4):367–379. https://doi.org/10.1080/10934529.2018.1553818

Shahid M, Singh UB, Khan MS, Singh P, Kumar R, Singh RN, Kumar A, Singh HV (2023) Bacterial ACC deaminase: insights into enzymology, biochemistry, genetics, and potential role in amelioration of environmental stress in crop plants. Front Microbiol 14:1132770. https://doi.org/10.3389/fmicb.2023.1132770

Sharma S, Khan F, Qazi G (2010) Molecular cloning and characterization of amylase from soil metagenomic library derived from Northwestern Himalayas. Appl Microbiol Biotechnol 86:1821–1828. https://doi.org/10.1007/s00253-009-2404-y

Shen L, Liu Y, Gu Z, Xu B, Wang N, Jiao N (2015) *Massilia eurypsychrophila* sp. nov. a facultatively psychrophilic bacteria isolated from ice core. Int J Syst Evol Microbiol 65:2124–2129. https://doi.org/10.1099/ijs.0.000229

Shrestha UB, Gautam S, Bawa KS (2012) Widespread climate change in the Himalayas and associated changes in local ecosystems. PloS One 7(5):e36741. https://doi.org/10.1371/journal.pone.0036741

Silva LI, Pereira MC, Carvalho AMX, Buttrós VH, Pasqual M, Dória J (2023) Phosphorus-solubilizing microorganisms: a key to sustainable agriculture. Agriculture 13(2):462. https://doi.org/10.3390/agriculture13020462

Singh BR, Singh O (2012) Study of impacts of global warming on climate change: rise in sea level and disaster frequency. Global warming—impacts and future perspective. InTech. https://doi.org/10.5772/50464

Singh J, Singh AV (2017) Microbial strategies for enhanced phytoremediation of heavy metals contaminated soils. In: Bharagava RN (ed) Environmental pollutants and their bioremediation approaches. Taylor & Francis, CRC Press London, New York, pp 249–264

Singh J, Singh AV, Prasad B, Shah S (2017) Sustainable agriculture strategies of wheat biofortification through microorganisms. In: Kumar et al (eds) Wheat a premier food crop. Kalyani, Publishers, New Delhi, India, pp 373–391

Singh J, Singh AV, Upadhayay VK, Khan A (2020) Comparative evaluation of developed carrier based bioformulations bearing multifarious PGP properties and their effect on shelf life under different storage conditions. Environ Ecol 38(1):96–103

Singh P, Hamid B, Ahmad Lone M, Ranjan K, Khan A, Chaurse V (2012) Evaluation of pectinase activity from the psychrophilic fungal strain *Truncatella angustata*-BPF5 for use in wine industry. Endocyt Cell Res 22:57–61

Soni R, Suyal DC, Agrawal K, Yadav A, Shouche Y, Goel R (2015) Differential proteomic expression of Himalayan psychrotrophic diazotroph *Pseudomonas palleroniana* N26 under low temperature diazotrophic conditions. Cryo Letters 36:74–82

Soni R, Suyal DC, Sai S, Goel R (2016) Exploration of *nifH* gene through soil metagenomes of the western Indian Himalayas. 3 Biotech 6:1–4. https://doi.org/10.1007/s13205-015-0324-3

Struvay C, Feller G (2012) Optimization to low temperature activity in psychrophilic enzymes. Int J Mol Sci 13:11643–11665. https://doi.org/10.3390/ijms130911643

Suarez C, Ratering S, Hain T, Fritzenwanker M, Goesmann A, Blom J, Chakraborty T, Bunk B, Sproer C, Overmann J, Schnell S (2019) Complete genome sequence of the plant growth-promoting bacterium *Hartmannibacter diazotrophicus* strain E19T. Int J Genomics. https://doi.org/10.1155/2019/7586430

Subramanian P, Joe MM, Yim WJ, Hong BH, Tipayno SC, Saravanan VS, Sa TM (2011) Psychrotolerance mechanisms in cold-adapted bacteria and their perspectives as plant growth-promoting bacteria in temperate agriculture. Korean J Soil Sci Fert 44:625–636. https://doi.org/10.7745/KJSSF.2011.44.4.625

Subramanian P, Kim K, Krishnamoorthy R, Mageswari A, Selvakumar G, Sa T (2016) Cold stress tolerance in psychrotolerant soil bacteria and their conferred chilling resistance in tomato (*Solanum lycopersicum* Mill.) under low temperatures. PloS One 11:8. https://doi.org/10.1371/journal.pone.0161592

Suyal DC, Joshi D, Kumar S, Soni R, Goel R (2019) Differential protein profiling of soil diazotroph *Rhodococcus qingshengii* S10107 towards low-temperature and nitrogen deficiency. Sci Rep 9(1):1–9. https://doi.org/10.1038/s41598-019-56592-8

Suyal DC, Yadav A, Shouche Y, Goel R (2014) Differential proteomics in response to low temperature diazotrophy of Himalayan psychrophilic nitrogen fixing *Pseudomonas migulae* S10724 strain. Curr Microbiol 68:543–550. https://doi.org/10.1007/s00284-013-0508-1

Tapia-Vázquez I, Sánchez-Cruz R, Arroyo-Domínguez M, Lira-Ruan V, Sánchez-Reyes A, del Rayo Sánchez-Carbente M, Folch-Mallol JL (2020) Isolation and characterization of psychrophilic and psychrotolerant plant-growth promoting microorganisms from a high-altitude volcano crater in Mexico. Microbiol Res 232:126394. https://doi.org/10.1016/j.micres.2019.126394

Tiryaki D, Aydın I, Atıcı Ö (2019) Psychrotolerant bacteria isolated from the leaf apoplast of cold-adapted wild plants improve the cold resistance of bean (*Phaseolus vulgaris* L.) under low temperature. Cryobiology 86:111–119. https://doi.org/10.1016/j.cryobiol.2018.11.001

Tomova I, Stoilova-Disheva M, Lazarkevich I, Vasileva-Tonkova E (2015) Antimicrobial activity and resistance to heavy metals and antibiotics of heterotrophic bacteria isolated from sediment and soil samples collected from two Antarctic islands. Front Life Sci 8(4):348–357. https://doi.org/10.1080/21553769.2015.1044130

Tribelli PM, López NI (2018) Reporting key features in cold-adapted bacteria. Life (Basel) 8(1):8. https://doi.org/10.3390/life8010008

Tribelli PM, Solar Venero EC, Ricardi MM, Gómez-Lozano M, Raiger Iustman LJ, Molin S, López NI (2015) Novel essential role of ethanol oxidation genes at low temperature revealed by transcriptome analysis in the Antarctic bacterium *Pseudomonas extremaustralis*. PloS One 10(12):e0145353. https://doi.org/10.1371/journal.pone.0145353

Turan M, Güllüce M, Çakmak R, Şahin F (2013) Effect of plant growth-promoting rhizobacteria strain on freezing injury and antioxidant enzyme activity of wheat and barley. J Plant Nutr 36:731–748. https://doi.org/10.1080/01904167.2012.754038

Vega-Celedón P, Bravo G, Velásquez A, Cid FP, Valenzuela M, Ramírez I, Vasconez IN, Álvarez I, Jorquera MA, Seeger M (2021) Microbial diversity of psychrotolerant bacteria isolated from wild flora of Andes Mountains and Patagonia of Chile towards the selection of plant growth-promoting bacterial consortia to alleviate cold stress in plants. Microorganisms 9(3):538. https://doi.org/10.3390/microorganisms9030538

Verma P, Yadav AN, Khannam KS, Kumar S, Saxena AK, Suman A (2016) Molecular diversity and multifarious plant growth promoting attributes of *Bacilli* associated with wheat (*Triticum aestivum* L.) rhizosphere from six diverse agro-ecological zones of India. J Basic Microbiol 56:44–58. https://doi.org/10.1002/jobm.201500459

Verma P, Yadav AN, Khannam KS, Panjiar N, Kumar S, Saxena AK, Suman A (2015) Assessment of genetic diversity and plant growth promoting attributes of psychrotolerant bacteria allied with wheat (*Triticum aestivum*) from the northern hills zone of India. Ann Microbiol 65:1885–1899. https://doi.org/10.1007/s13213-014-1027-4

Vicente D, Proença DN, Morais PV (2023) The role of bacterial polyhydroxyalkanoate (PHA) in a sustainable future: a review on the biological diversity. Int J Environ Res Public Health 20(4):2959. https://doi.org/10.3390/ijerph20042959

Wang H, Wang H, Shao H, Tang X (2016a) Recent advances in utilizing transcription factors to improve plant abiotic stress tolerance by transgenic technology. Front Plant Sci 7:67. https://doi.org/10.3389/fpls.2016.00067

Wang K, Ke S, Fang W, Liu F, Zhang Z (2023) Agroactive volatile organic compounds from microbes: chemical diversities and potentials of application in crop protection. Adv Agrochem 2:39–57. https://doi.org/10.1016/j.aac.2022.12.004

Wang R, Gao M, Ji S, Wang S, Meng Y, Zhou Z (2016b) Carbon allocation, osmotic adjustment, antioxidant capacity and growth in cotton under long-term soil drought during flowering and boll-forming period. Plant Physiol Biochem 107:137–146. https://doi.org/10.1016/j.plaphy.2016.05.035

Wolf D, Rippa V, Mobarec JC, Sauer P, Adlung L, Kolb P, Bischofs IB (2016) The quorum-sensing regulator ComA from Bacillus subtilis activates transcription using topologically distinct DNA motifs. Nucleic Acids Res 44(5):2160–2172

Xu X, Liu W, Tian S, Wang W, Qi Q, Jiang P, Yu H (2018) Petroleum hydrocarbon-degrading bacteria for the remediation of oil pollution under aerobic conditions: a perspective analysis. Front Microbiol 9:2885. https://doi.org/10.3389/fmicb.2018.02885

Yadav AN, Kour D, Sharma S, Sachan SG, Singh B, Chauhan VS, Saxena AK (2019) Psychrotrophic microbes: biodiversity, mechanisms of adaptation, and biotechnological implications in alleviation of cold stress in plants. In: Sayyed RZ et al (eds) Plant growth promoting rhizobacteria for sustainable stress management. Springer, Singapore. https://doi.org/10.1007/978-981-13-6536-2_12

Yadav AN, Sachan SG, Verma P, Saxena AK (2015) Prospecting cold deserts of north western Himalayas for microbial diversity and plant growth promoting attributes. J Biosci Bioeng 119:683–693. https://doi.org/10.1016/j.jbiosc.2014.11.006

Yadav AN, Verma P, Kumar V, Sachan SG, Saxena AK (2017a) Extreme cold environments: a suitable niche for selection of novel psychrotrophic microbes for biotechnological applications. Adv Biotechnol Microbiol 2:1–4. https://doi.org/10.19080/AIBM.2017.02.555584

Yadav AN, Verma P, Sachan SG, Kaushik R, Saxena AK (2016) Microbiome of Indian Himalayan regions: molecular diversity, phylogenetic profiling and biotechnological applications. *In:* Proceeding of 86th annual session of NASI & the symposium on "Science, Technology and Entrepreneurship for Human Welfare in The Himalayan Region" (Vol. 58).

Yadav AN, Verma P, Sachan SG, Saxena AK (2017b) Biodiversity and biotechnological applications of psychrotrophic microbes isolated from Indian Himalayan regions. EC Microbiol Eco 1:48–54

Zhang DC, Busse HJ, Liu HC, Zhou YG, Schinner F, Margesin R (2011) *Sphingomonas glacialis* sp. nov. a psychrophilic bacterium isolated from alpine glacier cryoconite. Int J Syst Evol Microbiol 61:587–591. https://doi.org/10.1099/ijs.0.023135-0

Zhao Q, Li M, Jia Z, Liu F, Ma H, Huang Y, Song S (2016) AtMYB44 positively regulates the enhanced elongation of primary roots induced by N-3-oxo-hexanoyl-homoserine lactone in *Arabidopsis thaliana*. Int Soc Mol Plant Microbe Interact 29(10):774–785. https://doi.org/10.1094/MPMI-03-16-0063-R

Zhou Z, Jiang F, Wang S, Peng F, Dai J, Li W (2012) *Pedobacter arcticus* sp. nov. a facultative psychrophile isolated from Arctic soil, and emended descriptions of the genus *Pedobacter*, *Pedobacter heparinus*, *Pedobacter daechungensis*, *Pedobacter terricola*, *Pedobacter glucosidilyticus* and *Pedobacter lentus*. Int J Syst Evol Microbiol 62:1963–1969. https://doi.org/10.1099/ijs.0.031104-0

Zubair M, Hanif A, Farzand A, Sheikh TMM, Khan AR, Suleman M, Ayaz M, Gao X (2019) Genetic screening and expression analysis of psychrophilic *Bacillus* spp. reveal their potential to alleviate cold stress and modulate phytohormones in wheat. Microorganisms 7(9):337. https://doi.org/10.3390/microorganisms7090337

Microbial Inoculants for Improved Soil Nutrient Accessibility and Maintenance of Soil Fertility

Vaibhav Dhaigude, Anuj Chaudhary, Pratap Gore, Jaskiran Kaur, Aditi Ghosh, S. T. M. Aravindharajan, Geeta Bhandari, Viabhav Kumar Upadhayay, and Sami Abou Fayssal

Abstract With a projected 10 billion global population, food security needs to be addressed urgently. For sustaining global food security, agricultural output must be increased to face changing climate conditions. To enhance the yield of crops, inorganic chemical fertilizers are widely utilized in conventional agricultural practices; however, they are associated with several detrimental impacts on the environment and human health. Bioinoculants are environmentally friendly and extremely effective and can be used as biopesticides that have no negative effects on plants. Microbial inoculants are soil-amended to improve various soil-related processes and plant growth; they enclose helpful microorganisms, i.e., bacteria, fungi, or archaea. These inoculants may be extremely important for enhancing nutrient accessibility and preserving soil fertility. These microbial inoculants' field efficacies are still variable because they are employed under various farming environments, climates, and management techniques. Finding answers to limitations and application challenges, as well as discovering additional microbial inoculants for unknown

V. Dhaigude · P. Gore · J. Kaur · A. Ghosh
Animal Nutrition Division, ICAR-National Dairy Research Institute, Karnal, Haryana, India

A. Chaudhary
School of Agriculture and Environmental Sciences, Shobhit University, Gangoh, Uttar Pradesh, India

S. T. M. Aravindharajan
Division of Microbiology, ICAR-Indian Agricultural Research Institute, New Delhi, India

G. Bhandari
Department of Biosciences, Himalayan School of Biosciences, Swami Rama Himalayan University, Dehradun, Uttarakhand, India

V. K. Upadhayay
Department of Microbiology, Dr. Rajendra Prasad Central Agricultural University, Pusa, Samastipur, Bihar, India

S. Abou Fayssal (✉)
Department of Agronomy, Faculty of Agronomy, University of Forestry, Sofia, Bulgaria

Department of Plant Production, Faculty of Agriculture, Lebanese University, Beirut, Lebanon
e-mail: sami.aboufaycal@st.ul.edu.lb

P. Chaudhary, A. Chaudhary (eds.), *Microbial Inoculants*, Interdisciplinary Biotechnological Advances, https://doi.org/10.1007/978-981-97-0633-4_12

areas, is thus a never-ending process. On the other hand, the quality of bioinoculants is the main key issue as private companies tend to give little care to quality standards. Therefore, to increase soil nutrient accessibility and fertility, and ensure the sustainability and productivity of field and horticultural crops, qualitative bioinoculants should be used.

Keywords Soil microbes · Soil nutrient · Soil fertility · Sustainable agriculture

1 Introduction

Microbial inoculants, sometimes referred to as bioinoculants or soil inoculants, are a type of agricultural modification that uses beneficial rhizospheric or endophytic microorganisms to improve plant health. In plant root system, these microorganisms create either autonomous or mutual relationships, contributing in nutrient availability and stress resistance (Joshi et al. 2019; Manzar et al. 2022). By transforming inaccessible nutrients into usable forms, they support plant growth (Rawat et al. 2019). These nutrient-utilizing inoculants, also known as plant growth-promoting microorganisms (PGPM), improve a plant's resistance to stressors that are abiotic or biotic in nature and function as biocontrol elements as well (Tyagi et al. 2022a, b).

Kaminsky et al. (2019) describe microbial inoculants as specific organisms administered into an ecosystem for a specific goal, like biocontrol or the growth of plants. Bioinoculants have no adverse impacts on plant and soil microbial populations. Microbial inoculants are now more important for creating sustainable agricultural production systems (Chaudhary et al. 2021g). The adoption of microbial inoculants has become essential in developing sustainable agricultural systems due to challenges such as soil erosion and degradation, salt accumulation, unwanted metal deposition, nutrient imbalances, and environmental stresses (Sammauria et al. 2020). Moreover, increased use of agrochemicals impairs the biological system, and in subsequent years' agriculture's reliance on them will worsen soil health and increase the risk of water pollution (Chaudhary et al. 2022a). The combined impact of these effects makes agriculture farming uncertain, which puts an immense strain on the economy. Microorganisms are quickly becoming a viable choice for preserving sustainable production for enterprises and the financial stability of economies. Therefore, microbial inoculants are alternative ways to fill current and upcoming nutritional demands for crop productivity and soil health (Pathak and Kumar 2016; Khati et al. 2017, 2018; Kukreti et al. 2020).

The specific function for a specific need of these microbial inoculants and their application in diverse combinations made them feasible and environmentally friendly solutions for maintaining and even improving the health of the soil (Kumari et al. 2020, 2021). In addition, several compatible microbial inoculation combinations considerably aid in controlling plant nutrients through transformation, solubilization, or fixation in soil (Chaudhary and Sharma 2019; Sammauria et al. 2020; Chaudhary et al. 2022e). It is firmly believed that to achieve global agricultural promotion, a comprehensive perspective on microbial inoculants and their unique applications for enriching soil fertility and nutrient availability improvements is required. In this book chapter, an overview of the ideas and uses of inoculants that aim to improve soil fertility and crop productivity is given, along with a discussion of the importance of different beneficial microorganisms in nutrient accessibility and maintenance of soil fertility.

2 Nitrogen-Fixing Microbes for Improving Soil Fertility

Nitrogen-fixing microbes are important for promoting the development of crops and improving the fertility of soil and because they have a distinctive capacity to transform nitrogen from the atmosphere into ammonia that plants can readily use, which is known as biological nitrogen fixation (BNF) (Aasfar et al. 2021). The group of microbes known as diazotrophs stabilize nitrogen biologically in combination with plant roots.

Green sulfur bacteria, cyanobacteria, actinomycetes, and all sub-divisions of *Proteobacteria* are categorized as nitrogen-fixing bacteria. Aerobic and anaerobic chemolithotrophs phototrophs are all capable of nitrogen fixation. The main enzyme responsible for turning di-nitrogen into ammonia during the nitrogen fixation process is nitrogenase (Koirala and Brözel 2021). Table 1 highlights the contribution of nitrogen-fixing microorganisms to the enhancement of soil fertility.

2.1 Microbes that Fix Nitrogen in a Symbiotic Relationship

Nitrogen-fixing bacteria are a valuable group of microbes that play a crucial role in enhancing soil fertility. They have the ability to convert atmospheric nitrogen gas (N2) into a form of ammonium nitrogen that plants can readily absorb and use for their growth. This process, known as nitrogen fixation, is essential for plant nutrition and overall soil health (Bai et al. 2022). The organic matter and ammonium nitrogen of soil can both be greatly enhanced by nitrogen-fixing bacteria. Nitrogen-fixing bacteria not only provide plants with nitrogen but also enhance soil structure and microbial diversity. This can lead to improved overall soil health, including increased water-holding capacity and nutrient availability (Sepp et al. 2023).

Various microbe species form symbiotic associations with legume roots. The first step is to select appropriate nitrogen-fixing bacteria strains for the specific soil and crop conditions. Common examples include species of *Rhizobium* for leguminous plants, such as beans, soybeans and peas, and species of *Azotobacter* and *Clostridium* for non-leguminous plants (Sindhu et al. 2019). The selected bacteria that fix nitrogen have a mutualistic relationship with the chosen host plants. For instance, leguminous plants, such as soybeans or clover, will develop nodules on their roots where *Rhizobium* bacteria can grow. Because of *Rhizobium* nitrogen

Table 1 Soil fertility improved by nitrogen-fixing microorganisms

Nitrogen-fixing microbes' name	Nitrogen-fixing microbes	Role in soil fertility improvement	References
Nitrogen-fixing endophytic bacteria	Endophytic bacteria	Improvement of ammonium and nitrate contents in soil	Thuc et al. (2022)
Frankia bacteria	Symbiotic bacteria	Increase in N_2 fixation	Ghodhbane-Gtari et al. (2021)
Azospirillum brasilense and *Bradyrhizobium japonicum*	Gram-negative and micro-symbiotic legume root-nodulating bacteria	Enhances N2 fixation and solubilization of phosphate	Gómez-Duque et al. (2022)
Saprobes fungi	Soil macrofungi	Organic matter decomposition and the release of nitrogen and carbon sources in soil aid in nutrient cycling	Frąc et al. (2018)
Glomus spp.	Arbuscular mycorrhizal fungi	Increased microbial activity, reduced soil erosion, heavy metal chelation, and formation of alkaline phosphatase for phosphorus mineralization in soil are all benefits of improved soil carbon sequestration	Fall et al. (2022)
Yeast spp.	Yeast	Upregulation of soil enzymes and stabilization of soil N, P, and K contents	Alzandi and Naguib (2022)
Yeast spp.	Yeast	Participation in soil nutrient cycling and mineralization and maintenance of aggregate formation	Hernández-Fernández et al. (2021)

fixation mechanism, legumes are less reliant on synthetic fertilizers than non-leguminous plants (Goyal et al. 2021). This not only leads to cost savings for farmers but also has environmental benefits by decreasing the potential for nitrogen runoff and its associated pollution.

The introduction of strains of *Rhizobium* into soybean cultivation has led to enhancements in various growth and yield characteristics. These improvements encompass an increase in the branches bearing pods on each plant, a higher overall pod count per plant, and an augmented quantity of seeds within each pod (Mfilinge et al. 2014). Furthermore, the utilization of *Rhizobium leguminosarum* strains in the cultivation of peas and lentils has resulted in enhanced nodulation, increased yield of pea seeds, and an improved shoot-to-root weight ratio (Bourion et al. 2017). Dual inoculation, for example, has been demonstrated to promote plant development and increase yield more than single inoculation techniques (Chaudhary and Sindhu 2016). For instance, when *Bradyrhizobium* spp. and *Azospirillum brasilense* were inoculated together rather than separately, the yield of soybeans increased considerably (Hungria et al. 2015).

2.2 Free-Living Nitrogen-Fixing Bacteria

Numerous studies have demonstrated the positive effects of free-living nitrogen-fixing bacteria on crop productivity. Research has shown that the inoculation of crops with these bacteria can lead to increased crop yields, improved nutrient uptake, and enhanced plant vigor (Soumare et al. 2020). Additionally, these bacteria contribute to the formation of organic matter in the soil, improving soil structure and water retention capacity.

Azotobacter stands out as a prominent example of diazotrophic bacteria (Aasfar et al. 2021). These bacteria have been identified across a spectrum of soil types, spanning from neutral to alkaline environments, and are commonly associated with the rhizospheres of non-leguminous crops, including wheat, rice, cotton, and various vegetables (Jain et al. 2021). Their ability to thrive in diverse soil conditions makes them valuable contributors to nitrogen fixation and soil fertility enhancement. Another noteworthy category of nitrogen-fixing organisms encompasses cyanobacteria commonly referred to as blue-green algae. These cyanobacteria are typically found in alkaline and moist soils. Additionally, they exist symbiotic relationships involving cyanobacteria, such as the *Azolla-Anabaena* system. Singh et al. (2016) reported that collectively, these cyanobacteria fix an impressive 4–6 billion kilograms of nitrogen annually, making a substantial contribution to nitrogen cycling in ecosystems.

A study conducted by Noar and Bruno-Barcena (2018) investigated the remarkable impact of microbial inoculation on wheat seed yield. The researchers examined the effects of inoculating wheat WH711 seeds with *Bacillus* SYB101, a beneficial bacterial strain. The results were striking, demonstrating a substantial seed yield increased by 32.6%, compared to the control group, which consisted of uninoculated seeds. This outcome underscores the potential of *Bacillus* SYB101 as a powerful agent for enhancing wheat yield, thereby holding promise for improving agricultural productivity. In a parallel aspect of the study, Noar and Bruno-Barcena (2018) also explored the influence of microbial inoculation on the Raj3765 variety of wheat. In this case, the inoculation led to a noteworthy 23.1% increase in seed yield when compared to the control group. This finding highlights the varietal differences in the response to microbial inoculation and suggests that Raj3765 possesses traits that can be harnessed for increased crop production. These findings emphasize the importance of selecting the appropriate microbial inoculant that is suited to specific crop varieties to maximize the benefits of microbial-assisted agriculture.

2.3 Associative Nitrogen-Fixing Microbes

Microbial species such as *Azospirillum* are known to produce various plant growth-promoting chemicals, including indole-3-acetic acid (IAA), gibberellins, and cytokinins. Among these species, *A. brasilense* and *A. lipoferum* have garnered significant attention and research focus (Rodrigues et al. 2015). These microorganisms establish associations with plant roots and play a crucial role in enhancing plant growth and nutrient uptake through the secretion of these bioactive compounds. One notable area of research involves the co-inoculation of *Rhizobium* species with *Azospirillum* and other plant growth-promoting rhizobacteria (PGPR) strains. This dual inoculation strategy has demonstrated remarkable effects, leading to a significant increase in total nitrogen content in various leguminous plants, nodule formation, and nitrogen fixation in plant biomass (Korir et al. 2017). These findings emphasize the potential of combined microbial inoculation approaches to enhance nitrogen fixation and overall plant productivity, surpassing the benefits of single *Rhizobium* inoculation or control conditions.

Furthermore, the impact of lectins produced by *Azospirillum* strains on the growth and development of host wheat plants under abiotic stress conditions was investigated by Alenkina and Nikitina (2021). Wheat seed germination can be adversely affected by stressors such as heavy metals, temperature fluctuations, salinity, and dehydration, each exerting unique effects. Interestingly, the study revealed that exposure to heavy metals had detrimental effects on *Azospirillum* strains' ability to support wheat germination. However, the application of lectins from these strains demonstrated a positive influence on root length and quantity in wheat seedlings, with the lectin from the endophytic strain showing greater efficacy compared to the epiphytic strain. These findings underscore the potential of microbial-derived compounds, such as lectins, in mitigating the adverse effects of abiotic stressors on plant growth.

3 Phosphate-Solubilizing Bacteria/Fungi for Improving Soil Fertility

The utilization of phosphate-solubilizing microorganisms, such as bacteria and fungi, offers a promising approach to enhance soil fertility (He et al. 2022). These microorganisms play a pivotal role in making phosphate nutrients more available to plants, ultimately contributing to improved soil fertility (Agri et al. 2021, 2022). These microbes have a unique ability to convert insoluble phosphates in the soil into a soluble form that plants can readily absorb. By doing so, they play a crucial role in increasing the bioavailability of phosphorus, an essential nutrient for plant growth and development (Qarni et al. 2021; Chaudhary et al. 2021a). This biofertilization strategy not only reduces the need for chemical phosphate fertilizers, which can have negative environmental impacts, but also promotes sustainable and eco-friendly agriculture.

Phosphorus is an essential element for the plant development and growth. Plants utilize orthophosphate, a form of phosphorus, by synthesizing siderophores, hydroxyl ions, organic acids, carbon dioxide, and protons. Microbes play a vital role in converting insoluble forms of phosphorus into this plant-accessible orthophosphate. The limited availability of soluble or inorganic phosphorus, particularly orthophosphate, in soil significantly impacts crop yields, leading to reduced agricultural productivity (Wang et al. 2017).

Various bacterial species have been identified for their ability to solubilize soil phosphorus, including *Bacillus, Escherichia freundii, Brevibacterium, Pseudomonas, Calothrix brauna, Tolypothrix ceylonica, Burkholderia, Erwinia, Corynebacterium, Xanthomonas*, and *Achromobacter* (Santoyo et al. 2021). Inoculating soil with these microorganisms can significantly contribute to meeting the phosphorus requirements of plants. This microbial-mediated process is of utmost importance in enhancing crop productivity and promoting sustainable agriculture (Chaudhary et al. 2021b, c, d). Figure 1 illustrates the mechanism employed by phosphate-solubilizing bacteria to break down phosphate bonds, elucidating their crucial role in phosphorus cycling and plant nutrition.

Sindhu et al. (2014) identified a diverse range of fungi, including *Aspergillus* sp., *Fusarium* sp., *Cephalosporium* sp., *Paecilomyces, Saccharomyces cerevisiae, Sclerotium rolfsii, Penicillium, Cylindrocladium* sp., and *Torula thermophila*. Their research underscores the significance of these fungal species in various ecological

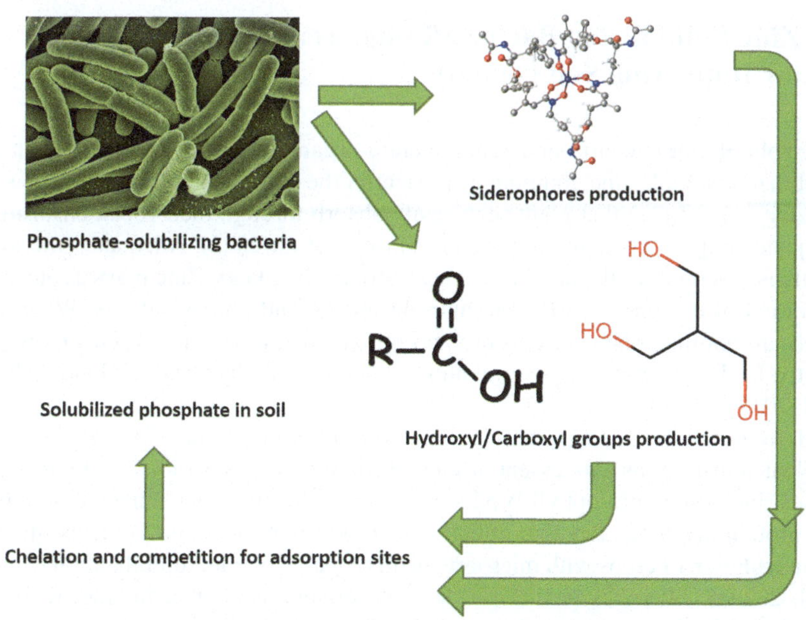

Fig. 1 Phosphate solubilization process by phosphate-solubilizing bacteria

processes, emphasizing their potential roles in soil health and nutrient cycling. It was observed that inoculating peanuts with the purple non-sulfur bacterium *Rhodopseudomonas palustris* had a remarkable impact on peanut cultivation (Wang et al. 2021). Over a 5-year period, this inoculation strategy led to a substantial increase in peanut yield, ranging from 8.1% to 19.5%. Moreover, it positively influenced the protein content and nitrogen absorption in peanut seeds, and it contributed to favorable genetic changes related to inorganic and organic phosphorus solubilization.

Another critical component in soil fertility enhancement is mycorrhizal fungi, particularly arbuscular mycorrhizae (AM). These fungi form symbiotic relationships with plant roots, fostering plant growth and health. Begum et al. (2019) conducted research demonstrating that mycorrhizal inoculation significantly boosts agricultural productivity, enhances soil quality and structure, and improves overall plant well-being. Furthermore, the intricate interplay between bacteria, fungi, and plants plays a pivotal role in phosphorus mobilization, nutrient cycling, and the sustainable availability of nutrients for plant uptake. Etesami et al. (2021) explored the synergistic effects of arbuscular mycorrhizal fungus (AMF), silicon addition, and phosphate-solubilizing bacteria inoculation on soil phosphorus availability. Their findings revealed that the combination of mycorrhizae and silicon proved highly effective in increasing soil phosphorus availability and subsequent plant uptake, illustrating the potential of such multifaceted approaches to enhance soil fertility and nutrient cycling.

4 Zinc-Solubilizing Bacteria/Fungi as an Agent for Improving Soil Fertility

Zinc-solubilizing bacteria and fungi secrete organic acids and enzymes that can solubilize insoluble zinc compounds present in the soil. This process releases soluble zinc ions (Zn2+) that plants can easily absorb through their roots, ensuring an adequate supply of this micronutrient (Masood et al. 2022). Adequate zinc availability is essential for the uptake of other nutrients by plants. Zinc plays a role in the activation of enzymes involved in nutrient uptake and photosynthesis. When zinc levels are optimal, plants can absorb and utilize other essential nutrients more efficiently, leading to improved overall nutrient uptake and plant health (Gontia-Mishra et al. 2017).

Hafeez et al. (2013) reported that low concentration of zinc is a prevalent issue in plant nutrition, as this essential element is crucial for various biological processes, including chlorophyll production, metabolic functions, enzyme activities, and protein synthesis. Another effective method to provide zinc to plants involves the inoculation of crops with microorganisms capable of zinc solubilization (Sindhu et al. 2019a). Microbes in the soil generate organic acids that interact with zinc complexes, leading to a reduction in pH levels and the sequestration of zinc cations.

Several bacterial species exhibit the capacity to solubilize zinc, such as *Thiobacillus thiooxidans, Saccharomyces* spp. and *Bacillus subtilis*. These microorganisms have the ability to dissolve insoluble zinc compounds that are commonly found in soil. Figure 2 illustrates the mechanism employed by zinc-solubilizing bacteria to break down zinc bonds, shedding light on the microbial contribution to zinc availability for plants.

Mumtaz et al. (2017) reported that inoculation of maize with zinc-solubilizing bacteria, specifically *B. aryabhattai*, was found to have a positive influence on plant growth. The research demonstrated that this microbial inoculation promoted the growth of maize. Similarly, Jha (2019) highlighted the growth-promoting effects of inoculating rice with zinc-dissolving bacteria, *Pseudomonas pseudoalcaligenes* and *B. pumilus*. This treatment resulted in increased dry weight and plant height in rice plants. Zinc-solubilizing *Bacillus* sp. influence was observed on maize (Hussain et al. 2020). Their findings revealed significant improvements, including a remarkable 90% increase in biomass production, overall growth, and chlorophyll content in maize plants.

According to a study conducted by Vaid et al. (2020), the utilization of a rhizobacterial consortium effectively mobilized zinc to rice grains. Over a 2-year period, rice production increased by 19.7–27.9% and 17.1–20.4%, respectively, showcasing the potential of microbial interventions to enhance crop yields. Furthermore, Bhatt and Maheshwari (2020) reported that zinc-solubilizing bacteria isolated from *Capsicum annuum* displayed the highest zinc-solubilizing capacity. Additionally, siderophores produced by *Leclercia adecarboxylata* were instrumental in zinc and

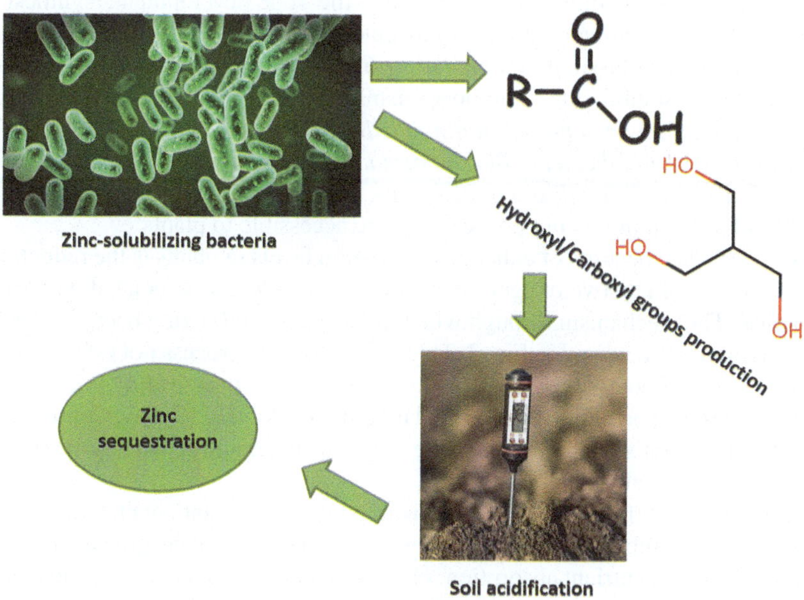

Fig. 2 Zinc solubilization process by zinc-solubilizing bacteria

silicate dissolution. This bacterium exhibited tolerance to elevated zinc levels in growth media and also synthesized a significant amount of indole-3-acetic acid (IAA). Kang et al. (2021) demonstrated that the inoculation of cucumber plants with the *L. adecarboxylata* MO1 strain not only reduced hydrogen peroxide (H_2O_2) levels but also enhanced zinc absorption by both shoots and roots. This microbial inoculation resulted in increased antioxidant activity in cucumber plants, including elevated levels of polyphenol peroxidase (PPO), glutathione (GSH), catalase (CAT), and peroxidase (POD).

5 Potassium-Solubilizing Bacteria/Fungi for Improving Soil Fertility

Potassium-solubilizing microorganisms can stimulate root growth and development in plants. Healthy and robust root systems enable plants to explore a larger volume of soil, resulting in better access to water and nutrients, including potassium (Meena et al. 2014). Potassium plays a significant role in plant stress tolerance, helping plants withstand drought, salinity, and other environmental stresses. Adequate potassium availability, facilitated by these microorganisms, can enhance a plant's resilience to adverse conditions (Khan et al. 2010).

Phosphorus and nitrogen are recognized as essential nutrients for plant growth, and potassium stands out as the third vital element in this regard (Ding et al. 2021; Patel et al. 2021). Potassium plays a pivotal role in promoting several critical physiological processes within plants, including the root development, synthesis of starch, and the regulation of stomatal movement (Gallegos-Cedillo et al. 2016).

The realm of potassium solubilization in soil is enriched by a diverse array of potassium solubilizing microorganisms (KSMs). These microorganisms, encompassing *Pseudomonas, Acidithiobacillus, Sphingomonas, Acidithiobacillus ferrooxidans, Aminobacter, Burkholderia, Paenibacillus* sp., *Paenibacillus glucanolyticus*, and *Bacillus circulans*, as documented by Parmar and Sindhu (2019), are helpful in making potassium more accessible to plants.

One of the key benefits of enhanced potassium levels in plants is the reduction in the production of reactive oxygen species which are closely associated with oxidative stress. The mechanism through which potassium-solubilizing bacteria mineralize potassium is elucidated in Fig. 3. This mineralization process not only facilitates the production of potassium but also plays a pivotal role in activating the adenosine triphosphate (ATP) synthase enzyme and modulating ATPase activity. Consequently, it bolsters the plant's resilience to a wide range of stresses, whether abiotic or biotic in nature (Khan et al. 2010).

Recent research findings have demonstrated the significant benefits of introducing potassium-solubilizing bacteria on various aspects of plant growth, including germination rates, nutrient absorption, plant development, and overall yield (Sindhu et al. 2016). These advantageous soil microorganisms, including actinomycetes,

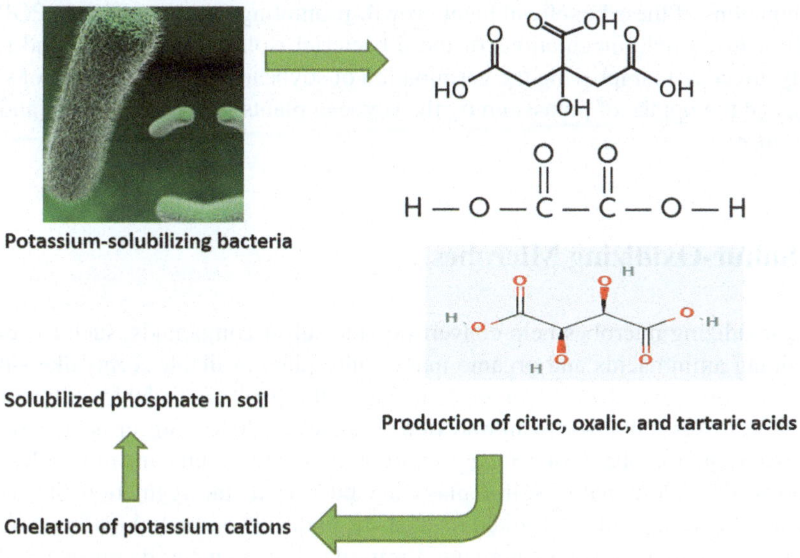

Fig. 3 Potassium solubilization process by potassium-solubilizing bacteria

bacteria, and fungi, collaborate in the chelation of potassium within the soil, enhancing its accessibility to plants. The process of making potassium available to plants involves a range of intricate mechanisms, encompassing chelation, acidolysis, polysaccharide formation, intricate chemical reactions, and exchange reactions.

The utilization of potassium-solubilizing microorganisms (KSMs) holds promising potential for elevating crop productivity, primarily due to the release of organic acids by these microorganisms, resulting in increased potassium dissolution (Sindhu et al. 2016). According to Prajapati and Modi (2012) and Zarjani et al. (2013), one of the most common agents responsible for solubilizing mineral potassium is the production of tartaric acid. Microorganisms generate these organic acids, which are subsequently released into the surrounding environment. This environmental acidification triggers the release of potassium ions through protonation and acidification (Meena et al. 2013). An illustrative example of the positive impact of inoculating potato plants with the potassium-solubilizing bacterium *Bacillus cereus* was studied by Ali et al. (2021), who observed notable increases in shoot dry weight, plant height, and branch number and improvements of approximately 26%, 15%, 27%, and, respectively.

In laboratory and pot experiments conducted by Bakhshandeh et al. (2020), the research team identified three specific bacteria, namely, *Bacillus cereus* (B1), *Trichoderma Simmons* (F2), and *Bacillus megaterium* (B2), that exhibited the capacity to stimulate plant growth. These bacteria were found to have a positive impact on soybean seed germination and the subsequent growth of seedlings. Through a series of trials, it was observed that the most effective strategies for enhancing seed germination in soyabean and potassium absorption involved combinations of these beneficial plant growth-promoting microorganisms (PGPM). Additionally, single inoculations of these bacterial isolates were also found to be highly effective in improving the germination of soybean seeds, the growth of seedlings, and the uptake of potassium by the soybean plants in both laboratory and pot experiments.

6 Sulfur-Oxidizing Microbes

Sulfur-oxidizing microbes help convert organic sulfur compounds, such as sulfur-containing amino acids and organic matter, into plant-available forms like sulfate ($SO4^{2-}$). This mineralization process increases the pool of available sulfur in the soil, which is essential for plant nutrition (Rana et al. 2020). Sulfur is an essential macronutrient for plants due to its incorporation into significant molecules like amino acids. Additionally, sulfur plays a vital role in the regulation of various enzymes, including but not limited to superoxide dismutase and glutathione reductase. In soil, both organic and inorganic forms of sulfur are naturally present. Sulfur-oxidizing bacteria (SOB), exemplified by species like *Alcaligenes, Pseudomonas, Bacillus, T. thiooxidans, Thiobacillus thioparus*, and *Xanthobacter*, play a pivotal role in converting organic sulfur compounds into the inorganic sulfur form, specifically sulfate (SO42), which can be readily utilized by plants for their nutritional requirements (Riaz et al. 2020).

In alkaline soils suffering from sulfur deficiency, the introduction of sulfur-oxidizing bacteria (SOBs) has demonstrated significant benefits, including a remarkable increase in grain yields, along with up to a fourfold enhancement in uptake of nitrogen (Anne et al. 2014). Furthermore, the application of sulfur-oxidizing bacteria has proven effective in augmenting the weight and width of garlic bulbs (Pourbabaee et al. 2020). Beyond these findings, the potential utility of sulfur-oxidizing bacteria as biofertilizers is noteworthy, particularly in alkaline soil conditions. Crops such as cauliflower, grape, ginger, oats, garlic, and onion have all exhibited promising responses to the use of SOBs as biofertilizers (Macik et al. 2020). This highlights the practical applicability of sulfur-oxidizing bacteria in improving growth and yield of various crops under alkaline soil conditions.

7 Plant Growth-Promoting Rhizobacteria

PGPRs are free-living bacteria which colonize the rhizosphere and have the ability to synthesize and secrete compounds that stimulate plant growth (Chaudhary et al. 2021d, e, f). Notably, plants that have been inoculated with PGPRs exhibit enhanced resilience and adaptability to environmental stressors, including challenges like drought and salt conditions (Bharti et al. 2013; Verma et al. 2016; Ilyas et al. 2020). The inoculation of plants with PGPRs has been found to contribute positively to soil fertility, seed germination, and overall plant growth. This is primarily achieved through the increased production of plant hormones such as gibberellins and auxins (Tahir et al. 2017). PGPRs encompass a diverse range of microbial genera, including but not limited *to Bacillus, Pseudomonas, Flavobacterium, Acinetobacter, Thiobacillus, Agrobacterium, Micrococcus, Azotobacter, Cellulomonas, Frankia, Actinoplanes, Rhizobium, Xanthomonas, Serratia*, and more, as outlined in Table 2. These bacteria play a crucial role in enhancing nutrient availability and uptake, as well as releasing hormones that benefit plant growth, all while mitigating soil-borne diseases (Khati et al. 2019). The cumulative effect of these actions leads to improved plant growth and increased crop yields (Santoyo et al. 2021; Chaudhary et al. 2022b, c, d).

In contrast to individual inoculations of plant growth-promoting rhizobacteria (PGPRs) such as *Rhizobium* or *Pseudomonas fluorescens*, co-inoculation of PSB and PGPR strains, including *B. polymyxa* and *P. fluorescens*, yielded significant improvements in nutrient uptake. Pellegrini et al. (2021) assessed the impact of bacterial strains such as *B. ambifaria, Gluconacetobacter diazotrophicus*, and *A. brasilense* on *Allium cepa* L. (onion) growth and soil health.

The application of bacterial strains had a positive influence on various aspects of plant growth, production, and composition. Specifically, it led to an 18% increase in plant height, a 42% rise in total chlorophyll content, a 13% boost in crop yields, and a 3% enhancement in bulb dry matter over control. Additionally, the treatment group exhibited a 25% increase in total phenolic content and a 20% improvement in antioxidant activities of the bulb extracts, showcasing the beneficial effects of the bacterial application (Tahir et al. 2017).

Furthermore, the bacterial consortia contributed to higher nutrient concentrations, including available phosphorus, organic matter, and total organic carbon, within the soil at the time of harvest as compared to control. This illustrates the potential of bacterial consortia in enhancing both plant performance and soil fertility. It's important to highlight that the utilization of synthetic microbial consortia, consisting of both bacterial and fungal components, as soil amendments, has gained widespread acceptance as a means to promote plant growth and development. Strategies for further refinement and engineering of such consortia have been implemented and are highly recommended (Chaudhary et al. 2023a, b).

8 Conclusion

In integrated agricultural management systems, microbial inoculants may augment control of plant disease, thus improving plant performance. They can aid in dropping the harmful environmental effects of pesticide abuse. Even though many of them have variable growth-promoting properties when utilized in inoculant formulations, the prospects of inoculants such as rhizobacteria, arbuscular mycorrhizal fungi, and some rhizobium outweigh the drawbacks in accomplishing the objective of sustainable agricultural output. In addition to encouraging plant growth, using

Table 2 Inoculation responses of beneficial microbes

Type of microbial inoculant	Bacterial used	Growth promoting activity	Impact on plant health parameters	References
Nitrogen-fixing biofertilizer	Strain Avi2 of (MCC 3432) *Azotobacter* sp. *Herbaspirillum seropedicae*, *Enterobacter cloacae*, *Bradyrhizobium* sp., *Rhizobium meliloti*	Nitrogen fixation, production of IAA (indole-3-acetic acid) and siderophore Nitrogen fixation	Improved vegetative and reproductive growth in rice Increased chlorophyll content, length of root and shoot, leaf, and root number of maize	Banik and Dangar (2019) Avila et al. (2020)
Phosphate-solubilizing/ mobilizing biofertilizer	*Pseudomonas stutzeri* (CSP03), *Bacillus subtilis* (TTP02), *Pseudomonas putida*, *Bacillus subtilis* LK14 *Alcaligenes faecalis*, *Pseudomonas frederiksbergensis*, *R. leguminosarum*, *Pseudomonas moraviensis*, *Pseudomonas hormaechei*, *Bacillus halotolerans*	Solubilization of P and zinc, IAA, N_2 fixation, and siderophore and NH_3 production Increased production of enzymes, pectinolytic and proteolytic, IAA and hydrogen cyanide	Enhanced plant growth in arid soils Increased height of the plant and weight of the root Increased weight of the root and overall plant growth	El-Sayed et al. (2014) Abdallah et al. (2016)
Potassium solubilizers	*Acidithiobacillus*, *Bacillus cereus*, *Bacillus mucilaginosus*, *Acidithiobacillus ferrooxidans*	These strains facilitate solubilization of potassium		Ali et al. (2021); Ansari et al. (2017)
Zinc-solubilizing biofertilizer	*Thiobacillus thiooxidans*	These biofertilizers solubilize zinc (Zn)	Enhanced growth of various crops including ginger, wheat, pulses, citrus, and paddy	Mehnaz (2016)
Sulfur-oxidizing biofertilizer	*Delftia acidovorans*, *Bradyrhizobium japonicum*	These biofertilizers are involved in sulfur oxidation	They contribute to improved growth in fiber crops, oilseeds, cereals, plantation crops, vegetables, ornamentals, medicinal crops, flowers and orchards	Mehnaz (2016)
Plant growth-promoting biofertilizer	*Azotobacter chroococcum*, *Pseudomonas fluorescens*	Promote overall plant growth	Improved most crop growth	Minaxi Saxena et al. (2013); Mehnaz (2016)

microbial inoculants may help address soil degradation issues. More investigation is needed to fully recognize the intricate connections between the soil, plants, and microbes to reach the objective of completely replacing the ecologically harmful agrochemicals with environmentally benefiting microbial inoculants. Numerous cases have shown the utility of a mix of microorganisms in inoculant formulations, and this value can be increased through cutting-edge research. Biotechnological research finding powerful microbes that are well-suited with crops and adapt well to the soil environment would also be extremely beneficial. Application of bioinoculants as a safe method for boosting crop output and promoting sustainable agriculture may be embraced.

References

Aasfar A, Bargaz A, Yaakoubi K, Hilali A, Bennis I, Zeroual Y, Meftah Kadmiri I (2021) Nitrogen-fixing *Azotobacter* species as potential soil biological enhancers for crop nutrition and yield stability. Front Microbiol 12:628379. https://doi.org/10.3389/fmicb.2021.628379

Abdallah RAB, Trabelsi BM, Nefzi A, Khiareddine HJ, Remadi MD (2016) Isolation of endophytic bacteria from *Withania somnifera* and assessment of their ability to suppress Fusarium wilt disease in tomato and to promote plant growth. J Plant Pathol Microbiol 7:352. https://doi.org/10.4172/2157-7471.1000352

Agri U, Chaudhary P, Sharma A (2021) In vitro compatibility evaluation of agriusable nanochitosan on beneficial plant growth-promoting rhizobacteria and maize plant. Natl Acad Sci Lett 44:555–559. https://doi.org/10.1007/s40009-021-01047-w

Agri U, Chaudhary P, Sharma A, Kukreti B (2022) Physiological response of maize plants and its rhizospheric microbiome under the influence of potential bioinoculants and nanochitosan. Plant Soil 474:451–468. https://doi.org/10.1007/s11104-022-05351-2

Alenkina SA, Nikitina VE (2021) Stimulating effect from lectins of associative bacteria of the genus Azospirillum on the germination and morphometric characteristics of spring wheat sprouts in simulated abiotic stress. Russ J Plant Physiol 68:315–321. https://doi.org/10.1134/S1021443721010027

Ali AM, Awad MY, Hegab SA, Gawad AMAE, Eissa MA (2021) Effect of potassium solubilizing bacteria (*Bacillus cereus*) on growth and yield of potato. J Plant Nutr 44(3):411–420. https://doi.org/10.1080/01904167.2020.1822399

Alzandi AA, Naguib DM (2022) Effect of yeast application on soil health and root metabolic status of corn seedlings under drought stress. Arch Microbiol 204:233. https://doi.org/10.1007/s00203-022-02843-8

Anne R, Joseph Suresh K, Kavimandan, Kolluru VBR, Lata N (2014) Response of canola and wheat to amendment of pyrite and Sulphur-oxidizing bacteria in soil. Arch Agron Soil Sci 60(3):367–375. https://doi.org/10.1080/03650340.2013.799275

Ansari MH, Hashemabadi D, Kaviani B (2017) Effect of cattle manure and Sulphur on yield and oil composition of pumpkin (Cucurbita pepo var. Styriaca) inoculated with *thiobacillus thiooxidans* in calcareous soil. Commun Soil Sci Plant Anal 48:2103–2118. https://doi.org/10.1080/00103624.2017.1406106

Avila JS, Ferreira JS, Santos JS, Rocha PAD, Baldani VL (2020) Green manure, seed inoculation with Herbaspirillumseropedicae and nitrogen fertilization on maize yield. Rev Bras Eng Agríc Ambient 24:590–595. https://doi.org/10.1590/1807-1929/agriambi.v24n9p590-595

Bai L, Yang Y, Shi Z, Zou Y, Zhou H, Jia J (2022) Improvement of low-fertility soils from a coal mining subsidence area by immobilized nitrogen-fixing bacteria. PRO 10(6):1185. https://doi.org/10.3390/pr10061185

Bakhshandeh E, Gholamhosseini M, Yaghoubian Y, Pirdashti H (2020) Plant growth promoting microorganisms can improve germination, seedling growth and potassium uptake of soybean under drought and salt stress. Plant Growth Regul 90(1):123–136. https://doi.org/10.1007/s10725-019-00556-5

Banik A, Dangar TK (2019) Application of rice (*Oryza sativa* L.) root endophytic diazotrophic *Azotobacter* sp. strain AviZ (MCC 3432) can increase rice yield under green house and field condition. Microbiol Res 219:56–65. https://doi.org/10.1016/j.micres.2018.11.004

Begum N, Qin C, Ahanger MA, Raza S, Khan MI, Ahmed N, Ashraf M, Zhang L (2019) Role of arbuscular mycorrhizal fungi in plant growth regulation: implications in abiotic stress tolerance. Front Plant Sci 10:1068. https://doi.org/10.3389/fpls.2019.01068

Bharti N, Yadav D, Barnawal D, Maji D, Kalra A (2013) *Exiguobacterium oxidotolerans*, a halotolerant plant-growth promoting rhizobacteria, improves yield and content of secondary metabolites in *Bacopa monnieri* (L.) Pennell under primary and secondary salt stress. J World Microbiol Biotechnol 29:379–387. https://doi.org/10.1007/s11274-012-1192-1

Bhatt K, Maheshwari DK (2020) Zinc solubilizing bacteria (*Bacillus megaterium*) with multifarious plant growth promoting activities alleviates growth in *Capsicum annuum* L. Biotechnology 10:36. https://doi.org/10.1007/s13205-019-2033-9

Bourion V, Heulin-Gotty K, Aubert V, Tisseyre P, Chabert-Martinello M, Pervent M, Delaitre C, Vile D, Siol M, Duc G, Brunel B (2017) Coinoculation of pea core-collection with diverse rhizobial strains shows competitiveness for nodulation and efficiency of nitrogen fixation are distinct traits in interaction. Front Plant Sci 8:2249–2263. https://doi.org/10.3389/fpls.2017.02249

Chaudhary A, Chaudhary P, Upadhyay A, Kumar A, Singh A (2022e) Effect of gypsum on plant growth promoting rhizobacteria. Environ Ecol 39(4A):1248–1256

Chaudhary A, Parveen H, Chaudhary P, Bhatt P, Khatoon H (2021g) Rhizospheric microbes and their mechanism. In: Bhatt P, Gangola S, Udayanga D, Kumar G (eds) Microbial Technology for Sustainable Environment. Springer, Singapore. https://doi.org/10.1007/978-981-16-3840-4_6

Chaudhary P, Agri U, Chaudhary A, Kumar A, Kumar G (2022c) Endophytes and their potential in biotic stress management and crop production. Front Microbiol 13:933017. https://doi.org/10.3389/fmicb.2022.933017

Chaudhary P, Chaudhary A, Agri U, Khatoon H, Singh A (2022d) Recent trends and advancements for agro-environmental sustainability at higher altitudes. In: Goel R, Soni R, Suyal DC, Khan M (eds) Survival strategies in cold-adapted microorganisms. Springer, Singapore. https://doi.org/10.1007/978-981-16-2625-8_19

Chaudhary P, Chaudhary A, Parveen H, Rani A, Kumar G, Kumar A, Sharma A (2021a) Impact of nanophos in agriculture to improve functional bacterial community and crop productivity. BMC Plant Biol 21:519. https://doi.org/10.1186/s12870-021-03298-7

Chaudhary P, Chen S, Rajput VD, Jaiswal DK (2023b) Editorial: bioinoculants with nanocompounds to improve soil health: a step toward sustainable agriculture. Front Environ Sci 11:1270002. https://doi.org/10.3389/fenvs.2023.1270002

Chaudhary P, Khati P, Chaudhary A, Gangola S, Kumar R, Sharma A (2021b) Bioinoculation using indigenous Bacillus spp. improves growth and yield of *Zea mays* under the influence of nanozeolite. 3 Biotech 11:11. https://doi.org/10.1007/s13205-020-02561-2

Chaudhary P, Khati P, Chaudhary A, Maithani D, Kumar G, Sharma A (2021c) Cultivable and metagenomic approach to study the combined impact of nanogypsum and *Pseudomonas taiwanensis* on maize plant health and its rhizospheric microbiome. PLoS One 16:e0250574. https://doi.org/10.1371/journal.pone.0250574

Chaudhary P, Khati P, Gangola S, Kumar A, Kumar R, Sharma A (2021d) Impact of nanochitosan and *Bacillus* spp. on health, productivity and defence response in Zea mays under field condition. 3 Biotech 11:237. https://doi.org/10.1007/s13205-021-02790-z

Chaudhary P, Parveen H, Gangola S, Kumar G, Bhatt P, Chaudhary A (2021e) Plant growth promoting rhizobacteria and their application in sustainable crop production. In: Bhatt P, Gangola S, Udayanga D, Kumar G (eds) Microbial Technology for Sustainable Environment. Springer, Singapore. https://doi.org/10.1007/978-981-16-3840-4_13

Chaudhary P, Sharma A (2019) Response of nanogypsum on the performance of plant growth promotory bacteria recovered from nanocompound infested agriculture field. Environ Ecol 37:363–372

Chaudhary P, Sharma A, Chaudhary A, Khati P, Gangola S, Maithani D (2021f) Illumina based high throughput analysis of microbial diversity of maize rhizosphere treated with nanocompounds and bacillus sp. Appl Soil Ecol 159:103836. https://doi.org/10.1016/j.apsoil.2020.103836

Chaudhary P, Singh S, Chaudhary A, Agri U, Bhandari G (2022a) Chapter 3 Agrochemicals and their effects on soil microbial population. In: Soni R, Suyal D, Goel R (eds) Plant protection: from chemicals to biologicals. De Gruyter, Berlin, Boston, pp 45–58. https://doi.org/10.1515/9783110771558-003

Chaudhary P, Singh S, Chaudhary A, Sharma A, Kumar G (2022b) Overview of biofertilizers in crop production and stress management for sustainable agriculture. Front Plant Sci 13:930340. https://doi.org/10.3389/fpls.2022.930340

Chaudhary P, Xu M, Ahamad L, Chaudhary A, Kumar G, Adeleke BS, Verma KK, Hu D-M, Širić I, Kumar P, Popescu SM, Abou Fayssal S (2023a) Application of synthetic consortia for improvement of soil fertility, pollution remediation, and agricultural productivity: a review. Agronomy 13:643. https://doi.org/10.3390/agronomy13030643

Chaudhary SR, Sindhu SS (2016) Growth stimulation of clusterbean (*Cyamopsis tetragonoloba*) by coinoculation with rhizosphere bacteria and *Rhizobium*. Legum Res 39(6):1003–1012. https://doi.org/10.18805/lr.v0iOF.8605

Ding Z, Ali EF, Almaroai YA, Eissa MA, Abeed AHA (2021) Effect of potassium solubilizing bacteria and humic acid on faba bean (*Vicia faba* L.) plants grown on sandy loam soils. J Soil Sci Plant Nutr 21:791–800. https://doi.org/10.1007/s42729-020-00401-z

El-Sayed WS, Akhkha A, El-Nagga M, Elbadry M (2014) In vitro antagonistic activity, plant growth promoting traits and phylogenetic affiliation of rhizobacteria associated with wild plants grown in arid soil. Front Microbiol 5:651. https://doi.org/10.3389/fmicb.2014.00651

Etesami H, Jeong BR, Glick BR (2021) Contribution of arbuscular mycorrhizal fungi, phosphate–solubilizing bacteria and silicon to P uptake by plant. Front Plant Sci 12:699618. https://doi.org/10.3389/fpls.2021.699618

Fall AF, Nakabonge G, Ssekandi J, Founoune-Mboup H, Apori SO, Ndiaye A, Badji A, Ngom K (2022) Roles of arbuscular mycorrhizal fungi on soil fertility: contribution in the improvement of physical, chemical, and biological properties of the soil. Front Fungal Biol 3:723892. https://doi.org/10.3389/ffunb.2022.723892

Frąc M, Hannula SE, Bełka M, Jędryczka M (2018) Fungal biodiversity and their role in soil health. Front Microbiol 9. https://doi.org/10.3389/fmicb.2018.00707

Gallegos-Cedillo VM, Urrestarazu M, Alvaro JE (2016) Influence of salinity on transport of nitrates and potassium by means of the xylem sap content between roots and shoots in young tomato plants. J Soil Sci Plant Nutr 16(4):991–998. https://doi.org/10.4067/S0718-95162016005000072

Ghodhbane-Gtari F, D'Angelo T, Gueddou A, Ghazouani S, Gtari M, Tisa LS (2021) Alone yet not alone: Frankia lives under the same roof with other bacteria in actinorhizal nodules. Front Microbiol 12:749760. https://doi.org/10.3389/fmicb.2021.749760

Gómez-Duque AM, Morales-Londoño CS, Hurtado-Salazar A, Ceballos-Aguirre N (2022) Nitrogen-fixing bacteria and nitrogen fertilization on economic feasibility of tomato. Revista Colombiana de Ciencias Hortícolas 16:e13623. https://doi.org/10.17584/rcch.2022v16i2.13623

Gontia-Mishra I, Sapre S, Tiwari S (2017) Zinc solubilizing bacteria from the rhizosphere of rice as prospective modulator of zinc biofortification in rice. Rhizosphere 3:185–190. https://doi.org/10.1016/j.rhisph.2017.04.013

Goyal RK, Mattoo AK, Schmidt MA (2021) Rhizobial–host interactions and symbiotic nitrogen fixation in legume crops toward agriculture sustainability. Front Microbiol 12:669404. https://doi.org/10.3389/fmicb.2021.669404

Hafeez B, Khanif YM, Saleem M (2013) Role of zinc in plant nutrition—a review. Am J Exp Agric 3(2):374–391. https://doi.org/10.9734/AJEA/2013/2746

He W-X, Wu Q-S, Hashem A, Abd-Allah EF, Muthuramalingam P, Al-Arjani A-BF, Zou Y-N (2022) Effects of symbiotic fungi on sugars and soil fertility and structure-mediated changes in plant growth of *Vicia villosa*. Agriculture 12:1523. https://doi.org/10.3390/agriculture12101523

Hernández-Fernández M, Cordero-Bueso G, Ruiz-Muñoz M, Cantoral JM (2021) Culturable yeasts as biofertilizers and biopesticides for a sustainable agriculture: a comprehensive review. Plan Theory 10(5):822. https://doi.org/10.3390/plants10050822

Hungria M, Nogueira MA, Araujo RS (2015) Soybean seed coinoculation with *Bradyrhizobium* spp. and *Azospirillum brasilense*: a new biotechnological tool to improve yield and sustainability. Acad J Plant Sci 6(6):8–11. https://doi.org/10.4236/ajps.2015.66087

Hussain A, Zahir ZA, Ditta A, Tahir MU, Ahmad M, Mumtaz MZ, Hayat K, Hussain S (2020) Production and implication of bio-activated organic fertilizer enriched with zinc-solubilizing bacteria to boost up maize (*Zea mays* L.) production and biofortification under two cropping seasons. Agronomy 10(1):39. https://doi.org/10.3390/agronomy10010039

Ilyas N, Mumtaz K, Akhtar N, Yasmin H, Sayyed R, Khan W, Enshasy HAE, Dailin DJ, Elsayed EA, Ali Z (2020) Exopolysaccharides producing bacteria for the amelioration of drought stress in wheat. J Sustain Agric Environ 12:8876. https://doi.org/10.3390/su12218876

Jain D, Sharma J, Kaur G, Bhojiya AA, Chauhan S, Sharma V, Maharjan E (2021) Phenetic and molecular diversity of nitrogen fixating plant growth promoting *Azotobacter* isolated from semiarid regions of India. Biomed Res Int 2021:1–9. https://doi.org/10.1155/2021/6686283

Jha Y (2019) The importance of zinc-mobilizing rhizosphere bacteria to the enhancement of physiology and growth parameters for paddy under salt-stress conditions. Jordan J Biol Sci 12(2):167–173. ISSN https://doi.org/1995-6673

Joshi D, Chandra R, Suyal DC, Kumar S, Goel R (2019) Impact of bioinoculants *Pseudomonas jesenii* MP1 and *Rhodococcus qingshengii* S10107 on *Cicer arietinum* yield and soil nitrogen status. Pedosphere 29(3):388–399. https://doi.org/10.1016/S1002-0160(19)60807-6

Kaminsky LM, Trexler RV, Malik RJ, Hockett KL, Bell TH (2019) The inherent conflicts in developing soil microbial inoculants. Trends Biotechnol 37(2):140–151. https://doi.org/10.1016/j.tibtech.2018.11.011

Kang S-M, Shahzad R, Khan MA, Hasnain Z, Lee K-E, Park H-S, Kim L-R, Lee I-J (2021) Ameliorative effect of indole-3-acetic acid- and siderophore- producing Leclerciaade carboxylata MO1 on cucumber plants under zinc stress. J Plant Interact 16:30–41. https://doi.org/10.1080/17429145.2020.1864039

Khan MS, Zaidi A, Ahemad M, Oves M, Wani PA (2010) Plant growth promotion by phosphate solubilizing fungi–current perspective. Arch Agron Soil Sci 56(1):73–98. https://doi.org/10.1080/03650340902806469

Khati P, Chaudhary P, Gangola S, Bhatt P, Sharma A (2017) Nanochitosan supports growth of *Zea mays* and also maintains soil health following growth. 3 Biotech 7:81. https://doi.org/10.1007/s13205-017-0668-y

Khati P, Chaudhary P, Gangola S, Sharma P (2019) Influence of Nanozeolite on plant growth promotory bacterial isolates recovered from Nanocompound infested agriculture field. Environ Ecol 37:521–527

Khati P, Parul BP, Nisha Kumar R, Sharma A (2018) Effect of nanozeolite and plant growth promoting rhizobacteria on maize. 3 Biotech 8:141. https://doi.org/10.1007/s13205-018-1142-1

Koirala A, Brözel VS (2021) Phylogeny of nitrogenase structural and assembly components reveals new insights into the origin and distribution of nitrogen fixation across bacteria and archaea. Microorganisms 9:1662. https://doi.org/10.3390/microorganisms9081662

Korir H, Mungai NW, Thuita M, Hamba Y, Masso C (2017) Co-inoculation effect of rhizobia and plant growth promoting rhizobacteria on common bean growth in a low phosphorus soil. Front Plant Sci 8:141. https://doi.org/10.3389/fpls.2017.00141

Kukreti B, Sharma A, Chaudhary P, Agri U, Maithani D (2020) Influence of nanosilicon dioxide along with bioinoculants on Zea mays and its rhizospheric soil. 3 Biotech 10:345. https://doi.org/10.1007/s13205-020-02329-8

Kumari H, Khati P, Gangola S, Chaudhary P, Sharma A (2021) Performance of plant growth promotory rhizobacteria on maize and soil characteristics under the influence of TiO2 nanoparticles. Pantnagar J Res 19:28–39

Kumari S, Sharma A, Chaudhary P, Khati P (2020) Management of plant vigor and soil health using two agriusable nanocompounds and plant growth promotory rhizobacteria in fenugreek. 3 Biotech 10:1–11. https://doi.org/10.1007/s13205-020-02448-2

Macik M, Gryta A, Frac M (2020) Biofertilizers in agriculture: an overview on concepts, strategies, and effects on soil microorganisms. Adv Agron 160:31. https://doi.org/10.1016/bs.agron.2020.02.001

Manzar N, Jayaswal D, Kishan G, Chauhan S, Joshi D, Singh J, Chaudhary P, Jamali H, Kushwaha P, Parveen H, Kashyap AS, Sahu PK, Singh UB (2022) Application of molecular ecology approaches in sustainable agriculture for a better understanding of plant–microbiome interactions. In: Singh UB, Sahu PK, Singh HV, Sharma PK, Sharma SK (eds) Rhizosphere microbes. Microorganisms for sustainability, vol 40. Springer, Singapore. https://doi.org/10.1007/978-981-19-5872-4_3

Masood F, Ahmad S, Malik A (2022) Role of rhizobacterial bacilli in zinc solubilization. In: Khan ST, Malik A (eds) Microbial Biofertilizers and Micronutrient Availability. Springer, Switzerland, pp 361–377

Meena OP, Maurya BR, Meena VS (2013) Influence of K-solubilizing bacteria on release of potassium from waste mica. Agric Sustain Dev 1(1):53–56. ISSN https://doi.org/2347-5358

Meena VS, Maurya BR, Verma JP (2014) Does a rhizospheric microorganism enhance K+ availability in agricultural soils. Microbiol Res 169(5–6):337–347. https://doi.org/10.1016/j.micres.2013.09.003

Mehnaz S (2016) An overview of globally available bioformulations. In: Arora NK (ed) Bioformulations: for sustainable agriculture. Springer, pp 267–281. https://doi.org/10.1007/978-81-322-2779-3_15

Mfilinge A, Mtei K, Ndakidemi P (2014) Effect of *Rhizobium* inoculation and supplementation with phosphorus and potassium on growth and total leaf chlorophyll (Chl) content of bush bean *Phaseolus vulgaris* L. J Agric Sci. https://doi.org/5:1413-1419

Minaxi Saxena J, Chandra S, Nain L (2013) Synergistic effect of phosphate solubilizing rhizobacteria and arbuscular mycorrhiza on growth and yield of wheat plants. J Soil Sci Plant Nutr 13:511–525. https://doi.org/10.4067/S0718-95162013005000040

Mumtaz MZ, Ahmad M, Jamil M, Hussain T (2017) Zinc solubilizing Bacillus spp. potential candidates for biofortification in maize. Microbiol Res 202:51–60. https://doi.org/10.1016/j.micres.2017.06.001

Noar JD, Bruno-Barcena JM (2018) *Azotobacter vinelandii*: the source of 100 years of discoveries and many more to come. Nat Microbiol 164:421–436. https://doi.org/10.1099/mic.0.0006

Parmar P, Sindhu SS (2019) The novel and efficient method for isolating potassium solubilizing bacteria from rhizosphere soil. Geomicrobiol J 36(2):130–136. https://doi.org/10.1080/01490451.2018.1514442

Patel SH, Viradiya MB, Prajapati BJ (2021) Effect of potassium and potassium mobilizing bacteria (KMB) with and without FYM on yield of wheat (*Triticum aestivum* L.). J Pharmacol Phytochem 10(1):1615–1620. https://doi.org/2349-8234

Pathak DV, Kumar M (2016) Microbial inoculants as biofertilizers and biopesticides. Microbial Inoculants in Sustainable Agricultural Productivity: Vol. 1: Research Perspectives 197–209. https://doi.org/10.1007/978-81-322-2647-5_11

Pellegrini M, Spera DM, Ercole C, Del Gallo M (2021) *Allium cepa* L. inoculation with a consortium of plant growth-promoting bacteria: effects on plants, soil and the autochthonous microbial community. Microorganisms 9(3):639. https://doi.org/10.3390/microorganisms9030639

Pourbabaee AA, Koohbori Dinekaboodi S, Seyed Hosseini HM, Alikhani HA, Emami S (2020) Potential application of selected sulfur-oxidizing bacteria and different sources of sulfur in plant growth promotion under different moisture conditions. Commun Soil Sci Plant Anal 51:735–745. https://doi.org/10.1080/00103624.2020.1729377

Prajapati KB, Modi HA (2012) Isolation and characterization of potassium solubilizing bacteria from ceramic industry soil. J Microbiol 1(2–3):8–14. ISSN: https://doi.org/2319-3867

Qarni A, Billah M, Hussain K, Shah SH, Ahmed W, Alam S, Khan N (2021) Isolation and characterization of phosphate solubilizing microbes from rock phosphate mines and their potential effect for sustainable agriculture. Sustain For 13(4):2151. https://doi.org/10.3390/su13042151

Rana K, Rana N, Singh B (2020) Applications of sulfur oxidizing bacteria. In: Physiological and biotechnological aspects of extremophiles. Academic Press, pp 131–136. https://doi.org/10.1016/B978-0-12-818322-9.00010-1

Rawat N, Sharma M, Suyal DC, Singh DK, Joshi D, Singh P, Goel R (2019) Psyhcrotolerant bio-inoculants and their co-inoculation to improve *Cicer arietinum* growth and soil nutrient status for sustainable mountain agriculture. J Soil Sci Plant Nutr 19(3):639–647. https://doi.org/10.1007/s42729-019-00064-5

Riaz U, Mehdi SM, Iqbal S, Khalid HI, Qadir AA, Anum W, Ahmad M, Murtaza G (2020) Bio-fertilizers: eco-friendly approach for plant and soil environment. Bioremediation and biotechnology. Springer, Berlin/Heidelberg, Germany, pp 189–213. https://doi.org/10.1007/978-3-030-35691-0_9

Rodrigues AC, Bonifacio A, de Araujo FF, Junior MAL, Figueiredo MVB (2015) *Azospirillum* sp. as a challenge for agriculture. In: Maheshwari DK (ed) Bacterial metabolites in sustainable agroecosystem. Springer International Publishing, pp 29–51. https://doi.org/10.1007/978-3-319-24654-3_2

Sammauria R, Kumawat S, Kumawat P, Singh J, Jatwa TK (2020) Microbial inoculants: potential tool for sustainability of agricultural production systems. Arch Microbiol 202(4):677–693. https://doi.org/10.1007/s00203-019-01795-w

Santoyo G, Guzman-Guzman P, Parra-Cota FI, de los Santos-Villalobos S, Orozco-Mosqueda MC, Glick BR (2021) Plant growth stimulation by microbial consortia. Agronomy 11:219. https://doi.org/10.3390/agronomy11020219

Sepp S-K, Vasar M, Davison J, Oja J, Anslan S, Al-Quraishy S, Bahram M, Bueno CG, Cantero JJ, Fabiano EC, Decocq G, Drenkhan R, Fraser L, Garibay Oriel R, Hiiesalu I, Koorem K, Kõljalg U, Moora M, Mucina L, Öpik M, Põlme S, Pärtel M, Phosri C, Semchenko M, Vahter T, Vasco Palacios AM, Tedersoo L, Zobel M (2023) Global diversity and distribution of nitrogen-fixing bacteria in the soil. Front Plant Sci 14:1100235. https://doi.org/10.3389/fpls.2023.1100235

Sindhu SS, Parmar P, Phour M, Sehrawat A (2016) Potassium solubilizing microorganisms (KSMs) and its effect on plant growth improvement. Springer-verlag, Berlin, Heidelberg, pp 171–185. https://doi.org/10.1007/978-81-322-2776-2_13

Sindhu SS, Phour M, Choudhary SR, Choudhary D (2014) Phosphorus cycling: prospects of using rhizosphere microorganisms for improving phosphorus nutrition of plants. In: Parmar N, Singh A (eds) Geomicrobiology and biogeochemistry. Springer-Verlag, Berlin, Heidelberg, pp 199–237. https://doi.org/10.1007/978-3-642-41837-2_11

Sindhu SS, Sharma R, Sindhu S, Phour (2019) Plant nutrient management through inoculation of zinc solubilizing bacteria for sustainable agriculture. In: Giri B, Prasad R, Wu QS, Verma A (eds) Biofertilizers for sustainable agriculture and environment. Springer Nature Springer, pp 173–201. https://doi.org/10.1007/978-3-030-18933-4_8

Sindhu SS, Sharma R, Sindhu S, Sehawat A (2019a) Soil fertility improvement by symbiotic rhizobia for sustainable agriculture. In: Panpatte DG, Jhala VK (eds) Soil fertility management for sustainable development. Springer Nature, Singapore, Pte Ltd., pp 101–166. https://doi.org/10.1007/978-981-13-5904-0_15

Singh JS, Kumar A, Rai AN, Singh DP (2016) Cyanobacteria: a precious bio-resource in agriculture, ecosystem, and environmental sustainability. Front Microbiol 7:529. https://doi.org/10.3389/fmicb.2016.00529

Soumare A, Diedhiou AG, Thuita M, Hafidi M, Ouhdouch Y, Gopalakrishnan S, Kouisni L (2020) Exploiting biological nitrogen fixation: a route towards a sustainable agriculture. Plan Theory 9:1011. https://doi.org/10.3390/plants9081011

Tahir HA, Gu Q, Wu H, Raza W, Hanif A, Wu L, Colman MV, Gao X (2017) Plant growth promotion by volatile organic compounds produced by *Bacillus subtilis* SYST2. J Front Microbiol 8:171. https://doi.org/10.3389/fmicb.2017.00171

Thuc LV, Huu TN, Ngoc TM, Hue NH, Quang LT, Xuan DT, Nhan TC, Xuan LNT, Thu LTM, Akagi I, Sakagami J-I, Khuong NQ (2022) Effects of nitrogen fertilization and nitrogen fixing endophytic bacteria supplementation on soil fertility, N uptake, growth, and yield of sesame (*Sesamum indicum* L.) cultivated on alluvial soil in dykes. Appl Environ Soil Sci 2022:1972585. https://doi.org/10.1155/2022/1972585

Tyagi J, Chaudhary P, Jyotsana BU, Bhandari G, Chaudhary A (2022a) Chapter 17 Impact of endophytic fungi in biotic stress management. In: Soni R, Suyal D, Goel R (eds) Plant protection: from chemicals to biologicals. De Gruyter, Berlin, Boston, pp 447–462. https://doi.org/10.1515/9783110771558-017

Tyagi J, Chaudhary P, Mishra A, Khatwani M, Dey S, Varma A (2022b) Role of endophytes in abiotic stress tolerance: with special emphasis on Serendipita indica. Int J Environ Res 16:62. https://doi.org/10.1007/s41742-022-00439-0

Vaid SK, Srivastava PC, Pachauri SP, Sharma A, Rawat D, Shankhadhar SC, Shukla AK (2020) Effective zinc mobilization to rice grains using rhizobacterial consortium. Israel J Plant Sci 67(3–4):145–157. https://doi.org/10.1163/22238980-20201085

Verma P, Yadav AN, Khannam KS, Kumar S, Saxena AK, Suman A (2016) Molecular diversity and multifarious plant growth promoting attributes of *Bacilli* associated with wheat (*Triticum aestivum* L.) rhizosphere from six diverse agro-ecological zones of India. J Basic Microbiol 56:44–58. https://doi.org/10.1002/jobm.201500459

Wang D, Lv S, Jiang P, Li Y (2017) Roles, regulation, and agricultural application of plant phosphate transporters. Front Plant Sci 8:817. https://doi.org/10.3389/fpls.2017.00817

Wang Y, Peng S, Hua Q, Qiu C, Wu P, Liu X, Lin X (2021) The long-term effects of using phosphate-solubilizing bacteria and photosynthetic bacteria as biofertilizers on peanut yield and soil bacteria community. Front Microbiol 12:693535. https://doi.org/10.3389/fmicb.2021.693535

Zarjani JK, Aliasgharzad N, Oustan S, Emadi M, Ahmadi A (2013) Isolation and characterization of potassium solubilizing bacteria in some Iranian soils. Arch Agron Soil Sci 59(12):1713–1723. https://doi.org/10.1080/03650340.2012.756977

Significance of Microbes in Heavy Metal Detoxification from Environment

Deeksha Kashyap, Minerva Sharma, Portia Sharma, Arshad Khan, and Lukman Ahamad

Abstract Contamination that began via heavy metals (HMs) has become a global issue in recent years as it affects the environment such as soil, plants, aquatic life, and human health. Heavy metals are widely known as environmental pollutants due to their toxicity, environmental persistence, and bioaccumulative nature. However, heavy metals (HMs) are introduced in the environment by various means including industrial and anthropogenic activities. Due to their toxicity and implications in the environment, it is necessary to detoxify them with the utilization of different microorganisms. Microbes that include fungi, bacteria, and algae played a crucial role in heavy metal detoxification through various mechanisms.

Keywords Heavy metals · Toxicity · Environment · Microbes · Detoxification

1 Introduction

The harmful impacts of long-term heavy metal contaminations pose a severe threat to all types of life-forms present in the environment. The heavy metals can accumulate in the food webs and mainly sensitive in low amounts, which puts the general public health at risk (Mitra et al. 2022). However, various organic contaminants and metals are nonbiodegradable and persist for a longer period in their surroundings. The conventional chemical and physical approaches of remediation are not cost-effective and generate significant quantity of chemical wastes in the environment. In

D. Kashyap · A. Khan · L. Ahamad (✉)
Section of Plant Pathology and Nematology, Department of Botany, Aligarh Muslim University, Aligarh, Uttar Pradesh, India

M. Sharma
Department of Biochemistry, School of Life Sciences, Dr. Bhimrao Ambedkar University, Agra, Uttar Pradesh, India

P. Sharma
Department of Botany, St. John's College, Agra, Uttar Pradesh, India

P. Chaudhary, A. Chaudhary (eds.), *Microbial Inoculants*, Interdisciplinary Biotechnological Advances, https://doi.org/10.1007/978-981-97-0633-4_13

recent years, there has been a significant and expanding interest in the balance of dangerous metals. The use of biosensors and microbes such as algae, fungi, and bacteria is economical and environmentally friendly to degrade or remove the pollutants from water and soil systems (Kumar et al. 2021a, b; Parveen et al. 2022; Dhaigude et al. 2023; Singh et al. 2023). Because of this, the microbes have a range of metal sequestration methods which have better metal biosorption capabilities. Heavy metals have harmful consequences including displacing and decreasing the availability of vital soil minerals, which can impact the growth and ultimately decrease crop productivity (Tarekegn et al. 2020).

By preserving sustainability, marine bacteria like *Pseudomonas putida*, *P. fluorescens*, *Alcanivorax borkumensis*, *Dechloromonas aromatica*, *Bacillus subtilis*, *Methylibium petroleiphilum*, and *Phanerochaete chrysosporium* play an important part in heavy metal remediation process. It was shown that these bacteria played a significant role in altering the concentration of heavy metals in the soil and water as well as assisting plants in surviving a concentration of heavy metals that had positive ecological effects (Husain et al. 2022). Unfortunately, the Black environment has a reputation for being a very dirty sea, with toxins from many sources contaminating the marine sediments. However, when exposed to salts of heavy metals, *Bacillus wiedmannii* was found to be the most resilient bacterial community (Kalkan 2022).

However, there are few data available on the application of metals tolerant of the rhizobiome of maize cultivation. It has been discovered that the root microbiome has greater metal-tolerant potential and has been employed for improving maize production in polluted soils. However, the significance of maize as a food crop and animal feed requires dearth of knowledge regarding the metal induced phytotoxicity and its prevention methods (Rizvi et al. 2022).

By oxidizing HMs, adhering to cell surfaces, immobilizing contaminants, and volatilizing or transforming contaminants, the natural microbiota participating in HM removal or effluent is polishing in the support environmental restoration (Greeshma et al. 2022). This is especially true when the HM concentration is between 1 and 50 mg/L. Contrary to conventional approaches, biosorption and bioaccumulation mechanisms used by algae, bacteria, and fungi to effectively remove HMs do not need energy or chemicals (Diep et al. 2018).

Microorganisms with small genomes, rapid evolution, and high capacity for multiplication over a short period of time and that have the ability to adapt to new, challenging environments are the best candidates for large-scale bioremediation of wastewater and pollutants (Kumar et al. 2021a, b; Chaudhary et al. 2023a, b, c, d). The elimination processes in each taxon are distinct according to the detoxification mechanisms examined for certain HMs, pointing to an improved impact when employed as a microbiome (Greeshma et al. 2022). Additionally, genetically modified microorganisms were shown to have strong bioremediation skills and great potential for removing hazardous substances (Sharma et al. 2022a, b). The majority of soil-dwelling species have high HM tolerance, which they probably developed through evolutionary adaptation to polluted environments. Due to variations in cellular metabolism, the fungi have better resistance to HMs in the environment than other microbes (Gajewska et al. 2022). Fungi are able to tolerate harsh environments because of high osmotic pressure in their cell structures. In addition, fungi can survive as chlamydospores, sclerotia, and other structures in soil that help microorganisms endure harsh circumstances. However, fungal activity increased at high Cu and Zn concentrations separately supplied to growth media in comparison to control (Gajewska et al. 2022). Through adsorption or cell inclusion, the algal biomass is employed for the bioremediation of heavy metals in contaminated wastewater. To remove heavy metals, phyco-remediation employs several kinds of algae and cyanobacteria that reduce or oxidize the toxicant. Numerous researchers came to the conclusion that the brown algae (*Sargassum*) are effective in adsorption and extraction of heavy metals such as Sm, Pr, Cd, Cr, Pd, Ni, and Cu because their cell walls contain active bio-absorption sites (Cardoso et al. 2017; Zaynab et al. 2022). Over the years, there has been significant development in the bioremediation of toxic metals. Microbial bio-sorbents are a viable solution for the cleanup of heavy metal-contaminated settings since they are economical and environmentally acceptable. Various metal sequestration methods that possess better metal bio-sorption capabilities are found in microorganisms. Utilizing living or dead biomass and its constituent parts, microbial bio-sorption aims to extract and/or recover metals and metalloids from liquids (Ayanbenro and Babalola 2017).

In this chapter, we highlighted the source of harmful heavy metals and microbial groups with bio-sorbent capabilities for heavy metal removal by bacteria, algae, fungi, and other microbes, as well as the mechanisms and enzymes involved in each. Some specific microbes and their role for removal of heavy metals (HMs) that are toxic are discussed. We hope the readers will benefit from this discussion.

2 General Information about Heavy Metals and its Types

Heavy metals (HMs) are the elements having atomic numbers of more than 20 and atomic densities greater than 5 g/cm^3. They have been associated with contamination and potential toxicity or eco-toxicity. The environmental pollution caused by heavy metals is continuously becoming a problem and has become a prime concern because of their adverse effects on the human health and environment around the world (Briffa et al. 2020). The heavy metals can enter soil, water, and air ecosystems by natural process and industrial and anthropogenic activities (Li et al. 2019). Additionally, heavy metals exert toxic effects on soil microorganisms which resulted in the change in the diversity, population size, and overall activities of microbial communities in soil (Table 1).

The most frequent heavy metals are nickel (Ni), copper (Cu), lead (Pb), chromium (Cr), mercury (Hg), arsenic (As), cadmium (Cd), and iron (Fe) (Bakshi et al. 2018). Among them, few heavy metals such as nickel (Ni) and iron (Fe) are essential for survival of all life-forms at low concentrations (Bakshi et al. 2018). However, heavy metals such as cadmium (Cd), lead (Pb), and mercury (Hg) are harmful to the living organisms at both higher and lower concentrations. In addition (Fig. 1), Fe, Al, Sb, Cr, Be, As, Cu, Cd, Hg, Pb, Se, Ni, Zn, and Ag are categorized as primary contaminant metals by the US Environmental Protection Agency (US EPA) (Patniak 2010).

Table 1 Harmful effects of heavy metals on the environment

Heavy metals	Effects on environment	Reference
Aluminum (Al^{3+})	Interaction of metal (Al^{3+}) with plasma membrane leads to toxicity and disturbs the cellular activity in plants	Gupta et al. (2013)
Mercury (Hg)	Heavy metal (Hg) reduced the % germination, plant length, flowering, and weight of fruits in tomato	Shekar et al. (2011)
Cd and Pb	Sheep tissues affected by heavy metals were greater than control animals, while Pb and Cd concentrations in hair and blood samples of residents were remarkably greater than in control samples	Shen et al. (2019)
Mercury (Hg)	Mercury heavy metal caused the abnormality in metabolic functions of organisms	Akanchise et al. (2020)
Lead (Pb)	Lead heavy metal affects the soil enzyme activities such as catalase, urease, acid phosphatase, and invertase	Bakshi et al. (2018)
Arsenic (As) and chromium (Cr)	Heavy metal exposure enlarges the size and incidence of tumor. It was regulated via signaling pathway (Wnt/beta catenin) induced through generation of ROS	Wang et al. (2012)
Cadmium (Cd)	The heavy metal decreased the richness, bioactivity, and microbial diversity with an increase in HM concentration	Xie et al. (2016)
Zn, Ni, Cd, and Cu	All the heavy metals negatively affect the structure and activity of microbial community in polluted soils	Jarosławiecka and Piotrowska-Seget (2022)

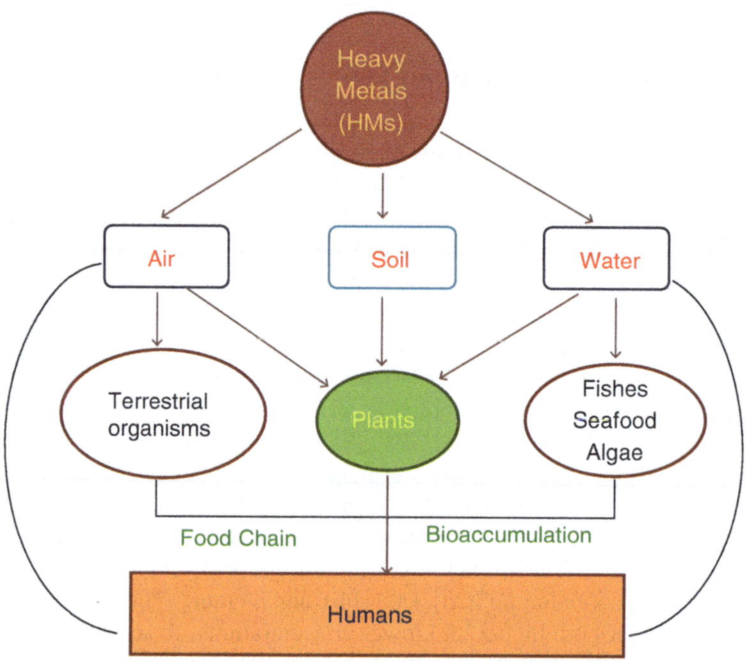

Fig. 1 Heavy metals and impacts on the environment

3 Role of Fungi in Detoxification of Heavy Metals

In many nations where industrial effluents often include hazardous metal ions, heavy metal pollution is a serious problem. The degree of usage, species of origin, bioavailability, ubiquity, and amount of absorption and diffusion all affect a metal's toxicity (Zaidi and Pal 2017). Heavy metal accumulation in the environment is largely posed by industrial and human activities. Excessive levels of heavy metals in soil have significant negative effect on fungal growth and the microbial-plant symbiotic relationship, which in turn affects soil fertility, mineral uptake, and the availability of nutrients to plants (Miransari 2016; Tyagi et al. 2022a, b).

3.1 Heavy Metal Toxicity in Fungi

Reactive oxygen species are produced as a result of excessive concentrations of accumulated metals, which damage microbial membranes, organelles, lipids, and organelles, as well as trigger cell death (Igiri et al. 2018). According to studies, heavy metal ions have a negative impact on the fungal cellular life process, which eventually results in growth suppression and death. Heavy metals mainly Mn, Cu, Pb, Cd, and Zn were shown to hinder the growth of certain species of the *Aspergillus* fungus and *Sterigmatomyces halophilus* (Bano et al. 2018), while Pb also suppresses the growth of another fungus, *Abortiporus biennis* (Grąz et al. 2011). Induced protein and nucleic acid damage from excessive Cd buildup in bacteria prevent transcription as well as cell division. Hg ions are also known to denature proteins, damage microbial cell membranes, and interfere with enzyme activity (Ayanbenro and Babalola 2017). The sporulation and viability of plant pathogenic fungi such as *Phytophthora infestans* (Gajewska et al. 2020), *P. capsica* (Liu et al. 2018), *Alternaria alternata*, *Botrytis cinerea* (Ouda 2014), and *Phanerochaete chrysosporium* (Chen et al. 2014) can also be affected by heavy metals, therefore creating stressful circumstances that dramatically enhance the production of reactive oxygen species (ROS) as peroxides and superoxides, and hydroxyl radicals, which harm the fungal cell and the structure of the organelle and change the metabolism. In conclusion, oxidative stress may be seen as the main cause of cell death caused by the heavy metals in microorganisms (Lin et al. 2020).

3.2 Heavy Metal Tolerance in Fungi

Prokaryotes also have natural defensive mechanisms that enable heavy metal ions to be absorbed and used as nutrition through absorptive and accumulative capability (Fig. 2). The ability of an eukaryotic microbial system like a fungus to adapt to

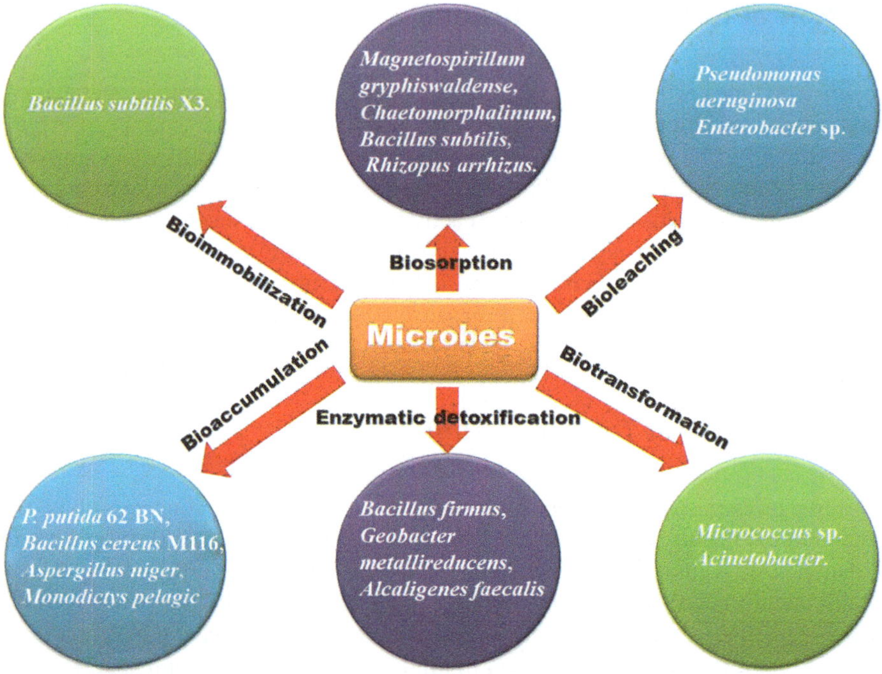

Fig. 2 Role of microbes in heavy metal detoxification from the environment

metal stress has also been demonstrated. The fungus uses a variety of strategies to live in the presence of metals, including binding with various microbial proteins, metal storage, and active efflux or exclusion of metals. Fungi have the capability to decrease metal ions into corresponding metallic forms when they attached or absorbed into the cell interior. However, the fungal enzymes, proteins, cofactors, and other metabolites play vital roles in the survival of organisms and also decrease the metal ions to their nano-particulate forms (Priyadarshini et al. 2021).

3.3 *Detoxification Mechanism of Heavy Metals*

The way that metals interact with fungi varies on several variables, which include the types and concentration of the metals, organisms, and structure of the polluted environment (Fig. 3). Diverse techniques, including biotransformation, extracellular precipitation, intracellular bioaccumulation, biosorption, and biomineralization, are used in the detoxification procedures (Wang et al. 2019).

The kind of heavy metal and its concentration, temperature, pH of the site or medium, exposure period, ratio of fungal surface to volume, and level of metabolic activity are only a few of the variables that affect the dynamics of myco-remediation (Khan et al. 2019). However, biosorption is a frequent process seen in microorganisms

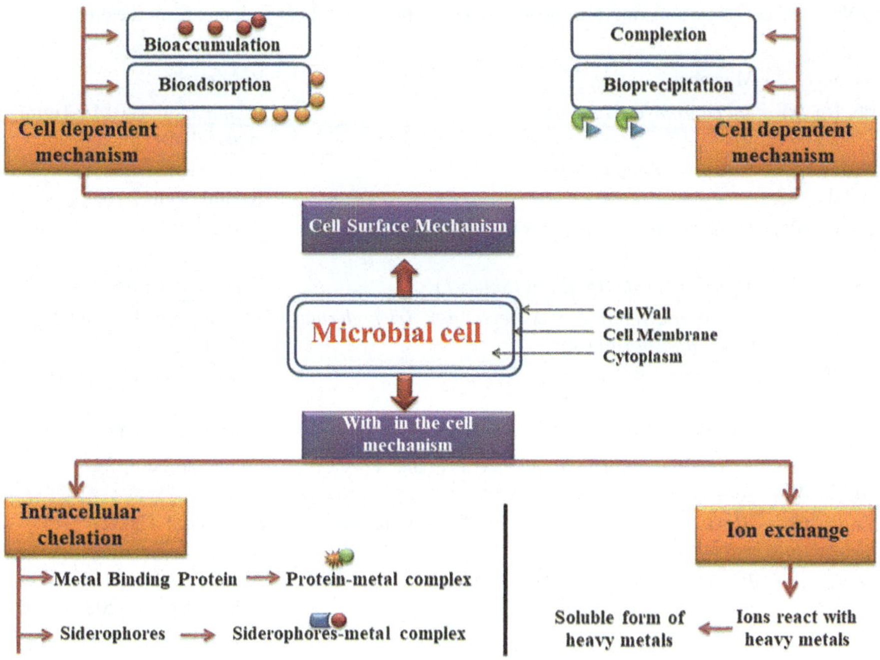

Fig. 3 Mechanism of action of microbes to detoxify heavy metals

and used to trap metals by ionic interaction between their sorption materials such as components of the cell wall or through the other polymeric compounds.

It is accomplished through the microorganisms' dead biomass, which is a non-metabolic mechanism. Environmental elements including pH, temperature, and the availability of certain metals or metal species are crucial throughout this process (Papirio et al. 2017). Among bacterial and fungal communities, interaction of metal species with functional groups of sorption components of cell walls, or with bacterial extracellular polysaccharides, is the major component of biosorption. As the first stage of biosorption, complexation, crystallization, and absorption, the fungal cell wall and its functional groups are strong ligand sites for the interaction with metal ions. This contact helps to establish the tolerance against heavy metal ions through biological persistence processes. Due to variations in the fungi's ability to absorb metal, the functional groups are altered as the mycelia grow (Siddiquee et al. 2015). Dead biomass is the main source of the electrostatic interaction's contribution to biosorption. Heavy metals interact with functional groups, changing their physical characteristics (Table 2). A variety of metabolic activities, including reduction, immobilization, and cell-bound or intracellular accumulation as well as concurrent detoxification processes, have been made possible by the alterations that occurred during contact (Gu et al. 2015).

Table 2 Effect of various microorganisms in detoxification of heavy metals from the environment

Heavy metals	Microbes	References
Pb, Hg, Ni	*Saccharomyces cerevisiae*	Infante et al. (2014); Tigini et al. (2010)
Heavy metals	*Cunninghamella elegans*	Paranthaman and Karthikeyan (2015)
Cu, Co, Pb, Cr	*Lysinibacillus sphaericus CBAM5*	Wu et al. (2015)
Fe	*Microbacterium profundi (Shh49T)*	Fazli et al. (2015)
Cd	*Aspergillus fumigatus*, *A. versicolor*, *Trichoderma* sp., *Cladosporium* sp.	Mirlahiji and Eisazadeh (2014)
Fe(III), U(VI)	*Geobacter* sp.	Priyalaxmi et al. (2014)
Cd	*Bacillus safensis*	Sinha et al. (2011)
Cu, U, Cr, Ni	*Aeromonas* sp., *Pseudomonas aeruginosa*	Sinha and Paul (2014)
Cr, Pb, Cd	*Rhodopseudomonas palustris*, *Aerococcus* sp.	Pande et al. (2022)
Pb, Cr, Cd	*B. megaterium*, *B. subtilis*, *Penicillium* spp., *A. niger*	Abioye et al. (2018) Tarekegn et al. (2020)

4 Role of Bacteria in Detoxification of Heavy Metals

The toxicities of cadmium, mercury, and arsenic metals have been linked with a number of diseases like Itai-itai, Minamata, and blackfoot disease, which are all examples of heavy metal poisoning diseases. Due to their continuous interactions with the water, air, and food for existence and survival, humans are particularly susceptible to environmental heavy metal exposure (Jaishankar et al. 2014). Regarding morphological alterations, there are numerous bacterial adaptive/resistance tactics against toxicity of heavy metals, including efflux pumps, the synthesis of metallothioneins, the generation of EPS, the development of siderophores, and the biotransformation of metal ions (Mathivanan et al. 2021). Regarding their environmental heavy metal detoxification, the most common bacteria are discussed.

4.1 Potential Biosorbent "Bacillus"

Many *Bacillus* species, including *B. thuringiensis*, *B. subtilis*, *B. megaterium*, *B. stearothermophilus*, *B. pumilus*, *B. jeotgali*, *B. cereus*, and *B. licheniformis* have been found to eliminate heavy metals more efficiently (Ghosh et al. 2022). The Zur family controls zinc uptake in *B. subtilis*, allowing Zn ions to be transported via two transporter proteins. Additionally, a third Zn ion absorption pathway in *B. subtilis* known as Zos A (Ptype ATPase) has been discovered, which was expressed under oxidative stress. However, a CPx-type ATPase efflux pump in *B. subtilis* called Cad

A facilitates the efflux of Zn at high concentrations. The molecular mechanisms of *Bacillus* spp. suggestively have transport, efflux, resistance, and detoxification systems, which function in both low and high concentrations of heavy metals to regulate and maintain the metal homeostasis at the cellular level by making them viable options for bioremediation (Alotaibi et al. 2021).

4.2 Pseudomonas aeruginosa

This bacterial strain is thought to be the most effective in removing heavy metal pollution and had various strategies for metal tolerance and resistance. There are three different ways that *Pseudomonas aeruginosa* resists heavy metals. Firstly, *Pseudomonas aeruginosa* is the only organism that has a method for actively extruding heavy metal ions from cells, which can be used to reduce the concentration of a specific ion rather than through interfering with absorption (Alfarras et al. 2022). Secondly, thiol-containing molecules can separate cations, particularly the "sulfur lovers" into complex compounds before expelling them from the cell. Thirdly, the complex enzymes and unique oxidation processes within the cells may also decrease some metal ions to a less lethal aerophilic form. Last but not the least, *Pseudomonas aeruginosa* effectiveness on various metals depends on a combination of two or three of the aforementioned fundamental processes. Pyoverdine, an extracellular substance with yellowish green fluorescence which is produced by *Pseudomonas aeruginosa*, serves as a by-product. Pyoverdine, formerly known as fluorescein, is produced simultaneously with pyochelin, another by-product, as well as other soluble pigments, including the blue pigment pyocyanin (Abdelbary et al. 2019).

4.3 Rhizosphere Bacteria

Areas with copper contamination put the health of living things at danger. Thus, remediation procedures are needed for both industrial and crop production. The promotion of copper phyto-extraction is thought to be greatly aided by three Cu-resistant isolates of bacteria from oats' rhizosphere (*Pseudomonas putida* A1, *Acinetobacter calcoaceticus* A6, and *Stenotrophomonas maltophilia* A2) (Andreazza et al. 2010). One of the most crucial functions of rhizosphere bacteria, such as rhizobia, is the defense of plant roots against toxic environments and other environmental stresses (Moretto et al. 2022). Redox processes that are mediated by microbes have a substantial impact on transformation of HMs into lesser or nontoxic forms. The extracellular electron conduit MtrABC, trans-outer membrane porin-cytochrome protein complex (Pcc), and outer membrane c-type cytochromes (OM c-Cyts) played vital roles in microbial metal reduction activities. The systems in *Shewanella* and *Geobacter* species have been thoroughly studied. Furthermore, certain enzymes may also oxidize HM. In the case of Cu efflux, multi-copper oxidases like CueO, CuiD, and/or CopR are substantially necessary. As opposed to Chr A of chromate reductase, this converts Cr^{+6} to Cr^{+3}. However, protein Mer A transforms Hg^{2+} into the less harmful form of Hg0 for Hg.

Under toxic HM stress, the expression of these proteins increases. HMT bacteria frequently contribute significantly to the detoxification of HM in plants (Mishra et al. 2017). In the treatment of heavy metals in polluted soils, the application of HMT-PGP (heavy metal-tolerant plant growth-promoting) microorganisms and their host plants can be both cost-efficient and beneficial to the environment. To put this technology into practice at a level where it can be commercialized, there is still a knowledge gap. When the contaminated sites are highly polluted, the accumulation of HMs in plant tissues frequently slows down the cleanup process. The application of HMT-PGP microorganisms with additives is proven to be the most effective in few situations where the soil is contaminated with numerous forms of heavy metals (Mishra et al. 2017).

4.4 Ricinus communis

One of these plant species, *Ricinus communis* (castor bean), has garnered a lot of attention due to its potential to accumulate metal ions and its ability to flourish in highly polluted soil (Boda et al. 2017). *R. communis* is also a great companion crop for rotation and an industrial crop with a variety of non-food uses. The ability to grow *R. communis* readily in various climates and use its biomass in biofuel production can both make heavy metal-contaminated soils productive and, albeit slowly, help to repair them. It denotes both a benefit to the economy and improved soil.

Under stress, the found bacterium produced a number of substances that aided plant growth, such as ligninolytic enzymes, siderophores, and indole acetic acid (IAA), which increased nutrient uptake and metal mineralization in plants (Banerjee et al. 2018). It is possible to effectively treat distillery wastewater using *R. communis* L. because of its capacity to grow in metals consisting of wastewater and remove organic pollutants and heavy metals. An environmentally favorable method for lowering organometallic contamination and safeguarding agricultural land is the application of *R. communis* L. (Tripathi et al. 2021).

Municipal solid waste (MSW) contributes to the emission of greenhouse gases (GHGs), the buildup of plastic in rivers, and other forms of environmental damage. Native microorganisms (*Pseudomonas*, *Flavobacterium*, *Bacillus*, *Nitrosomonas*, etc.) can provide possible methods for the remediation of contaminants by using new metabolic routes and channels (Sharma et al. 2022a, b).

4.5 Cyanobacteria

Different strategies are used by cyanobacteria to combat the harmful buildup of heavy metals. Metal homeostasis and resistance are significantly influenced by phytochelatins (PCs), which are high-affinity metal-binding proteins. Additionally, the cyanobacteria can tolerate HM stress via both enzymatic (catalase, superoxide dismutase, peroxidase) and nonenzymatic (carotenoids, glutathione) redox machinery. Numerous cyanobacteria have so far been identified for their potential use in heavy metal bioremediation including *Anabaena*, *Nostoc*, *Synechocystis*, *Synechococcus*, *Phormidium*, and *Oscillatoria* (Chakdar et al. 2022). Additionally, exo-polysaccharide (EPS) is one of the many parts of the cyanobacterial cell wall, which is the primary mechanism by which this blue-green algae biosorbs the metal ions (Al-Amin et al. 2021).

The usage of non-neutral, negatively charged EPS (EPS packed with numerous anionic functional groups) to be included as a suitable biosorbent is essential for heavy metal remediation by bacterial EPS (Gupta and Diwan 2017). Some of the commercially available bacteria that have been reported to have the necessary anionicity include gellan (*Sphingomonas paucimobilis*), alginate (*Azotobacter vinelandii*), hyaluronan (*Pseudomonas aeruginosa* and *Streptococci* attenuated strains), and xanthan (*Xanthomonas*). Because they are stress-tolerant, cyanobacteria including *Anabaena*, *Oscillatoria*, and *Spirogyra* can naturally grow in wastewater contaminated with heavy metals. Microbial treatment of metals offers a good and fresh viewpoint for the prevention of environmental pollution at the molecular level (Sharma et al. 2021).

4.6 Probiotics

Probiotics are living cells of bacteria that have qualities that are advantageous to human health. Although *Lactococcus*, *Pediococcus*, *Bacillus*, and yeasts are included as probiotics, *Lactobacillus* and *Bifidobacterium* are the two main probiotic families. Probiotics have a critical role in maintaining bodily health, according to research. Probiotics were noted for their potent ability to bind a variety of targets and expel them with feces. These substances could be arsenic, lead, cadmium, or aluminum (Abdel-Megeed 2021). Among gut microorganisms, *Lactobacilli* and *Bifidobacteria* species are mainly recognized to have the highest antioxidant activities. By modifying physiological conditions and intestinal permeability and increasing enzymes for metabolizing heavy metals, the gut microbiota seeks to detoxify the heavy metals (Arun et al. 2021).

The lactic acid bacteria (LABs), *Bifidobacterium* sp., and yeast (*Saccharomyces*) are the most often utilized probiotics (*S. cerevisiae* var. *boulardii*). These strains have a number of advantageous traits that can be crucial in the detoxification of xenobiotics (Średnicka et al. 2021). They have strong capability to bind, tolerate, or detoxify a high tolerance to acid and bile, a strong adhesion to the gut mucosa, and strong antioxidant or immunoregulatory capacities that allow them to adapt the changes in the gut environment brought through xenobiotics. The majority of peptidoglycans with minor quantity of lipoteichoic and teichoic acids, polysaccharides, and exo-polysaccharides make up the cell wall of LABs (Średnicka et al. 2021). Further investigation is needed to illustrate formerly unidentified molecular mechanisms of resistance that can help in explaining the role of *Bacillus* spp. in bioremediation process for the effective elimination of contaminants (Pal et al. 2021).

5 Role of Algae in Detoxification of Heavy Metals

The most covered technology for using plants to eliminate or detoxify toxins from the environment has emerged as phytoremediation (Bhandari et al. 2023). A wide variety of pollutants, including some of the most significant ones, such as pentachlorophenol (PCP), metals, radionuclides, fertilizers, petroleum hydrocarbons, and chlorinated solvents, are potentially amenable to phytoremediation (PAHs). Growing in popularity are phytoremediation techniques, which can be applied to clean up hazardous metal-contaminated locations for a much lower cost than more traditional techniques including soil replacement, solidification, and washing techniques (Chekroun and Baghour 2013; Bhandari et al. 2023).

Due to the large concentrations of sulfated polysaccharides and alginates in their cell walls, which metals have a significant affinity for, brown algae (Phaeophyta) are particularly effective metal accumulators. Exclusion, compartmentalization, the formation of complexes, and the synthesis of binding proteins like metallothioneins (MTs) or phytochelatins (PCs) and their translocation into vacuoles are just a few of the defense mechanisms that organisms employ in response to heavy metal stress (Table 2). As possible ligands for heavy metals, carboxylic and amino acids including malate, citrate, oxalate, histidine (His), citrate, nicotianamine (NA), and phosphate derivatives (phytate) have been linked to tolerance and detoxification (Singh and Chauhan 2011).

Although algae have a great deal of potential for HM accumulation, they are constrained by factors including low biomass and sensitivity to high HM concentrations. Some cutting-edge therapies, such as genetic alterations that increase biomass output, may be able to alleviate these issues (Ankit et al. 2022). The most prominent algae are discussed with respect to their heavy metal detoxification in the environment.

*5.1 Green Algae (*Scenedesmus obliquus*)*

It belongs to the family Chlorophyceae and is a genus of green algae. One of the most prevalent freshwater algae, *Scenedesmus*, is frequently cultivated for uses like oil extraction or treatment of wastewater (Abdel-Raouf et al. 2012). The cell wall of *Scenedesmus* species is incredibly thick, making them exceedingly resistant to digestion and predatory activity. The negatively charged groups on the surface of the algal cell wall can interact with positively charged metal ions at binding sites on functional groups at the outer layer of the algal cell wall. Thus, understanding systems like biosorption requires an understanding of their characteristics. The quick and reversible binding of ions from aqueous solutions to functional groups found on the surface of biomass is known as biosorption. According to reports, one of the main participants in the adsorption of heavy metals is the carboxyl group. The heavy metals (Cd^{2+}) can be bound by the carboxylic moieties in one of two ways: either by substituting H^+ ion in the carboxy group (type-I) or after prolonged exposure, by forming carboxylates (type-II). Numerous methods, including as Fourier transform infrared spectroscopy (FTIR) and X-ray photoelectron spectroscopy (XPS), demonstrate the interaction with heavy metals during adsorption process and have been used to confirm the influence of these groups (Spain et al. 2021).

Synthetic biology is frequently studied in microalgae as a method to increase biomass and chemical output (Sattayawat et al. 2021). *Scenedesmus obliquus* is an extremely adaptable microalga used as a starting point for the generation of biofuels. *S. obliquus* may flourish in industrial wastewater from many sources, demonstrating good adaptability. *Scenedesmus* has been used as a bioremediation agent to enhance the quality of polluted water by removing inorganic nutrients. Various studies have shown that the microalga *S. obliquus* is an efficient agent for bio-removal of heavy metals, making it one of the most significant producers of aquatic biomass and potentially a substantial source of renewable energy (bioremediation). *S. obliquus* has the capacity to rapidly synthesize huge amounts of carbohydrates. *S. obliquus* is a potential feedstock choice for the production of bioethanol. Nitrogen and phosphorus removal from wastewater by microalgae has a lot of potential.

Whereas wastewater treatment using algae supplies the oxygen for aerobic bacteria, mechanical aeration used in wastewater treatment requires significant energy costs. Aerobic bacteria need oxygen to consume the organic components in the wastewater. These are exemplary species for bioethanol production as microalgae used as source of carbohydrates for the production fermentable sugars (Hamouda et al. 2016). *Chlorella vulgaris*, *S. obliquus*, *Spirulina* sp., and *Porphyridium purpureum* have the potential for the production of high levels of carbohydrates in lieu of lipids as reserve polymers.

Artificial sweeteners (ASs) are growing organic polar contaminants that can chelate with heavy metals in the environment. Currently, the artificial sweeteners have been detected widely in aquatic environments. It is unclear whether ASs influence the toxicity and mobility of heavy metals associated with sediments. Additionally, heavy metal discharges from the lakes' sediment and their toxicity to *Scenedesmus obliquus* (green algae) were observed in two commonly detected artificial sweeteners (acesulfame/sucralose and humic acid) (Liu et al. 2020).

*5.2 Brown Algae (*Padina gymnospora*)*

The medical value of seaweeds has been widely documented by numerous researchers from around the world. They may provide the biologically active metabolites to pharmaceutical and drug industries that are not present in other organisms (Lomartire and Gonçalves 2022). Due to the unfavorable environmental conditions, they dwell in their body tissues, thus creating these antioxidant chemicals. Since *Padina* extracts have hepatoprotective, hypolipidemic, antioxidant, and anti-inflammatory properties, they can be utilized to treat hepatotoxicity.

They serve as reliable predictors of changes in many variables, including light, temperature, nutrients, and numerous other pollutants in the aquatic environment. With rising Cd concentrations, *Padina* (Phaeophyta) displayed a decline in specific growth rate and chlorophyll levels, indicating Cd pollution. They can be used as biological indicators of environmental contamination, and in the restoration of the coastal marine ecosystem, macroalgae like *Padina* are significant ecologically (Ansari et al. 2019). The brown algae can accumulate significant quantities of metals, making them useful as biosorbents and heavy metal biomonitors. The primary chemical mechanism for the metal sequestration has been identified as cation exchange by cell wall polysaccharides. When exposed to high metals, *P. gymnospora* overproduces cell wall polysaccharides as a defense mechanism (Andrade et al. 2010). After the majority of its components were extracted using 70% methanol, the residue of *P. gymnospora* (RPG) was assessed as a potentially effective adsorbent for the elimination of Cd^{2+} and Cr^{3+} from wastewater (Mohamed et al. 2019). In comparison to other algal species, the brown algae represent a class of biosorbents that are substantially more effective. Fortunately, they are less expensive, have superior metal ion uptake, and are a viable treatment for algae in wastewater bioremediation, which can be used in a variety of beneficial applications. As a result, the cultivation of vast algal biomass and the treatment of it with industrial wastewater became possible through the use of algae (Sreevani and Anierudhe 2022).

*5.3 Red Algae (***Palmaria palmata***)*

The removal of copper ions from aqueous solutions can be accomplished by using red algae and beer draff. The pH level and starting Cu (II) concentration had a significant impact on the materials' ability to adsorb Cu (II). For both materials, the ideal pH range was between 5 and 6. Hydrogen ions have participated in the copper adsorption process, as evidenced by the correlation between final pH and adsorption capacity (Li et al. 2011). A popular and highly prized red macroalga is *Palmaria palmata*, sometimes known as dulse. When nutrients are scarce, *P. palmata* can store nitrogen, phosphate, and carbon (it can use both CO_2 and HCO_3) to maintain growth. In addition to food accessibility, irradiance is crucial for nutrient storage in *P. palmata*. When compared to plants cultivated under high irradiation, individuals adapted to low irradiance had higher nitrogen and phosphate levels but lower soluble carbohydrate contents. A portion of nitrogen is accumulated as light-absorbing compound phycoerythrin to promote photosynthesis in low light situations and to boost growth in nutrient-rich environments. In order to guard against oxidative stress and the breakdown of biomolecules (DNA, proteins), *P. palmata* also produces and accumulates photoprotective compounds, especially mycosporine-like amino acids (Stevant et al. 2023).

5.4 Spirulina platensis

Spirulina platensis dry biomass is just as capable as other familiar algae employed in the removal of heavy metals and has enormous potential for removing cadmium ions from aqueous solution. One of the most hazardous elements present in aquatic habitats is cadmium. Many ailments are brought on by this metal, which tends to accumulate in fish and plants that use photosynthesis for energy. In order to decrease any health hazards, it must be removed from our surroundings. *S. platensis*, a microalga (cyanobacterium), was employed as a biosorbent to remove cadmium ions (Cd^{2+}) from aqueous solutions. A particularly strong contender for removing heavy metals from aquatic environments is *S. platensis*. The procedure is doable, dependable, and environmentally safe (Al-Homaidan et al. 2015). To improve the water quality and separate the heavy metals contaminated in the wastewater, *S. platensis* can be cultured on the wastewater. *S. platensis* may be able to discard heavy metals (mercuric chloride) from water. Given that mercuric poisoning is one of the main causes of Minamata disease, which first affect fish and then humans, the current study offers a simple and affordable solution to the problem of water pollution. Additional research can be used to better understand the mechanism behind *S. platensis*'s mercuric chloride absorption and phytoremediation (Sharma et al. 2019).

5.5 Algae as a Green Technology for Heavy Metals

Algae offer a sustainable and eco-friendly alternative for the remediation of heavy metals (Ankit et al. 2022). However, the algal biomass could be considered as alternative to the traditional adsorbent materials for treatment of heavy metals due to the following:

1. Algae can flourish in various environmental factors.
2. They have rapid growth rates as a result of their quick cell cycles.
3. When compared to other biomass species, they require minimal nutrient concentrations.
4. They may be cultivated without using farmland.
5. Even though wastewater requirements can be reduced, algal cultivation can be accomplished.
6. They can also be applied to different tasks, such as producing biofuel. For the conversion of wastewater into biomass that can be utilized to produce biofuel, they can use both inorganic and organic matter (Maurice 2022).

The heavy metal ions in wastewater surrounding the algae are replaced with elemental ions (Na^+, Ca^{2+}, and K^+) held on the cell surface. The viability of this method is dependent on critical elements like metal selectivity and regeneration potential (Ankit et al. 2022). Heavy metal ions bind to the cell surface by physicochemical exchanges, and selectivity in biosorption is typically limited. The elemental ions retained on the cell surface of the algae, like Ca^{2+}, Na^{+}, and K^{+}, exchange with heavy metal ions in the wastewater surrounding the algae. However, chemical modification of biomass, such as cross-linking with epichlorohydrin, can boost selectivity. Commercially accessible algal biomass is not made for phyco-remediation applications; hence, its performance might not be at its best. When dealing with lower concentrations of heavy metals, use dead biomass conditions based on the ability of living cultures to perform phyco-remediation (Salama et al. 2019).

5.6 *Microalgae Show Various Mechanisms against Heavy Metals*

Algae have unique mechanisms for protecting themselves from the toxicity of heavy metals. They carry out complications on ion exchange and chelation and create reducing agents, thus resulting in immobilization of HMs (Chugh et al. 2022). With the aid of these mechanisms, microalgae are able to combat the harmfulness of heavy metals, which impair cellular metabolism, improper physiological and structural development, nucleic acid faults, and cell membrane flaws. These self-defense systems aid algae in surviving in wastewater that contains greater concentrations of heavy metals and aid in the removal of hazardous HMs. The four primary forms of self-defense mechanisms used by algae are biosorption, compartmentalization, biotransformation, and metal exclusion (Greeshma et al. 2022).

5.6.1 Extracellular Uptake (Biosorption)

In comparison to other algae, brown algae are found to remove most of the heavy metals. Various functional groups such as alginic acid, cellulose, and other polymers make up the cell wall of brown algae. The hydroxyl and carboxyl group present on the surface of brown algae (*Fucus vesiculosus*) enables it for zinc extraction (Chugh et al. 2022). Instead, alkaline pH encourages the uptake of metals. As nonliving and living biomass can perform this activity, it is very efficient and economical and does not produce any hazardous by-products. Furthermore, it has no impact on any aspect of cellular metabolism or the buildup of coproduct in algae.

5.6.2 Intracellular Uptake (Bioaccumulation and Compartmentalization)

This activity depends on metabolism and can only be carried out by live cells. As a result, microalgal cells are using a gradual intracellular active bioaccumulation self-protection approach (Amaro et al. 2023). These heavy metal ions are easily absorbed by microalgal cells because of their small size, which corresponds to high surface to volume ratio. Sulfur (S), iron (Fe), and nitrogen (N) are found to aid in the bioaccumulation of metal ions. Potential bioindicators and bioremediators include this species. Both *Skeletonema costatum* and the green alga *Tetraselmis suecica* have demonstrated the capacity to bioaccumulate cadmium and copper (Chugh et al. 2022).

5.6.3 Metal Detoxification

The biotransformation, which involves chelators that can change heavy metals into nontoxic molecules, is the third type of self-defense mechanism displayed by microalgal cells (Danouche et al. 2021). Chelating agents are organic substances that generate complexes by joining with metal ions. The term "complexation" also refers to this process. It has been noted that algae release chelating compounds to protect themselves from hazardous metals. The metal ions are joined by these chelating agents and changed into a nontoxic form. *Phaeodactylum tricornutum* produced metallothioneins that could detoxify copper ions. *P. tricornutum* releases metallothioneins and combines cadmium and sulfide ions to generate these complexes. *Chlamydomonas reinhardtii* releases ROS and hydrophilic antioxidants as a defence mechanism against cadmium and chromium poisoning (Chugh et al. 2022).

5.6.4 Metal Exclusion

In this, the microalgal cells use efflux transport to eliminate heavy metals. The microalgal cell employs this defence mechanism in addition to others. With this method, the concentration of heavy metals inside the cell is maintained. In microalgae, this active transport system aids in the removal of heavy metals that are hazardous to cells. Ion-selective transporters are used to carry out this exclusion transfer. In order to lessen the influx of heavy metal, it has also been discovered that algae change the cell membrane. Other microbial organisms, such as bacteria or fungi, also frequently employ this self-defense strategy, but additional research is needed to fully understand how heavy metals are removed from algae (Chugh et al. 2022).

6 Conclusion

Nature is capable of renewing itself in multiple ways. The actions of microorganisms through heavy metal detoxification can manage environmental contamination, especially heavy metal pollution. Bacteria, fungi, and algae are the most prevalent detoxifying microorganisms in heavy remediation. For the purpose of cleaning up polluted places, microbe-based bioremediation of hazardous organics is by far the most cost-effective, effective, sustainable, and environmentally friendly method. For resistance to heavy metal stress, microbes have evolved various active defense mechanisms like bioaccumulation, efflux transport, metal chelation, and biosorption. Nowadays, numerous researchers have focused on determining the level of tolerance of various microbes to the heavy metals through different mechanisms. Therefore, a wide range of microbes played a significant role in decontamination of heavy metals from the environment.

References

Abdelbary S, Elgamal MS, Farrag A (2019) Trends in heavy metals tolerance and uptake by *P. aeruginosa*. 122. https://doi.org/10.5772/InTech.77869

Abdel-Megeed RM (2021) Probiotics: a promising generation of heavy metal detoxification. Biol Trace Elem Res 199:2406–2413. https://doi.org/10.1007/s12011-020-02350-1

Abdel-Raouf N, Al-Homaidan AA, Ibraheem IBM (2012) Microalgae and wastewater treatment. Saudi J Biol Sci 19:257–275. https://doi.org/10.1016/j.sjbs.2012.04.005

Abioye OP, Oyewole OA, Oyeleke SB, Adeyemi MO, Orukotan AA (2018) Biosorption of lead, chromium and cadmium in tannery effluent using indigenous microorganisms. Brazilian J Biol Sci 5:25–32. https://doi.org/10.21472/bjbs.050903

Al-Amin A, Parvin F, Chakraborty J (2021) Cyanobacteria mediated heavy metal removal: a review on mechanism, biosynthesis, and removal capability. Environ Technol Rev 10:44–57. https://doi.org/10.1080/21622515.2020.1869323

Akanchise T, Boakye S, Borquaye LS, Dodd M, Darko G (2020) Distribution of heavy metals in soils from abandoned dump sites in Kumasi, Ghan. Sci Afr 10:e00614. https://doi.org/10.1016/j.sciaf.2020.e00614

Alfarras FA, Al-Fahdawi HM, Albayaty MK (2022) Heavy metal resistance ability of *Pseudomonas* species isolated from sludge and sewage in Iraq. Arch Razi Institute 77(3):1041–1047. https://doi.org/10.22092/ARI.2022.357399.2036

Al-Homaidan AA, Alabdullatif JA, Al-Hazzani AA, Al-Ghanayem AA, Alabbad AF (2015) Adsorptive removal of cadmium ions by *Spirulina platensis* dry biomass. Saudi J Biol Sci 22:795–800. https://doi.org/10.1016/j.sjbs.2015.06.010

Alotaibi BS, Khan M, Shamim S (2021) Unravelling the underlying heavy metal detoxification mechanisms of *Bacillus* species. Microorganisms 9(8):1628. https://doi.org/10.3390/microorganisms9081628

Amaro HM, Salgado EM, Nunes OC, Pires JCM, Esteves AF (2023) Microalgae systems - environmental agents for wastewater treatment and further potential biomass valorisation. J Environ Manag 337:117678. https://doi.org/10.1016/j.jenvman.2023.117678

Andrade LR, Leal RN, Noseda M, Duarte ME, Pereira MS, Mourão PA, Farina M, Amado Filho GM (2010) Brown algae overproduce cell wall polysaccharides as a protection mechanism against the heavy metal toxicity. Mar Pollut Bull 60:1482–1488. https://doi.org/10.1016/j.marpolbul.2010.05.004

Andreazza R, Okeke BC, Lambais MR, Bortolon L, Bastos de Melo GW, Camargo FAO (2010) Bacterial stimulation of copper phytoaccumulation by bioaugmentation with rhizosphere bacteria. Chemosphere 81:1149–1154. https://doi.org/10.1016/j.chemosphere.2010.09.047

Ankit, Bauddh K, Korstad J (2022) Phycoremediation: use of algae to sequester heavy metals. Hydrobiology 1(3):288–303. https://doi.org/10.3390/hydrobiology1030021

Ansari AA, Ghanem SM, Naeem M (2019) Brown Alga *Padina*: a review. Int J Bot Stud 4:1–3

Arun KB, Madhavan A, Sindhu R, Emmanual S, Binod P, Pugazhendhi A, Sirohi R, Reshmy R, Awasthi MK, Gnansounou E, Pandey A (2021) Probiotics and gut microbiome - prospects and challenges in remediating heavy metal toxicity. J Hazard Mater 420:126676. https://doi.org/10.1016/j.jhazmat.2021.126676

Ayanbenro AS, Babalola OO (2017) A new strategy for heavy metal polluted environments: a review of microbial biosorbents. Int J Environ Res Public Health 14(1):94. https://doi.org/10.3390/ijerph14010094

Bakshi S, Banik C, He Z (2018) The impact of heavy metal contamination on soil health. In: Reicosky D (ed) Managing soil health for sustainable agriculture, vol 2. Burleigh Dodds Science Publishing, pp 63–95

Banerjee A, Jhariya MK, Yadav DK, Raj A (2018) Micro-remediation of metals: a new frontier in bioremediation. In: Hussain C (ed) Handbook of environmental materials management. Springer, Cham. https://doi.org/10.1007/978-3-319-58538-3_10-1

Bano A, Hussain J, Akbar A, Mehmood K, Anwar M, Hasni MS, Ullah S, Sajid S, Ali I (2018) Biosorption of heavy metals by obligate halophilic fungi. Chemosphere 199:218–222. https://doi.org/10.1016/j.chemosphere.2018.02.043

Bhandari G, Dhasmana A, Chaudhary P, Gupta S, Gangola S, Gupta A, Rustagi S, Shende SS, Rajput VD, Minkina T, Malik S, Slama P (2023) A perspective review on green nanotechnology in agro-ecosystems: opportunities for sustainable agricultural practices & environmental remediation. Agriculture 13:668. https://doi.org/10.3390/agriculture13030668

Boda RK, Majeti NVP, Suthari S (2017) *Ricinus communis* L. (castor bean) as a potential candidate for revegetating industrial waste contaminated sites in peri-urban Greater Hyderabad: remarks on seed oil. Environ Sci Pollut Res 24(24):19955–19964. https://doi.org/10.1007/s11356-017-9654-5

Briffa J, Sinagra E, Blundell R (2020) Heavy metal pollution in the environment and their toxicological effects on humans. Heliyon 6(9):e04691. https://doi.org/10.1016/j.heliyon.2020.e04691

Cardoso SL, Costa CSD, Nishikawa E, da Silva MGC, Vieira MGA (2017) Biosorption of toxic metals using the alginate extraction residue from the brown algae *Sargassum filipendula* as a natural ion-exchanger. J Clean Prod 165:491–499. https://doi.org/10.1016/j.jclepro.2017.07.114

Chakdar H, Thapa S, Srivastava A, Shukla P (2022) Genomic and proteomic insights into the heavy metal bioremediation by cyanobacteria. J Hazard Mater 424:127609. https://doi.org/10.1016/j.jhazmat.2021.127609

Chaudhary P, Ahamad L, Chaudhary A, Kumar G, Chen F-S, Chen S (2023b) Nanoparticle-mediated bioremediation as a powerful weapon in the removal of environmental pollutants. J Environ Chem Eng 11:109591. https://doi.org/10.1016/j.jece.2023.109591

Chaudhary P, Chen S, Rajput VD, Jaiswal DK (2023c) Editorial: bioinoculants with nano-compounds to improve soil health: a step toward sustainable agriculture. Front Environ Sci 11:1270002. https://doi.org/10.3389/fenvs.2023.1270002

Chaudhary P, Chen S, Rajput VD, Jaiswal DK (eds) (2023d) Bioinoculants with nano-compounds to improve soil health: a step toward sustainable agriculture. Frontiers Media SA, Lausanne. https://doi.org/10.3389/978-2-8325-3287-4

Chaudhary P, Xu M, Ahamad L, Chaudhary A, Kumar G, Adeleke BS, Verma KK, Hu D-M, Širić I, Kumar P, Popescu SM, Abou Fayssal S (2023a) Application of synthetic consortia for improvement of soil fertility, pollution remediation, and agricultural productivity: a review. Agronomy 13:643. https://doi.org/10.3390/agronomy13030643

Chekroun KB, Baghour M (2013) The role of algae in phytoremediation of heavy metals: a review. J Mater Environ Sci 4(6):873–880

Chen A, Zeng G, Chen G, Liu L, Shang C, Hu X, Lu L, Chen M, Zhou Y, Zhang Q (2014) Plasma membrane behavior, oxidative damage, and defense mechanism in *Phanerochaete chrysosporium* under cadmium stress. Process Biochem 49:589–598. https://doi.org/10.1016/j.procbio.2014.01.014

Chugh M, Kumar L, Shah MP, Bharadvaja N (2022) Algal bioremediation of heavy metals: an insight into removal mechanisms, recovery of by-products, challenges, and future opportunities. Energy Nexus 7:100129. https://doi.org/10.1016/j.nexus.2022.100129

Danouche M, Ghachtouli NE, Arroussi HE (2021) Phycoremediation mechanisms of heavy metals using living green microalgae: physicochemical and molecular approaches for enhancing selectivity and removal capacity. Heliyon 7:e07609. https://doi.org/10.1016/j.heliyon.2021.e07609

Dhaigude V, Naliyapara HB, Chaudhary P, Fayssal SA, Nadda AK, Minkina T, Rajput VD (2023) Role of Nanobiosensors in agriculture advancement. In: Chaudhary P, Chaudhary A, Nadda

AK, Khati P (eds) Advances in nanotechnology for smart agriculture: techniques and applications, 1st edn. CRC Press. https://doi.org/10.1201/9781003345565-9
Diep P, Mahadevan R, Yakunin AF (2018) Heavy metal removal by bioaccumulation using genetically engineered microorganisms. Front Bioeng Biotechnol 6:157. https://doi.org/10.3389/fbioe.2018.00157
Fazli MM, Soleimani N, Mehrasbi M, Darabian S, Mohammadi J, Ramazani A (2015) Highly cadmium tolerant fungi: their tolerance and removal potential. J Environ Health Sci Eng 13(19):1–9. https://doi.org/10.1186/s40201-015-0176-0
Gajewska J, Azzahra NA, Bingöl ÖA, Izbiańska-Jankowska K, Jelonek T, Deckert J, Floryszak-Wieczorek J, Arasimowicz-Jelonek M (2020) Cadmium stress reprograms ROS/RNS homeostasis in *Phytophthora infestans* (Mont.) de Bary. Int J Mol Sci 21:8375. https://doi.org/10.3390/ijms21218375
Gajewska J, Floryszak-Wieczoreket J, Sobieszczuk-Nowicka E, Mattoo A, Arasimowicz-Jelonek M (2022) Fungal and oomycete pathogens and heavy metals: an inglorious couple in the environment. IMA Fungus 13(1):1–20. https://doi.org/10.1186/s43008-022-00092-4
Ghosh S, Bhattacharya J, Nitnavare R, Webster TJ (2022) Heavy metal removal by *Bacillus* for sustainable agriculture. Bacilli Agrobiotechnol 454:1–30. https://doi.org/10.1007/978-3-030-85465-2_1
Grąz M, Pawlikowska-Pawlęga B, Jarosz-Wilkołazka A (2011) Growth inhibition and intracellular distribution of Pb ions by the white-rot fungus *Abortiporus biennis*. Int Biodeterior Biodegradation 65(1):124–129
Greeshma K, Kim H-S, Ramanan R (2022) The emerging potential of natural and synthetic algae-based microbiomes for heavy metal removal and recovery from wastewaters. Environ Res 215:114238. https://doi.org/10.1016/j.envres.2022.114238
Gu Y, Xu W, Liu Y, Zeng G, Huang J, Tan X, Jian H, Hu X, Li F, Wang D (2015) Mechanism of Cr (VI) reduction by *Aspergillus niger*: enzymatic characteristic, oxidative stress response, and reduction product. Environ Sci Pollut Res 22(8):6271–6279. https://doi.org/10.1007/s11356-014-3856-x
Gupta N, Gaurav SS, Kumar A (2013) Molecular basis of aluminium toxicity in plants: a review. Am J Plant Sci 4:21–37. https://doi.org/10.4236/ajps.2013.412A3004
Gupta P, Diwan B (2017) Bacterial exopolysaccharide mediated heavy metal removal: a review on biosynthesis, mechanism and remediation strategies. Biotechnol Rep 13:58–71. https://doi.org/10.1016/j.btre.2016.12.006
Hamouda RA, Yeheia DS, Hussein MH, Hamzah HA (2016) Removal of heavy metals and production of bioethanol by green alga Scenedesmus obliquus grown in different concentrations of wastewater. Sains Malaysiana 45(3):467–476
Husain R, Vikram N, Yadav G (2022) Chapter 9—Microbial bioremediation of heavy metals by Marine bacteria. Development in wastewater treatment research and processes 177–203. DOI: https://doi.org/10.1016/B978-0-323-85839-7.00014-1
Igiri BE, Okoduwa SIR, Idoko GO, Akabuogu EP, Adeyi AO, Ejiogu IK (2018) Toxicity and bioremediation of heavy metals contaminated ecosystem from tannery wastewater, a review. J Toxicol 2018:2568038. https://doi.org/10.1155/2018/2568038
Infante JC, De Arco RD, Angulo ME (2014) Removal of lead, mercury and nickel using the yeast Saccharomyces cerevisiae. Revista MVZ Cordoba 19(2):4141–4149
Jaishankar M, Tseten T, Anbalagan N, Mathew BB, Beeregowda KN (2014) Toxicity, mechanism and health effects of some heavy metals. Interdiscip Toxicol 7(2):60–72
Jarosławiecka AK, Piotrowska-Seget Z (2022) The effect of heavy metals on microbial communities in industrial soil in the area of Piekary Śląskie and Bukowno (Poland). Microbiol Res 13:626–642. https://doi.org/10.3390/microbiolres13030045
Kalkan S (2022) Heavy metal resistance of marine bacteria on the sediments of the Black Sea. Mar Pollut Bull 179:113652. https://doi.org/10.1016/j.marpolbul.2022.113652

Khan I, Ali M, Aftab M, Shakir S, Qayyum S, Haleem KS, Tauseef I (2019) Mycoremediation: a treatment for heavy metal-polluted soil using indigenous metallotolerant fungi. Environ Monit Assess 191:622. https://doi.org/10.1007/s10661-019-7781-9

Kumar A, Sharma A, Chaudhary P, Gangola S (2021a) Chlorpyrifos degradation using binary fungal strains isolated from industrial waste soil. Biologia 10:3071–3081. https://doi.org/10.1007/s11756-021-00816-8

Kumar G, Lal S, Maurya SK, Bhattacherjee AK, Chaudhary P, Gangola S, Rajan S (2021b) Exploration of *Klebsiella pneumoniae* M6 for paclobutrazol degradation, plant growth attributes, and biocontrol action under subtropical ecosystem. PLoS One 16:e0261338. https://doi.org/10.1371/journal.pone.0261338

Li C, Zhou K, Qin W, Tian C, Qi M, Yan X, Han W (2019) A review on heavy metals contamination in soil: effects, sources, and remediation techniques. Soil Sediment Contam 28(4):380–394

Li Y, Helmreich B, Horn H (2011) Biosorption of Cu (II) ions from aqueous solution by red alga (*Palmaria palmata*) and beer draff. Mater Sci Appl 2(2):70–80. https://doi.org/10.4236/msa.2011.22010

Lin Y, Xiao W, Ye Y, Wu C, Hu Y, Shi H (2020) Adaptation of soil fungi to heavy metal contamination in paddy fields—A case study in eastern China. Environ Sci Pollut Res 27(22):27819–27830. https://doi.org/10.1007/s11356-020-09049-9

Liu P, Wei M, Zhang J, Wang R, Li B, Chen Q, Weng Q (2018) Changes in mycelia growth, sporulation, and virulence of *Phytophthora capsici* when challenged by heavy metals (Cu^{2+}, Cr^{2+} and Hg^{2+}) under acid pH stress. Environ Pollut 235:372–380. https://doi.org/10.1016/j.envpol.2017.12.100

Liu Y, Hu H, Wang Y, Wang L, Feng Y (2020) Effects of heavy metals released from sediment accelerated by artificial sweeteners and humic acid on a green algae *Scenedesmus obliquus*. Sci Total Environ 729:138960. https://doi.org/10.1016/j.scitotenv.2020.138960

Lomartire S, Gonçalves AMM (2022) An overview of potential seaweed-derived bioactive compounds for pharmaceutical applications. Mar Drugs 20(2):141. https://doi.org/10.3390/md20020141

Mathivanan K, Chandirika JU, Vinothkanna A, Yin H, Liu X, Meng D (2021) Bacterial adaptive strategies to cope with metal toxicity in the contaminated environment—a review. Ecotoxicol Environ Saf 226:112863. https://doi.org/10.1016/j.ecoenv.2021.112863

Maurice N (2022) Algal biofuel production from municipal waste waters. Technological advancement in algal biofuels production. Technol Adv Algal Biofuels Prod 63:193–236. https://doi.org/10.1007/978-981-19-6806-8

Miransari M (2016) Soybean production and heavy metal stress. In: Abiotic and biotic stresses in soybean production, 1st edn. Academic Press, pp 197–216

Mirlahiji SG, Eisazadeh K (2014) Bioremediation of uranium via *Geobacter* spp. J Res Dev 1(12):52–58

Mishra J, Singh R, Arora NK (2017) Alleviation of heavy metal stress in plants and remediation of soil by rhizosphere microorganisms. Front Microbiol 8:1706. https://doi.org/10.3389/fmicb.2017.01706

Mitra S, Chakraborty AJ, Tareq AM, Emran TB, Nainu F, Khusro A, Idris AM, Khandaker MU, Osman H, Alhumaydhi FA, Simal-Gandara J (2022) Impact of heavy metals on the environment and human health: novel therapeutic insights to counter the toxicity. J King Saud Unive Sci 34:101865. https://doi.org/10.1016/j.jksus.2022.101865

Mohamed HS, Soliman NK, Abdelrheem DA, Ramadan AA, Elghandour AH, Ahmed SA (2019) Adsorption of Cd^{2+} and Cr^{3+} ions from aqueous solutions by using residue of *Padina gymnospora* waste as promising low-cost adsorbent. Heliyon 5(3):e01287. https://doi.org/10.1016/j.heliyon.2019.e01287

Moretto C, Castellane TCL, Leonel TF, Campanharo JC, de Macedo Lemos EG (2022) Bioremediation of heavy metal-polluted environments by non-living cells from rhizobial isolates. Environ Sci Pollut Res 29:46953–46967

Ouda SM (2014) Antifungal activity of silver and copper nanoparticles on two plant pathogens. *Alternaria alternata* and *Botrytis cinerea*. Res J Microbiol 9:34–42

Pal A, Bhattacharjee S, Saha J, Sarkar M, Mandal P (2021) Bacterial survival strategies and responses under heavy metal stress: a comprehensive overview. Crit Rev Microbiol 48:327–355. https://doi.org/10.1080/1040841X.2021.1970512

Pande V, Pandey SC, Sati D, Bhatt P, Samant M (2022) Microbial interventions in bioremediation of heavy metal contaminants in agroecosystem. Front Microbiol 13:824084. https://doi.org/10.3389/fmicb.2022.824084

Papirio S, Frunzo L, Mattei MR, Ferraro A, Race M, D'Acunto B, Pirozzi F, Esposito G (2017) Heavy metal removal from wastewaters by biosorption: mechanisms and Modeling. In: Rene E, Sahinkaya E, Lewis A, Lens P (eds) Sustainable heavy metal remediation. Environmental chemistry for a sustainable world, vol 8. Springer, Cham. https://doi.org/10.1007/978-3-319-58622-9_2

Paranthaman SR, Karthikeyan B (2015) Bioremediation of heavy metal in paper mill effluent using *Pseudomonas* spp. J Microbiol Biomed Res 1:1–5

Parveen H, Chaudhary A, Chaudhary P, Sultana R, Khati P, Kumar G, Khati P, Rana M, Bhatt P (2022) Omics approaches for the production of the microbial enzymes and applications. In: Bhatt P (ed) Industrial applications of microbial enzymes. Ist edition. CRC Press. https://doi.org/10.1201/9781003202998-16

Patniak P (2010) Handbook of environmental analysis. Chemical pollutants in air, water, soil and solid wastes, 3rd edn. CRC Press Taylor and Francis Group, Boca Raton, p 770

Priyadarshini E, Priyadarshini SS, Cousins BG, Pradhan N (2021) Metal-fungus interaction: review on cellular processes underlying heavy metal detoxification and synthesis of metal nanoparticles. Chemosphere 274:129976. https://doi.org/10.1016/j.chemosphere.2021.129976

Priyalaxmi R, Murugan A, Raja P, Raj KD (2014) Bioremediation of cadmium by *Bacillus safensis* (JX126862), a marine bacterium isolated from mangrove sediments. Int J Curr Microbiol App Sci 3(12):326–335

Rizvi A, Ahmed B, Khan MS, Rajput VD, Umar S, Minkina T, Lee J (2022) Maize associated bacterial microbiome linked mitigation of heavy metal stress: a multidimensional detoxification approach. Environ Exp Bot 200:104911. https://doi.org/10.1016/j.envexpbot.2022.104911

Salama ES, Roh HS, Dev S, Khan MA, Abou-Shanab RAI, Chang SW, Jeon BH (2019) Algae as a green technology for heavy metals removal from various wastewater. World J Microbiol Biotechnol 35(5):75. https://doi.org/10.1007/s11274-019-2648-3

Sattayawat P, Yunus IS, Noirungsee N, Mukjang N, Pathom-Aree W, Pekkoh J, Pumas C (2021) Synthetic biology-based approaches for microalgal bio-removal of heavy metals from wastewater effluents. Front Environ Sci. https://doi.org/10.3389/fenvs.2021.778260

Sharma A, Manjari, Kaur KP, Chatli AS (2019) Bioremediation of heavy metal (HgCl2) from water by *Spirulina platensis*. Int J Sci Dev Res 4:2455–2631

Sharma P, Dutta D, Udayan A, Singh P, Singh B, Singh RP (2022a) Role of microbes in bioaccumulation of heavy metals in municipal solid waste: impacts on plant and human being. Environ Pollut 305:119248. https://doi.org/10.1016/j.envpol.2022.119248

Sharma P, Kumar S, Pandey A (2021) Bioremediated techniques for remediation of metal pollutants using metagenomics approaches: a review. J Environ Chem Eng 9(6):105684. https://doi.org/10.1016/j.jece.2021.105684

Sharma P, Parakh SK, Singh SP, Parra-Saldívar R, Kim SH, Varjani S, Tong YW (2022b) A critical review on microbes-based treatment strategies for mitigation of toxic pollutants. Sci Total Environ 834:155444. https://doi.org/10.1016/j.scitotenv.2022.155444

Shekar CHC, Sammaiah D, Shasthree T, Reddy KJ (2011) Effect of mercury on tomato growth and yield attributes. Int J Pharm Bio Sci 2:B358–B364

Shen X, Chi Y, Xiong K (2019) The effect of heavy metal contamination on humans and animals in the vicinity of a zinc smelting facility. PLoS One 14(10):e0207423. https://doi.org/10.1371/journal.pone.0207423

Siddiquee S, Rovina K, Azad SA, Naher L, Suryani S, Chaikaew P (2015) Heavy metal contaminants removal from wastewater using the potential filamentous fungi biomass: a review. J Microb Biochem Technol 7:384–395. https://doi.org/10.4172/1948-5948.1000243

Singh D, Chauhan SK (2011) Organic acids of crop plants in aluminium detoxification. Curr Sci 100:1509–1515

Singh S, Chaudhary P, Sharma A (2023) Efficient decolourization of toluidine blue and methylene blue using *Bacillus cereus*. Vegetos. https://doi.org/10.1007/s42535-023-00747-6

Sinha SN, Biswas M, Paul D, Rahaman S (2011) Biodegradation potential of bacterial isolates from tannery effluent with special reference to hexavalent chromium. Biotechnol Bioinformatics Bioeng 1(3):381–386

Sinha SN, Paul D (2014) Heavy metal tolerance and accumulation by bacterial strains isolated from waste water. J Chem Biol Phys Sci 4(1):812–817

Spain O, Plöhn M, Funk C (2021) The cell wall of green microalgae and its role in heavy metal removal. Physiol Plant 173(2):526–539. https://doi.org/10.1111/ppl.13405

Średnicka P, Juszczuk-Kubiak E, Wójcicki M, Akimowicz M, Roszko ML (2021) Probiotics as a biological detoxification tool of food chemical contamination: a review. Food Chem Toxicol 153:112306. https://doi.org/10.1016/j.fct.2021.112306

Sreevani K, Anierudhe VV (2022) Reduction of heavy metal pollution using brown algae: a review. J Nanosci Res Rep 4:140. https://doi.org/10.47363/JNSRR/2022(4)140

Stevant P, Schmedes PS, Gall LL, Wegeberg S, Dumay J, Rebours C (2023) Concise review of the red macroalga dulse, *Palmaria palmata* (L.) Weber & Mohr. J Appl Phycol 35:523–550. https://doi.org/10.1007/s10811-022-02899-5

Tarekegn MM, Salilih FZ, Ishetu AI (2020) Microbes used as a tool for bioremediation of heavy metal from the environment. Cogent Food Agric 6. https://doi.org/10.1080/23311932.2020.1783174

Tigini V, Prigione V, Giansanti P, Mangiavillano A, Pannocchia A, Varese GC (2010) Fungal biosorption, an innovative treatment for the decolourisation and detoxification of textile effluents. Water 2(3):550–565. https://doi.org/10.3390/w2030550

Tripathi S, Sharma P, Purchase D, Chandra R (2021) Distillery wastewater detoxification and management through phytoremediation employing *Ricinus communis* L. Bioresour Technol 333:125192. https://doi.org/10.1016/j.biortech.2021.125192

Tyagi J, Chaudhary P, Jyotsana, Bhagwati U, Bhandari G, Chaudhary A (2022a) Chapter 17 impact of endophytic fungi in biotic stress management. In: Soni R, Suyal D, Goel R (eds) Plant protection: from chemicals to biologicals. De Gruyter, Berlin, Boston, pp 447–462. https://doi.org/10.1515/9783110771558-017

Tyagi J, Chaudhary P, Mishra A, Khatwani M, Dey S, Varma A (2022b) Role of endophytes in abiotic stress tolerance: with special emphasis on *Serendipita indica*. Int J Environ Res 16:62. https://doi.org/10.1007/s41742-022-00439-0

Wang X, Mandal AK, Saito H, Pulliam JF, Lee EY, Ke ZJ, Lu J, Ding S, Li L, Shelton BJ, Tucker T, Evers BM, Zhang Z, Shi X (2012) Arsenic and chromium in drinking water promote tumorigenesis in a mouse colitis-associated colorectal cancer model and the potential mechanism is ROS mediated Wnt/β-catenin signaling pathway. Toxicol Appl Pharmacol 262(1):11–21. https://doi.org/10.1016/j.taap.2012.04.014

Wang Y, Yi B, Sun X, Yu L, Wu L, Liu W, Wang D, Li Y, Jia R, Yu H, Li X (2019) Removal and tolerance mechanism of Pb by a filamentous fungus: a case study. Chemosphere 225:200–208. https://doi.org/10.1016/j.chemosphere.2019.03.027

Wu YH, Zhou P, Cheng H, Wang CS, Wu M, Xu XW (2015) Draft genome sequence of *Microbacterium profundi* Shh49T, an Actinobacterium isolated from deep-sea sediment of a polymetallic nodule environment. Genome Announcement 3(3):1–2. https://doi.org/10.1128/genomeA.00642-15

Xie Y, Fan J, Zhu W, Amombo E, Lou Y, Chen L, Fu J (2016) Effect of heavy metals pollution on soil microbial diversity and bermudagrass genetic variation. Front Plant Sci 7:755. https://doi.org/10.3389/fpls.2016.00755

Zaidi J, Pal VA (2017) Review on heavy metal pollution in major lakes of India: remediation through plants. Afr J Environ Sci Technol 11(6):255–265. https://doi.org/10.5897/AJEST2017.2299
Zaynab M, Al-Yahyai R, Ameen A, Sharif Y, Ali L, Fatima M, Khan KA, Li S (2022) Health and environmental effects of heavy metals. J King Saud Univ Sci 34(4):101653

Occurrence, Fate, and Biodegradation of Microplastics in Marine Environments

Geeta Bhandari, Sanjay Gupta, Saurabh Gangola, Mayank Bahuguna, and Shalu Chaudhary

Abstract The ubiquitous and recalcitrant nature of synthetic polymers contributes extensively toward xenobiotic pollutants entering the marine environments. Microplastics (≤5-mm fragments) are extensively distributed emerging contaminants present in the marine ecosystems. In general, microplastics comprise polyvinyl chloride, polyethylene terephthalate, polystyrene, nylon, etc. Due to deficiency of effective treatment technologies, the amount of microplastics is expanding at a dismaying rate, thus causing hazardous impact on the marine ecosystem and raising a global concern. The minute particle size of microplastics enables for its enhanced pathways in the aquatic food web and element cycles. Further, other organic and inorganic pollutants present in the marine environment can readily adhere to microplastics, which then function as active carriers for a wide range of contaminants in aquatic food webs. Processes such as weathering, degradation (microbial, oxidative, and hydrolytic), aggregation, and biofilm production have been reported to impose significant impact on transport of microplastics in the aquatic ecosystem. This chapter provides a deep insight into sources, distribution, fate, and toxic impact of microplastics in the environment. In addition, the role of marine microbes in the degradation of microplastics has also been explored.

Keywords Biodegradation · Marine · Microplastics · Microorganisms

G. Bhandari (✉) · S. Gupta · M. Bahuguna · S. Chaudhary
Department of Biosciences, Himalayan School of Biosciences, Swami Rama Himalayan University, Dehradun, Uttarakhand, India

S. Gangola
School of Agriculture, Graphic Era Hill University, Bhimtal Campus, Bhimtal, Uttarakhand, India

P. Chaudhary, A. Chaudhary (eds.), *Microbial Inoculants*, Interdisciplinary Biotechnological Advances, https://doi.org/10.1007/978-981-97-0633-4_14

1 Introduction

Plastic is the most commonly used synthetic material across the globe due to its durability, lightweight, and cost-effectiveness. In accordance with statistical data, 360 million tons of plastic items were manufactured in 2018, with plastic manufacture expected to reach 500 million tons by 2025. Nonetheless only 60% of the total plastic produced annually is collected (Yuan et al. 2020; Skoczinski et al. 2021). Subsequently the collected waste is generally not treated appropriately, and thus plastic waste inescapably contaminates the surrounding environment and is converted into big plastics (>2.5 cm), medium plastics (0.5–2.5 cm), microplastics (MPs) (1 μm–5 mm), and nanoplastics (NPs) (<1 μm) by different physicochemical processes, including weathering, UV radiation, and man-made activities (Lin et al. 2022). MPs account for 92.4% plastic waste in the marine environment and are abundantly found in different aquatic matrices such as soil, ocean, river, and lake (Yuan et al. 2020). Microplastics generally comprise polystyrene (PS), polyethylene (PE), polyethylene terephthalate (PET), polypropylene (PP), polyurethane (PU), polyvinyl chloride (PVC), polyamide nylon 6, and polyacrylonitrile (PAN) (Santana et al. 2016). MPs also comprise other hazardous compounds that are added, including pigments, plasticizers, and flame retardants, or obtained while transporting plastics in the ecosystem (pathogenic organisms, organic compounds, and metals) (Miri et al. 2022).

MPs are recalcitrant and have long half-lives and are stable in the environment causing detrimental impacts on different components of the ecosystem (Lin et al. 2022). MPs generally enter surrounding ecosystems through the food chain and subsequently get bioaccumulated. The occurrence of MPs in the diverse environmental matrices imposes adverse impacts on living beings and the ecosystem. Studies suggest that consumption of MPs by different organisms results in reproductive complications, false satiety, pathological stress, and impaired growth (Batel et al. 2016). MPs are readily ingested by aquatic organisms, causing physical blockages, and entangle the digestive system resulting in chronic toxicity owing to the adsorbed MPs (Frydkjaer et al. 2017). Furthermore, MP's electric charge, large specific surface area, and hydrophobic nature enable them to be perfect transporters of other recalcitrant organic contaminants, heavy metals, and antibiotics that can pollute water and soil (Lin et al. 2022). Thus, the aquatic organisms become particularly vulnerable to exposure to other toxic contaminants, such as pharmaceutically active compounds, personal care products, endocrine disruptors, polybrominated diphenyl ethers, and persistent organic pollutants that may enter the food chain and result in adverse impacts on the environment and human health (Huang et al. 2020). Different chemicals added during manufacture of plastics as additives are also detrimental to aquatic organisms since they release toxic molecules and resins in high concentrations. The recent progress for development of biodegradable plastic products by using renewable natural resources also causes production of MP after biodegradation (Miri et al. 2022). MPs have been proven to have a direct effect on plants by inhibiting nutrient uptake and accumulation of MP in roots, shoots, and leaves, as well as an indirect effect on soil qualities, physicochemical properties, and soil health (Khalid et al. 2020). MP accumulation in plants causes oxidative damage by activating the antioxidative enzyme activity (Yu et al. 2020). Ingestion and inhalation of MP promote an inflammatory response by activation of MAP-kinase pathway, neurotoxic effects by inhibiting acetylcholinesterase activity, and cancer development in humans. Owing to the highly toxic impact of MP on flora and fauna, its removal from the environment is of utmost importance.

Several different methods are being explored for treating MP from contaminated environments. Physical treatment processes (sedimentation, coagulation, filtration) physically segregate MPs without degradation and thus they persist in the sludge during the wastewater treatment. Furthermore, the employment of waste sludge in agricultural practices as organic fertilizers eventually releases MPs into the soil (Klein et al. 2018). Thus, there is an immense need for acquiring detailed insight of occurrence and distribution of MPs in different environmental compartments (Lambert and Wagner 2018). In the past decade, bioremediation has gained attention from researchers across the world as an eco-friendly and sustainable alternative for minimizing toxic contaminants in the environment (Gangola et al. 2018; Bhatt et al. 2022a; Chaudhary et al. 2023a). Microbes have an inherent capability of adaptation to different habitats and have the capability to biodegrade different toxic pollutants (Bhatt et al. 2021a, b; Bhandari et al. 2021; Kumar et al. 2022; Singh et al. 2023). The employment of microorganisms for the degradation of MPs in a green manner makes it a promising and sustainable strategy for clean-up of contaminated environments without imparting hazardous impact (Bhatt et al. 2022b).

Nonetheless, presently only few MP-metabolizing functional microbes are reported, and the complete interaction and degradation mechanism is yet to be elucidated (Yuan et al. 2020). Thus, it is significant to review and assess the present literature in order to figure out methods to acquire more functional microorganisms, enhance the MP-degrading activity of bacteria, and encourage a comprehensive insight into microbial metabolism of MPs.

2 Microplastics: Source, Distribution, and Toxicity

MPs have been reported from primary and secondary sources, respectively. Primary MPs are manufactured as small plastics that include polymeric material (pellet, flake, powder), plastic fiber fragments, powders utilized in molding, micro-exfoliants, microbeads employed in cosmetics, drug delivery system, and plastic nanoparticles generally composed of PS, PE, PET, PP, PU, PAN, PVC, and polyamide nylon (Miri et al. 2022). The employment of plastic in air blasting technique and the petroleum industry for drilling fluids and abrasives also produces primary MPs (Haider et al. 2019). The secondary MPs are generated when large plastics from textiles, tires, and paints are improperly treated and released into the environment which then break down or weather to produce small-sized plastics by physical (abrasion, high temperature, UV, ozone, oxidation, and pressure), chemical, and

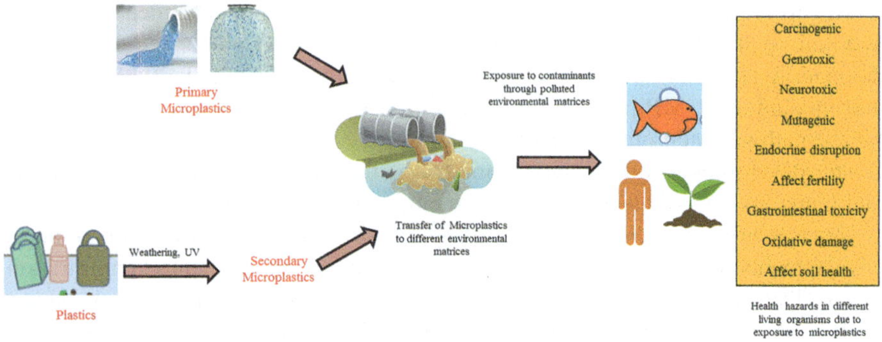

Fig. 1 Source, distribution, and transport of different types of microplastics in environment and their related health hazards

enzymatic processes (Chen et al. 2022). In natural conditions, the thermal and photooxidation reactions damage the plastic integrity, thereby also resulting in MP formation. It is a general consideration that majority of MPs entering into the environment are mostly due to secondary sources since massive amounts of macroplastics are released in the environment without any treatment. After being released or created in the environment, MPs will go through weathering and will be transported to different components of the ecosystem, thereby contaminating the water, atmosphere, and soil (Sanchez 2019) (Fig. 1). The unrestricted release and improper treatment of primary and secondary plastics causes contamination of freshwater ecosystem with subtle particles of MPs (Driedger et al. 2015; Lambert and Wagner 2018). Another potent source of entry of macro- and microplastics into the ecosystem is through urban water ecosystems (Prata 2018). The industrial and wastewater treatment plant effluent is another source of MPs in the aquatic system (Eerkes-Medrano et al. 2015; Chubarenko et al. 2016). MP can accidently enter the food chain on consumption by aquatic organisms especially various zooplanktons. The ingested MP can result in intestinal blockage, loss in nutrient absorption, hormonal imbalance, endocrine disruption, neurological and immunological impairment, and reproductive complications (Anand et al. 2023). Residual monomers and additives of MP exposure in marine animals can cause inflammation, oxidative damage, change in gene expression, and immune and central nervous system impairment (Vethaak and Martinez-Gomez 2020). Continuous increase in MP in aquatic bodies will also inevitably affect the marine biodiversity and aquaculture industry (Anand et al. 2023).

3 Biodegradation of Microplastics

The degradation of MP in nature usually occurs through physical, chemical, and biological processes (Ariza-Tarazona et al. 2020). Mechanical breakdown of large plastics in the water bodies forms micro- and nanoplastics having smaller particle sizes and higher surface areas, resulting in better degradation owing to high reactivity. Thermal degradation is not significant for freshwaters since the temperature in aquatic systems is not high enough to initiate chemical changes in synthetic polymers. On a molecular level, photooxidation (UV) or hydrolysis typically starts the deterioration of synthetic polymers in the environment, which is subsequently followed by chemical oxidation. MPs remain stable for ages in natural environment; however, specific microbes are capable of decomposing them. Owing to their better adaptation, microbes adapt to different habitats and possess capability to degrade various toxic organic contaminants, including MPs (Oliveira et al. 2020). MPs function as distinct ecological niche for microorganisms by providing supporting material for bacterial colonization and growth by working as a carbon source. The prime mechanism for degradation involves attachment of microbes to the polymer surface with subsequent colonization to the exposed surface. The biodegradation of MPs is initiated by degradation of large polymers to small-sized particles, oligomers, dimers, and monomers by extracellular enzymes. The initial enzymatic depolymerization occurs by hydrolytic breakdown of amide, ester, or urethane bonds. The small MP units are then easily absorbed by microbes and subsequently mineralized. In some cases, MPs are used as substrates for the growth of microbial biofilms and the produced biofilm causes pits and cracks that weaken the MPs' structure (Du et al. 2021) (Fig. 2).

Thereafter microbial enzymes attack the weakened MP fragments either selectively or inadvertently during the next stage of biodeterioration in a rate-limiting step. As soon as MP fragments are lighter than 600 kDa, bacteria in the biofilm can quickly assimilate them. Eventually, the assimilated units are further mineralized to small molecules (CO_2, N_2, CH_4, H_2O, H_2S) by microbial metabolism (Du et al. 2021). Light and oxygen assist in the physicochemical breakdown of MPs and also improve their bioavailability by photo -and thermooxidative degradation and hydrolysis for microbial degradation (Niu et al. 2020). However, light and oxygen are under limiting condition in deeper sediments, and thus redox environment is a vital parameter for MP biodegradation in these environments. In anaerobic environment, microbes use MPs or organic contaminants as electron donors (Rogers et al. 2020). Till date various anaerobes and facultative anaerobes capable of plastic degradation have been reported, thus creating possibility of occurrence of MP-degrading microbes in anoxic sediments. In the anoxic environment, organic compounds are usually degraded by denitrification, fermentative pathways, sulfate reduction, and methanotrophic process (Wurzbacher et al. 2017). The degree of colonization on the MP is determined from the surface characterization in terms of electrostatic interaction, roughness, topographical features, free energy, and hydrophobicity. The hydrophilic surfaces are found to be more easily colonized by microorganisms in

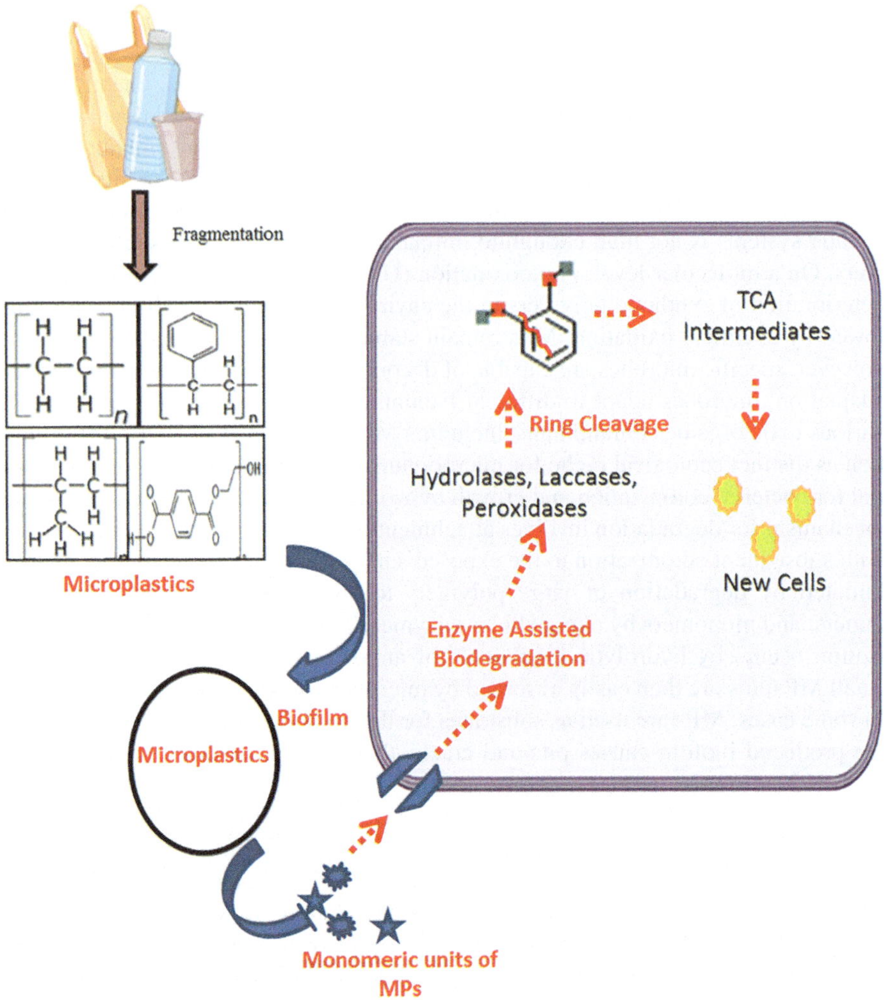

Fig. 2 Mechanism of biodegradation of microplastics

comparison to hydrophobic surfaces (Restrepo-Flórez et al. 2014). Crystallinity also affects the pace of biodegradation; the more crystalline a material is, the more difficult it will be for microorganisms to degrade it. As a result, the amorphous areas are degraded initially, since the monomeric units are loosely arranged in this region. The molecular mass is another determinant of biodegradation, since the complex MPs having side groups are slowly mineralized in comparison to linear MPs. Thus, the increase in molecular mass reduces the biodegradation rate of the MPs (Oliveira et al. 2020). Additionally, environmental conditions such as aeration, salinity, temperature, humidity, and light have a significant effect on the MP biodegradation. The additives in the polymer enhance the biodegradability by modifying the chemical and thermal sensitiveness and capability to absorb UV of MP and result in obliteration of stable character which are responsible for microbial attachment (Yang et al. 2021).

3.1 Bacterial Degradation of Microplastics

Microbes are capable of metabolically adsorbing, desorbing, and degrading MPs (Auta et al. 2018) (Table 1). Some bacterial phyla such as *Bacteroidetes*, *Cyanobacteria*, *Firmicutes*, and *Proteobacteria* colonize MPs more readily in comparison to other phyla, suggesting that distinct genera of microorganisms contemplate MPs as an ecologically beneficial habitat and a potent material for metabolic adaptation (adherence, chemotaxis, additive resistance, and degradation). Microbial-mediated MP biodegradation consists of successive steps including biodeterioration, biofragmentation, assimilation, and mineralization. Bacterial consortia of *Pseudomonas stutzeri*, *P. chlororaphis*, and *Vibrio* sp. are capable of utilizing PP as a carbon source with subsequent degradation of PP (Arkatkar et al. 2010). In a subsequent investigation, *Achromobacter xylosoxidans* produced a thin biofilm on high-density PE with subsequent degradation of the MP (Kowalczyk et al. 2016). Auta et al. (2017) investigated *Bacillus cereus* and *Bacillus gottheilii's* ability to break down MPs. The weight loss percentages by *Bacillus cereus* are 1.6%, 6.6%, and 7.4% for PE, PET, and PS, respectively. For PE, PET, PP, and PS, the weight loss percentages promoted by *B. gottheilii* are 6.2%, 3.0%, 3.6%, and 5.8%, respectively. Surprisingly, certain MPs had rough surfaces and several cracks and grooves in it (Auta et al. 2017). The FTIR results of PE revealed the disappearance of carbonyl band after incubation with *Bacillus cereus* and *Bacillus gottheilii* and occurrence of new peaks at 3738 and 3419 cm^{-1} which were correlated to N-H and O-H bonds, respectively. Additionally, Auta et al. (2018) assessed the capability of two pure microbial strains reported from mangrove sediment for biodegradation of MPs. Within 40 days, *Bacillus* sp. and *Rhodococcus* sp. increased PP weight loss by 4.0% and 6.4%, respectively. Moreover, several abnormalities on the PP surface were discovered, implying that PP may function as support for attachment and colonization by pure bacterial isolates with subsequent degradation of PP. These reports suggest that microbes can modify not only the surface features of MPs but also their functional groups and other characteristics.

According to Jeon and Kim (2013), on incubating with *Stenotrophomonas maltophilia* LB 2–3, the molecular mass and tensile strength of polylactic acid (PLA) rapidly reduced, implying that microbes impact these properties of MPs. *Pseudomonas aeruginosa* biodegraded PS-PLA nanocomposites and the highest degradation rate was 10% (Shimpi et al. 2012). Mohan et al. (2016) studied the capability of novel *Bacillus* strains to biodegrade brominated high-impact PS and the biodegradation rate was 23% in 30 days. According to these findings, microbes have comparatively better degradation impact on MP and their modified forms. Moreover, various reports also indicate that the employment of bacterial consortia

Table 1 Bacterial strains capable of degrading microplastics

Source	Bacterial strain	Type of MP	Degradation efficiency	References
Polymeric sheet under degradation	*Pseudomonas aeruginosa*	PS-PLA Nanocomposites	9.9% within 0.94 months	Shimpi et al. (2012)
Compost	*Chelatococcus* E1	PE	–	Jeon and Kim (2013)
Gut of mealworm	*Exiguobacterium* sp. strain YT2	PS	7.4 ± 0.4% within 60 days	Yang et al. (2015)
Contaminatedsite	*Stenotrophomonas panacihumi* PA3–2	PP	20.3 ± 1.39% within 90 days	Jeon and Kim (2016)
Marine	*Salinibacterium* sp. *Lysinibacillus* sp.	PE	19% within 6 months	Syranidou et al. (2017)
Gut of *Lumbricus terrestris*	*Bacillus simplex*, *Bacillus* sp.	LDPE	60% degradation of LDPE within 21 days	Huerta Lwanga et al. (2018)
Compost	*Bacillus thuringiensis*	PP and poly- L-lactide	12% degradation within 15 days	Jain et al. (2018)
Compost	*Bacillus licheniformis*	PP and poly- L-lactide	10% degradation within 15 days	Jain et al. (2018)
Contaminated site	*Streptomyces* sp.	LDPE	15–19% weight loss within 30 days	Alaa (2019)
Contaminated sediments	*Bacillus* sp., *Paenibacillus* sp.	PP	14.7% within 60 days	Park and Kim (2019)
Sludge	*Thermus*, *Bacillus*, and *Geobacillus*	PS	43.7% within 45 days	Chen et al. (2020)
Cow dung	Consortia of *Pseudomonas aeruginosa*, *Enterobacter* sp., *Enterobacter cloacae*	LDPE and PP	64.25 ± 2% and 63.00 ± 2% within 160 days	Skariyachan et al. (2021)
Hot spring soil	*Anoxybacillus flavithermus* Gecek4s	PE	Colonization of the polymer surface and surface damage	Akarsu et al. (2022)
Compost	*Bacillus cereus*, *Bacillus licheniformis*, *Bacillus thuringiensis*	PP and poly-L-lactide	–	Jain et al. (2022)
Hot spring	*Anoxybacillus flavithermus ST3* and *Anoxybacillus* sp. *ST6*	PE	Colonization of the polymer surface and surface damage	Özdemir et al. (2022)
Mangrove	Indigenous mangrove bacteria	PET and PS	18% weight loss within 90 days	Auta et al. (2022)

is more suitable for the absolute mineralization of toxic intermediates. Generally, the microbial strains of the consortia possess a mutual symbiotic and synergistic interactions that enable them to have higher tolerance and biodegradability (Chaudhary et al. 2023b). Park and Kim (2019), investigated the efficacy of a mesophilic bacterial consortia comprising *Paenibacillus* sp. and *Bacillus* sp. for biodegradation of PE, and the results indicated a decrease in MP dry weight by 14.7% within 60 days. A novel consortium of *Enterobacter* and *Pseudomonas* obtained from cow dung exhibits improved degradation of PP and PE and demonstrates 15% dry weight reduction after 120 days (Skariyachan et al. 2021). *Anoxybacillus flavithermus* Gecek4s and *Bacillus firmus* ST5 were investigated for PE degradation, and results revealed that *A. flavithermus* Gecek4s colonizes the polymer surface and causes surface damage. Additionally, dimensional changes and cracks were noticed on the PE surface (Akarsu et al. 2022). Assemblages of *Mycosphaerella* sp., *Phaeosphaeria spartinicola*, and *P. halima* capable of degrading complex MPs produce a biofilm on the polymer surface with subsequent ingestion of MPs (Kawai et al. 2019). A diverse range of plastics have been found to be degraded in the insect gut such as insect larvae of mealworms, Indian meal moth, and wax moths by nonspecific oxidative breakdown process (Dey et al. 2023). Yang et al. (2015) reported that waxworms can consume PE films, and two microbial strains (*Enterobacter asburiae* YT1 and *Bacillus* sp. YP1) that were able to degrade PE were then recovered from the gut of waxworms. Better understanding of the molecular intricacies through which insect and gut microbiome degrade MPs is needed, which then could be useful in the future for enzyme-based techniques for MP waste management (Wierckx et al. 2018; Kong et al. 2019).

3.2 Fungal-Mediated Microplastic/Plastic Degradation

Moreover, fungi are also capable of adhering to and utilizing MP (Table 2). The hydrophobicity of MPs can be reduced by fungi through formation of several chemical linkages, including carbonyl, carboxyl, and ester functional groups. In fungi, extracellularly and intracellularly synthesized enzymes initially decompose the polymeric plastics into monomeric units and generate CO_2, H_2O, and CH_4 in anaerobic environment and CO_2 and H_2O in aerobic conditions. Moreover, several fungal strains also release hydrophobins (surface proteins), which contribute in the biodegradation process by increasing substrate mobility and availability. Fungal strains enhance the degradation and circulation of toxic contaminants (Chen et al. 2016; Bhandari 2022). Additionally, pre-treating MPs with NaOH and HNO_3 is reported to improve the PE degradation efficacy of *Aspergillus niger* (Nwachukwu et al. 2010). In recent times, numerous studies have investigated the fungal-mediated biodegradation of MPs, and fungi have used MPs as a sole carbon source. Several different genera of fungi are reported to degrade different kinds of MPs (PE, PP, PET) including *Aspergillus fumigatus*, *Aspergillus niger*, *Aspergillus tubingensis*, *Fusarium solani*, *Penicillium simplicissimum*, *Pestalotiopsis microspora*, and

Table 2 Fungal strains capable of degrading microplastics

Source	Fungal strain	Type of MP	Degradation efficiency	References
Commercially obtained	*Phanerochaete chrysosporium* NCIM 1170	PP	57% weight loss within 1 year	Jeyakumar et al. (2013)
Contaminated site	*Engyodontium album* MTP091	PP	79% weight loss within 1 year	Jeyakumar et al. (2013)
Contaminated site	*Aspergillus*, *Fusarium*	LDPE	10.3 and 9% within 60 days	Das and Kumar (2014)
Endophytic microbes isolated from *Humboldtia brunonis*	*Aspergillus*, *Lasiodiplodia theobromae*	PP	Laccase production	Sheik et al. (2015)
Marine sediments	*Bjerkandera adusta*	Gamma-irradiated PP	Production of ligninase	Butnaru et al. (2016)
Contaminated site	*Aspergillus niger*, *Fusarium oxysporum*	LDPE and PU	Reduction in percent elongation and tensile strength of LDPE and PU after 90 days	Raghavendra et al. (2016)
Contaminated site	*Aspergillus tubingensis*	PPU	Complete degradation of PPU film into small fragments within 60 days	Khan et al. (2017)
Contaminated site	*Penicillium oxalicum*, *Penicillium chrysogenum*	HDPE and LDPE	Surface morphology of PE changed after 60 days	Ojha et al. (2017)
Marine sediments	*Zalerion maritimum*	PE	Reduction in both the size and mass of the PE within 28 days	Paço et al. (2017)
Dumping sites of mangrove rhizosphere soil	*Aspergillus terreus*, *Aspergillus sydowii*	PE	Significant reduction in weight and tensile strength of PE within 60 days	Sangale et al. (2019)
Commercially obtained	*Chaetomium globosum*	PVC	Attachment and growth of *C. globosum* fertile structures on PVC within 28 days	Vivi et al. (2019)
Gut of wax moth	*Aspergillus flavus*	PE	3.9025 ± 1.18% weight loss within 28 days	Zhang et al. (2020)
--	*Aspergillus* sp., *Penicillium* sp.	Polypropylene/ butylene adipate-co-terephthalate	Modification in morphological features and hydrophilicity of polymer, attachment of fungi on polymer	de Oliveira et al. (2020)

Zalerion maritimum (Devi et al. 2015; Muenmee et al. 2016; Espinosa et al. 2020). Fungal strains, *A. tubingensis* VRKPT1 and *A. flavus* VRKPT2, isolated from PE-contaminated sites from marine environment, demonstrated efficient HDPE biodegradation capabilities of 6.02 0.2% and 8.51 0.1%, respectively (Devi et al. 2015). Kumar et al. (2013) also observed degradation of LDPE by *Aspergillus* spp. and *Fusarium* spp. Kunlere et al. (2019) investigated biodegradation of LDPE by *Mucor circinelloides* and *Aspergillus flavus* obtained from a municipal landfill site (Kunlere et al. 2019). PE-degrading fungus *Aspergillus flavus* PEDX3 was obtained from the gut contents of wax moth *Galleria mellonella* (Zhang et al. 2020). The study suggested that high-density polyethylene MP was transformed into lower molecular weight MP by the strain PEDX3 within 28 days. The FTIR results revealed the presence of carbonyl groups and ether groups validating the degradation of PE. The expression of two laccase-like multicopper oxidase (LMCO) genes, AFLA_006190 and AFLA_053930, was induced during the biodegradation (Zhang et al. 2020).

4 Enzymatic Basis of MP Degradation

Microbial extracellular enzymes (manganese peroxidases, lignin peroxidases, esterases, lipases, poly-3-hydroxybutyrate depolymerases, cutinases, and laccases) enhance the hydrophilic nature of MPs by transforming them to smaller units containing carbonyl or alcoholic groups, which may improve bacterial adherence to MP and subsequent degradation (Taniguchi et al. 2019; Sol et al. 2020). These enzymes attack certain bonds in MP or its side chain during the process of biodegradation, thus enhancing chain cleavage. However, owing to their size, the enzymes are unable to diffuse into the polymeric structure; therefore, the breakdown mainly occurs on the polymeric surface causing cracks. Nonetheless, intercellular enzymes from microbes are yet to be explored, and the process of monomer transport across the microbial membrane deserves more investigation (Sol et al. 2020; Parveen et al. 2022). Santo et al. (2013) identified that *Rhodococcus ruber* C208 is capable of biodegrading PE by synthesis of extracellularly secreted laccase. Copper ions were necessary for the induction and activity of the synthesized laccase enzyme. The amorphic portion of PE films was oxidized by laccases resulting in generation of readily accessible carbonyl groups, validated by FTIR results, and a drop in the molecular mass of the PE sheets (Santo et al. 2013). By assessing the transcription levels of alkB1 and alkB2 qRT-PCR, Jeon and Kim (2016) examined the PE degradation capability of two alkane monooxygenases (AlkB1 and AlkB2) produced in *Pseudomonas aeruginosa* E7. When *P. aeruginosa* E7 was cultured in growth media supplemented with PE as the only carbon source, alkB2 transcription levels were higher than alkB1. Yoshida et al. (2016) reported that *Ideonella sakaiensis* was able to degrade PET and demonstrated that it attaches to the polymer surface and synthesizes two enzymes: PETase and MHETase. PETase is secreted extracellularly and causes hydrolytic breakdown of PET to main intermediates (mono(2-hydroxyethyl) terephthalic acid (MHET) and terephthalic acid (TPA)). Thereafter, MHETase breaks down MHET into TPA and ethylene glycol, which are monomeric units of PET obtained from crude petroleum. TPA is typically delivered to the cytoplasmic region by TPA transporters, which is then subsequently metabolized by oxidation and degradation of the ring. In a subsequent study by Taniguchi et al. (2019), the PET degradation pathway by *Ideonella sakaiensis* was described. Extracellularly secreted enzyme PETase hydrolyzes PET to MHET as a major intermediated and TPA, which was then transformed by MHETase into TPA and EG. TPA enters the cytoplasm via TPA transporters and TPA-binding proteins, where it enters the tricarboxylic acid (TCA) cycle via protocatechuic acid (PCA) (Taniguchi et al. 2019).

A diverse range of fungal enzymes are capable of catalyzing MP degradation reactions. Intracellularly secreted enzymes play an important function in fungal adaptability and degradation of contaminants (Schwartz et al. 2018). Conjugation, desulfuration, deamination, dehalogenation, epoxidation, oxidation, and sulfoxidation of MP are all carried out by the cytochrome P450 family epoxidase and transferase (Shin et al. 2018). The cytochrome P450 enzyme assists in preserving the hyphal wall integrity and the development of spore walls by utilizing cofactors such as heme, $NADPH^+$, H^+, and FAD. Fungal hydrolase, cutinase, lipase, laccase, serine esterase, and p-nitrophenyle sterase are extracellular enzymes that are responsible for catalyzing the breakdown of MPs and increasing the solubility of contaminants, hence minimizing bioaccumulation (Olicón-Hernández et al. 2017). Class II peroxidase (Mn peroxidases and lignin peroxidases), laccase, and dye-decolorizing peroxidase can oxidize a diverse range of contaminants and thus may be utilized as effective environment cleaners. Laccase is produced by lignin-degrading fungal strains that catalyze the oxidation of aromatic and nonaromatic compounds (Kumar et al. 2021a, b). The thermostable nature of these enzymes may encourage their usage in large-scale reactors where PP degradation can take place at high temperatures, allowing for high kinetics responses. Fungal cutinases and lipases work together to hydrolyze PET, generating TPA and EG. Because MHET works as a competitive inhibitor, cutinase can hydrolyze PET until MHET (an intermediate) is produced. Lipase, on the other hand, has no limitations; it totally transforms PE to TPA (Anbalagan et al. 2022).

5 Biotechnological Tools for Microplastic Biodegradation

The development of many genetic engineering technologies has allowed us to alter their microbial genetic components and improve their biodegradation efficacy. Different approaches utilizing recombinant DNA technologies, gene cloning, genetic manipulation, and metabolic engineering can improve microbial biodegradation in contaminated environments (Kumar et al. 2020, 2021a, b; Gangola et al. 2022) (Fig. 3). Nonetheless, relatively few studies have been undertaken to date on the use of genetic engineering to create a superior strain for MP biodegradation. Genetic changes have also been made to increase PVC capture inside the bacterial biofilm (Liu et al. 2021). *Pseudomonas aeruginosa* was genetically modified by

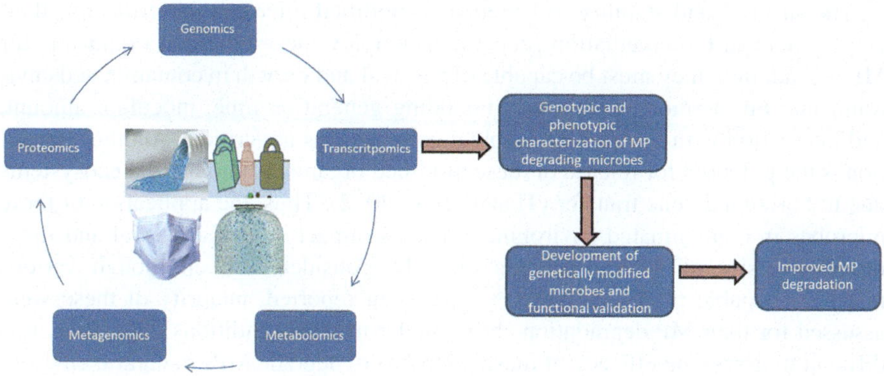

Fig. 3 Application of Omics approach for Microplastic degradation

deletion of the *wspF* gene in order to stimulate the synthesis of sticky exopolysaccharides, which improves the bacteria's ability to concentrate MPs in its biofilm. Furthermore, the *yhjH* gene was created and inserted into the bacterium under the regulation of an arabinose-induced promoter. The role of the *yhjH* gene is to downregulate the expression of cyclic dimeric GMP level and upregulate the expression of the gene inhibiting biofilm development sufficiently to allow the release of trapped MPs. This "catch and release" technique allows the development of efficient MP scavenger for aquatic ecosystem bioremediation. Lameh et al. (2022) performed in silico site-specific mutagenesis of carboxylesterases to create BTA hydrolases in *Archaeoglobus fulgidus* to improve its capability to breakdown polyethylene terephthalate.

The primary enzyme responsible for MP biodegradation was synthesized by the genetically modified *E. coli* and *S. cerevisiae* BY 4741 strains, such as Mn peroxidases, and laccases were synthesized by the genetically engineered strains of *E. coli BL21* (Paço et al. 2019). These genetically engineered enzymes can degrade polyethylene terephthalate more effectively. Genetically engineered yeast synthesizes microbial cutinases, by adding sugars at distinct positions that allow the enzyme to remain folded at even high temperatures and by developing physical barriers. Microbial cutinases developed by this method degrade polyester linkage and may be employed for degradation of polyethylene terephthalate at an ideal temperature of 75 °C (Shirke et al. 2018). Oda et al. (2018) developed a genetically modified calcium-dependent cutinases (Cut190) by using *Saccharomonospora viridis* AHK190. The modified enzymes degraded PET more efficiently than the original enzymes. According to Islam et al. (2019), genetically altered cutinases decrease the biodegradation duration from 41.8 to 6.2 h as compared to the wild strain. Yan et al. (2021) identified a thermostable cutinase from the plant compost metagenome that can degrade PET at 70 °C. *Clostridium thermocellum* was utilized to clone the thermostable cutinase, and a strong promoter of a constitutively expressed gene was employed to achieve a higher amount of gene expression and signal peptide of exoglucanase Cel48S for extracellularly releasing the enzyme.

The survival and stability of genetically modified microbes are critical in their employment in bioremediation process. To employ these modified organisms for MP degradation, they must be capable of survival and growth in contaminated environments. Furthermore, parameters including generation time, inoculum amount, and interaction with the native microbial communities are critical. Another restriction is the potential for release of these modified organisms into natural ecosystems due to horizontal gene transfer (Thakur et al. 2022). Thus, the application of these microbes in contaminated environment and its impact on the structural and functional diversity of natural habitats must be considered. Even though various microbes capable of degrading MPs have been reported, majority of these were assessed for their MP degradation ability under in vitro conditions only. Thus, it is difficult to assess the efficacy of these microbes to degrade MPs in natural environments. As a result, new approaches for improving survivability of these modified organisms in contaminated environments must be developed, and our understanding of the biodegradation mechanism must be expanded.

6 Limitations

To enhance the efficacy of MP biodegradation and make it a viable alternative for managing MP pollution, different limitations need to be investigated. Even though there are numerous benefits of using microorganisms for MP degradation, the effect of environmental factors including pH, substrate properties, temperature, and biosurfactants needs to be assessed. Since majority of the MPs are exceedingly recalcitrant and not readily biodegradable, microbial-mediated breakdown is generally slow, with major alterations observable only within 60–90 days. Thus, advanced approaches are needed to improve growth conditions and reduce the biodegradation duration. Furthermore, the degradation of MPs by a single culture produces toxic intermediates limiting the growth of microbes. To promote the biodegradation of MPs, different pre-treatments with UV and HNO_3 are also advised. In addition, various novel fungal strains with improved organic contaminant degradation have been reported in recent decades. However, very few studies report fungal-mediated MP degradation, signaling that fungal isolates with better MP biodegradation ability through ectopic selection are difficult (Yuan et al. 2020). The degradation of MPs by fungus in diverse environmental conditions is still an active field study, with considerable progress made.

7 Conclusion

Microplastics are gaining popularity owing to their pervasiveness and harmful influence on human health and the environment. The exposure of MPs can occur by inhalation, digestion, and skin absorption resulting in neurotoxic and carcinogenic effect, endocrine disruption, and metabolic disturbances. As a result, effective methods of MP remediation are required. The processes currently being utilized for the treatment of MPs include landfills and incineration that cause secondary pollution by releasing more toxic by-products in the environment. Microbes have the capability to biodegrade MPs in a sustainable manner. However, there are numerous obstacles and limits in the use of microorganisms for MP biodegradation that can be solved by developing genetically modified organisms. However, majority of genetically engineered bacteria have only been verified in the lab studies, and their effectiveness in the field is yet to be explored. Furthermore, understanding various MP degradative pathways and enzymes is substantially inadequate. The current advancement in metagenome studies, engineering of uncultivated bacterial communities from polluted environments, and gene editing in silico tools can help develop novel bioremediation processes, and culture-independent techniques can open up new avenues for the discovery of novel metabolic pathways and enzymes.

References

Akarsu C, Özdemir S, Ozay Y, Acer O, Dizge N (2022) Investigation of two different size microplastic degradation ability of thermophilic bacteria using polyethylene polymers. Environ Technol. https://doi.org/10.1080/09593330.2022.2071638

Alaa S (2019) Biodegradation of polyethylene LDPE plastic waste using locally isolated *Streptomyces* sp. J Pharm Sci Res 11:1333–1339

Anand U, Dey S, Bontempi E, Ducoli S, Vethaak AD, Dey A, Federici S (2023) Biotechnological methods to remove microplastics: a review. Environ Chem Lett. https://doi.org/10.1007/s10311-022-01552-4

Anbalagan S, Venkatakrishnan HR, Ravindran J, Sathyamoorthy J, Rangabashyam KA, Ragini YP, Sathasivam J, Sureshbabu K (2022) Hydrolytic degradation of polyethylene terephthalate by cutinase enzyme derived from fungal biomass– molecular characterization. Biointerface Res Appl Chem 12(1):653–667. https://doi.org/10.33263/BRIAC121.653667

Ariza-Tarazona MC, Villarreal-Chiu JF, Hernandez-Lopez JM, Rivera De la Rosa J, Barbieri V, Siligardi C, Cedillo-Gonzalez EI (2020) Microplastic pollution reduction by a carbon and nitrogen-doped TiO2: effect of pH and temperature in the photocatalytic degradation process. J Hazard Mater 395:122632

Arkatkar A, Juwarkar AA, Bhaduri S, Uppara PV, Doble M (2010) Growth of Pseudomonas and Bacillus biofilms on pretreated polypropylene surface. Int Biodeterior Biodegradation 64(6):530–536

Auta HS, Abioye OP, Aransiola SA, Bala JD, Chukwuemeka VI, Hassan A, Aziz A, Fauziah SF (2022) Enhanced microbial degradation of PET and PS microplastics under natural conditions in mangrove environment. J Environ Manag 304:114273. https://doi.org/10.1016/j.jenvman.2021.114273

Auta HS, Emenike CU, Fauziah SH (2017) Screening of *Bacillus* strains isolated from mangrove ecosystems in Peninsular Malaysia for microplastic degradation. Environ Pollut 231:1552–1559

Auta HS, Emenike CU, Jayanthi B, Fauziah SH (2018) Growth kinetics and biodeterioration of polypropylene microplastics by *Bacillus* sp. and *Rhodococcus* sp. isolated from mangrove sediment. Mar Pollut Bull 127:15–21. https://doi.org/10.1016/j.marpolbul.2017.11.036

Batel A, Linti F, Scherer M, Erdinger L, Braunbeck T (2016) Transfer of benzo [α] pyrene from microplastics to artemia nauplii and further to zebrafish via a trophic food web experiment: CYP1A induction and visual tracking of persistent organic pollutants. Environ Toxicol Chem 35(7):1656–1666

Bhandari G (2022) Characterization and screening of pesticide-degrading indigenous fungi from soil and water. In: Udayanga D, Bhatt P, Manamgoda D, Saez JM (eds) Mycoremediation protocols. Springer protocols handbooks. Humana, New York, NY. https://doi.org/10.1007/978-1-0716-2006-9_3

Bhandari G, Bagheri AR, Bhatt P, Bilal M (2021) Occurrence, potential ecological risks, and degradation of endocrine disrupter, nonylphenol, from the aqueous environment. Chemosphere 275:130013. https://doi.org/10.1016/j.chemosphere.2021.130013

Bhatt P, Bhandari G, Bilal M (2022b) Occurrence, toxicity impacts and mitigation of emerging micropollutants in the aquatic environments: recent tendencies and perspectives. J Environ Chem Eng 10(3):107598. https://doi.org/10.1016/j.jece.2022.107598

Bhatt P, Gangola S, Bhandari G, Zhang W, Maithani D, Mishra S, Chen S (2021a) New insights into the degradation of synthetic pollutants in contaminated environments. Chemosphere 268:128827. https://doi.org/10.1016/j.chemosphere.2020.128827

Bhatt P, Gangola S, Joshi C, Chaudhary P, Kumar G, Bhandari G, Kumar K, Joshi S, Kumar A, Bhandari NS, Tewari S (2021b) Recent advancements and mechanism of microbial enzymes in sustainable agriculture. In: Bhatt P, Gangola S, Udayanga D, Kumar G (eds) Microbial technology for sustainable environment. Springer, Singapore. https://doi.org/10.1007/978-981-16-3840-4_15

Bhatt P, Pandey SC, Joshi S, Chaudhary P, Pathak VM, Huang Y, Wu X, Zhou Z, Chen S (2022a) Nanobioremediation: a sustainable approach for the removal of toxic pollutants from the environment. J Hazard Mater 427:128033. https://doi.org/10.1016/j.jhazmat.2021.128033

Butnaru E, Darie-Nita RN, Zaharescu T, Balaes T, Tanase C, Hitruc G, Vasile C (2016) Gamma irradiation assisted fungal degradation of the polypropylene/biomass composites. Radiat Phys Chem 125:134–144

Chaudhary P, Ahamad L, Chaudhary A, Kumar G, Chen F-S, Chen S (2023a) Nanoparticle-mediated bioremediation as a powerful weapon in the removal of environmental pollutants. J Environ Chem Eng 11:109591. https://doi.org/10.1016/j.jece.2023.109591

Chaudhary P, Xu M, Ahamad L, Chaudhary A, Kumar G, Adeleke BS, Verma KK, Hu D-M, Širić I, Kumar P, Popescu SM, Abou Fayssal S (2023b) Application of synthetic consortia for improvement of soil fertility, pollution remediation, and agricultural productivity: a review. Agronomy 13:643. https://doi.org/10.3390/agronomy13030643

Chen J, Wu J, Sherrell PC, Chen J, Wang H, Zhang WX, Yang J (2022) How to build a microplastics-free environment: strategies for microplastics degradation and plastics recycling. Adv Sci 9(6):2103764. https://doi.org/10.1002/advs.202103764

Chen Y, Stemple B, Kumar M, Wei N (2016) Cell surface display fungal laccase as a renewable biocatalyst for degradation of persistent micropollutants bisphenol a and sulfamethoxazole. Environ Sci Technol 50:8799–8808. https://doi.org/10.1021/acs.est.6b01641

Chen Z, Zhao W, Xing R, Xie S, Yang X, Cui P, Lü J, Liao H, Yu Z, Wang S, Zhou S (2020) Enhanced in situ biodegradation of microplastics in sewage sludge using hyperthermophilic composting technology. J Hazard Mater 384:121271. https://doi.org/10.1016/j.jhazmat.2019.121271

Chubarenko I, Bagaev A, Zobkov M, Esiukova E (2016) On some physical and dynamical properties of microplastic particles in marine environment. Mar Pollut Bull 108:105–112. https://doi.org/10.1016/j.marpolbul.2016.04.048

Das PM, Kumar S (2014) Microbial deterioration of low density polyethylene by Aspergillus and Fusarium sp. Int J ChemTech Res 6(1):299–305

Devi RS, Kannan VR, Nivas D, Kannan K, Chandru S, Antony AR (2015) Biodegradation of HDPE by *Aspergillus* spp. from marine ecosystem of gulf of Mannar, India. Mar Pollut Bull 96(1–2):32–40

Dey S, Anand U, Kumar V, Kumar S, Ghorai M, Ghosh A, Kant N, Suresh S, Bhattacharya S, Bontempi E, Bhat SA, Dey A (2023) Microbial strategies for degradation of microplastics generated from COVID-19 healthcare waste. Environ Res 216(1):114438. https://doi.org/10.1016/j.envres.2022.114438

Driedger AG, Dürr HH, Mitchell K, Van Cappellen P (2015) Plastic debris in the Laurentian Great Lakes: a review. J Gt Lakes Res 41(1):9–19

Du H, Xie Y, Wang J (2021) Microplastic degradation methods and corresponding degradation mechanism: research status and future perspectives. J Hazard Mater 418:126377. https://doi.org/10.1016/j.jhazmat.2021.126377

Eerkes-Medrano D, Thompson RC, Aldridge DC (2015) Microplastics in freshwater systems: a review of the emerging threats, identification of knowledge gaps and prioritisation of research needs. Water Res 75:63–82

Espinosa MJC, Blanco AC, Schmidgall T, Atanasoff-Kardjalieff AK, Kappelmeyer U, Tischler D, Pieper DH, Heipieper HJ, Eberlein C (2020) Toward biorecycling: isolation of a soil bacterium that grows on a polyurethane oligomer and monomer. Front Microbiol 11:404. https://doi.org/10.3389/fmicb.2020.00404

Frydkjaer CK, Iversen N, Roslev P (2017) Ingestion and egestion of microplastics by the *Cladoceran Daphnia magna*: effects of regular and irregular shaped plastic and sorbed phenanthrene. Bull Environ Contam Toxicol 99(6):655–661

Gangola S, Bhatt P, Kumar AJ, Bhandari G, Joshi S, Punetha A, Bhatt K, Rene ER (2022) Biotechnological tools to elucidate the mechanism of pesticide degradation in the environment. Chemosphere 296:133916. https://doi.org/10.1016/j.chemosphere.2022.133916

Gangola S, Sharma A, Bhatt P, Khati P, Chaudhary P (2018) Presence of esterase and laccase in *Bacillus subtilis* facilitates biodegradation and detoxification of cypermethrin. Sci Rep 8:12755. https://doi.org/10.1038/s41598-018-31082-5

Haider TP, Volker C, Kramm J, Landfester K, Wurm FR (2019) Plastics of the future? The impact of biodegradable polymers on the environment and on society. Angew Chem Int Ed 58:50–62

Huang D, Tao J, Cheng M, Deng R, Chen S, Yin L, Li R (2020) Microplastics and nanoplastics in the environment: macroscopic transport and effects on creatures. J Hazard Mater 407:124399

Huerta Lwanga E, Thapa B, Yang X, Gertsen H, Salanki T, Geissen V, Garbeva P (2018) Decay of low-density polyethylene by bacteria extracted from earthworm's guts: a potential for soil restoration. Sci Total Environ 624:753–757. https://doi.org/10.1016/j.Scitotenv.2017.12.144

Islam S, Apitius L, Jakob F, Schwaneberg U (2019) Targeting microplastic particles in the void of diluted suspensions. Environ Int 123:428–435. https://doi.org/10.1016/j.envint.2018.12.029

Jain K, Bhunia H, Reddy MS (2018) Degradation of polypropylene- poly-L-lactide blend by bacteria isolated from compost. Biorem J 22(3–4):73–90. https://doi.org/10.1080/10889868.2018.1516620

Jain K, Bhunia H, Reddy MS (2022) Degradation of polypropylene-poly-L-lactide blends by *Bacillus* isolates: a microcosm and field evaluation. Biorem J 26(1):64–75. https://doi.org/10.1080/10889868.2021.1886037

Jeon HJ, Kim MN (2013) Biodegradation of poly (L-lactide) (PLA) exposed to UV irradiation by a mesophilic bacterium. Int Biodeterior Biodegradation 85:289–293. https://doi.org/10.1016/j.ibiod.2013.08.013

Jeon HJ, Kim MN (2016) Comparison of the functional characterization between alkane monooxygenases for low-molecular-weight polyethylene biodegradation. Int Biodeterior Biodegradation 114:202–208

Jeyakumar D, Chirsteen J, Doble M (2013) Synergistic effects of pretreatment and blending on fungi mediated biodegradation of polypropylenes. Bioresour Technol 148:78–85. https://doi.org/10.1016/j.biortech.2013.08.074

Kawai F, Kawabata T, Oda M (2019) Current knowledge on enzymatic PET degradation and its possible application to waste stream management and other fields. Appl Microbiol Biotechnol 103:4253–4268

Khalid N, Aqeel M, Noman A (2020) Microplastics could be a threat to plants in terrestrial systems directly or indirectly. Environ Pollut 267:115653. https://doi.org/10.1016/j.envpol.2020.115653

Khan S, Nadir S, Shah ZU, Shah AA, Karunarathna SC, Xu J, Khan A, Hasan F (2017) Biodegradation of polyester polyurethane by *Aspergillus tubingensis*. Environ Pollut 225:469–480

Klein S, Dimzon IK, Eubeler J, Knepper TP (2018) Analysis, occurrence, and degradation of microplastics in the aqueous environment. Freshwater Microplastics. Springer, Cham, pp 51–67

Kong HG, Kim HH, Chung JH, Jun J, Lee S, Kim HM, Jeon S, Park SG, Bhak J, Ryu CM (2019) The Galleria mellonella hologenome supports microbiota-independent metabolism of long-chain hydrocarbon beeswax. Cell Rep 26(9):2451–2464. https://doi.org/10.1016/j.celrep.2019.02.018

Kowalczyk A, Chyc M, Ryszka P, Latowski D (2016) *Achromobacter xylosoxidans* as a new microorganism strain colonizing high-density polyethylene as a key step to its biodegradation. Environ Sci Pollut Res 23(11):11349–11356

Kumar A, Sharma A, Chaudhary P, Gangola S (2021a) Chlorpyrifos degradation using binary fungal strains isolated from industrial waste soil. Biologia 10:3071–3081. https://doi.org/10.1007/s11756-021-00816-8

Kumar G, Lal S, Maurya SK, Bhattacherjee AK, Chaudhary P, Gangola S, Rajan S (2021b) Exploration of *Klebsiella pneumoniae* M6 for paclobutrazol degradation, plant growth attributes, and biocontrol action under subtropical ecosystem. PLoS One 16:e0261338. https://doi.org/10.1371/journal.pone.0261338

Kumar G, Lal S, Soni SK, Maurya SK, Shukla PK, Chaudhary P, Bhattacherjee AK, Garg N (2022) Mechanism and kinetics of chlorpyrifos co-metabolism by using environment restoring microbes isolated from rhizosphere of horticultural crops under subtropics. Front Microbiol 13:891870. https://doi.org/10.3389/fmicb.2022.891870

Kumar M, Xiong X, He M, Tsang DC, Gupta J, Khan E, Bolan NS (2020) Microplastics as pollutants in agricultural soils. Environ Pollut 265:114980. https://doi.org/10.1016/j.envpol.2020.114980

Kumar S, Das MP, Rebecca LJ, Sharmila S (2013) Isolation and identification of LDPE degrading fungi from municipal solid waste. J Chem Pharm Res 5(3):78–81

Kunlere IO, Fagade OE, Nwadike BI (2019) Biodegradation of low density polyethylene (LDPE) by certain indigenous bacteria and fungi. Int J Environ Stud 76(3):428–440. https://doi.org/10.1080/00207233.2019.1579586

Lambert S, Wagner M (2018) Microplastics are contaminants of emerging concern in freshwater environments: an overview. Freshwater microplastics. Springer, Cham, pp 1–23

Lameh F, Baseer AQ, Ashiru AG (2022) Comparative molecular docking and molecular-dynamic simulation of wild-type-and mutant carboxylesterase with BTA-hydrolase for enhanced binding to plastic. Eng Life Sci 22(1):13–29. https://doi.org/10.1002/elsc.202100083

Lin Z, Jin T, Zou T, Xu L, Xi B, Xu L, He J, Xiong L, Tang C, Peng J, Zhou Y, Fei J (2022) Current progress on plastic/microplastic degradation: fact influences and mechanism. Environ Pollut 304:119159. https://doi.org/10.1016/j.envpol.2022.119159

Liu SY, Leung MML, Fang JKH, Chua SL (2021) Engineering a microbial 'trap and release' mechanism for microplastics removal. Chem Eng J 404:127079. https://doi.org/10.1016/j.cej.2020.127079

Miri S, Saini R, Davoodi SM, Pulicharla R, Brar SK, Magdouli S (2022) Biodegradation of microplastics: better late than never. Chemosphere 286:131670. https://doi.org/10.1016/j.chemosphere.2021.131670

Mohan AJ, Sekhar VC, Bhaskar T, Nampoothiri KM (2016) Microbial assisted High Impact Polystyrene (HIPS) degradation. Bioresour Technol 213:204–207. https://doi.org/10.1016/j.biortech.2016.03.021

Muenmee S, Chiemchaisri W, Chiemchaisri C (2016) Enhancement of biodegradation of plastic wastes via methane oxidation in semi-aerobic landfill. Int Biodeterior Biodegradation 113l:244–255. https://doi.org/10.1016/j.ibiod.2016.03.016

Niu L, Li Y, Li Y, Hu Q, Wang C, Hu J, Zhang W, Wang L, Zhang C, Zhang H (2020) New insights into the vertical distribution and microbial degradation of microplastics in urban river sediments. Water Res. https://doi.org/10.1016/j.watres.2020.116449

Nwachukwu S, Obidi O, Odocha C (2010) Occurrence and recalcitrance of polyethylene bag waste in Nigerian soils. Afr J Biotechnol 9:6096–6104

Oda M, Yamagami Y, Inaba S, Oida T, Yamamoto M, Kitajima S, Kawai F (2018) Enzymatic hydrolysis of PET: functional roles of three Ca^{2+} ions bound to a cutinase-like enzyme, Cut190*, and its engineering for improved activity. Appl Microbiol Biotechnol 102:10067–10077

Ojha N, Pradhan N, Singh S, Barla A, Shrivastava A, Khatua P, Rain V, Bose S (2017) Evaluation of HDPE and LDPE degradation by fungus, implemented by statistical optimization. Sci Rep 7:39515

Olicón-Hernández DR, González-López J, Aranda E (2017) Overview on the biochemical potential of filamentous fungi to degrade pharmaceutical compounds. Front Microbiol 8:1792. https://doi.org/10.3389/fmicb.2017.01792

Oliveira J, Belchior A, da Silva VD, Rotter A, Petrovski Ž, Almeida PL, Lourenço ND, Gaudêncio SP (2020) Marine environmental plastic pollution: mitigation by microorganism degradation and recycling valorization. Front Mar Sci 7:567126. https://doi.org/10.3389/fmars.2020.567126

de Oliveira TA, Barbosa R, Mesquita AB, Ferreira JH, de Carvalho LH, Alves TS (2020) Fungal degradation of reprocessed PP/PBAT/ thermoplastic starch blends. J Mater Res Technol 9(2):2338–2349. https://doi.org/10.1016/j.jmrt.2019.12.065

Özdemir S, Akarsu C, Acer Ö, Fouillaud M, Dufossé L, Dizge N (2022) Isolation of thermophilic bacteria and investigation of their microplastic degradation ability using polyethylene polymers. Microorganisms 10:2441. https://doi.org/10.3390/microorganisms10122441

Paço A, Duarte K, da Costa JP, Santos PSM, Pereira R, Pereira ME, Freitas AC, Duarte AC, Rocha-Santos TAP (2017) Biodegradation of polyethylene microplastics by the marine fungus *Zalerion maritimum*. Sci Total Environ 586:10–15

Paço A, Jacinto J, da Costa JP, Santos PS, Vitorino R, Duarte AC, Rocha-Santos T (2019) Biotechnological tools for the effective management of plastics in the environment. Crit Rev Environ Sci Technol 49(5):410–441. https://doi.org/10.1080/10643389.2018.1548862

Park SY, Kim CG (2019) Biodegradation of micro-polyethylene particles by bacterial colonization of a mixed microbial consortium isolated from a landfill site. Chemosphere 222:527–533. https://doi.org/10.1016/j.chemosphere.2019.01.159

Parveen H, Chaudhary A, Chaudhary P, Sultana R, Khati P, Kumar G, Khati P, Rana M, Bhatt P (2022) Omics approaches for the production of the microbial enzymes and applications. In: Bhatt P (ed) Industrial applications of microbial enzymes. Ist edition. CRC Press. https://doi.org/10.1201/9781003202998-16

Prata JC (2018) Airborne microplastics: consequences to human health? Environ Pollut 234:115–126

Raghavendra VB, Uzma M, Govindappa M, Vasantha RA, Lokesh S (2016) Screening and identification of polyurethane (PU) and low density polyethylene (LDPE) degrading soil fungi isolated from municipal solid waste. Int J Curr Res 8:34753–34761

Restrepo-Flórez J, Bassi A, Thompson MR (2014) Microbial degradation and deterioration of polyethylene—a review. Int Biodeterior Biodegradation 88:83–90. https://doi.org/10.1016/j.ibiod.2013.12.014

Rogers KL, Carreres-Calabuig JA, Gorokhova E, Posth NR (2020) Micro-by-micro interactions: how microorganisms influence the fate of marine microplastics. Limnol Oceanogr Lett 5(2):18–36

Sanchez PA (2019) Properties and management of soils in the tropics. Cambridge University Press, Cambridge. https://doi.org/10.1017/9781316809785

Sangale MK, Shahnawaz M, Ade AB (2019) Potential of fungi isolated from the dumping sites mangrove rhizosphere soil to degrade polythene. Sci Rep 9:5390

Santana MF, Ascer LG, Custodio MR, Moreira FT, Turra A (2016) Microplastic contamination in natural mussel beds from a Brazilian urbanized coastal region: rapid evaluation through bioassessment. Mar Pollut Bull 106:183–189

Santo M, Weitsman R, Sivan A (2013) The role of the copper-binding enzyme—laccase—in the biodegradation of polyethylene by the actinomycete *Rhodococcus Ruber*. Int Biodeterior Biodegradation 84:204–210

Schwartz M, Perrot T, Aubert E, Dumarçay S, Favier F, Gérardin P, Gelhaye E (2018) Molecular recognition of wood polyphenols by phase II detoxification enzymes of the white rot *Trametes versicolor*. Sci Rep 8(1):1–11. https://doi.org/10.1038/s41598-018-26601-3

Sheik S, Chandrashekar KR, Swaroop K, Somashekarappa HM (2015) Biodegradation of gamma irradiated low density polyethylene and polypropylene by endophytic fungi. Int Biodeterior Biodegradation 105:21–29

Shimpi N, Borane M, Mishra S, Kadam M (2012) Biodegradation of polystyrene (PS)-poly (lactic acid) (PLA) nanocomposites using *Pseudomonas aeruginosa*. Macromol Res 20:181–187. https://doi.org/10.1007/s13233-012-0026-1

Shin J, Kim JE, Lee YW, Son H (2018) Fungal cytochrome P450s and the P450 complement (CYPome) of *Fusarium graminearum*. Toxins 10(3):112. https://doi.org/10.3390/toxins10030112

Shirke AN, White C, Englaender JA, Zwarycz A, Butterfoss GL, Linhardt RJ, Gross RA (2018) Stabilizing leaf and branch compost cutinase (LCC) with glycosylation: mechanism and effect on PET hydrolysis. Biochemistry 57:1190–1200. https://doi.org/10.1021/acs.biochem.7b01189

Singh S, Chaudhary P, Sharma A (2023) Efficient decolourization of toluidine blue and methylene blue using *Bacillus cereus*. Vegetos. https://doi.org/10.1007/s42535-023-00747-6

Skariyachan S, Taskeen N, Kishore AP, Krishna BV, Naidu G (2021) Novel consortia of *Enterobacter* and *Pseudomonas* formulated from cow dung exhibited enhanced biodegradation of polyethylene and polypropylene. J Environ Manag 284:112030. https://doi.org/10.1016/j.jenvman.2021.112030

Skoczinski P, Krause L, Raschka A, Dammer L, Carus M (2021) Chapter one—current status and future development of plastics: solutions for a circular economy and limitations of environmental degradation. In: Weber G, Bornscheuer UT, Wei R (eds) Methods in enzymology, vol 648. Academic Press, pp 1–26. https://doi.org/10.1016/bs.mie.2020.11.001

Sol D, Laca A, Laca A, Díaz M (2020) Approaching the environmental problem of microplastics: importance of WWTP treatments. Sci Total Environ 140016. https://doi.org/10.1016/j.scitotenv.2020.140016

Syranidou E, Karkanorachaki K, Amorotti F, Repouskou E, Kroll K, Kolvenbach B, Corvini PF, Fava F, Kalogerakis N (2017) Development of tailored indigenous marine consortia for the degradation of naturally weathered polyethylene films. PLoS One 25;12(8):e0183984. https://doi.org/10.1371/journal.pone.0183984

Taniguchi I, Yoshida S, Hiraga K, Miyamoto K, Kimura Y, Oda K (2019) Biodegradation of PET: current status and application aspects. ACS Catal 9:4089–4105

Thakur S, Mathur S, Patel S, Paital B (2022) Microplastic accumulation and degradation in environment via biotechnological approaches. Water 14(24):4053. https://doi.org/10.3390/w14244053

Vethaak AD, Martinez-Gomez C (2020) Micro and nanoplastics in the aquatic environment with special reference to synthetic fibers. Synthetic Nano and Microfibers:111–130

Vivi VK, Martins-Franchetti SM, Attili-Angelis D (2019) Biodegradation of PCL and PVC: *Chaetomium globosum* (ATCC 16021) activity. Folia Microbiol 64:1–7

Wierckx N, Narancic T, Eberlein C, Wei R, Drzyzga O, Magnin A, Ballerstedt H, Kenny ST, Pollet E, Averous L (2018) Plastic biodegradation: challenges and opportunities. In: Steffan R (ed) Handbook of hydrocarbon and lipid microbiology consequences. Springer International Publishing, pp 1–30

Wurzbacher C, Fuchs A, Attermeyer K, Frindte K, Grossart HP, Hupfer M, Casper P, Monaghan MT (2017) Shifts among Eukaryota, bacteria, and archaea define the vertical organization of a lake sediment. Microbiome 5(41). https://doi.org/10.1186/s40168-017-0255-9

Yan F, Wei R, Cui Q, Bornscheuer UT, Liu YJ (2021) Thermophilic whole-cell degradation of polyethylene terephthalate using engineered *Clostridium Thermocellum*. Microb Biotechnol 14:374–385. https://doi.org/10.1111/1751-7915.13580

Yang H, Chen G, Wang J (2021) Microplastics in the marine environment: sources, fates, impacts and microbial degradation toxics. 9:41. https://doi.org/10.3390/toxics9020041

Yang Y, Yang J, Wu WM, Zhao J, Song Y, Gao L, Yang R, Jiang L (2015) Biodegradation and mineralization of polystyrene by plastic-eating mealworms: part 2. Role of gut microorganisms. Environ Sci Technol 49:12087–12093

Yoshida S, Hiraga K, Takehana T, Taniguchi I, Yamaji H, Maeda Y, Toyohara K, Miyamoto K, Kimura Y, Oda K (2016) A bacterium that degrades and assimilates poly (ethylene terephthalate). Science 351:1196–1199

Yu H, Zhang X, Hu J (2020) Ecotoxicity of polystyrene microplastics to submerged carnivorous *Utricularia vulgaris* plants in freshwater ecosystems. Environ Pollut 265:114830. https://doi.org/10.1016/j.envpol.2020.11483

Yuan J, Ma J, Sun Y, Zhou T, Zhao Y, Yu F (2020) Microbial degradation and other environmental aspects of microplastics/plastics. Sci Total Environ 715:136968. https://doi.org/10.1016/j.scitotenv.2020.136968

Zhang J, Danling G, Quanhao L, Yixuan Z, Li L, Hanfeng L, Qirui B, Yucheng Z (2020) Biodegradation of polyethylene microplastic particles by the fungus *Aspergillus flavus* from the guts of wax moth *Galleria mellonella*. Sci Total Environ 704:135931. https://doi.org/10.1016/j.scitotenv.2019.135931

Role of Microbes in Dye Degradation

Shivani Singh, Parul Chaudhary, Geeta Bhandari, Durgesh Kumar Jaiswal, Viabhav Kumar Upadhayay, Ashish Kumar, Neha Saini, and Anita Sharma

Abstract Modern industrialization has high demand for dyes to impart better color to cosmetic, paper, textile and pharmaceutical industry. Dyes are of several kinds depending on the base of their chromophore groups, chemical structures, methods of application on the substrate, and dissociation properties in aqueous solution. The wastewater created during dyeing process, application, and operation contains residual dyes, mordants, and organic and inorganic contaminants, which are having drastic negative effect on the valuable environment and its living creatures. Various kinds of technologies are needed for the absolute degradation of dyes before the safely discharge of effluents. Among them, physiochemical methods are mostly used but are not environment friendly and cost-effective and are producing secondary hazards that required further treatment. So, there is a growing demand for alternative environment and budget-friendly biological approaches that involve the use of microbes, fungi, algae, and plants. These biological methods involve several techniques such as biosorption, biodegradation, and biomineralization. This chapter discusses different types of dyes and their degradation using microorganisms.

S. Singh · A. Kumar · N. Saini · A. Sharma
Department of Microbiology, Govind Ballabh Pant University of Agriculture and Technology, Pantnagar, Uttarakhand, India

P. Chaudhary (✉)
School of Agriculture, Graphic Era Hill University, Dehradun, Uttarakhand, India

G. Bhandari
Department of Biosciences, Himalayan School of Biosciences, Swami Rama Himalayan University,
Dehradun, Uttarakhand, India

D. K. Jaiswal
Department of Biotechnology, Graphic Era Deemed to be University, Dehradun, Uttarakhand, India

V. K. Upadhayay
Department of Microbiology, College of Basic Sciences and Humanities, Dr. Rajendra Prasad Central Agricultural University, Pusa, Samastipur, Bihar, India

P. Chaudhary, A. Chaudhary (eds.), *Microbial Inoculants*, Interdisciplinary Biotechnological Advances, https://doi.org/10.1007/978-981-97-0633-4_15

Keywords Dye degradation · Bacteria · Fungi · Bioremediation · Biological methods

1 Introduction

Rapid urban population growth brought on by industrial expansion leads to a tremendous increase in solid and wastewater, which ultimately degrades the quality of aquatic resources and adds to the burden of serious human diseases. According to the Ministry of Environment, Forest and climate change (MOEFCC), paint industry is placed under "Red Category" which causes severe environmental pollution. Textile industries are a major source of water pollution globally (Gaikwad et al. 2014).

A colorant can be defined as dye or pigment. A dye is a water- or oil-soluble colored unsaturated organic substance that chemically binds to the substrate to which it is applied. It imparts color to the fabric by absorbing wavelength from visible spectrum Ardila-Leal et al. (2021). Dyes should have a chromophore group (unsaturated ring that absorbs light in the visible spectrum and exhibits resonance of electron in conjugated ring) and an auxochrome group that is responsible for the intensity of color and influences the solubility of dye and a conjugated aromatic ring (benzene, naphthalene, anthracene, etc.). This property distinguishes dyes from pigments which do not chemically bind to the material they color and are insoluble. However, pigment is a water-insoluble substance and is used as a coloring agent (Benkhaya et al. 2020).

Textile industries are chemically intensive industries in India and consume a large volume of water. They are major polluters of aquatic bodies and groundwater. These industries also consume high amounts of fuels and chemicals. In India, textile industries are mostly situated in Mumbai, Ahmedabad, Ludhiana, and Kanpur. Textile industries do not only pollute water but they are also responsible for air pollution as they release oxides of sulfur, dust, volatile organic composites, and nitrogen. Water pollution is mainly caused by discharge of untreated effluent (Bhatia et al. 2017). Color effluents not only deteriorate the quality of water but also deteriorate the quantity of water (Lellis et al. 2019).

In the textile industry, three different types of fibers are used such as cellulose fiber (plant origin: cotton, rayon, hemp, linen), protein fiber (animal origin: wool, cashmere, angora, and silk), and synthetic fiber (man-made: nylon, polyester, polypropylene) (Ghaly et al. 2014). Nowadays, most of the textile industries use synthetic dyes that have complex molecular structures and are difficult to degrade. Worldwide production of dye is about 7×10^5 metric tons. Nearly 10–15% of dyes is lost during their synthesis and dyeing processes (Azanaw et al. 2022). Among various classes of dyes, azo dyes are widely used in textile industries. Depending on the class of dyes and application, 2% of basic dyes and 50% of reactive dyes are directly lost with effluents and lead to severe pollution of the surface and groundwater (Singh et al. 2012). Industrial effluents show a drastic effect on water bodies by decreasing water quality by increasing BOD and COD. They inhibit penetration of light in water and impair photosynthesis, inhibit plant growth, enter our food chain, and bioaccumulate in higher organisms (Lellis et al. 2019).

Industrial wastewaters usually have considerably higher toxicity and need a pretreatment before they are sent to municipal wastewater treatment plants (Ahmed et al. 2021). Many industries use physical and chemical methods to treat wastewater. Physiochemical procedures are quite expensive and generate a lot of sludge. Biological methods have many advantages over physiochemical methods as they are of low cost, eco-friendly, and easy to use and convert toxic compounds into less toxic compounds. A variety of biological techniques such as biosorption and biodegradation have been proposed for the removal of textile effluent (Karim et al. 2018).

A number of microorganisms like bacteria, fungi, algae, and yeasts have been reported to decolorize or detoxify the industrial effluents. These microorganisms detoxify and mineralize dye with the help of different enzymes under different environmental conditions. Different decolorizing bacteria have been reported such as *Pseudomonas* sp., *Bacillus* sp., *Staphylococcus* sp., *Klebsiella* sp., *Aeromonas* sp., etc. (Fareed et al. 2022). Microbial efficiency to decolorize dye depends upon the adaptability of microorganisms for dyes. Microbial consortia have many advantages over single strain for decolorization. Mixture of bacterial strains attacks dye at different positions and metabolizes dye completely into CO_2, H_2O, and salt of organic origin without leaving any harmful product (Mahmood et al. 2015; Parmar and Shukla 2019). Mixed microbial culture completely mineralized the dyes by the synergistic metabolic activities (Sultana 2017).

Dye decolorization process depends on several enzymes of microbial origin like laccase, tyrosinase, lignin peroxidase, manganese peroxidase, and veratryl alcohol oxidase (Kurade et al. 2012; Karim et al. 2018). Extracellular peroxidase enzymes include laccase, lignin, and manganese peroxidase, versatile peroxidase, and dye decolorization peroxidase. These enzymes have wide application in lignin degradation, removal of phenolic compounds, and synthetic dye decolorization (Falade et al. 2017). Laccase is a copper comprising enzyme formed by different plants and fungi that oxidize anilines and phenols in the presence of oxygen (Li et al. 2014). Bacterial laccases have more applications over fungal laccases because of their highly efficient expression, much higher thermostability, and tolerance to alkaline condition (Falade et al. 2017). Many bacterial species are reported for lignin degradation, color removal, and toxicity reduction (Zainith et al. 2019). According to Kannan et al. (2013), *Pseudomonas putida* decolorizes Remazol black B dye at pH 7 through degradation mechanisms instead of adsorption on the cell surface. Microorganisms used for decolorization are chosen on the basis of several factors such as structure of the dye, concentration of the dye to be decolorized, adsorption of the dye on the cell membrane, permeability of the dye through the cell membrane, pH, temperature, and salt concentration (Parmar and Shukla 2019).

Dyes are photolytically constant and insistent in natural surroundings for a long. Dyes can be classified on the basis of their chromophore groups, chemical

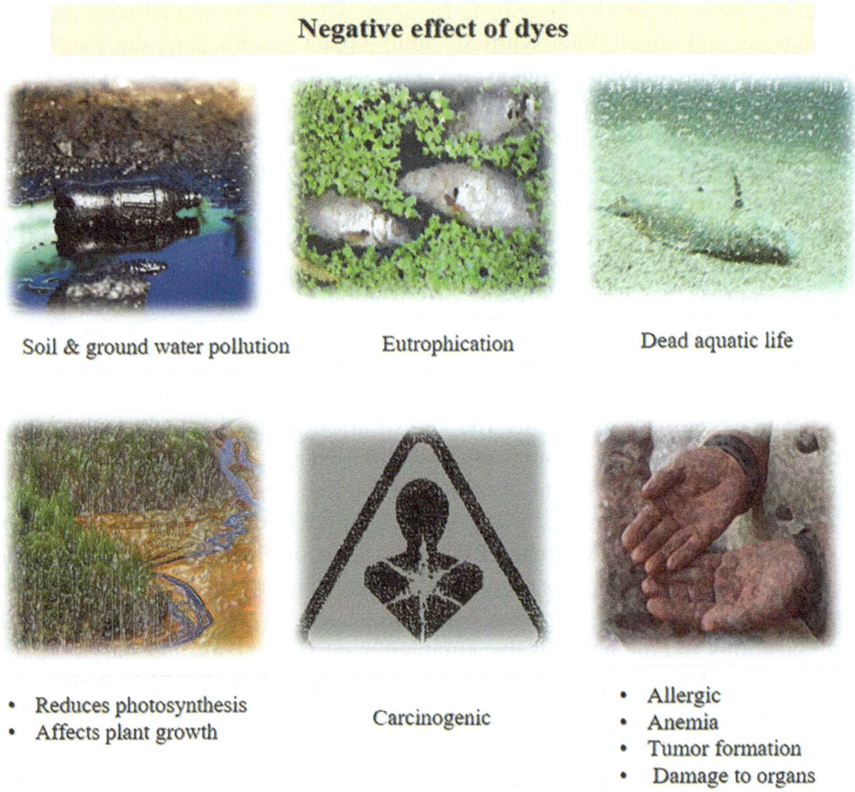

Fig. 1 Toxic effects of dyes

structures, methods of application on the substrate, and dissociation properties in aqueous solution (acidic dye, basic dye, and direct dye) (Singh et al. 2012) (Fig. 1).

2 Synthetic Dyes

The first synthetic dye was revealed by William Henry Perkin. It was mauveine or aniline purple that had a heterocyclic structure. Synthetic dyes were originally derived from organic molecules (Ardila-Leal et al. 2021). Azo dye, a type of synthetic dye, is used in large amount as compared to natural dyes due to its sustainable nature, high coloring power in a short time, and cost-effectiveness. Azo dyes are nonbiodegradable and persist in the environment due to their greater stability toward light, temperature, water, detergent, chemicals, and bleaching (Ali et al. 2023a, b). Dyes may be classified into different types on the basis of various factors such as chromophore structure (like nitro and nitroso dye, azo dye, triaryl methane, indigo, and anthraquinone dye) and methods of application such as acidic, basic, vat, mordant disperse, azo, reactive, and direct dye.

2.1 Acid (Anionic) Dyes

Acidic dyes, the acidified form of basic dyes, are used in the form of sulfonic acid, carboxylic acid, or phenol. These dyes are mainly applied on protein fibers (wool, leather, and silk), nylon, and acrylics. Anionic dyes are applied at low pH where they form ionic bond between the NH_2 group of protein fiber and the acidic group of the dye. Rose bengal, bromocresol green, Congo red, Indian ink, and Nigrosine are some examples of anionic dyes (Cerboneschi et al. 2015).

2.2 Basic (Cationic) Dyes

Basic dyes are also known as cationic dyes as they have positive charge on chromophore. They contain basic group such as ($-NH_2$) or ($-NR_2$). These dyes are water-soluble and electrostatically attracted to the negatively charged group of the substrate. They are used to dye acrylic fibers and modified nylons, polyesters, wool, cotton, leather, paper, etc. (Benkhaya et al. 2020). Malachite green, toluidine blue, methylene blue, and crystal violet are some examples of basic dyes.

2.3 Direct Dyes

Direct dyes are directly applied on the substrate under neutral or alkaline medium. They bind to the fiber through hydrogen bonding. They are classified on the basis of the chromophoric group such as azo, phthalocyanine, etc. They give very bright shades to the fabrics without using any mordant. The first direct dye discovered by Paul Bottiger was Congo red (Benkhaya et al. 2020). These dyes bind to cotton, paper, wool, leather, and nylon. Congo red and Martius yellow are examples of direct dyes.

2.4 Vat Dyes

Vat dyes are derivatives of indigo which are natural dyes and are now produced synthetically (Ramasany et al. 2012). They are insoluble in water but soluble in hot water and in the presence of a little amount of sodium carbonate. Halogenated derivatives of indigo provide vat dyes of other class, i.e., thioindigo and anthraquinone. These dyes are used to color rayon and cellulose. Alizarin is an example of vat dye.

2.5 *Reactive Dyes*

Reactive dyes constitute the second largest class of dye. They attach themselves through reactive groups to bind to fibers irreversibly. These dyes bind with covalent bond to hydroxyl amine and the sulfhydryl group of protein fibers. Poor fixation on the fiber is one of the major drawbacks of this dye. They are used to dye cotton, wool, or silk but lost during washing (Benkhaya et al. 2020). Remazol Brilliant Blue R and Procion Red are examples of reactive dyes.

2.6 *Sulfur Dyes*

These dyes are water-insoluble and have the highest molecular weight. They have the dullest range of all colors from all the classes. They are generally used to dye cellulosic fibers (Benkhaya et al. 2020). Indophenol and sulfur black are example of sulfur dyes.

2.7 *Azo Dyes*

Azo dyes belong to the largest class of dye. Unique features of azo dyes are the presence of azo linkage (N=N) along with aromatic heterocyclic units which are directly attached with nitrogen (Fareed et al. 2022). These dyes are used in cosmetics, paper, paint, and food industries. Congo red, Sudan stain, and Azure B are some examples of azo dyes.

3 Decolorization

Colored effluents are released from various textile industries such as dyestuff, paper, and cosmetic industries (Markandeya and Shukla, 2022). Polluted water containing dyes shows less degradability due to high molecular weight, high water solubility, and conjugated unsaturated ring structure, which reduce their permeability through the microbial cell membrane. Dye-containing effluents have very high BOD and presence of metal ions. Three methods are commonly used to decolorize textile effluents, namely, physical, chemical, and biological methods. Physical and chemical methods are expensive and produce high amounts of sludge after treatment which have to be further processed. Biological approaches are cost-effective and eco-friendly and produce nontoxic products such as CO_2 and H_2O in the environment after degradation (Singh et al. 2017).

3.1 *Physical Method*

3.1.1 Membrane Filtration

Membrane filtration is also known as membrane technology. Membrane technology consists of four different membrane groups, such as microfiltration (MF), ultrafiltration (UF), nanofiltration (NF), and reverse osmosis (RO). Membranes are made up of inorganic and polymeric materials. Membrane filtration is very selective and consumes low energy. Some modern membranes are also available which are highly resistant to heat, alkaline, acidic condition, harmful chemical, and microbial attack. Separation is made by membrane by creating a boundary between different bulk materials as the flow rate of different solutes and solvent is different (Thamaraiselvan and Noel 2015; Ben et al. 2018).

The membrane separation method, which has been deemed as a promising approach for heavy metal removal, possesses some of its best qualities, including high efficiency, simple operation, and limited space requirements (Xiang et al. 2022). Devis et al. (2023) reported the potential of graphene oxide membranes for the removal of the combination of organic dyes (MB and CR) and exhibited good water penetrability with a rate of 183 L m^{-2} h^{-1} bar^{-1} emphasizing its high possibility for purification of water.

3.1.2 Coagulation

Coagulation is a process that promotes aggregation of dissolved and suspended solids. Immediate formation of flocs in the effluents containing coagulants is known as flocculation. It is one of the most efficient and low-cost methods for the removal of color, chemical oxygen demand, and pigments. Partial removal of dye depends upon the coagulant used (Moghaddam et al. 2010; Singh et al. 2017). Coagulants are of different types such as inorganic and synthetic organic polymers and natural coagulants. Inorganic and synthetic organic polymer coagulants are iron- and aluminum-based (aluminum sulfate, alum, lime). Natural coagulants are hyacinth bean, cactus, *Tamarindus indica*, acacia, neem, etc. These coagulants have many advantages over chemical coagulants. Natural coagulants are biodegradable and have low toxicity and low residual sludge production and are environment friendly (Hoong and Ismail 2018; Kolhe and Ingole 2020). Removal of reactive dyes is mostly carried out by coagulants (Hynes et al. 2020).

3.1.3 Electrochemical Treatment

Electrochemical process involves generation of metal ions (Fe^{2+} or Al^{3+}) from iron or aluminum anodes. These metal ions are hydrolyzed into $Fe(OH)_2$ or $Al(OH)_3$ which have higher adsorption flocculation activity and can effectively remove colloidal particles of the dyes and other impurities (Palanisamy et al. 2020). Generation of hydrogen gas on a cathode helps coagulated matter to float. The electrochemical process decomposes organic compounds present in the wastewater into low molecular weight compound or directly into CO_2. This method involves oxidation, floatation, flocculation and micro-electrolysis. It does not produce secondary pollution. It requires a compact equipment and has high removal efficiency at low temperature, but it is a cost-intensive, high-energy-consuming, and time-consuming process and erosion of the electrodes also takes place. This method has more advantages than others (Chen et al. 2018; Beulah and Muthukumaran 2020; Palanisamy et al. 2020). A new electrochemical approach to degrade organic dyes in water is presented. This will be an effective approach in real water samples (Sarfo et al. 2023).

3.1.4 Adsorption

The process of transferring organic compounds from wastewater to the surface of the adsorbent is called adsorption. This is the most effective and highly efficient method for the removal and decolorization of dye (Beulah and Muthukumaran 2020; Mavinkattimath et al. 2023). Absorbent is the material on which adsorption process occurs and adsorbate is the material that binds to the adsorbent due to the potential for diffusion process. Presence of pores, more surface area, contact time, and temperature are characteristics of the adsorbent. The common absorbents are activated carbon, metal oxides, metal hydroxides, and clay (Chong and Tam 2020). It is a low-cost easily available method and used to remove dye and waste (Beulah and Muthukumaran 2020).

3.1.5 Ion Exchange Method

Removal of cations and anions from industrial effluents can be done by passing the effluent through the bed of resins, where anions and cations replace hydroxyl and sodium ions, respectively. This method is known as ion exchange method (Vijayaraghavan et al. 2013). Anion exchange resins are used for reactive dyes, whereas ion exchange containing acrylic resin is used to remove other classes of dyes. Effluents are passed over the ion exchange resins until all the exchange sites are not saturated. There is no loss of absorbent. On the other hand, this method uses expensive organic solvents (Fig. 2).

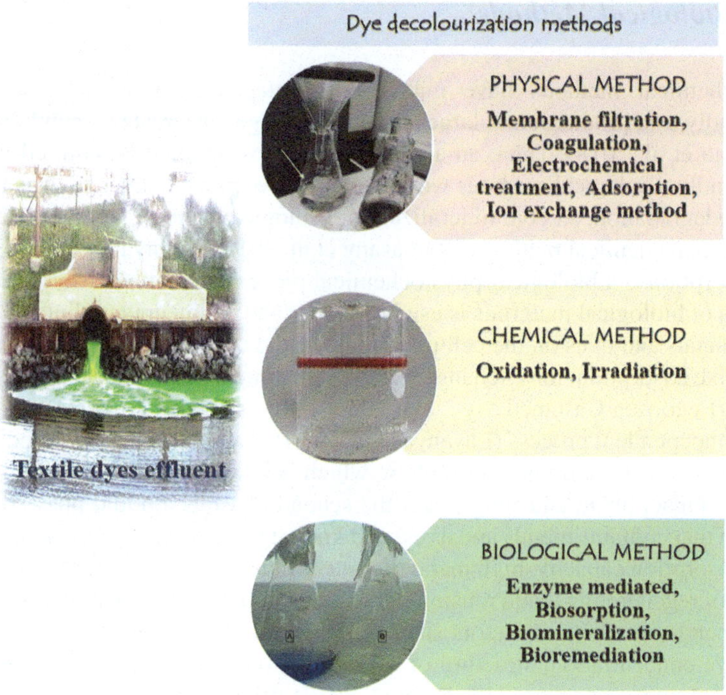

Fig. 2 Methods of dye decolorization

3.2 *Chemical Methods*

3.2.1 Oxidation

The aim of any oxidation process is to use hydroxyl free radicals as strong oxidants that react with organic contamination of the effluents and then destroy the compounds. Using advanced oxidation Fenton and ultraviolet/hydrogen peroxide processes is investigated to remove color and COD (Márquez et al. 2023; Slama et al. 2023).

3.2.2 Irradiation

Irradiation uses gamma radiation or electron beam which induces oxidation of the effluents. It is used in the laboratory to eliminate a wide variety of organic compounds and harmful organisms. This method is efficient in degradation of a large variety of pollutants. It only removes colorants at very low level. The disadvantage of the method is the requirement of a lot of dissolved oxygen during this process (Lavanya et al. 2014).

3.3 Biological Methods

Physiochemical methods have many disadvantages such as high cost, non-ecofriendly, and production of large amount of sludge that creates secondary pollutants (Gao et al. 2018). Thua, an alternative low-cost method is required which is eco-friendly. Biological methods which include several techniques such as biosorption, biodegradation, and biomineralization. Biological methods are also 5–20 times cheaper than chemical methods (Ramasany et al. 2012; Nachiyar et al. 2023).

Biosorption is a biological-physiochemical process where inactive, dead, or live biomass of biological materials is used to absorb and disintegrate pollutants such as heavy metals and dyes on the cell membrane or the cell wall through different processes like adsorption, ion exchange, electrostatic interaction, and chemical precipitation (Legorreta-Castañeda et al. 2020; Mushtaq et al. 2023). It is an energy-independent process (Lavanya et al. 2014). Two phases are required for biosorption. The first phase is a solid phase which is known as biosorbent (biomass of bacteria, algae, plant, and fungi), and the second phase is a liquid phase (aqueous solution that contains metals and dyes) and known as sorbate. Biosorbent must have high biosorption capacity, high metal binding efficiency, sustainability, and the possibility of regenerating it and reusing it (Ali Redha 2020). The feasibility of biosorption depends on different factors such as temperature, pH, ionic quality, exposure time, dye concentration, structure of dyes, functional groups present in dyes, and the type of microorganisms (Bhatia et al. 2017). Biomineralization is the chemical modification of some pollutants by microorganisms. Pollutants are transformed into CO_2, H_2O, or NH_3. This process is known as biotransformation. It also describes the formation of composite materials in organisms (Fenner et al. 2021).

Bioremediation is an energy-dependent process that requires living microorganisms to break down hazardous substances into less toxic inorganic compounds (Ayilara and Babalola 2023). It is a green technique to remove dyes. This action is based upon microbial enzymes such as laccase, lignin peroxidase, tyrosinase, and manganese peroxidase (Bhatia et al. 2017). A complete biodegradation process is known as mineralization that produces water, carbon dioxide, or any simpler inorganic molecules from the pollutants. This process is separated into two stages. The first phase takes place under anaerobic condition (pyrolysis and reduction of dye occur), and in the second phase, by-products that are formed during the first phase are employed by aerobic bacteria and converted into small molecules (CO_2 and NH_3) (Gao et al. 2018). This process is generally followed during reduction of azo dyes. Production of aromatic amines after anaerobic treatment is a major disadvantage of this process. Amines are readily bio-transformed under aerobic condition (Vijayaraghavan et al. 2013). It is efficient in reducing COD and turbidity of the effluents.

3.3.1 Bacterial Remediation

3.3.1.1 Decolorization by Microbial Cultures

Bacterial decolorization is normally faster than fungal decolorization. Microbes remediate organic contamination and can be used as a pollution indicator for numerous substances in wastewater (Siregar et al. 2020; Singh et al. 2022). Microbes have easy culturing methods and show fast growth. The in-between products produced in dye decolorization are also reduced by enzymes like oxygenases and hydroxylases. Various gram-positive and gram-negative bacteria that show decolorization of dyes are *Bacillus subtilis* and *Proteus* sp. *Klebsiella pneumonia*, *Pseudomonas aeruginosa*, *Pseudomonas putida*, and *Escherichia coli* exhibit auspicious decolorizing efficiency of dyes (Table 1). *Bacillus circulans* strain decolorizes methyl orange with the help of azo reductase and laccase enzymes. It is also reported that this bacterium could also degrade the aromatic amines, thereby reducing dye toxicity (Gao et al. 2018). John et al. (2020) stated that *Salinivibrio kushneri* is involved in Coomassie brilliant blue degradation within 48 h due to the creation of NADH azoreductase enzyme which causes deamination and isomerization of diverse complexes. Congo red dye (25 ppm) was degraded by using the *Bacillus* sp. in 5 days (Hanis et al. 2020). *Bacillus* sp. and *Staphylococcus* degraded mordant black dye at acidic pH due to genetic heterogenicity, resistance, and interactions between microbes and azo dyes (Tahir et al. 2021). Bacterial isolates such as *Mesobacillus fusiformis* and *Lysinibacillus fusiformis* showed Congo red degradation in the presence of azoreductase enzymes (Subramaniam et al. 2023). Immobilized culture of *Bacillus cereus* is involved in methylene decolorization and showed presence of dye-degrading enzymes (Singh et al. 2023). *Ochrobactrum intermedium* is also involved in methyl orange degradation up to 89% (Mishra et al. 2023). Advantages

Table 1 Role of bacteria in dye degradation

S. no.	Microbes	Dye removed	References
1.	*Oerskovia paurometabola*	Azo dyes	Franca et al. (2020)
2.	*Bacillus pumilus*, *B. thuringiensis consortium*	Azo dyes	Afrin et al. (2021)
3.	*Enterobacter* sp.	Crystal violet	Roy et al. (2018)
4.	*Bacillus pumilus*	Textile dyes	Liu et al. (2023)
5.	*Pseudomonas* sp.	Dyes	Cagide et al. (2023)
6.	*Lactococcus lactis*	Azo dyes	Oliveira et al. (2023)
7.	*Alteribacillus*	Azo dye	Tian et al. (2023)
8.	*Microbial consortium*	Pharmaceutical dyes	Poddar et al. (2023)
9.	*Streptomyces coelicolor*	Azo dyes	Preethi et al. (2023)
10.	*Bacillus* sp.	Congo red	Ellafi et al. (2023)
11.	*Pseudomonas alcaliphila*	Methylene blue	Ghoniem et al. (2023)
12.	*Ralstonia eutropha*	Methylene blue	Habibi and Mehrabi (2017)

of using microbial cultures for dye decolorization are: they are environment friendly and cost-effective, they produce less sludge and end products that are nontoxic as compared to other methods (Saratale et al. 2010; Avci et al. 2023).

3.3.1.2 Decolorization by Extremophilic Bacteria

Extremophiles are the organisms that have the ability to grow and survive under extreme/adverse conditions such as highly acidic pH, alkaline pH, high or low temperature, and high salt concentrations. The unique enzymes of extremophiles and physiological conditions allow them to survive under harsh conditions (Giovanella et al. 2020). Under extreme conditions, the clean-up process of textile effluents is not much successful by physiochemical methods. Extremophiles are adapted to extreme conditions for their survival by using unique biological pathways. During the process of adaptation, extremophiles also evolve the capability to convert more unstable toxic compounds to less toxic compounds which are further used for their metabolism (Jeong and Choi 2020) (Fig. 3).

Acidithiobacillus ferrooxidans NCIMB 8455 is a halophilic bacterium and found very effective for bioremediation of U(IV) from mine effluents (Romero-Gonzalez et al. 2016). There have been several reports on marine bacteria that live in high salt concentration and decontaminate the seawater from toxic heavy metals. Abd-Elnaby et al. (2011) reported that *Vibrio harveyi* shows high absorption capacity for cadmium cations. *Nesterenkonia lacusekhoensis* EMLA3 is a halotolerant and

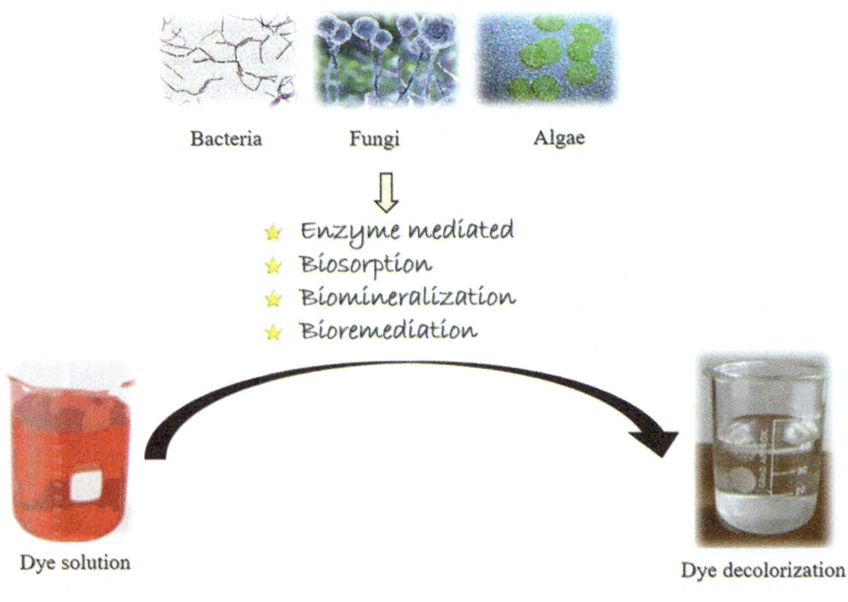

Fig. 3 Biological method of dye decolorization using bacteria, fungi, and algae

alkaliphilic bacterium that degraded methyl red under alkaline condition (pH 11.5) by 97% during the initial 16 h (Bhattacharya et al. 2017). *Marinobacter* spp. were able to transform aromatic compounds (naphthalene, biphenyl, phenanthrene) into simple carbon sources in high salt-containing environment (Ramalingam and Devi 2017).

3.3.1.3 Decolorization by Engineered Bacteria

The dye decolorization capability of the microbes can be enhanced by using biotechnological tactics. The recombinant technology helps in inserting gene segment, responsible for the production of enzymes involved in the degradation process in bacterial chromosomes. Wang et al. (2012) created a whole cell biocatalyst by engineering a bacterial laccase enzyme on the cell surface of *Pseudomonas putida*. This engineered cell was useful to decolorize the acid green dye and acid red 18. The engineered bacteria showed high activity toward acid green dye having decolorization percentage 91.2% starting in 3 h and 97.1% in 18 h, and for acid red, the organism showed 80.5% starting in 3 h and 89% after 18 h. No significant deterioration in decolorization effectiveness was seen at the first two rounds, and then the reaction equilibrium was extended and the remaining laccase activity ultimately reduced to low stages. This study shows that *P. putida* with surface-immobilized laccase was successful and could be used as regenerable biocatalyst for dye degradation. In *Bacillus subtilis*, a gene Cot A encoding laccase enzyme can also be transferred into *E. coli*, which can decolorize azo and anthraquinone dyes (Lellis et al. 2019). Liu et al. (2011) introduced *ars M* gene from *Rhodopseudomonas palustris* into indigenous *Sphingomonas desiccabilis* and *Bacillus idriensis*. This gene was responsible for removal of arsenic from arsenic-contaminated soil and environment by biovolatilization. Dye-degrading peroxidase enzyme of *Serratia* was cloned which effectively decolorized synthetic dyes (Xuejiao et al. 2023).

3.3.1.4 Decolorization by Immobilized Bacteria

Immobilization of microbial cultures is done to protect the microbial cells from dyes and toxic chemicals and from changes in environmental condition (Prasad and Aikat 2014). Immobilization is of two types: entrapment (microorganisms are enclosed with fibrous or porous material) and attachment (absorption of microorganisms on support material) (Wang et al. 2012). Two types of gels (natural and synthetic gels) are used for immobilization. Natural gels are soluble and biodegradable and have low mechanical strength, while synthetic gels are less soluble and have high mechanical strength. Some commonly used natural dyes are alginate, collagen, cellulose, keratins, chitin, chitosan, and carrageen (Zdarta et al. 2018). Bacterial cells can be covalently bonded to a support matrix. Covalent cross-linkages help in binding between cell to cell and cell to matrix. Prasad and Aikat (2014) isolated *Enterobacter cloacae SXCR* from petroleum-contaminated soil and immobilized bacterial cells in calcium alginate beads through entrapment. They showed that immobilized bacterial cells were able to reduce azo bond in Congo red dye enzymatically.

3.3.2 Myco-Remediation and Phyco-Remediation

Remediation by fungal cultures is known as myco-remediation (Table 2). For remediation, fungi must have characteristic features such as fast growth and large amount of biomass production and must produce some extracellular enzymes such as lignin

Table 2 Role of fungi and algae in dye decolorization

S. no.	Microbes	Dye removed	References
1.	*Aspergillus flavus*	Acid blue	Ameen et al. (2021)
2.	*Trametes hirsuta*	Anthraquinone dyes	Alam et al. (2021)
3.	*Myrothecium verrucaria*	Anthraquinone	Agrawal and Verma (2020)
4.	*Aspergillus niger*, *Trichoderma viride*	Azo dyes	Ali et al. (2023a, b)
5.	*Pestalotiopsis* sp.	Remazol Brilliant Blue	Syafiuddin and Fulazzaky (2021)
6.	*Bjerkandera adusta*	Malachite green, crystal violet	Gao et al. (2020)
7.	*Ralstonia pickettii* and *T. viride consortium*	Methylene blue	Nabilah et al. (2023)
8.	*Aspergillus quadrilineatus*	Dimethyl yellow	Yusuf et al. (2021)
9.	*Trichoderma asperellum*	Malachite green	Ting et al. (2021)
10.	*Aspergillus niger*	Congo red	Singh and Dwivedi (2022)
11.	*Phanerochaete chrysosporium*	Azo dyes	de Almeida et al. (2021)
12.	*Bjerkandera adusta*	Triphenylmethane	Gao et al. (2020)
13.	*Aspergillus* and *Chlorella sorokiniana consortium*	Disperse red	Tang et al. (2019)
14.	*Chromochloris zofingiensis*	Methylene blue	Bayat et al. (2023)
15.	*Eucheuma spinosum*	Azo dye	Mokhtar et al. (2017)
16.	*Scenedesmus*	Methylene blue	Afshariani and Roosta (2019)
17.	*Ulva fasciata*, *Pterocladia capillacea*	Synthetic dyes	El-Agawany et al. (2023)
18.	*Chlorella pyrenoidosa*	Methylene blue	Lebron et al. (2018)
19.	*Shewanella*	Textile dye	Chaieb et al. (2023)
20.	*Laminaria digitata*	Methylene blue	Bouzikri et al. (2023)
21.	*Penicillium italicum*	Erythrosine B	Çetinkaya et al. (2023)
22.	*Schizophyllum commune*	Reactive dye	Raees et al. (2023)
23.	*Aspergillus oryzae*	Rhodamine B	Souza et al. (2020)
24.	*Fusarium oxysporum*	Synthetic dyes	Thao et al. (2023)

peroxidases, manganese peroxidases, laccase, and tyrosinase (Bhatia et al. 2017). Due to higher cell surface ratio of fungi than single-celled organisms, fungi have greater physical contact with dye and produce more enzymes (Wang et al. 2012; Gul et al. 2023). White rot fungi, *Phanerochaete chrysosporium*, is a much studied fungus (Saratale et al. 2011). *Trametes flavida* is involved in degradation of Congo red dye and showed presence of degrading enzymes (Sharma et al. 2023). Alam et al. (2023) reported that *T. flavida* is involved in azo dye degradation due to the presence of laccase enzyme.

Remediation process that involves algae or microalgae is termed as phycoremediation. Algae, fresh or marine water organisms, can be extensively used in biosorption and bioaccumulation processes (Sachan et al. 2022; Sivalingam and Gopal 2023). Biosorption of methylene blue using *Lentinus crinitus* was reported by de Freitas et al. (2023). Algae are gaining attention in remediation processes because of the presence of large surface area, binding ability, simple nutrition requirements, high metal removal capacity, and ability to regenerate (Bhatia et al. 2017; Chandra et al. 2020). El-Naggar et al. (2021) reported that methyl orange dye was degraded using *Fucus vesiculosus* algae. Decolorization of dye by algae involves three different stages. Algal species that degrade various synthetic dyes are *Chlorella vulgaris*, *Lyngbya lagerlerimi*, and *Chroococcus minutes*. Remediation process through cyanobacteria is known as cyanoremediation. Cyanobacteria such as *Nostoc*, and *Anabaena* sp. have been studied extensively (Priya 2018). Bacterial-algal combination (*P. putida*, *Chlorella*, *L. plantarum*) were used in degradation of CI reactive blue dye (Ayed et al. 2021).

3.4 Lignin- and Manganese Peroxide-Degrading Enzymes

Various enzymes are involved in dye decolorization such as laccase, manganese peroxidases (MnP), lignin peroxidases (LiP), etc. Laccases, commonly referred to as phenol oxidases, are an example of an oxidoreductase. This multicopper oxidase enzyme may remove one electron to oxidize phenols and anilines (Bhatia et al. 2017). With the conversion of molecular oxygen to water, laccase oxidizes not only aromatic amines but also phenolic compounds (Sheikhi et al. 2012). Laccase has a molecular weight that is between 60 and 390 kDa (Saratale et al. 2011). The azo dye degradation process is carried out by laccase enzyme via a nonspecific free radical mechanism, which avoids the formation of dangerous aromatic amines. Four adjacent copper atoms are visible in the laccase's molecular structure and they are dispersed across various binding sites. They are divided into copper I, II, and III types. They are distinguished from one another based on particular traits that enable them to contribute significantly to the enzyme catalytic mechanism. Several industrial effluents, including textiles, can be detoxified using laccases (Lellis et al. 2019). Blánquez et al. (2019) found that *Streptomyces* degraded acetosyringone due to the laccase enzyme production at 55 °C. Palazzolo et al. (2019) reported thar *Ganoderma lucidum* (fungal species) is involved in degradation of azo dyes and showed the presence of laccase enzyme. The white rot fungus *Pseudolagarobasidium acaciicola* produced laccase enzyme and showed its application in decolorization of dyes (Adak et al. 2016). Laccase enzyme produced by *Oudemansiella canarii* is involved in degradation of Congo red dye as reported by Iark et al. (2019). Azo dye decolorization was improved by enhancing lignin-modifying enzymes by *Ganoderma lucidum* fungi (Sun et al. 2023). Laccase-producing *Pseudomonas stutzeri* and *Acinetobacter baumannii* and their consortium showed Congo red decolorization up to 89% and 97%, respectively (Kuppusamy et al. 2017).

Lignin peroxidase (LiP) was initially identified based on the need for hydrogen peroxide to catalyze the C–C cleavage of lignin-based compounds. It was later demonstrated that this enzyme could also depolymerize methylated lignin. These are glycoproteins, which range in molecular weight from 38 to 46 kDa. Many resistant aromatic chemicals, including polychlorinated biphenyls and polyaromatic hydrocarbons, have been mineralized using LiP. LiP produces a radical at carbon carrying azo linkages by oxidizing the phenolic group, which then oxidizes the azo dyes (Lellis et al. 2019). Ligninolytic microbes produce LiP that is involved in dye decolorization. *Bacillus paramycoides* showed production of lignin peroxidase and degrade Congo red and alizarin red S up to 82.79% and 40.51%, respectively (Rahayu et al. 2023). Pham et al. (2022) reported that the LiP enzyme in *Bacillus velezensis* showed reactive decolorization of up to 98% within 48 h . LiP producing yeast *Sterigmatomyces halophilus* showed decolorization of reactive black dye (Al-Tohamy et al. 2020).

Manganese peroxidase is a member of class oxidoreductase. It utilizes ions to target phenolic chemicals through intermediary redox processes (Bhatia et al. 2017). H_2O_2 oxidizes Mn peroxidase to create an intermediate molecule, which then oxidizes Mn^{2+} to Mn^{3+}. Inorganic acids and Mn^{3+} cations interact to create a complex that can oxidize a variety of substrates, including lignin, phenols, and textile dyes (Lellis et al. 2019). White rot fungus produces oxidoreductase enzymes such as lignin, peroxidase, and laccase and can break down contaminants such as dyes, lignocelluloses, and aromatic hydrocarbons (Alam et al. 2021). Rekik et al. (2019) isolated the manganese peroxidase enzyme from *Trametes pubescens* and *P. chrysosporium* which showed the decolorization of dyes. Manganese peroxidase producing consortium of yeast is also involved in azo dye degradation (Samir Ali et al. 2022). Peroxidase enzyme from *Phlebia radiata* fungus is involved in dye degradation as reported by Bouacem et al. (2022). The brown rot fungi *Fomitopsis pinicola* showed the presence of manganese peroxidase enzyme and showed dye degradation (Mahmood et al. 2023).

3.5 Conclusion

Microbes are increasingly being used in dye degradation due to their unique well-developed metabolic mechanism and enzymes which allow them to mineralize and totally decolorize the dye molecules under appropriate environments. Thorough research on the enzymes, genes, and metabolic pathways responsible for microbial dye decolorization is required for application in effluent treatment. The potential interactions of merging such technologies can be carefully evaluated by employing different consortia inaccessible from dye-contaminated areas. More research focusing on selecting a range of bacteria present at contaminated environments and then using them for complete breakdown and mineralization of hazardous dyes would undoubtedly be a sustainable and environmentally beneficial solution.

References

Abd-Elnaby H, Abou-Elela G-M, El-Sersy NA (2011) Cadmium resisting bacteria in Alexandria Eastern Harbor (Egypt) and optimization of cadmium bioaccumulation by *Vibrio harveyi*. Afr J Biotechnol 10(17):3412–3423. https://doi.org/10.5897/AJB10.1933

Adak A, Tiwari R, Singh S, Sharma S, Nain L (2016) Laccase production by a novel white-rot fungus *Pseudolagarobasidium acaciicola* LA 1 through solid-state fermentation of Parthenium biomass and its application in dyes decolorization. Waste Biomass Valorization 7(6):1427–1435. https://doi.org/10.1007/s12649-016-9550-0

Afrin S, Shuvo HR, Sultana B, Islam F, Abu Rus'd A, Begum S, Hossain M (2021) The degradation of textile industry dyes using the effective bacterial consortium. Heliyon 7:e08102. https://doi.org/10.1016/j.heliyon.2021.e08102

Afshariani F, Roosta A (2019) Experimental study and mathematical modeling of biosorption of methylene blue from aqueous solution in a packed bed of microalgae *Scenedesmus*. J Clean Prod 225:133–142. https://doi.org/10.1016/j.jclepro.2019.03.275

Agrawal K, Verma P (2020) Myco-valorization approach using entrapped Myrothecium verrucaria ITCC-8447 on synthetic and natural support via column bioreactor for the detoxification and degradation of anthraquinone dyes. Int Biodeterior Biodegradation 153:105052. https://doi.org/10.1016/j.ibiod.2020.105052

Ahmed D-A-E-A, Gheda S-F, Ismail G-A (2021) Efficacy of two seaweeds dry mass in bioremediation of heavy metal polluted soil and growth of radish (*Raphanus sativus* L.) plant. Environ Sci Pollut Res 28:12831–12846

Alam R, Ardiati FC, Solihat NN, Alam MB, Lee SH, Yanto DHY, Watanabe T, Kim S (2021) Biodegradation and metabolic pathway of anthraquinone dyes by *Trametes hirsuta* D7 immobilized in light expanded clay aggregate and cytotoxicity assessment. J Hazard Mater 405:124176. https://doi.org/10.1016/j.jhazmat.2020.124176

Alam R, Mahmood R-A, Islam S, Ardiati F-C, Solihat N-N, Alam M-B, Kim S (2023) Understanding the biodegradation pathways of azo dyes by immobilized white-rot fungus, *Trametes hirsuta* D7, using UPLC-PDA-FTICR MS supported by in silico simulations and toxicity assessment. Chemosphere 313:137505. https://doi.org/10.1016/j.chemosphere.2022.137505

Ali E, Amjad I, Rehman A (2023a) Evaluation of azo dyes degradation potential of fungal strains and their role in wastewater treatment. Saudi J Biol Sci 30:103734. https://doi.org/10.1016/j.sjbs.2023.103734

Ali AE, Chowdhury ZZ, Devnath R, Ahmed MM, Rahman MM, Khalid K, Wahab YA, Badruddin IA, Kamangar S, Hussien M (2023b) Removal of azo dyes from aqueous effluent using bio-based activated carbons: toxicity aspects and environmental impact. Separations 10(9):506. https://doi.org/10.3390/separations10090506

Ali Redha A (2020) Removal of heavy metals from aqueous media by biosorption. Arab J Basic Appl Sci 27(1):183–193. https://doi.org/10.1080/25765299.2020.1756177

Al-Tohamy R, Sun J, Fareed MF, Kenawy E-R, Ali SS (2020) Ecofriendly biodegradation of reactive Black 5 by newly isolated *Sterigmatomyces Halophilus* SSA1575, valued for textile azo dye wastewater processing and detoxification. Sci Rep 10:12370–12386. https://doi.org/10.1038/s41598-020-69304-4

Ameen F, Dawoud TM, Alshehrei F, Alsamhary K, Almansob A (2021) Decolorization of acid blue 29, disperse red 1 and Congo red by different indigenous fungal strains. Chemosphere 271:129532. https://doi.org/10.1016/j.chemosphere.2021.129532

Ardila-Leal LD, Poutou-Piñales RA, Pedroza-Rodríguez AM, Quevedo-Hidalgo BE (2021) A brief history of colour, the environmental impact of synthetic dyes and removal by using laccases. Molecules 26(13):3813. https://doi.org/10.3390/molecules26133813

Avci A, Yıldırım A, Cerit I, Keskin SY, Hamk M, Keskin CS, Demirkol O (2023) Influence of culture age and environmental conditions on the decolorization and biodegradation of Reactive Orange 13 by *Bacillus* sp. SBT8. Biomass Convers Biorefinery 13:4353–4363. https://doi.org/10.1007/s13399-022-03157-x

Ayed L, Ladhari N, El Mzoughi R, Chaieb K (2021) Decolorization and phytotoxicity reduction of reactive blue 40 dye in real textile wastewater by active consortium: anaerobic/aerobic algal-bacterial-probiotic bioreactor. J Microbiol Methods 181:106129. https://doi.org/10.1016/j.mimet.2020.106129

Ayilara MS, Babalola OO (2023) Bioremediation of environmental wastes: the role of microorganisms. Front Agron 5:1183691. https://doi.org/10.3389/fagro.2023.1183691

Azanaw A, Birlie B, Teshome B, Jemberie M (2022) Textile effluent treatment methods and eco-friendly resolution of textile wastewater. Case Stud Chem Environ Eng 6:100230. https://doi.org/10.1016/j.cscee.2022.100230

Bayat M, Salehi E, Mahdieh M (2023) *Chromochloris zofingiensis* microalgae as a potential dye adsorbent: adsorption thermo-kinetic, isothermal, and process optimization. Algal Res 71:103043. https://doi.org/10.1016/j.algal.2023.103043

Ben AR, Oun A, Cerneaux S, Cretin M, Khemakhem S (2018) Decolorization of dyeing effluent by novel ultrafiltration ceramic membrane from low-cost natural material. J Membr Sci Res 4(2):101–107. https://doi.org/10.22079/jmsr.2017.69818.1154

Benkhaya S, Mrabet S, El Harfi A (2020) A review on classifications, recent synthesis and applications of textile dyes. Inorg Chem Commun 115:107891. https://doi.org/10.1016/j.inoche.2020.107891

Beulah S-S, Muthukumaran K (2020) Methodologies of removal of dyes from wastewater: a review. Int Res J Pure Appl Chem 50:68–78. https://doi.org/10.9734/irjpac/2020/v21i1130225

Bhatia D, Sharma N-R, Singh J, Kanwar R-S (2017) Biological methods for textile dye removal from wastewater: a review. Crit Rev Environ Sci Technol 47(19):1836–1876. https://doi.org/10.1080/10643389.2017.1393263

Bhattacharya A, Goyal N, Gupta A (2017) Degradation of azo dye methyl red by alkaliphilic, halotolerant *Nesterenkonia lacusekhoensis* EMLA3: application in alkaline and salt-rich dyeing effluent treatment. Extremophiles 21(3):479–490

Blánquez A, Rodríguez J, Brissos V, Mendes S, Martins LO, Ball AS, Arias ME, Hernández M (2019) Decolorization and detoxification of textile dyes using a versatile *Streptomyces* laccase-natural mediator system. Saudi J Biol Sci 26:913–920. https://doi.org/10.1016/j.sjbs.2018.05.020

Bouacem K, Allala F, Zaraî Jaouadi N, Hamdi S, Mechri S, Ighilahriz K, Jaouadi B (2022) A novel peroxidase from white-rot Agaricomycetes fungus *Phlebia radiata* strain KB-DZ15: its purification, characterisation, and potential application for dye-decolorisation and lignin-biodegradation. Biocatal Biotransformation 40(5):365–377. https://doi.org/10.1080/10242422.2021.1939315

Bouzikri S, Ouasfi N, Bentiss F, Jama C, Khamliche L (2023) The brown marine algae "*Laminaria digitata*" for the outstanding biosorption of methylene blue and reactive blue 19 dyes: kinetics, equilibrium, thermodynamics, regeneration, and mechanism studies. Nanotechnol Environ Eng 8:317–332. https://doi.org/10.3390/pr11061793

Cagide C, Marizcurrena JJ, Vallés D, Alvarez B, Castro-Sowinski S (2023) A bacterial cold-active dye-decolorizing peroxidase from an Antarctic *Pseudomonas* strain. Appl Microbiol Biotechnol 107:1707–1724. https://doi.org/10.1007/s00253-023-12405-7

Cerboneschi M, Corsi M, Bianchini R, Bonanni M, Tegli S (2015) Decolorization of acid and basic dyes: understanding the metabolic degradation and cell-induced adsorption/precipitation by *Escherichia coli*. Appl Microbiol Biotechnol 19:8235–8245. https://doi.org/10.1007/s00253-015-6648-4

Çetinkaya S, Eyupoglu V, Çetintaş H-İ, Yenidünya A-F, Kebabcı Ö, Tüzün B (2023) Removal of erythrosine B dye from wastewater by *Penicillium italicum*: experimental, DFT, and molecular docking studies. J Biomol Struct Dyn 41(23):14212. https://doi.org/10.1080/07391102.2023.2186704

Chaieb K, Kouidhi B, Ayed L, Hosawi SB, Abdulhakim JA, Hajri A, Altayb HN (2023) Enhanced textile dye removal from wastewater using natural biosorbent and *Shewanella* algae B29: application of Box Behnken design and genomic approach. Bioresour Technol 374:128755. https://doi.org/10.1016/j.biortech.2023.128755

Chandra S-P, Sanyal D, Dasgupta S, Banik A (2020) Cadmium biosorption and biomass production by two freshwater microalgae *Scenedesmus acutus* and *Chlorella pyrenoidosa*. An integrated approach. Chemosphere 269:128755. https://doi.org/10.1016/j.chemosphere.2020.128755

Chen J, Xiaohui W, Hao W, Qi J (2018) Study on decolorization of dyeing wastewater by electrochemical treatment. IOP Conf Ser Earth Environ Sci 113:1315–1755. https://doi.org/10.1088/1755-1315/113/1/012207

Chong M-Y, Tam Y-J (2020) Bioremediation of dyes using coconut parts via adsorption: a review. SN Appl Sci 2(2):1–16. https://doi.org/10.1007/s42452-020-1978-y

de Almeida AP, Macrae A, Ribeiro BD, do Nascimento RP (2021) Decolorization and detoxification of different azo dyes by *Phanerochaete chrysosporium* ME-446 under submerged fermentation. Braz J Microbiol 52:727–738. https://doi.org/10.1007/s42770-021-00458-7

de Freitas J-D-S, Delgado Bertéli M-B, Neto J-C, Gasparotto E-S, Celso Gonçalves A, do Valle J-S, Colauto N-B (2023) Biosorption of methylene blue by residue from *Lentinus crinitus* mushroom cultivation. World J Microbiol Biotechnol 39(5):1–11. https://doi.org/10.1007/s11274-023-03562-y

Devis A-M, Sreekumar R, Das S, Jayasankar K (2023) Fabrication of cross-linked highly stable graphene oxide membrane for dye removal. Int J Environ Anal Chem:1–17. https://doi.org/10.1080/03067319.2022.2164716

El-Agawany NI, Abdel-Razik SAR, Abdel-Kareem MS, Kaamoush MA (2023) Environmental performance of two marine algae *Ulva fasciata* and *Pterocladia capillacea* in the biological treatment of four reactive dyes from aqueous solutions based on fresh and dried biomass. Appl Water Sci 13:82. https://doi.org/10.1007/s13201-023-01893-7

El-Naggar N-E-A, Hamouda R-A, Saddiq AA, Alkinani MH (2021) Simultaneous bioremediation of cationic copper ions and anionic methyl orange azo dye by brown marine alga *Fucus vesiculosus*. Sci Rep 11(1):3555. https://doi.org/10.1038/s41598-021-82827-8

Ellafi A, Dali A, Mnif S, Ben Younes S (2023) Microbial enzymatic degradation, spectral analysis and phytotoxicity assessment of Congo red removal by *Bacillus* spp. Catal Lett 153:3620–3633. https://doi.org/10.1007/s10562-023-04272-8

Falade A-O, Eyisi O-A, Mabinya L-V, Nwodo U-U, Okoh A-I (2017) Peroxidase production and ligninolytic potentials of fresh water bacteria *Raoultella ornithinolytica* and *Ensifer adhaerens*. Biotechnol Rep 16:12–17. https://doi.org/10.1016/j.btre.2017.10.001

Fareed A, Zaffar H, Bilal M, Hussain J, Jackson C, Naqvi TA (2022) Decolorization of azo dyes by a novel aerobic bacterial strain *Bacillus cereus* strain ROC. PLoS One 17(6):e0269559. https://doi.org/10.1371/journal.pone.0269559

Fenner K, Elsner M, Lueders T, McLachlan MS, Wackett LP, Zimmermann M, Drewes JE (2021) Methodological advances to study contaminant biotransformation: new prospects for understanding and reducing environmental persistence? CS EST Water 7:1541–1554. https://doi.org/10.1021/acsestwater.1c00025

Franca RDG, Vieira A, Carvalho G, Oehmen A, Pinheiro HM, Crespo MTB, Lourenço ND (2020) *Oerskovia paurometabola* can efficiently decolorize azo dye Acid Red 14 and remove its recalcitrant metabolite. Ecotoxicol Environ Saf 191:110007. https://doi.org/10.1016/j.ecoenv.2019.110007

Gaikwad G-L, Wate S-R, Ramteke D-S, Kunal R (2014) Development of microbial consortia for the effective treatment of complex wastewater. J Bioremed Biodegr 5(4):1

Gao Y, Yang B, Wang Q (2018) Biodegradation and decolorization of dye wastewater: a review. IOP Conf Ser Earth Environ Sci 178:012013

Gao T, Qin D, Zuo S, Peng Y, Xu J, Yu B, Dong J (2020) Decolorization and detoxification of triphenylmethane dyes by isolated endophytic fungus, *Bjerkandera adusta* SWUSI4 under non-nutritive conditions. Bioresour Bioprocess 7:1–12. https://doi.org/10.1186/s40643-020-00340-8

Ghaly A-E, Ananthashankar R, Alhattab M-V-V-R, Ramakrishnan V-V (2014) Production, characterization and treatment of textile effluents: a critical review. J Chem Eng Process Technol 5(1):1–18. https://doi.org/10.4172/2157-7048.1000182

Ghoniem A-A, Moussa Z, Alenzi A-M, Alotaibi A-S, Fakhry H, El-Khateeb A-Y, Elsayed A (2023) *Pseudomonas alcaliphila* NEWG-2 as biosorbent agent for methylene blue dye: optimization, equilibrium isotherms, and kinetic processes. Sci Rep 13(1):3678. https://doi.org/10.1038/s41598-023-30462-w

Giovanella P, Vieira G-A, Otero I-V-R, Pellizzer E-P, de Jesus Fontes B, Sette L-D (2020) Metal and organic pollutants bioremediation by extremophile microorganisms. J Hazard Mater 382:121024. https://doi.org/10.1016/j.jhazmat.2019.121024

Gul R, Sharma P, Kumar R, Umar A, Ibrahim A-A, Alhamami M-A, Baskoutas S (2023) A sustainable approach to the degradation of dyes by fungal species isolated from industrial wastewaters: performance, parametric optimization, kinetics and degradation mechanism. Environ Res 216:114407. https://doi.org/10.1016/j.envres.2022.114407216

Habibi A, Mehrabi Z (2017) Aerobic degradation of methylene blue from colored effluents by *Ralstonia eutropha*. Pollution 3(3):363–375. https://doi.org/10.7508/pj.2017.03.002

Hanis KKA, Nasri AM, Farahiyah WW, Rabani MM (2020) Bacterial degradation of azo dye Congo red by Bacillus sp. Int J Phys Conf Ser 1529(2):022048

Hoong H-N-J, Ismail N (2018) Removal of dye in wastewater by adsorption-coagulation combined system with *Hibiscus sabdariffa* as the coagulant. In: MATEC web of conferences, vol 152, p 01008. https://doi.org/10.1051/matecconf/201815201008

Hynes N-R-J, Kumar J-S, Kamyab H, Sujana J-A-J, Al-Khashman O-A, Kuslu Y, Suresh B (2020) Modern enabling techniques and adsorbents based dye removal with sustainability concerns in textile industrial sector—a comprehensive review. J Clean Prod 272:122636. https://doi.org/10.1016/j.jclepro.2020.122636

Iark D, dos Reis Buzzo AJ, Garcia JAA, Côrrea VG, Helm CV, Corrêa RCG, Peralta RA, Moreira RFPM, Bracht A, Peralta RM (2019) Enzymatic degradation and detoxification of azo dye Congo red by a new laccase from *Oudemansiella canarii*. Bioresour Technol 289:121655. https://doi.org/10.1016/j.biortech.2019.121655

Jeong S-W, Choi Y-J (2020) Extremophilic microorganisms for the treatment of toxic pollutants in the environment. Molecules 25(21):4916. https://doi.org/10.3390/molecules25214916

John J, Dineshram R, Hemalatha KR, Dhassiah MP, Gopal D, Kumar A (2020) Bio-decolorization of synthetic dyes by a halophilic bacterium *Salinivibrio* sp. Front Microbiol 11:594011. https://doi.org/10.3389/fmicb.2020.594011

Kannan S, Dhandayuthapani K, Sultana M (2013) Original research article decolorization and degradation of azo dye-remazol Black B by newly isolated *Pseudomonas putida*. Int J Curr Microbiol Appl Sci 2(4):108–116

Karim M-E, Dhar K, Hossain M-T (2018) Decolorization of textile reactive dyes by bacterial monoculture and consortium screened from textile dyeing effluent. J Genet Eng Biotechnol 16(2):375–380. https://doi.org/10.1016/j.jgeb.2018.02.005

Kolhe P-P, Ingole N-W (2020) Colour removal from textile industry wastewater by coagulation-flocculation technique. J Water Resour Eng Pollut Stud 5(1)

Kuppusamy S, Sethurajan M, Kadarkarai M, Aruliah R (2017) Biodecolourization of textile dyes by novel, indigenous *Pseudomonas Stutzeri* MN1 and *Acinetobacter Baumannii* MN3. J Environ Chem Eng 5:716–724. https://doi.org/10.1016/j.jece.2016.12.021

Kurade M-B, Waghmode T-R, Kagalkar A-N, Govindwar S-P (2012) Decolorization of textile industry effluent containing disperse dye Scarlet RR by a newly developed bacterial-yeast consortium BL-GG. Chem Eng J 184:33–41. https://doi.org/10.1016/j.cej.2011.12.058

Lavanya C, Rajesh D, Sunil C, Sarita S (2014) Degradation of toxic dyes: a review. Int J Curr Microbiol Appl Sci 3(6):189–199

Lebron YAR, Moreira VR, Santos LVS, Jacob RS (2018) Remediation of methylene blue from aqueous solution by *Chlorella pyrenoidosa* and *Spirulina maxima* biosorption: equilibrium, kinetics, thermodynamics and optimization studies. J Environ Chem Eng 6:6680–6690. https://doi.org/10.1016/j.jece.2018.10.025

Legorreta-Castañeda AJ, Lucho-Constantino CA, Beltrán-Hernández RI, Coronel-Olivares C, Vázquez-Rodríguez GA (2020) Biosorption of water pollutants by fungal pellets. Water 12:1155. https://doi.org/10.3390/w12041155

Lellis B, Fávaro-Polonio C-Z, Pamphile J-A, Polonio J-C (2019) Effects of textile dyes on health and the environment and bioremediation potential of living organisms. Biotechnol Res Innov 3(2):275–290. https://doi.org/10.1016/j.biori.2019.09.001

Li H, Zhang R, Tang L, Zhang J, Mao Z (2014) Evaluation of *Bacillus* sp. MZS10 for decolorizing Azure B dye and its decolorization mechanism. J Environ Sci 26(5):1125–1134. https://doi.org/10.1016/S1001-0742(13)60540-9

Liu S, Zhang F, Chen J, Sun G (2011) Arsenic removal from contaminated soil via bio volatilization by genetically engineered bacteria under laboratory conditions. J Environ Sci 23(9):1544–1550. https://doi.org/10.1016/S1001-0742(10)60570-0

Liu J, Li B, Li Z, Yang F, Chen B, Chen J, Li H, Jiang Z (2023) Deciphering the alkaline stable mechanism of bacterial laccase from *Bacillus pumilus* by molecular dynamics simulation can improve the decolorization of textile dyes. J Hazard Mater 443:130370. https://doi.org/10.1016/j.jhazmat.2022.130370

Mahmood R, Sharif F, Ali S, Hayyat M-U (2015) Enhancing the decolorizing and degradation ability of bacterial consortium isolated from textile effluent affected area and its application on seed germination. Sci World J 2015:*628195*. https://doi.org/10.1155/2015/628195

Mahmood RT, Asad MJ, Hadri SH, El-Shorbagy MA, Mousa AAA, Dara RN, Awais M, Tlili I (2023) Bioremediation of textile industrial effluents by Fomitopsis pinicola IEBL-4 for environmental sustainability. Hum Ecol Risk Assess Int J 29:285–302. https://doi.org/10.1080/10807039.2022.2057277

Markandeya MD, Shukla SP (2022) Hazardous consequences of textile mill effluents on soil and their remediation approaches. Cleaner Engineering and Technology 7:100434. https://doi.org/10.1016/j.clet.2022.100434

Márquez A-A, Coreño O, Nava J-L (2023) Abatement of a complex mixture of dyes in the presence of chlorides by electrocoagulation and active chlorine-based photoelectro-Fenton-like processes. Process Saf Environ Prot 169:579–591. https://doi.org/10.1016/j.psep.2022.11.050

Mavinkattimath R-G, Shetty Kodialbail V, Srinikethan G (2023) Continuous fixed-bed adsorption of reactive azo dye on activated red mud for wastewater treatment-evaluation of column dynamics and design parameters. Environ Sci Pollut Res 30(19):57058–57075. https://doi.org/10.1007/s11356-023-26210-2

Mishra A, Singh D, Singh R, Singh D, Mishra V, Giri B (2023) Effect of noble bacteria *Ochrobactrum intermedium* (Alhpa-22) on decolorization of methyl orange dye in a bioreactor. Int J Chem React Eng 21:1517. https://doi.org/10.1515/ijcre-2023-0047

Moghaddam S-S, Moghaddam M-A, Arami M (2010) Coagulation/flocculation process for dye removal using sludge from water treatment plant: optimization through response surface methodology. J Hazard Mater 175(1–3):651–657. https://doi.org/10.1016/j.jhazmat.2009.10.058

Mokhtar N, Aziz EA, Aris A, Ishak WFW, Ali NSM (2017) Biosorption of azo-dye using marine macro-alga of *Euchema spinosum*. J Environ Chem Eng 5:5721–5731. https://doi.org/10.1016/j.jece.2017.10.043

Mushtaq S, Bareen F-E, Tayyeb A (2023) Equilibrium kinetics and thermodynamic studies on biosorption of heavy metals by metal-resistant strains of *Trichoderma* isolated from tannery solid waste. Environ Sci Pollut Res 30(4):10925–10954. https://doi.org/10.1007/s11356-022-22860-w

Nabilah B, Purnomo AS, Prasetyoko D, Rohmah AA (2023) Methylene blue biodecolorization and biodegradation by immobilized mixed cultures of *Trichoderma viride* and *Ralstonia pickettii* into SA-PVA-Bentonite matrix. Arab J Chem 16(8):104940. https://doi.org/10.1016/j.arabjc.2023.104940

Nachiyar C-V, Rakshi A-D, Sandhya S, Jebasta N-B-D, Nellore J (2023) Developments in treatment technologies of dye-containing effluent: a review. Case Stud Chem Environ Eng 7:100339. https://doi.org/10.1016/j.cscee.2023.100339

Oliveira JMS, Poulsen JS, Foresti E, Nielsen JL (2023) Microbial communities and metabolic pathways involved in reductive decolorization of an azo dye in a two-stage AD system. Chemosphere 310:136731. https://doi.org/10.1016/j.chemosphere.2022.136731

Palanisamy S, Nachimuthu P, Awasthi M-K, Ravindran B, Chang S-W, Palanichamy M, Nguyen D-D (2020) Application of electrochemical treatment for the removal of triazine dye using aluminium electrodes. J Water Supply Res Technol AQUA 69(4):345–354. https://doi.org/10.2166/aqua.2020.109

Palazzolo MA, Postemsky PD, Kurina-Sanz M (2019) From agro-waste to tool: biotechnological characterization and application of Ganoderma lucidum E47 laccase in dye decolorization. 3 Biotech 9(6):213. https://doi.org/10.1007/s13205-019-1744-2

Parmar N-D, Shukla S-R (2019) Decolourization of dye wastewater by microbial methods—a review. Indian J Chem Technol 25(4):315–323. http://nopr.niscpr.res.in/handle/123456789/44961

Pham VHT, Kim J, Chang S, Chung W (2022) Biodegradation of methylene blue using a novel lignin peroxidase enzyme producing bacteria, named *Bacillus* sp. React3, as a promising candidate for dye-contaminated wastewater treatment. Fermentation 8(5):190. https://doi.org/10.3390/fermentation8050190

Poddar K, Sarkar D, Sarkar A (2023) Construction of bacterial consortium for efficient degradation of mixed pharmaceutical dyes. Environ Sci Pollut Res 30(10):25226–25238. https://doi.org/10.1016/j.envpol.2023.121718

Prasad S-S, Aikat K (2014) Study of bio-degradation and bio-decolourization of azo dye by *Enterobacter* sp. SXCR. Environ Technol 35(8):956–965. https://doi.org/10.1080/09593330.2013.856957

Preethi PS, Vickram S, Das R, Hariharan NM, Rameshpathy M, Subbaiya R, Govarthanan M (2023) Bioprospecting of novel peroxidase from *Streptomyces coelicolor* strain SPR7 for carcinogenic azo dyes decolorization. Chemosphere 310:136836. https://doi.org/10.1016/j.chemosphere.2022.136836

Priya S (2018) Development of immobilized microbial consortia for decolorization of synthetic industrial dyes (Doctoral dissertation, GB Pant University of Agriculture and Technology, Pantnagar)

Raees A, Bhatti H-N, Alshehri S, Aslam F, Al-Fawzan F-F, Alissa S-A, Nazir A (2023) Adsorption potential of *Schizophyllum commune* white rot fungus for degradation of reactive dye and condition optimization: a thermodynamic and kinetic study. Adsorp Sci Technol 2023:1. https://doi.org/10.1155/2023/4725710

Rahayu F, Mustafa I, Marjani RF, Qazi RA, Zeb K, Ullah N (2023) Newly isolated ligninolytic bacteria and its applications for multiple dye degradation. Water Air Soil Pollut 234:359. https://doi.org/10.1007/s11270-023-06377-7

Ramalingam P, Devi V (2017) Biodegradation and decolourization of azo dyes using marine bacteria. Int J Adv Biotechnol Res 7:1–12

Ramasany R, Ahmed H-A-M, Karthik S-S (2012) Development of microbial consortium for the biodegradation and biodecolorization of textile effluents. J Urban Environ Eng 6(1):36–41. https://doi.org/10.4090/juee.2012.v6n1.036041

Rekik H, Zaraî Jaouadi N, Bouacem K, Zenati B, Kourdali S, Badis A, Annane R, Bouanane-Darenfed A, Bejar S, Jaouadi B (2019) Physical and enzymatic properties of a new manganese peroxidase from the white-rot fungus *Trametes Pubescens* strain I8 for lignin biodegradation and textile-dyes biodecolorization. Int J Biol Macromol 125:514–525. https://doi.org/10.1016/j.ijbiomac.2018.12.053

Romero-Gonzalez M, Nwaobi B-C, Hufton J-M, Gilmour D-J (2016) Ex-situ bioremediation of U(VI) from contaminated mine water using *Acidithiobacillus ferrooxidans* strains. Front Environ Sci 4:39. https://doi.org/10.3389/fenvs.2016.00039

Roy DC, Biswas SK, Saha AK, Sikdar B, Rahman M, Roy AK, Prodhan ZH, Tang SS (2018) Biodegradation of crystal violet dye by bacteria isolated from textile industry effluents. PeerJ 6:e5015. https://doi.org/10.7717/peerj.5015

Sachan D, Ghosh A, Das G (2022) Valorization of aquatic weed *Salvinia minima* to value-added eco-friendly biosorbent: preferential removal of dye and heavy metal. Int J Environ Sci Technol 20:3703–3712. https://doi.org/10.1007/s13762-022-04126-7

Samir Ali S, Al-Tohamy R, Khalil MA, Ho S-H, Fu Y, Sun J (2022) Exploring the potential of a newly constructed manganese peroxidase-producing yeast consortium for tolerating lignin degradation inhibitors while simultaneously decolorizing and detoxifying textile azo dye wastewater. Bioresour Technol 351:126861–126874. https://doi.org/10.1016/j.biortech.2022.126861

Saratale R-G, Saratale G-D, Chang J-S, Govindwar S-P (2010) Decolorization and biodegradation of reactive dyes and dye wastewater by a developed bacterial consortium. Biodegradation 21:999–1015. https://doi.org/10.1007/s10532-010-9360-1

Saratale R-G, Saratale G-D, Chang J-S, Govindwar S-P (2011) Bacterial decolorization and degradation of azo dyes: a review. J Taiwan Inst Chem Eng 42(1):138–157. https://doi.org/10.1016/j.jtice.2010.06.006

Sarfo DK, Kaur A, Marshall DL, O'Mullane AP (2023) Electrochemical degradation and mineralisation of organic dyes in aqueous nitrate solutions. Chemosphere 316:137821. https://doi.org/10.1016/j.chemosphere.2023.137821

Sharma B, Tiwari S, Kumar R, Kumar M, Tewari L (2023) Eco-friendly detoxification of hazardous Congo red dye using novel fungal strain Trametes flavida WTFP2: deduced enzymatic biomineralization process through combinatorial in-silico and in-vitro studies. J Hazard Mater 455:131503. https://doi.org/10.1016/j.jhazmat.2023.131503

Sheikhi F, Ardakani M-R, Enayatizamir N, Rodriguez-Couto S (2012) The determination of assay for laccase of *Bacillus subtilis* WPI with two classes of chemical compounds as substrates. Indian J Microbiol 52(4):701–707. https://doi.org/10.1007/s12088-012-0298-3

Singh G, Dwivedi S-K (2022) Mechanistic, adsorption kinetics and confirmatory study of Congo red dye removal by native fungus *Aspergillus niger*. Biomass Convers Biorefinery:1–19. https://doi.org/10.1007/s13399-022-03369-1

Singh P, Iyengar L, Pandey A (2012) Bacterial decolorization and degradation of azo dyes. In: Microbial degradation of xenobiotics. Springer, Berlin, pp 101–133

Singh K, Kumar P, Srivastava R (2017) An overview of textile dyes and their removal techniques: Indian perspective. Pollut Res 36:790–797

Singh D, Goswami RK, Agrawal K, Chaturvedi V, Verma P (2022) Bio-inspired remediation of wastewater: a contemporary approach for environmental clean-up. Curr Res Green Sustain Chem 5:100261. https://doi.org/10.1016/j.crgsc.2022.100261

Singh S, Chaudhary P, Sharma A (2023) Efficient decolourization of toluidine blue and methylene blue using *Bacillus cereus*. Vegetos. https://doi.org/10.1007/s42535-023-00747-6

Siregar R-A-N, Sanjaya A, Lucy J, Pinontoan R (2020) Methylene blue decolorizing bacteria isolated from water sewage in Yogyakarta, Indonesia. Biodiversitas 21(3):1136–1141. https://doi.org/10.13057/biodiv/d210338

Sivalingam S, Gopal V (2023) Coagulation study on extracted algal alginate from red algae as natural coagulant for remediation of textile dye Congo red. Environ Nat Resour J 21(2):90–101. https://doi.org/10.32526/ennrj/21/202200221

Slama N-E-H, Masmoudi G, Fizer M, Mariychuk R, Dhaouadi H (2023) Optimization of dark blue reactive dye degradation with Fenton reaction based on experimental design methodology and DFT calculations. Chem Papers 77:4425–4437. https://doi.org/10.1007/s11696-023-02792-6

Souza F-H, Leme V-F, Costa G-O, Castro K-C, Giraldi TR, Andrade G-S (2020) Biosorption of rhodamine B using a low-cost biosorbent prepared from inactivated *Aspergillus oryzae* cells: kinetic, equilibrium and thermodynamic studies. Water Air Soil Pollut 231(5):242. https://doi.org/10.1007/s11270-020-04633-8

Subramaniam S, Velu G, Palaniappan M, Jackson KM (2023) Decolourization of azo dyes using immobilized bacterial isolates from termite mound ecosystem. Total Environ Res Themes 6:100041. https://doi.org/10.1016/j.totert.2023.100041

Sultana B (2017) Degradation of dyes used in textile industries by bacteria isolated from local textile effluents (Doctoral dissertation, BRAC University)

Sun S, Liu P, Ullah M (2023) Efficient azo dye biodecolorization system using lignin-co-cultured white-rot fungus. J Fungi (Basel) 9(1):91. https://doi.org/10.3390/jof9010091

Syafiuddin A, Fulazzaky MA (2021) Decolorization kinetics and mass transfer mechanisms of Remazol Brilliant Blue R dye mediated by different fungi. Biotechnol Rep 29:1–14. https://doi.org/10.1016/j.btre.2020.e00573

Tahir U, Nawaz S, Khan UH, Yasmin A (2021) Assessment of biodecolorization potentials of biofilm forming bacteria from two different genera for Mordant Black 11 dye. Biorem J 25:252–270. https://doi.org/10.1080/10889868.2021.1911920

Tang W, Xu X, Ye BC, Cao P, Ali A (2019) Decolorization and degradation analysis of Disperse Red 3B by a consortium of the fungus *Aspergillus* sp. XJ-2 and the microalgae *Chlorella sorokiniana* XJK. RSC Adv 9(25):14558–14566. https://doi.org/10.1039/C9RA01169B

Thamaraiselvan C, Noel M (2015) Membrane processes for dye wastewater treatment: recent progress in fouling control. Crit Rev Environ Sci Technol 45(10):1007–1040. https://doi.org/10.1080/10643389.2014.900242

Thao T-T-P, Nguyen Thi M-L, Chung N-D, Ooi C-W, Park S-M, Lan T-T, Huy N-D (2023) Microbial biodegradation of recalcitrant synthetic dyes from textile-enriched wastewater by *Fusarium oxysporum*. Chemosphere 325:138392. https://doi.org/10.1016/j.chemosphere.2023.138392

Tian F, Wang Y, Guo G, Ding K, Yang F, Wang C, Yan M (2023) Meta-genome analysis of a newly enriched azo dyes detoxification halo-thermophilic bacterial consortium. Environ Res 237:116828. https://doi.org/10.1016/j.envres.2023.116828

Ting ASY, Cheng CKW, Santiago KAA (2021) Decolourization of malachite green dye by endolichenic fungi from the lichen Usnea sp.: a novel study on their dye removal potential. J King Saud Univ Sci 33(7):101579. https://doi.org/10.1016/j.jksus.2021.101579

Vijayaraghavan J, Basha S-S, Jegan J (2013) A review on efficacious methods to decolorize reactive azo dye. J Urban Environ Eng 7(1):30–47. https://www.jstor.org/stable/26203388

Wang W, Zhang Z, Ni H, Yang X, Li Q, Li L (2012) Decolorization of industrial synthetic dyes using engineered *Pseudomonas putida* cells with surface-immobilized bacterial laccase. Microb Cell Factories 11(1):1–14. https://doi.org/10.1186/1475-2859-11-75

Xiang H, Min X, Tang C-J, Sillanpää M, Zhao F (2022) Recent advances in membrane filtration for heavy metal removal from wastewater: a mini review. J Water Process Eng 49:103023. https://doi.org/10.1016/j.jwpe.2022.103023

Xuejiao A, Yi C, Ningjian L, Shulin Z, Liuwei W, Qinghua Z (2023) Green and sustainable approach for decoloration and detoxification of azo dye by bacterial-mediated thermo-halotolerant dye-decolorizing peroxidase (DyP). Process Biochem 130:334–346. https://doi.org/10.1016/j.procbio.2023.03.037

Yusuf F, Ahmad FA, Muhammad F, Shehu U, Yakasai HM (2021) Profiling the effects of pH and temperature on azo dye-decolourisation by *Aspergillus quadrilineatus* strain BUK_BCH_BTE1 isolated from textile effluents. ChemSearch J 12(2):64–69

Zainith S, Purchase D, Saratale G-D, Ferreira L-F-R, Bilal M, Bharagava R-N (2019) Isolation and characterization of lignin-degrading bacterium *Bacillus aryabhattai* from pulp and paper mill wastewater and evaluation of its lignin-degrading potential. 3 Biotech 9(3):92. https://doi.org/10.1007/s13205-019-1631-x

Zdarta J, Meyer A-S, Jesionowski T, Pinelo M (2018) A general overview of support materials for enzyme immobilization: characteristics, properties, practical utility. Catalysts 8(2):92. https://doi.org/10.3390/catal8020092

Microbial Inoculants in Amelioration of Stress by Regulation of Signalling Cascades in Plant

Manoj Kumar Prajapati, Anamita Sen, Gopinath V, and Viabhav Kumar Upadhayay

Abstract In agriculture, rapid manipulation and exploitation of PGPR (plant growth-promoting rhizobacteria) is rising rapidly due to vast functionality of plant-protective and growth-related benefits in plants against biotic and abiotic stresses. The interaction between host plants and the microbial community in the rhizospheric zone is a very complex and dynamic process which regulates the effective communication system in a coordinated manner. This interaction might happened at multiple levels, i.e., either intraspecies, interspecies, or tritrophic interaction. PGPR-assisted plant growth outcome and stress reduction outcome are triggered by means of the signal transduction pathway during plant–microbe interaction. Rapid advancement in technology has opened a new dimension for exploring the functionality of specific signal molecules and related chemical cues which participate in the communication process as well as induction of defense mechanism. The precise and comprehensive gene expression mechanism and communication between plant and microbes could provide enough strength to understand the mechanism behind activation of plant's immune response. The activation of plant PRR and NLR by microbial effector proteins activates the MAPK and CDPK pathways through transphosphorylation. These signalling pathways play a vital role as regulators of gene expression, influencing various functions such as plant immunity, virulence properties, and plant–microbe communications. Over the past decades, the use of PGPR has emerged as a promising tactic to improve tolerance to abiotic and biotic stresses in host plants. Therefore, it is imperative to understand the mechanism overview of the microbe–plant interaction phenomenon and understand its importance. This chapter focuses on elucidating the effects of defense signalling on the

M. K. Prajapati (✉) · A. Sen · G. V
Department of Plant Pathology and Nematology, PGCA, Dr. Rajendra Prasad Central Agricultural University, Pusa, Samastipur, Bihar, India

V. K. Upadhayay
Department of Microbiology, College of Basic Sciences and Humanities, Dr. Rajendra Prasad Central Agricultural University, Pusa, Samastipur, Bihar, India

P. Chaudhary, A. Chaudhary (eds.), *Microbial Inoculants*, Interdisciplinary Biotechnological Advances, https://doi.org/10.1007/978-981-97-0633-4_16

root microbiome. It also explains the signalling molecules produced by both microbes and plants, which contribute to the defense mechanisms and tolerance responses of host plants against biotic and abiotic stresses.

Keywords Biotic stress · Plant signalling · Microbial inoculants · Plant–microbe interaction

1 Introduction

Growing concerns over global warming and population growth have increased the demand for food, fiber, and energy. Meeting this demand for global food security and sustainable crop production is now a major challenge (Malhi et al. 2021). Apart from that, high consumption rate of fossil fuels along with rapid industrialization associated with urban activities, deforestation, and modern crop production techniques has increased the emission of greenhouse gases, which has played a significant role in the contribution of climate change (Foley et al. 2011). Consequently, change in temperature and rainfall pattern is one of the major constraints that will progressively affect food production in this century. Furthermore, it has been reported that abiotic stress accounts for around 50% of crop loss, while biotic stress is responsible for approximately 30% of losses (Malhi et al. 2021). Therefore, one of the drastic challenges for sustainable agriculture is to increase yield and mitigate crop losses using precise techniques and strategies under changing climate scenarios and economic constraint (Seleiman et al. 2021; Chaudhary et al. 2022b, c).

Similarly, massive use of chemical pesticides is being experienced in modern era, which has created serious issues in terms of plant, animal, soil, and environmental health concerns (Chaudhary et al. 2022a). Moreover, it might have a long-lasting adverse effect on fertility status of soil by affecting the soil microbial diversity present in it (Chaudhary et al. 2020, 2022d).

Significant efforts have been undertaken to reduce reliance on chemicals in crop production and replace them with bioinoculants like PGPR. These bioinoculants hold promise in addressing agricultural challenges and promoting stability and sustainability in crop production (Chaudhary et al. 2021a, b). Although soils harbor a diverse range of microorganisms, our understanding of them remains limited, despite their crucial role in maintaining the biological and physiochemical stability of soil ecosystems (Chaudhary et al. 2021c, d). The use of PGPR may provide a long-term solution to the problems that arise when chemical fertilizers are applied to soil in order to boost its fertility status (Upadhayay et al. 2018, 2022a, b; Chaudhary et al. 2021e, f). PGPR could directly trigger plant growth by secreting certain bioactive novel secondary metabolites and by escalating the enhanced mineral nutrient availability in the rhizospheric region of roots or circuitously make possible plant development by eliminating other pathogenic effects through antagonistic behavior toward other phytopathogens (Agri et al. 2021; Singh et al. 2021; Chaudhary et al. 2022a). The microbes that interact with their host plant are proven to be advantageous for both autotroph and heterotroph. Through processes like biological nitrogen fixation, nutrient solubilization (making nutrients like phosphorus, potassium, and zinc more readily available), chelation (of iron molecules via siderophore production), hydrogen cyanide (HCN) production, etc., plant growth can be stimulated (Glick 2014; Rawat et al. 2019; Kaviya et al. 2019; Khan et al. 2019, 2023a, b; Upadhayay et al. 2021, 2023b; Chaudhary et al. 2022e). Firstly, microorganisms act as plant stimulators by triggering plant growth through the release of hormones. Secondly, microbes are used as bioinoculants to enhance plant growth (Roshani et al. 2020; Upadhayay et al. 2022c, d; Chaudhary et al. 2022e; Khan et al. 2023b). Thirdly, microbes help maintain the health of host plants against abiotic stress (Singh et al. 2022; Upadhayay et al. 2023a). Additionally, microbes play a role in managing biotic stress by inhibiting the growth of phytopathogens (Tyagi et al. 2022a). They achieve this through the production of antimicrobial compounds such as antibiotics, chitinase, β-1,3-glucanase, and lytic enzymes and by activating plant defense responses against pathogens, among others (Beneduzi et al. 2012). Numerous reports about PGPR have revealed remarkable role in plant growth and on eliminating abiotic stresses (Singh et al. 2019a, b, c).

This chapter elucidates the symbiotic relationship between microbes and plants, showcasing their collaborative efforts to enhance sustainable crop production in agriculture. It specifically focuses on the utilization of PGPR as an environmentally friendly approach to mitigating excessive use of agrochemicals, including pesticides and fertilizers. The chapter explores the economic significance and diverse applications of PGPR, emphasizing their role within rhizospheric regions. Special attention is given to the signal transduction pathway and the activation of cellular signalling molecules, which play a crucial role in shaping plant responses to various stressors. Additionally, the chapter delves into the regulation of defense mechanisms at both the physiochemical and molecular levels, addressing biotic and abiotic stress conditions comprehensively.

2 PGPR and Its Rhizospheric Interaction

The rhizosphere, which encompasses the soil surrounding a plant's roots, harbors a diverse range of microbes (Santoyo et al. 2021). It serves as a habitat for various microorganisms, including bacteria, fungi, protozoa, algae, and actinomycetes (Poria et al. 2021). Plant roots possess the remarkable ability to produce, store, and release a wide array of chemical compounds (Canarini et al. 2019). These compounds are well recognized as "root exudates" and act as chemical attractants for the microbial communities in the soil. Root exudates modify physiochemical behavior of soil and thus influence soil microbiome of rhizospheric milieu. The dynamic interplay between the root exudates of plants and the signalling molecules produced by both the plants and microbes plays an imperative role in shaping the composition and arrangement of the microbiota within the rhizospheric zone (Jalmi 2020; Upadhayay et al. 2022b). This intricate communication network, involving a variety of nonhormonal signal compounds, regulates numerous aspects of growth and metabolism in plants and their associated microbiomes (holobionts). Plant roots synthesize, accumulate, and release numerous secondary metabolites, such as terpenes, N- and sulfur compounds, and pyrazines, which act as chemoattractants in the rhizosphere (Mohanty et al. 2021; Ullah et al. 2021). These chemoattractants possess the ability to influence plant growth and modulate the health of both plants and the soil (Agri et al. 2022; Chaudhary et al. 2023a, b, c). PGPR can be classified into two categories based on their interaction with the plant root: extracellular PGPR (ePGPR) and intracellular PGPR (iPGPR). ePGPR fosters the development of bacterial communities in the rhizospheric milieu and root cortex, encompassing species such as "*Azotobacter*," "*Arthrobacter*," "*Bacillus*," "*Burkholderia*," "*Erwinia*," "*Pseudomonas*," "*Serratia*," and "*Micrococcus*." On the other hand, iPGPR supports rhizobacterial communities within the intercellular space, including *Allorhizobium* and *Azorhizobium*. A variety of metabolically active substances are produced by PGPR that reside in plant tissue, communicate with one another, and associate with the surface of the root (Backer et al. 2018).

3 Multifunction and Growth-Promoting Activity Opted by PGPR

PGPR employs a range of mechanisms to improve the overall health of plants, both directly and indirectly. These mechanisms encompass activities such as biofertilization, phytostimulation, and biocontrol or biopesticide activities (Khati et al. 2018; Kukreti et al. 2020; Kumar et al. 2021a, b, c). By engaging in these processes, PGPR assists host plants in combating both biotic and abiotic stressors, resulting in enhanced plant health and resilience (Leontidou et al. 2020). Direct mechanism of growth enhancement activities involves biological nitrogen fixation (Joshi et al. 2019; Kusale et al. 2021), phosphate solubilization activity (Sham et al. 2019; Kusale et al. 2021), Zn and K solubilization activity (Baba et al. 2021), ACC deaminase production (Kapadia et al. 2022), synthesis of phytohormone (Ahemad and Khan 2012; Khati et al. 2019), and siderophore production (Rehman et al. 2020). Indirect mechanism of growth-promoting activities opted by PGPR which include antibiosis, nutritional competition, parasitism, inhibition of pathogen's toxin, and induced resistance plays a crucial role in plant disease control against a wide range of phytopathogens and triggers a defense response in a host plant (Elnahal et al. 2022). PGPR can keep out the threat of phytopathogens such as fungi, bacteria, and virus through competition for food and space, exclusion of niche, synthesis of antimicrobial metabolites, secretion of cell wall-degrading enzymes like chitinase and β-1,3-glucanase, protease, lipase, and other lytic enzymes and trigger induced systemic resistance (Rehman et al. 2020; Riaz et al. 2021; Elnahal et al. 2022; Parveen

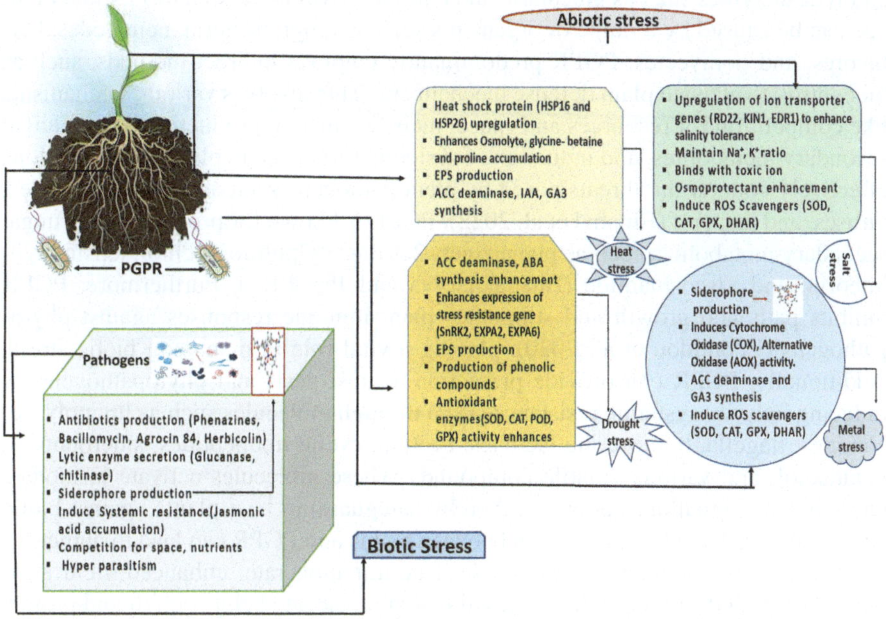

Fig. 1 Mechanisms involved in biotic and abiotic stress management in plants by PGPR

et al. 2022). A wide range of the antifungal metabolites that are secreted by PGPR such as "phenazines," "2,4-diacetylphloroglucinol," "HCN," "iturin," "zwittermicin A," "bacillomycin D," "pyoluteorin," and "viscosinamide" directly or indirectly degrade or lyse the fungal cell wall, mycelium, and their propagules (Bhattacharyya and Jha 2012). The activation of the plant's innate immune system response (ISR) can be initiated by a range of substances, including lipopolysaccharides, cyclic lipopeptides, siderophores, homoserine lactones, flagella, acetoin, and 2,3-butanediol (Prakash et al. 2022). This, in turn, leads to the production of jasmonate and ethylene signalling. Antibiotic synthesis is another important indirect mechanism (Keswani et al. 2020; Elnahal et al. 2022); antibiotics are compounds with low molecular weight which has the capability to kill or decrease the growth of pathogens. The mechanisms involved in biotic and abiotic stress management in plants by PGPR have been mentioned in Fig. 1.

4 PGPR Against Biotic Stress

PGPR exhibits various beneficial traits that contribute to its ability to mitigate plant diseases, enhance plant resistance to abiotic stress, improve nutrient uptake, and promote plant growth (Kumari et al. 2020; Tyagi et al. 2022b). These traits include increased production of phytohormones, siderophores, volatile organic compounds, and lytic enzymes such as glucanase and chitinase (Kumari et al. 2021). Plant damage can be caused by a range of organisms such as fungi, bacteria, nematodes, herbivores, and oomycetes. PGPR predominantly employs indirect methods, such as biocontrol, to activate plant defense mechanisms. This involves various mechanisms like competition for resources and space, niche exclusion, production of antifungal secondary metabolites, and induction of systemic resistance in plants. These actions effectively counter the threats posed by plant pathogens such as fungi, oomycetes, viruses, and bacteria (Elnahal et al. 2022). PGPR is known for producing antifungal secondary metabolites such as phenazines, 2,4-diacetylphloroglucinol, tensin, pyoluteorin, and viscosinamide (Bhattacharyya and Jha 2012). Furthermore, PGPR inhibits pathogen growth and stimulates plant immune responses against phytopathogens (Leontidou et al. 2020), playing a vital role in managing biotic stress. Additionally, PGPR can provide protection against pests and phytopathogens by inducing induced systemic resistance (ISR) through molecules such as lipopolysaccharides, flagella, homoserine lactones, acetoin, cyclic lipopeptides, siderophores, butanediol, and various volatile compounds. These molecules activate jasmonate and ethylene signalling pathways, thereby safeguarding host plants against biotic stress (Glick 2012). The interaction between plants and PGPR can lead to numerous positive effects, including improved seed germination rate, enhanced yield attributes, stimulated root growth, increased leaf area, elevated chlorophyll and protein content, enhanced nutrient uptake, higher germination percentage, increased root weight and length, improved hydraulic activity, increased abiotic stress tolerance, and biocontrol activity (Pérez-Montaño et al. 2014; Jochum et al. 2019; Kumari et al. 2020).

5 PGPR Against Abiotic Stress

Drought, high salinity, cold, heat, oxidative stress, flooding, and the presence of toxic heavy metals in the soil are among the abiotic stresses that plants commonly encounter (Upadhayay et al. 2023a). These stresses, similar to biotic stresses, adversely affect various aspects of plant growth, fertility, development, metabolism, photosynthesis, and immune response. As a result, plant yield and quality are significantly decreased in the field due to these detrimental effects (Hayes et al. 2021). The primary and direct effect of drought stress, frost stress, salt stress, and heat stress is the induction of water deficit conditions within cells, which is then followed by an increase in biochemical, molecular, and phenotypic expressions in resistance to stresses (Tyagi et al. 2022b). Reduced leaf water potential and stomatal opening, smaller leaves, stifled root growth, fewer and smaller seeds, reduced viability, delayed flowering, and stunted growth and productivity have all been linked to prolonged water stress in plants (Xu et al. 2016). Heat stress causes modifications to the structure, physiology, and biochemistry of plants. Understanding the mechanisms of plant heat tolerance and their physiological responses is crucial for improving crop production strategies. High temperatures negatively impact seed germination, photosynthesis, respiration, and cell membrane permeability (Xu et al. 2014). Some of the major responses of plants to heat stress include alterations in hormone, primary, and secondary metabolite levels, increased expression of heat shock proteins and related proteins, and increased production of reactive oxygen species (ROS). Plants employ a heat stress mitigation strategy by inducing MAPK and CDPK signalling cascades and thereby helping to maintain and stabilize cell membrane permeability. Several parallel processes help cells endure heat stress, including the removal of ROS, the synthesis of antioxidant compounds and compatible solutes, the activation of chaperone signalling and transcriptional intonation of genes encoding heat shock proteins, and the induction of a thermotolerant response (Moumbock et al. 2021). Microbes possess the remarkable ability to withstand various abiotic stresses, and when they possess multiple plant growth-promoting traits, they can be utilized as bioinoculants to alleviate abiotic stress in plants. PGPR has a potential benefits which makes the plant to withstand in abiotic stress condition by adopting diverse mechanisms like osmotic adjustment (synthesis of osmolytes) through regulation of osmolytes and stress responsive genes; increased antioxidant activity; synthesis of ROS scavenging agents such as SOD and CAT; synthesis of drought stress amino acids such as proline, glycine, and betaine; and synthesis of phytohormones, especially ethylene, VOCs, ACCD, exopolysaccharides, and many more (Khan et al. 2020; Rehman et al. 2020). The pictorial representation of the role of PGPR against abiotic stress management has been mentioned in Fig. 1. Additionally, Table 1 provides an overview of the application of bioinoculants in combating abiotic stress.

Toxicity of heavy metals especially Cd and Hg has also posed serious threat to crop productivity due to long persistence and least degradability in the soil environment that directly affects soil pH and texture, reducing crop growth and development by interfering the cellular as well as physio-biochemical processes (Riyazuddin et al. 2021). Many indirect mechanisms (such as glutathione reduction or binding to proteins' SH groups) and direct effects (such as preventing anti-oxidative enzymes, thereby triggering ROS-producing enzymes, i.e., NADPH) contribute to oxidative stress in plants that is caused by heavy metal stress. The most effective and widely used PGPRs, including *Pseudomonas*, *Streptomyces*, *Methylobacterium*, and *Bacillus*, have the powerful capability to lessen the negative impact of heavy metals on crops by altering the mobilization process of the metals' translocation to different parts of plants through complexation, precipitation, redox processes, chelation, and adsorption (Lengai et al. 2020). The application of PGPR against biotic and abiotic stresses has been mentioned in Tables 2 and 3.

5.1 PGPR and Acetyl-CoA Deamine Synthase

The molecular architecture of bacterial ACC deaminase enzymes has been illuminated by recent studies that have effectively identified their crystal structures. All ACC deaminase enzymes exhibit a significant shared trait of displaying an affinity

Table 1 Effect of bioinoculants in mitigating abiotic stress in various plant species

Stress type	Host	Micro organism	Effects/mechanism	References
Heat	Tomato (*Lycopersicon esculentum*)	*Bacillus cereus*	– Increases exopolysaccharide production – Cleavage of ACC deaminase	Mukhtar et al. (2020)
		Bacillus safensis SCAL1	Enhances activity of ACC deaminase, IAA, GA, and kinetin	Mukhtar et al. (2022)
	Soybean (*Glycine max*)	*Bacillus aryabhattai*	Enhances ABA production	Park et al. (2017)
	Sorghum (*Sorghum bicolor*)	*Pseudomonas aeruginosa*	Increases sugar, amino acid, and proline content	Saleem et al. (2018)
	Rice (*Oryza sativa*)	*Brevibacterium linens*	– Upregulates HSP16 and HSP26 – Enhances ACC deaminase activity	Choi et al. (2022)
	Wheat (*Triticum aestivum*)	*Bacillus safensis*	– Enhance expression of heat shock proteins (HSPs) – Increases antioxidant enzyme synthesis	Sarkar et al. (2021)
		Pseudomonas brassicacearum	– Upregulates ADC1 and ADC2 genes – Enhances phenol content, increases catalase and oxidation production	Ashraf et al. (2019)
Drought	Rice (*Oryza sativa*)	*Bacillus megaterium*	Decreases COX1 gene expression, increases EXP1 gene expression	Omar et al. (2021)
		Pseudomonas azotoformans	Increases expression of NRAMP6, NAM, and DHN gene	
		Rhizobium spp.	Increases expression of AP2-EREBP, EXP2, and EXP3 gene	
	Wheat (*Triticum aestivum*)	*Streptomyces pactum*	– Increases proline accumulation – Increases ABA content – Enhances expression of stress resistance gene SnRK2, EXPA2, and EXPA6	Li et al. (2020a, b)
		Bacillus megaterium	– Enhances seed germination – Increases IAA concentration, hence improving plant growth (shoot and root) – Increases proline and sugar content – Stimulates water stress-related proteins	Rashid et al. (2022)

	Wheat (*Triticum aestivum*)	*Serratia Marcescens*	– Increases exopolysaccharide production – Enhances osmolyte accumulation – Enhances Zn and Fe content in plants	Khan and Singh (2021)
		Bacillus subtilis	– Increases RWC, chlorophyll content, and antioxidant enzyme activity (SOD, CAT, POD)	Sood et al. (2020)
	Maize (*Zea mays*)	*Bacillus* sp., *Enterobacter* sp., *Pseudomonas fluorescence 1*	– Increases activity of SOD, GPX, and CAT – Increases chlorophyll a and b and carotenoid activity by producing ACC deaminase	Joshi et al. (2020)
	Soybean (*Glycine max*)	*Bacillus cereus AKAD A1–1*	– Enhances IAA production, hence improving plant biomass – Increases proline, soluble carbohydrate, and protein content Reduces H_2O_2 level	Dubey et al. (2021)
	Arabidopsis (*Arabidopsis thaliana*)	*Pseudomonas* SP2	– Improves SP2 colonization – Increases osmolyte, glycine betaine, and proline accumulation	Yasmin et al. (2022)
	Okra (*Abelmoschus esculentus*)	*Bacillus subtilis*	– Increases sugar and proline accumulation – Increases enzymatic (SOD, CAT, DHAR, APX) and nonenzymatic (GSH, ascorbate Asa) antioxidants	Puthiyottil and Akkara (2021)
	Pepper (*Capsicum annum*)	*Bacillus butanolivorans* KJ40	– Activates peroxidase and glutathione peroxidase activity – Enhances expression of Shsp, Cadhn, and NCED1 gene	Kim et al. (2022)

(continued)

Table 1 (continued)

Stress type	Host	Micro organism	Effects/mechanism	References
Salinity	Arabidopsis (*Arabidopsis thaliana*)	*Streptomyces* sp.	– Enhances oxidase and catalase production – Improves phosphate solubilization – Upregulates nitrogen assimilation (GS1) of proteins – Increases expression of cFBPase gene related to carbohydrate metabolism	Manullang and Chuang (2020)
	Arabidopsis (*Arabidopsis thaliana*)	*Bacillus oryzicola* YC7007	– Stem and root of seedling releases stress-related genes (RD22, KIN1, EDR1) and raises salt tolerance in plants	Chu et al. (2019)
	Soybean (*Glycine max*)	*Pseudomonas simiae*	– Increases vegetative storage protein (VSP), gamma-glutamyl hydrolase (GGH), and RuBisCo proteins – Reduces Na, increases K and P content in plants	Ma et al. (2019)
Salinity	Sunflower (*Helianthus annus*)	*Bacillus pumilus*	S1 and S3 genes are involved which enhance ACC deaminase, proline, and IAA production	Sultana et al. (2020)
	White clover (*Trifolium repens*)	*Bacillus subtilis*	Improves the potassium to sodium ratio by lowering sodium levels in plant tissues and roots	Shah et al. (2021)
	Wheat (*Triticum aestivum*)	*Halobacillus* sp. SL3	Enhances IAA and siderophore production by activating acdS gene	El-Esawi et al. (2018)
	Okra (*Abelmoschus esculentus*)	*Enterobacter* sp. UPMR18	– Activation of X55749 gene – ROS pathway upgradation	Habib et al. (2016)
Low temperature	Rice (*Oryza sativa*)	*Bacillus amyloliquefaciens*	Enhances production of ABA, SA, JA, and ET	Tiwari et al. (2017)
	Tomato (*Solanum lycopersicum*)	*Funneliformis mosseae*	Increases ROS activity	Caradonia et al. (2019)

Heavy metals	Cd	Rice (*Oryza sativa*)	*Enterobacter aerogenes*	IAA, ET	Pramanik et al. (2018)
	Ni, Cd, Cu, Pb	Cord grass (*Spartina maritima*)	*Bacillus methylotrophicus* *Bacillus licheniformis*	Induces cytochrome oxidase (COX), alternative oxidase (AOX) activity	Mesa-Marín et al. (2018)
	As	Soybean (*Glycine max*)	*Bradyrhizobium japonicum*, *Azospirillum brasilense*	Increases IAA production	Armendariz (2019)
	Cr	Indian mustard (*Brassica juncea*)	*Pseudomonas aeruginosa* *Burkholderia gladioli*	– Reduces expression of RBOH1 gene – Increases activities of SOD, APOX, GPOX, and DHAR	Sharma et al. (2017)

Table 2 Microbes/PGPR-mediated resistance mechanisms in plants against biotic stress

Microbial strains	Target pathogen	Diseases	Resistance mechanism	References
Bacillus amyloliquefaciens Ba13	Tomato leaf curl virus	Tomato leaf curl disease	– PR1, PR2, and PR3 genes enhance PAL activities (plant growth and stress tolerance) – Enhances the activities of beta-1,3-glucanase as well as chitinase	Guo et al. (2019)
Bacillus cereus AR156	*Pseudomonas syringae pv tomato*	Bacterial speck in tomato	Blocking miR825 and stimulating related defense gene expression	Nie et al. (2019)
Bacillus atrophaeus GBSC56	*Meloidogyne incognita*	Root knot disease	– Protection from oxidative stress and control of antioxidant enzymes – Enhances accumulation of volatile metabolites-dimethyl disulfide which exhibits resistance against root knot nematode	Ayaz et al. (2021)
Pseudomonas fluorescens PTA-CT2	*Plasmopara viticola*, *Botrytis cinerea*	Downy mildew and gray mold diseases	– The SA, JA, and ABA defensive pathways are all activated by this – Induces HR	Lakkis et al. (2019)
Pseudomonas fluorescens WCS417	–	Broad spectrum	Regulates iron uptake responses and activates the transcription factor MYB72	Desrut et al. (2020)
Streptomyces lydicus M01	*Alternaria alternata*	Foliar diseases of cucumber	The development of reactive oxygen species (ROS) and the subsequent suppression of mycelial growth and spore germination	Morcillo et al. (2020)
Acrophialophora jodhpurensis	*Rhizoctonia solani*	Tomato root and crown rot	ROS formation, enhancement of activities of β-1,3-glucanase, PAL, chitinase, peroxidase, and siderophore production	Daroodiet al. (2021)
Streptomyces pactum	Tomato yellow leaf curl virus	Tomato yellow leaf curl disease	– Inhibition of microbial growth and spore germination by activities of ROS, β-1,3-glucanase, and chitinase – Activation of PR-1, PR-2, and PR-5 genes Activation of SAR	Li et al. (2019)
Mortierella hyalina	*Alternaria brassicae*	Alternaria leaf spot of crucifers	– Jasmonic acid (JA) accumulation which induces response to biological stresses – Ca^{2+} ion influxes regulate plant cell membrane permeability	Johnson et al. (2019)

(continued)

Table 2 (continued)

Microbial strains	Target pathogen	Diseases	Resistance mechanism	References
Trichoderma atroviride	*Botrytis cinerea*	–	– Activation of glutamate: glyoxylate aminotransferase GGAT1 – Induces defense response to biotic stresses by activating WRKY transcription factors	González-López et al. (2021)
Trichoderma harzianum	*Bipolaris sorokiniana, Rhizoctonia solani*	Spot blotch, wilt	Increased phenylpropanoid activities lead to less cell wall disruption and more lignification in plant cells. Production of harzianic acid (secondary metabolite) which induces expression of several defense-related genes	Singh et al. (2019a, b, c)

for the substrate ACC within the millimolar range. Numerous functional bacterial ACC deaminase genes have been identified through comprehensive investigations, thereby facilitating a more profound comprehension of their involvement in diverse biological processes (Glick 2014). In addition, through sequence alignment, researchers have identified conserved amino acid residues that play a crucial part in the enzymatic function of ACC deaminase, which includes significant residues like Lys51 and Cys162. Several rhizobacteria that are recognized for their ability to enhance plant growth contain the enzyme 1-aminocyclopropane-1-carboxylate (ACC) deaminase (Duan et al. 2021). This enzyme aids in the degradation of the ethylene precursor ACC into a-ketobutyrate and ammonia; the presence of ACC deaminase-containing bacteria plays a crucial role in regulating ethylene production, thereby impeding root growth and promoting optimal plant development. In addition, the enzymatic process serves a crucial function in the regular development of plants and confers resistance to diverse abiotic stressors (Pandey et al. 2017). In the past few years, considerable endeavors have been directed toward the identification and delineation of several genes linked with ACC deaminase activity in bacteria. The comparison of amino acid sequences of these genes has led to the identification of conserved residues that are presumed to have a crucial impact on the enzyme's overall functionality and catalytic efficacy.

6 Rhizosphere: Cross Talk in Plant–Microbe Signalling

The rhizospheric zone is a highly intriguing environment that facilitates the growth of a diverse range of microorganisms, which play a significant role in plant–microbe communication through the secretion of chemical signals, specialized metabolites,

Table 3 PR protein- and gene-mediated resistance responses in plants against various biotic and abiotic stresses

PR protein/ genes	Source	Transgenic plant	Effects	References
VpPR10.1	*Vitis vinifera*	*Vitis vinifera*	Induces HR-mediated defense response against the downy mildew pathogen *Plasmopara viticola*	Ma et al. (2018)
BjPR proteins	*Brassica juncea*	*Brassica juncea*	– Activation of SA- and JA-dependent pathways during *Erysiphe cruciferarum* infection – Activation of JA and SA signature genes during salinity and drought stress	Ali et al. (2017)
JIOsPR10	*Oryza sativa*	*Oryza sativa*	– Reduces susceptibility to rice blast pathogen – Enhances salt and drought tolerance compared to control	Wu et al. (2016)
TaLr35PR5 gene (thaumatin-like)	*Allium cepa*, *Triticum aestivum*	*Triticum aestivum*	Provide resistance against *Puccinia triticina*	Zhang et al. (2018)
GmPRP	*Glycine max*	*Glycine max Nicotiana tabacum*	Enhances resistance against *Phytophthora nicotiana* and *Phytophthora sojae*	Jiang et al. (2015)
OSamiRNAs	*Oryza sativa*	*Oryza sativa*	Enhances resistance against *Magnaporthe oryzae*, *Xanthomonas oryzae pv oryzae*, rice stripe virus, and brown plant hopper	Kumar et al. (2022)
JsPRP1	*Juglans sigillata* Dode	*Nicotiana tabacum*	– High tolerance to drought, $CdCl_2$, and *Colletotrichum gloeosporioides* infection – ROS accumulation is lower in transgenic tobacco compared to wild tobacco	Liu et al. (2018)
PR10 proteins	–	*Arabidopsis thaliana*	Enhances resistance to salt stress	Han et al. (2017)

and root exudates in response to specific stimuli. These secretions trigger various regulatory pathways, leading to alterations in plant growth, root–root interaction, inhibition of soil pathogens, nutrient acquisition, formation of biofilm, development of microbial resistance to pathogens, and the accumulation of soil microbes (Rosier et al. 2016; Sasse, et al. 2018; Manzar et al. 2022). Chemical signal molecules or metabolites serve as mediators in the communication between plants and microbes, as well as between microbes themselves, eliciting plant responses in the process. The schematic representation of signalling cascade in plant–PGPR interaction has been mentioned in Fig. 2. The microbial communication in the vicinity of

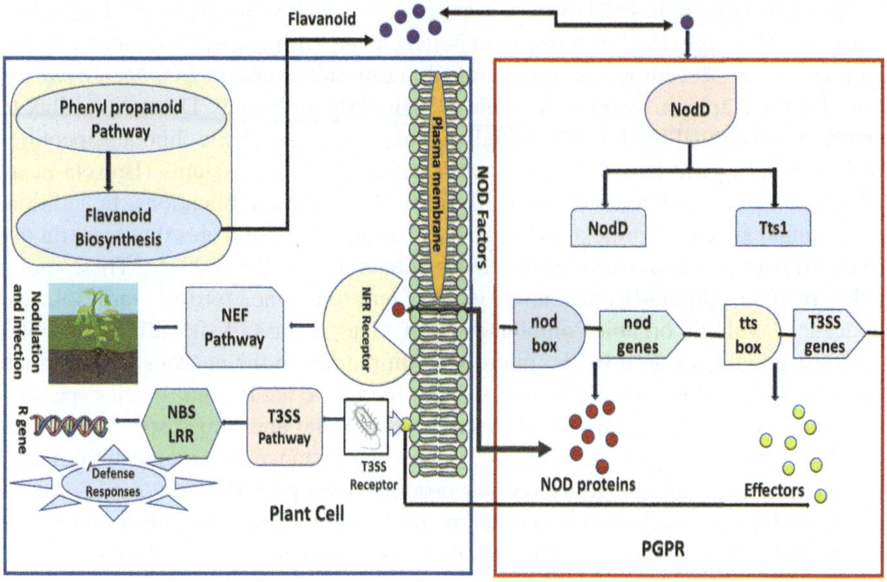

Fig. 2 Deciphering the signaling dynamics: schematic representation of plant-microbe interactions and signal activation

rhizospheric regions is regulated by a diverse group of signalling molecules which is produced during rhizomicrobiome interaction. Broadly speaking, the communication among the microbial community has been classified into two major groups, i.e., intra-/interspecies and interkingdom signalling between plants and microbes.

6.1 Intra- or Interspecies Signalling in Microbes

Quorum sensing (QS) is a specialized mechanism used for inter- and intraspecies communication in the rhizosphere. Through the reception of a cognate signal and a subsequent change in gene expression pattern, PGPR colonization in plant roots is fundamentally involved (Venturi and Keel 2016). Sequence-specific synthesis, release, and detection of signal molecules ensure that microbial communities can coordinate and build up their behavioral response through a process called quorum sensing (QS) (Helman and Chernin 2015). QS is facilitated by a wide variety of chemical signal molecules, including auto-inducer peptides (AIPs) and *N*-acyl homoserine lactones (AHLs). Biofilm formation, surface adhesion, cell motility, propagation, virulence property, conjugation, rhizospheric competition, synthesis of lytic enzymes and secondary metabolites, growth and metabolism, and symbiotic association are just some of the bacterial functions that are triggered by the recognition of these chemical signals by the recipient cell (Checcucci and Marchetti 2020).

Multiple studies have shown that antibiotics function as QS signalling molecules, which improves communication between and within species (Andersson and Hughes 2014). Recent research has linked antibiotics found in *Lysobacter capsici* found in tobacco's rhizosphere to a role as signalling molecules. Due to its ability to produce antifungal and antibacterial signals, such as cyclic lipodepsipeptides (c-LDP) and cyclic as well as polycyclic tetramate macrolactams (Brescia et al. 2020), this strain of bacteria may be useful for biocontrol applications. In addition, most fungal species secrete a wide variety of signalling molecules that bacteria use to coordinate and respond to each other's behaviors (Barriuso 2015). These molecules include gamma-heptalactone, gamma-butyrolactone, tyrosol, farnesol, and dodecanol. Volatile organic compounds (VOCs) are a class of signalling molecules synthesized and released by the microbial community in the gaseous form and are characterized by their low molecular weight, lipophilic nature, and species specificity. They are synthesized by a wide variety of bacterial and fungal spp. via distinct metabolic pathways and are responsible for long-distance communication among microbes. Fungal and bacterial species produce a wide variety of volatile organic compound classes, including alkanes, terpenoids, alkenes, ketones, sulfur-containing compounds, and alcohols. Because of their antimicrobial activity, they serve as a chemical arsenal, but they also affect the expression of genes involved in hormone signalling, defense, and microbial stress tolerance, virulence, and biofilm development (Audrain et al. 2015).

Furthermore, it has been established that VOCs' signalling molecules released by microbial communities can stimulate plant growth by altering root architecture and hormonal signalling and can boost plant resistance to biotic and abiotic stresses (Bitas et al. 2013). More research is needed to determine the precise role that microbial VOCs play in signalling. Therefore, the rhizomicrobiome structure is shaped in large part by inter- and intraspecies communication among rhizosphere microbes and their complex signalling mechanisms.

6.1.1 Cross Talk: Microbe–Plant Signalling

Microbial and host plant symbiotic interactions are triggered by the release of signals by microbes. The signals that emanate from the rhizosphere exhibit an exceptional capacity to elicit substantial modifications in the transcriptome of the plant (Sharma et al. 2020). Various strains of rhizobacteria, including *Azospirillum* spp., *B. amyloliquefaciens*, *B. muralis*, *B. thuringiensis*, *Rhizobium* spp., and *Pseudomonas* spp., have been recognized as significant producers of a wide range of phytohormone signals. Plant hormones are essential regulators of physiochemical processes and play a pivotal role in inducing resistance against biotic and abiotic stressors. Current scientific investigations have thoroughly explored the advantageous symbiotic relationships between microorganisms and plants, encompassing the facilitation of plant growth and the initiation of plant defense responses against diverse stressors. Notable rhizospheric microorganisms that confer benefits to plants are mycorrhiza, rhizobia, and plant growth-promoting bacteria and fungi (PGPR or

PGPF) (Venturi and Keel 2016). The microorganisms are capable of synthesizing substances that demonstrate both structural and functional similarities to the plant's intrinsic regulatory and defense mechanisms (Fahad et al. 2015). As a result, they exert significant influence over plant growth, hormonal signalling, immune responses, and overall development. Notable substances produced by these microbes include cytokinins (CK), IAA, GA3, JA, SA, and ABA.

The recognition of microbial-associated molecular patterns (MAMPs) such as flagellin, chitin, peptidoglycan, and lipopolysaccharides are attributed to receptor-like proteins known as pattern recognition receptors (PRRs). Upon detection, the organisms activate specific defense mechanisms via hormonal signalling pathways, as described by Pieterse et al. (2014). The discernment of quorum sensing (QS) signals resulting from the host plant by PRRs has a significant impact on root development, stress responses, plant defense stimulation, metabolic regulation, and hormonal balance. This suggests that AHLs play a role in interkingdom communication via QS signals (Schikora et al. 2016; Hassan et al. 2016). Apart from cyclo-dipeptides, DSF, and diketopiperazines, bacteria produce various other QS molecules. It is well-established that different plant PRRs can detect DSF QS signals and utilize them to trigger innate immune responses (Kakkar et al. 2015).

Research focusing on AHL mimicry within the microbial community (PGPR) has demonstrated that mimicry molecules, known as AHLs (AHL analogues), specifically disrupt quorum sensing regulatory pathways. These molecules have the ability to either stimulate or inhibit the expression of target genes in the original QS receptor or host. This process, known as quorum quenching, reduces pathogenicity by interfering with microbial QS signal molecules through the use of mimicry molecules (Kusari et al. 2015).

In a recent study exploring quorum quenching mechanisms, researchers made an intriguing discovery of a novel bacterial strain, *Acinetobacter* sp. XN-10 (Zhang et al. 2020). This strain exhibited a protective role against tissue maceration in *Solanum tuberosum*, *Brassica campestris*, and *Daucus carota*. Notably, *Acinetobacter* sp. XN-10 effectively eliminated QS molecules belonging to the AHL family, particularly those produced by *Pectobacterium carotovorum* subsp. *carotovorum*, a known causative agent of soft rot (Zhang et al. 2020).

Another study investigating the quorum quenching (qq) defense mechanism revealed an interesting phenomenon in plants infected with *Pseudomonas aeruginosa*. The infected plants secreted QQ molecules, such as rosmarinic acid, which acted as mimics and competitors to the pathogen-derived QS C4-HSL signalling molecules (Jamil et al. 2022). Consequently, these QQ molecules triggered QS-mediated responses, protecting the plants from virulence factors and inhibiting biofilm development. Furthermore, rhizospheric *Pseudomonas* spp. were found to produce antimicrobial signalling compounds like pyocyanin and diacetyl phloroglucinol. These compounds exhibited the remarkable ability to induce systemic resistance in plants against a broad spectrum of bacterial and fungal phytopathogens by modulating hormonal signalling pathways (Weller et al. 2012). In addition, a diverse group of plant growth-promoting rhizobacteria (PGPR) and plant growth-promoting fungi (PGPF) play a pivotal role in stimulating plant growth and enhancing systemic resistance against biotic stress through the release of volatile organic compounds (VOCs) (Kanchiswamy et al. 2015). Certain VOCs, including pentadecane, endecanone, undecanone, and heptanol, serve as recognizable signals for plants. For instance, specific rhizobacterial strains such as *B. subtilis* and *B. amyloliquefaciens* produce VOCs like 2-heptanol and 2-undecanone, respectively, which significantly promote the growth of *Arabidopsis thaliana* when the plants are inoculated with these strains (Wu et al. 2019).

6.1.2 Cross Talk: Plant–Microbe Signalling

Microbial communities predominantly reside on plants and engage in interactions with their host plants through mutualism, commensalism, or parasitism. Root exudates are plant-secreted molecules that specifically respond to microbial signals. These exudates can comprise various components, including organic acids, carbohydrates, aliphatic acids, fatty acids, amino acids, flavonoids, and low molecular weight secondary metabolites (Sasse et al. 2018; Manzar et al. 2022). The molecules the plant secretes attract rhizospheric microbes, which then form a parasitic or symbiotic relationship with the roots. The symbiotic relationship between *Rhizobium* spp. and the root nodule of legumes is the best-known example of plant–microbe interaction. *Rhizobia* and legumes form a symbiotic relationship, thanks to a signalling pathway that aids in nutrient uptake by the host plant. *Rhizobium* makes available the reduced form of N_2 to plants through biological nitrogen fixation and in exchange, the plants supply carbohydrates to the *Rhizobium* (Poole et al. 2018). The mutualistic relationship between *Rhizobium* and host plants is initiated through secretion of root exudates of the host plant, i.e., flavonoid derivatives of 2-phenyl-1,4-benzopyrone as a signal molecule. The flavonoids function as chemical attractants (signalling molecules) for *Rhizobium* cells to trigger nodulation process upon the activation of nod genes of *Rhizobium* cells which in turn synthesize the NOD factor termed as lipo-chitooligosaccharides (LCOs) (Buendia et al. 2019). LCOs are major signalling molecules and disperse through rhizobium and switch on certain nodulation genes which are responsible for not only nodule formation but also lateral root development through auxin homeostasis regulation process (Limpens et al. 2015; Buendia et al. 2019). Thc bacterial association lysin motif receptor-like kinases (LysM-RLKs), which contain an oligosaccharide-binding LysM site, help plant receptor-like kinase molecules (found in the root epidermis of the host plant) recognize LCO molecules. The nodulation process begins when LCOs bind to LysM, which activates signalling pathways. To easily recognize LCOs/Nod-like factors (NFRs) like NFR1 and NFR5 in the case of legume crops, Nod factor receptors are present (Maillet et al. 2011). The schematic representation of the symbiotic-signal pathway indicating the signal activation during plant–microbe interaction has been depicted in Fig. 2.

Occasionally, these pathways can be more efficiently triggered by the amalgamation of LCOs and chitiologosaccarides (Cos). When the interaction between *Cos* molecules and LYK/CERK1 receptors which in turn activate plant defense mechanism through activation of the Ca^{2+} channel in which influx of Ca^{2+} and H^+ and efflux of K^+ and Cl^- occurs across the cell membrane, synthesis of ROS molecules by NAPDH oxidase (RobHD) and by activating MAPK pathway through phosphorylation (Ge et al. 2016). On the other hand, in the presence of phytopathogens, the development of ROS can be suppressed by LCOs. Apart from the LCO proteins, the rhizobia also produce an effector molecule termed as NopL (nodulation outer protein L) as in the case of pathogenic effector proteins that are produced by plant pathogenic bacteria, i.e., T3 effector protein through type T3SS secretion system (coded by hrp genes) during infection cycle. The transcription of PR proteins like chitinase and glucanase is suppressed through interaction between NopL and MAPK signalling, and nodule senescence is delayed as a result. However, NopL phosphorylation can be blocked by Ser/Thr kinase inhibitors and MAPKK inhibitors, indicating that NopL is a MAPK substrate (Ge et al. 2016). Several other signalling molecules, including chalcones, betaines, vanillin, JA, and aldonic acid, have been reported to activate nod genes and thereby regulate the gene expression of careful partner symbionts for efficient mutualism. Enhanced nodulation, rapid N and P absorption, and higher total protein content in chickpea are just some of the PGP effects seen after co-inoculation with PGPRs such as *Mesorhizobium ciceri, Bacillus* sp., and *Enterobacter aerogenes* (Benjelloun et al. 2021).

In addition, the foregoing, numerous studies on nitric oxide (NO) have demonstrated NO's essential role as a signalling molecule in the signal transduction pathway between the PGPR and the host plant. In the presence of oxygen, the bacterial nitric oxide synthase (bNOS) gene in a number of PGPR can oxidize L-arginine to L-citrulline, resulting in the formation of nitric oxide (NO) (Grover et al. 2021). Several signalling pathways, including stress-mediated and developmental pathways, can involve nitric oxide (NO) (Grover et al. 2021). Nitrosoglutathione and sodium nitroprusside are two NO-related compounds that, like nitric oxide (NO), can be applied to plants to stimulate root growth (Chakraborty and Acharya 2017). While NO can promote ROS production in response to stress, it also plays a wide variety of roles in plant development via interactions with reactive oxygen species (ROS), plant hormones, posttranslational modifications of a wide variety of proteins, and Ca^{2+}-regulated pathways.

Similar to the induction of LCOs in rhizosphere fungi by root exudates, the activation of Myc signaling molecules in rhizosphere fungi is triggered by the presence of arbuscular mycorrhizal fungi (AMF) (Bukhat et al. 2020). Roots secrete primary signalling molecules like strigolactones (act as branching factors) and cutin monomers, which are recognized by AMF and cause the release of mycorrhizal lipochito-oligosaccharides (Myc factors), which are responsible for the establishment of a mutualistic relationship between the two. Mycorrhizae alter the hormonal signalling of their host plant by releasing proteins called mutualistic nonpathogenic effectors (MNEs), which include small secreted proteins (SSPs). When the ectomycorrhizal fungus *Laccaria bicolor* interacts with a root, it releases a 7-kDa protein called MiSSP7. This small protein enters host cells and blocks the effects of jasmonic acid by interacting with co-receptors for plant hormones (Plett et al. 2014).

Recent studies on the tripartite interaction of arbuscular mycorrhizal fungi (AFM), rhizobia (*Rhizobium*), and plants (ARP) have shed light on the fact that this type of symbiotic relationship improves biological nitrogen fixation, increases root-soluble sugar content and dry weight, and increases the total number of nodules (Hao et al. 2019). As Selvaraj et al. (2020) point out, this tripartite symbiosis mechanism is regarded in the literature as an effective interaction for plant growth.

7 Regulatory Network for Regulation of Plant Immunity Against Biotic Stress (Phytopathogens)

Insects, fungi, bacteria, viruses, viroids, protists, nematodes, and other biotic agents are responsible for about 30% and 15% of global yield losses and food losses, respectively (Haggag et al. 2015). Host plants respond to a wide variety of plant pathogens with an intrinsic immune response triggered by plant pathogen perception and response (PGPR) receptors. This response consists of two stages: (a) regulation of defense responses via activation of defense-related signalling pathways. Plants respond to threats by first mounting a local immune defense response, which is then translated into a systemic defense response via hormonal signalling pathways (Denancé et al. 2013). The signalling network and pathway that plants have evolved to transmit extracellular and intracellular signals is highly sophisticated, complex, and diverse. Gaining tolerance to stress in the host plant is the job of signalling modules like Ca^{2+} calcineurin B-like (CBL), CIPKs, and MAPKs, as well as stress-responsive transcription factors (Tang et al. 2020a, b). Changes in hormone levels and associated signalling pathways mediate the growth and stress response developed in plant by virtue of the action of PGPR. Roots of plants can form various associations with microbes through the use of these signalling molecules. Numerous signalling pathways, including MAPK and CDPK pathways, are active within the plant system based on stimulus or the signalling apparatus (Bredow and Monaghan 2019). The significance of these signalling pathways in response to abiotic or biotic stress is well-established. In addition, the PGPR indirectly boosts crop stress tolerance, which is governed by signal chemicals like phytohormone (i.e., auxins, GA3, CKs, brassinosteroids, and other specialized signal molecules) that facilitate host-to-phytomicrobiome-to-host interaction as a result of the regulation and coordination of multiple defense responses and developmental stages and may also contribute to plant immunity (Lyu et al. 2020; Rady et al. 2021). The mitogen-activated protein kinases (MAPKs) are a family of highly conserved signalling molecules consisting of the MAPKKK, MAPKK, and MAPK, which relay signals via a phosphorylation process. Many different signals can activate MAPKs, but in the end, they usually produce the same result (Sözen et al. 2020). The schematic representation of defense mechanism activation in plant through perception of endogenous and exogenous elicitors in response to biotic and abiotic stress conditions has been mentioned in Fig. 3.

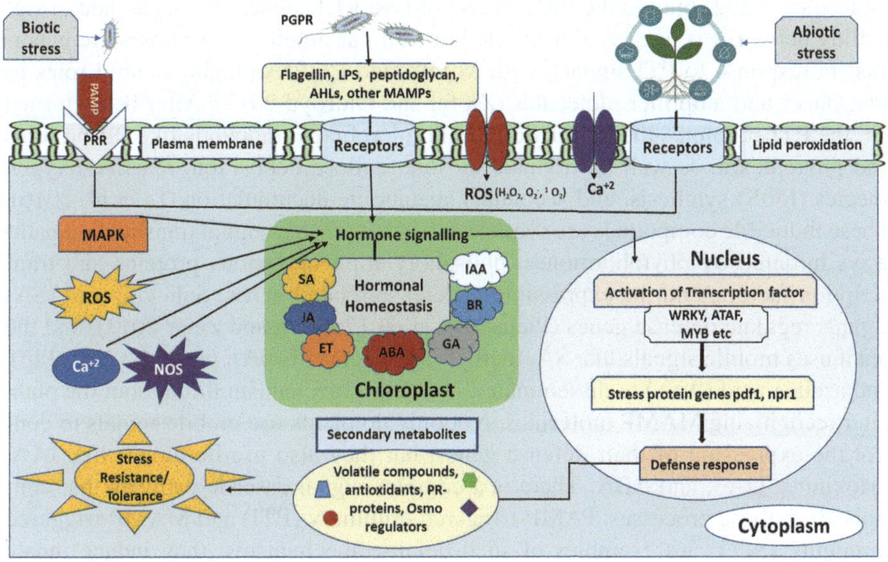

Fig. 3 Schematic representation of defense mechanism activation in plant through perception of endogenous and exogenous elicitors in response to biotic and abiotic stress condition

8 Perception of Stress Stimulus and Activation of Defense Response

8.1 PAMP-Triggered Immunity (PTI)

In plants, defense mechanisms are triggered by the molecular interaction of PRR molecules, which recognize pathogen-derived conserved molecules like PAMPs and MAMPs (Hou et al. 2019). Enhanced plant immunity is the result of a cascade of defense signalling modules, such as mitogen-activated protein kinases (MAPKs) and cyclic AMP-dependent protein kinases (CDPKs), being activated in response to the recognition of pathogen-associated molecular patterns (PAMPs) or pathogen-associated molecular patterns (MAMPs). Well-known examples of MAMPs include components of the cell walls of bacteria and fungi, such as chitosan, glucan, cutin oligomers, glycoproteins, peptidoglycans, chitin, LPS layer, flg-22, and several other molecules (Barreto-Bergter and Figueiredo 2014).

Biosurfactants, AHLs, siderophores, antibiotics like pyocyanin and DAPG, and other molecules produced by beneficial bacteria are all known to activate the plant defense system and are therefore considered MAMPs. Additional examples of DAMPs or endogenous elicitors include the lytic enzymes synthesized by plant pathogens to breach or degrade plant tissues. Both hormonally dependent processes, such as MAPKs and CDPKs, and hormonally independent processes govern the regulation of the signal cascade (Tena et al. 2011; Pandey et al. 2016). When PAMP molecules (flg22) bind to the PRR-FLS-2 of host RLK molecules, a cascade of signalling networks is triggered. Multiple intracellular defensive responses are generated in response to PTI signals, with MAPK and CDPKs playing central roles as transducer and amplifier molecules (Zipfel and Oldroyd 2017). After being turned on, the PTI promotes the production of antimicrobial compounds like PR proteins and phytoalexin, as well as stomatal closure, callose deposition, reactive oxygen species (ROS) synthesis, and secondary metabolite accumulation (Li et al. 2016). These inducible compounds are synthesized in response to signal transduction pathways initiated by phytohormones, phosphorylation of various proteins and transcription factors, and the expression of defense-related genes. Salicylic acid (SA) signals regulate defense genes (Pieterse et al. 2012; Shah and Zeier 2013), and the plant uses mobile signals like SA, methyl-jasmonate (Me-JA), pipecolic acid (Pip), and azelaic acid (AzA) to disseminate the defense mechanism throughout the plant after recognizing MAMP molecules. Not only do plants use mobile signals to control the expression of their defense genes, but they also use hormones like IAA, cytokinins, GA3, and ABA. There is a complex signal cascade pathway that controls all of these processes. PAMP-triggered immunity (PTI) and MAMP-triggered immunity (MTI) are examples of such defense mechanisms; they induce innate immunity and stop the pathogens from multiplying and colonizing (Bakker et al. 2014; Couto and Zipfel 2016; Li et al. 2016). Plant growth regulators (PGPR) are essential for the induction and accumulation of hormones such as SA and ABA that play a role in plant defense mechanisms like the regulation of stomatal movement, transpiration, and osmoregulation by abscisic acid and the protection of plants from pathogenic organisms by inducing the SA pathway (Bashir et al. 2020).

8.2 Effector-Triggered Immunity (ETI)

In addition to PTI, the pathogen employs numerous countermeasure mechanisms that disrupt PTI signalling and prevent detection by the host plant by triggering effector-triggered susceptibility (ETS) (Pel and Pieterse 2013; Gimenez-Ibanez et al. 2016). Due to the activation of ETS, plants have developed a second line of defense called ETI, which is regulated by the reorganization of pathogenic effector molecules via a specialized intracellular protein kinase domain called NB-LRR domains (Pieterse et al. 2014). Resistance to specific phytopathogens such as bacteria, fungi, virus, oomycete fungi, and phytopathogenic nematodes is encoded by a large class of plant resistance genes (R gene family), which includes NB-LRR. Either by directly binding to the effector molecules or by shifting its protein binding configuration site in response to effector binding, it is able to identify them. *Phytophthora infestans* is able to invade and colonize plants because of effector proteins that dampen hormonal signalling (i.e., SA and JA signals) (Boevink et al. 2016). When a wide variety of phytopathogens are detected, the PTI and ETI are activated, setting off a cascade of defense signals that spread throughout the plant and limit further infection. SAR refers to a type of long-distance defense response triggered by phytopathogens.

The SAR signalling pathway requires SA, which is produced by a large number of PGPR and used as a signal molecule to activate and regulate defense-related genes such as PR genes, PTI, and ETI (Shah and Zeier 2013; Adam et al. 2018). Salicylic acid (SA) is synthesized by PGPR, and an increase in SA level alters the cytoplasmic redox potential, which in turn regulates the transport of NPR1 (non-expressor PR1 genes) oligomers to the nucleus in response to a nuclear localization signal. After being transported to the nucleus, NPR1 monomers act as co-transcription factors by interacting with TGA TF (transcription factors). Upregulation and expression of genes such as WRKY70, which belong to a family of transcription factors, and activation of the SA-dependent pathway that encodes for antimicrobial compounds such phytoalexin, PR proteins, and others are initiated by this complex (Wu et al. 2012; Fu and Dong 2013). In order to activate the PR gene, SA binds to NPR1 and causes conformational changes in its structure (Wu et al. 2012). ISR refers to the activation of a defense signalling pathway against phytopathogens and pests by means of beneficial microbes, such as PGPR (Doornbos et al. 2011; Pieterse et al. 2014). The signalling pathways of plant hormones like jasmonic acid (JA) and ethylene (ET) control the SA-independent and PGPR-based ISR. An intricate defense signalling network is formed by ROS and RNOS, which promote the biosynthesis of JA, ET, and SA. Gene expression of ethylene (ET) and jasmonic acid (JA) as a defense mechanism against *Alternaria alternata* and *Botrytis cinerea* is triggered by the RSL1 (recombinant *Solanum lycopersicum*) protein (Singh et al. 2019a, b, c). Thus, it can be suggested that PGPR controls multiple stress responses by several ways through induction of ISR, pathogenesis-related (PR) protein-coding genes, etc. (Singh and Jha 2017), increased antioxidative enzymatic activities, reduction in reactive oxygen species (ROS) generation (Younas et al. 2021), increased ACC deaminase activity, reduced C2H4 biosynthetic pathway, etc. (Meena et al. 2015) during alleviation of biotic and abiotic stresses as well.

9 Conclusion and Future Perspective

Plant experiences a wide array of environmental stresses including metal toxicity, heat stress, salinity, drought stress, phytopathogens, and hormonal and nutritional imbalance which in turn alters biochemical and physiological properties and ultimately hampers the yield potential and quality of produce. However, plants can actively defend themselves against these stresses through chemical communication and various signalling cascade pathways during plant–microbiome interaction in rhizospheric regions and established either mutualistic or parasitic relationship. Plants and PGPR create a dynamic network of interaction and communication that improves nutrient acquisition and the regulation of phytohormones, as well as the production of the antioxidant enzymes superoxide dismutase, catalase, and peroxidase. Over the past decade, quorum sensing has been explained in interkingdom signalling between plants and PGPR and in the gene expression of beneficial and harmful interactions between inter- and intramicrobial species. Most plant metabolites have been characterized using *Arabidopsis thaliana* as a model. Metabolites and chemical cues may be found in other plants, according to the literature. New chemical cues to increase plant productivity via mutualism, resistance to biotic and abiotic stress, and phytopathogen defense can be developed from studying rhizosphere communities. However, there is lacunae in this area which need to be addressed, such as host specificity, quorum sensing, quorum quenching, biofilm formation, signalling pathway, bacterial motility, commercial formulations, and inconsistency under field condition. The manufacturing of various combinations of either bacterial strains with fungal strains or both bacterial strains to target the phytopathogens requires in-depth research into the identification of different pathways and their mechanism of action of PGPR. There are a number of biotechnological avenues that could be pursued in order to create synthetic bacterial consortiums with enhanced biological efficacy and longer shelf-life periods. PGPR-based bioformulation, such as biopesticides and nanopesticides, along with their selection, identification, screening, trait characterization, biosafety and compatibility assessment, and efficient delivery methods, is seen as an essential future component for sustainable agriculture and plant stress management due to its environmental friendliness, lack of toxicity, and low cost.

Additionally, it is important to place a greater emphasis on the novel approach "omics" technology, which includes genomics, transcriptomics, metagenomics, and metabolomics, and which has the potential to generate multiple pieces of knowledge regarding new chemical cues, cellular mechanisms, signal transduction pathway during plant microbes interaction, genes involved in SAR and ISR induction, and the discovery of virulence specificity of pathogen effectors. The expression of genes and metabolic actions accompanied with tripartite interactions involving arbuscular mycorrhizal fungi, rhizobia, and plants must be better understood. Recognizing and understanding the part played by each partner in the mutually beneficial relationships within the rhizospheric zone would be greatly aided by metagenomics. Further study on rhizosphere engineering is needed to figure out how specific plant–microbe communications can boost sustainable agriculture.

References

Adam AL, Nagy ZA, Katay G, Mergenthaler E, Viczian O (2018) Signals of systemic immunity in plants: progress and open questions. Int J Mol Sci 19:1146. https://doi.org/10.3390/ijms19041146

Agri U, Chaudhary P, Sharma A (2021) In vitro compatibility evaluation of agriusable nanochitosan on beneficial plant growth-promoting rhizobacteria and maize plant. Natl Acad Sci Lett 44:555–559. https://doi.org/10.1007/s40009-021-01047-w

Agri U, Chaudhary P, Sharma A, Kukreti B (2022) Physiological response of maize plants and its rhizospheric microbiome under the influence of potential bioinoculants and nanochitosan. Plant Soil 474:451–468. https://doi.org/10.1007/s11104-022-05351-2

Ahemad M, Khan MS (2012) Ecological assessment of biotoxicity of pesticides towards plant growth promoting activities of pea (*Pisum sativum*)-specific *Rhizobium* sp. strain MRP1. Emir J Sci Food Agric 24:334–343

Ali S, Mir ZA, Tyagi A, Bhat JA, Chandrashekar N, Papolu PK, Rawat S, Grover A (2017) Identification and comparative analysis of *Brassica juncea* pathogenesis-related genes in response to hormonal, biotic and abiotic stresses. Acta Physiol Plant 39:268. https://doi.org/10.1007/s11738-017-2565-8

Andersson DI, Hughes D (2014) Microbiological effects of sublethal levels of antibiotics. Nat Rev Microbiol 12:465–478. https://doi.org/10.1038/nrmicro3270

Armendariz AL, Talano MA, Olmos Nicotra MFO, Escudero L, Breser ML, Porporatto C, Agostini E (2019) Impact of double inoculation with *Bradyrhizobium japonicum* E109 and *Azospirillum brasilense* Az39 on soybean plants grown under arsenic stress. Plant Physiol Biochem 138:26–35. https://doi.org/10.1016/j.plaphy.2019.02.018

Ashraf A, Bano A, Ali SA (2019) Characterisation of plant growth promoting rhizobacteria from rhizosphere soil of heat-stressed and unstressed wheat and their use as bio-inoculant. Plant Biol 21:762–769. https://doi.org/10.1111/plb.12972

Audrain B, Farag MA, Ryu CM, Ghigo JM (2015) Role of bacterial volatile compounds in bacterial biology. FEMS Microbiol Rev 39:222–233. https://doi.org/10.1093/femsre/fuu013

Ayaz M, Ali Q, Farzand A, Khan AR, Ling H, Gao X (2021) Nematicidal volatiles from *Bacillus atrophaeus* GBSC56 promote growth and stimulate induced systemic resistance in tomato against *Meloidogyne incognita*. Int J Mol Sci 22:5049. https://doi.org/10.3390/ijms22095049

Baba ZA, Hamid B, Sheikh TA, Alotaibi SH, El Enshasy HA, Ansari MJ, Sayyed RZ (2021) Psychrotolerant *Mesorhizobium* sp. isolated from temperate and cold desert regions solubilize potassium and produces multiple plant growth-promoting metabolites. Molecules 26:5758. https://doi.org/10.3390/molecules26195758

Backer R, Rokem JS, Ilangumaran G, Lamont J, Praslickova D, Ricci E, Subramanian S, Smith DL (2018) Plant growth-promoting rhizobacteria: context, mechanisms of action, and roadmap to commercialization of biostimulants for sustainable agriculture. Front Plant Sci 9:1473. https://doi.org/10.3389/fpls.2018.01473

Bakker PAHM, Ran L, Mercado-Blanco J (2014) Rhizobacterial salicylate production provokes headaches. Plant Soil 382:1–16. https://doi.org/10.1007/s11104-014-2102-0

Barreto-Bergter E, Figueiredo RT (2014) Fungal glycans and the innate immune recognition. Front Cell Infect Microbiol 4:145. https://doi.org/10.3389/fcimb.2014.00145

Barriuso J (2015) Quorum sensing mechanisms in fungi. AIMS Microbiol 1:37–47. https://doi.org/10.3934/microbiol.2015.1.37

Bashir T, Naz S, Bano A (2020) Plant growth promoting rhizobacteria in combination with plant growth regulators attenuate the effect of drought stress. Pak J Bot 52:783–792. https://doi.org/10.30848/PJB2020-3(17)

Beneduzi A, Ambrosini A, Passaglia LMP (2012) Plant growth-promoting rhizobacteria (PGPR): their potential as antagonists and biocontrol agents. Genet Mol Biol 35:1044–1051. https://doi.org/10.1590/s1415-47572012000600020

Benjelloun I, Thami Alami I, El Khadir M, Douira A, Udupa SM (2021) Co-inoculation of *Mesorhizobium ciceri* with either *Bacillus* sp. or *Enterobacter aerogenes* on chickpea improves growth and productivity in phosphate-deficient soils in dry areas of a Mediterranean region. Plants (Basel) 10:571. https://doi.org/10.3390/plants10030571

Bhattacharyya PN, Jha DK (2012) Plant growth-promoting rhizobacteria (PGPR): emergence in agriculture. World J Microbiol Biotechnol 28:1327–1350. https://doi.org/10.1007/s11274-011-0979-9

Bitas V, Kim HS, Bennett JW, Kang S (2013) Sniffing on microbes: diverse roles of microbial volatile organic compounds in plant health. Mol Plant-Microbe Interact 26:835–843. https://doi.org/10.1094/MPMI-10-12-0249-CR

Boevink PC, Wang X, McLellan H, He Q, Naqvi S, Armstrong MR, Birch PR (2016) A *Phytophthora infestans* RXLR effector targets plant PP1c isoforms that promote late blight disease. Nat Commun 7:10311. https://doi.org/10.1038/ncomms10311

Bredow M, Monaghan J (2019) Regulation of plant immune signaling by calcium-dependent protein kinases. Mol Plant-Microbe Interact 32:6–19. https://doi.org/10.1094/MPMI-09-18-0267-FI

Brescia F, Marchetti-Deschmann M, Musetti R, Perazzolli M, Pertot I, Puopolo G (2020) The rhizosphere signature on the cell motility, biofilm formation and secondary metabolite production of a plant-associated *Lysobacter* strain. Microbiol Res 234:126424. https://doi.org/10.1016/j.micres.2020.126424

Buendia L, Maillet F, O'Connor D, van de Kerkhove Q, Danoun S, Gough C (2019) Lipo-chitooligosaccharides promote lateral root formation and modify auxin homeostasis in *Brachypodium distachyon*. New Phytol 221:2190–2020. https://doi.org/10.1111/nph.15551

Bukhat S, Imran A, Javaid S, Shahid M, Majeed A, Naqqash T (2020) Communication of plants with microbial world: exploring the regulatory networks for PGPR mediated defense signaling. Microbiol Res 238:126486. https://doi.org/10.1016/j.micres.2020.126486

Canarini A, Kaiser C, Merchant A, Richter A, Wanek W (2019) Corrigendum: root exudation of primary metabolites: mechanisms and their roles in plant responses to environmental stimuli. Front Plant Sci 10:420. https://doi.org/10.3389/fpls.2019.00420

Caradonia F, Francia E, Morcia C, Ghizzoni R, Moulin L, Terzi V, Ronga D (2019) Arbuscular mycorrhizal fungi and plant growth promoting rhizobacteria avoid processing tomato leaf damage during chilling stress. Agronomy 9:299. https://doi.org/10.3390/agronomy9060299

Chakraborty N, Acharya K (2017) "NO way"! Says the plant to abiotic stress. Plant Gene 11:99–105. https://doi.org/10.1016/j.plgene.2017.05.001

Chaudhary T, Dixit M, Gera R, Shukla AK, Prakash A, Gupta G, Shukla P (2020) Techniques for improving formulations of bioinoculants. 3 Biotech 10:199. https://doi.org/10.1007/s13205-020-02182-9

Chaudhary P, Chaudhary A, Parveen H, Rani A, Kumar G, Kumar A, Sharma A (2021a) Impact of nanophos in agriculture to improve functional bacterial community and crop productivity. BMC Plant Biol 21:519. https://doi.org/10.1186/s12870-021-03298-7

Chaudhary P, Khati P, Chaudhary A, Gangola S, Kumar R, Sharma A (2021b) Bioinoculation using indigenous *Bacillus* spp improves growth and yield of *Zea mays* under the influence of nanozeolite. 3 Biotech 11:11. https://doi.org/10.1007/s13205-020-02561-2

Chaudhary P, Khati P, Chaudhary A, Maithani D, Kumar G, Sharma A (2021c) Cultivable and metagenomic approach to study the combined impact of nanogypsum and *Pseudomonas taiwanensis* on maize plant health and its rhizospheric microbiome. PLoS One 16:e0250574. https://doi.org/10.1371/journal.pone.0250574

Chaudhary P, Khati P, Gangola S, Kumar A, Kumar R, Sharma A (2021d) Impact of nanochitosan and *Bacillus* spp. on health, productivity and defence response in *Zea mays* under field condition. 3 Biotech 11:237. https://doi.org/10.1007/s13205-021-02790-z

Chaudhary P, Parveen H, Gangola S, Kumar G, Bhatt P, Chaudhary A (2021e) Plant growth promoting rhizobacteria and their application in sustainable crop production. In: Bhatt P, Gangola S, Udayanga D, Kumar G (eds) Microbial technology for sustainable environment. Springer, Singapore. https://doi.org/10.1007/978-981-16-3840-4_13

Chaudhary P, Sharma A, Chaudhary A, Khati P, Gangola S, Maithani D (2021f) Illumina based high throughput analysis of microbial diversity of maize rhizosphere treated with nanocompounds and *Bacillus* sp. Appl Soil Ecol 159:103836. https://doi.org/10.1016/j.apsoil.2020.103836

Chaudhary A, Chaudhary P, Upadhyay A, Kumar A, Singh A (2022a) Effect of gypsum on plant growth promoting rhizobacteria. Environ Ecol 39(4A):1248–1256

Chaudhary P, Agri U, Chaudhary A, Kumar A, Kumar G (2022b) Endophytes and their potential in biotic stress management and crop production. Front Microbiol 13:933017. https://doi.org/10.3389/fmicb.2022.933017

Chaudhary P, Chaudhary A, Agri U, Khatoon H, Singh A (2022c) Recent trends and advancements for agro-environmental sustainability at higher altitudes. In: Goel R, Soni R, Suyal DC, Khan M (eds) Survival strategies in cold-adapted microorganisms. Springer, Singapore. https://doi.org/10.1007/978-981-16-2625-8_19

Chaudhary P, Singh S, Chaudhary A, Agri U, Bhandari G (2022d) Chapter 3 Agrochemicals and their effects on soil microbial population. In: Soni R, Suyal D, Goel R (eds) Plant protection: from chemicals to biologicals. De Gruyter, Berlin, pp 45–58. https://doi.org/10.1515/9783110771558-003

Chaudhary P, Singh S, Chaudhary A, Sharma A, Kumar G (2022e) Overview of biofertilizers in crop production and stress management for sustainable agriculture. Front Plant Sci 13:930340. https://doi.org/10.3389/fpls.2022.930340

Chaudhary P, Chen S, Rajput VD, Jaiswal DK (2023a) Editorial: bioinoculants with nano-compounds to improve soil health: a step toward sustainable agriculture. Front Environ Sci 11:1270002. https://doi.org/10.3389/fenvs.2023.1270002

Chaudhary P, Xu M, Ahamad L, Chaudhary A, Kumar G, Adeleke BS, Verma KK, Hu D-M, Širić I, Kumar P, Popescu SM, Abou Fayssal S (2023b) Application of synthetic consortia for improvement of soil fertility, pollution remediation, and agricultural productivity: a review. Agronomy 13:643. https://doi.org/10.3390/agronomy13030643

Chaudhary P, Chen S, Rajput VD, Jaiswal DK (eds) (2023c) Bioinoculants with nano-compounds to improve soil health: a step toward sustainable agriculture. Frontiers Media SA, Lausanne. https://doi.org/10.3389/978-2-8325-3287-4

Checcucci A, Marchetti M (2020) The rhizosphere talk show: the rhizobia on stage. Front Agron 2:591494. https://doi.org/10.3389/fagro.2020.591494

Choi J, Roy Choudhury A, Walitang DI, Lee Y, Sa T (2022) ACC deaminase-producing *Brevibacterium linens* RS16 enhances heat-stress tolerance of rice (*Oryza sativa* L.). Physiol Plant 174:e13584. https://doi.org/10.1111/ppl.13584

Chu TN, Tran BTH, Van Bui L, Hoang MTT (2019) Plant growth-promoting rhizobacterium *Pseudomonas* PS01 induces salt tolerance in *Arabidopsis thaliana*. BMC Res Notes 12:11. https://doi.org/10.1186/s13104-019-4046-1

Couto D, Zipfel C (2016) Regulation of pattern recognition receptor signalling in plants. Nat Rev Immunol 16:537–552. https://doi.org/10.1038/nri.2016.77

Denancé N, Sánchez-Vallet A, Goffner D, Molina A (2013) Disease resistance or growth: the role of plant hormones in balancing immune responses and fitness costs. Front Plant Sci 4:155. https://doi.org/10.3389/fpls.2013.00155

Desrut A, Moumen B, Thibault F, Le Hir R, Coutos-Thévenot P, Vriet C (2020) Beneficial rhizobacteria *Pseudomonas simiae* WCS417 induce major transcriptional changes in plant sugar transport. J Exp Bot 71:7301–7315. https://doi.org/10.1093/jxb/eraa396

Doornbos RF, Geraats BP, Kuramae EE, Van Loon LC, Bakker PA (2011) Effects of jasmonic acid, ethylene, and salicylic acid signaling on the rhizosphere bacterial community of *Arabidopsis thaliana*. Mol Plant-Microbe Interact 24:395–407. https://doi.org/10.1094/MPMI-05-10-0115

Duan B, Li L, Chen G, Su-Zhou C, Li Y, Merkeryan H, Liu W, Liu X (2021) 1-Aminocyclopropane-1-carboxylate deaminase-producing plant growth-promoting rhizobacteria improve drought stress tolerance in grapevine (*Vitis vinifera* L.). Front Plant Sci 12:706990. https://doi.org/10.3389/fpls.2021.706990

Dubey A, Saiyam D, Kumar A, Hashem A, Abd Allah EF, Khan ML (2021) Bacterial root endophytes: characterization of their competence and plant growth promotion in soybean (*Glycine max* (L.) Merr.) under drought stress. Int J Environ Res Public Health 18:931

El-Esawi MA, Alaraidh IA, Alsahli AA, Alzahrani SM, Ali HM, Alayafi AA, Ahmad M (2018) *Serratia liquefaciens* KM4 improves salt stress tolerance in maize by regulating redox potential, ion homeostasis, leaf gas exchange and stress-related gene expression. Int J Mol Sci 19:3310. https://doi.org/10.3390/ijms19113310

Elnahal ASM, El-Saadony MT, Saad AM, Desoky EM, El-Tahan AM, Rady MM, El-Tarabily KA (2022) The use of microbial inoculants for biological control, plant growth promotion, and sustainable agriculture: a review. Eur J Plant Pathol 162:759–792. https://doi.org/10.1007/s10658-021-02393-7

Habib SH, Kausar H, Saud HM (2016) Plant Growth-Promoting Rhizobacteria Enhance Salinity Stress Tolerance in Okra through ROS-Scavenging Enzymes. Biomed Res Int. 2016;6284547. doi: 10.1155/2016/6284547.

Fahad S, Hussain S, Bano A, Saud S, Hassan S, Shan D, Huang J (2015) Potential role of phytohormones and plant growth-promoting rhizobacteria in abiotic stresses: consequences for changing environment. Environ Sci Pollut Res 22:4907–4921. https://doi.org/10.1007/s11356-014-3754-2

Foley JA, Ramankutty N, Brauman KA, Cassidy ES, Gerber JS, Johnston M, Zaks DP (2011) Solutions for a cultivated planet. Nature 478:337–342. https://doi.org/10.1038/nature10452

Fu ZQ, Dong X (2013) Systemic acquired resistance: turning local infection into global defense. Annu Rev Plant Biol 64:839–863. https://doi.org/10.1146/annurev-arplant-042811-105606

Ge YY, Xiang QW, Wagner C, Zhang D, Xie ZP, Staehelin C (2016) The type 3 effector NopL of *Sinorhizobium* sp. strain NGR234 is a mitogen-activated protein kinase substrate. J Exp Bot 67:2483–2494. https://doi.org/10.1093/jxb/erw065

Gimenez-Ibanez S, Chini A, Solano R (2016) How microbes twist jasmonate signaling around their little fingers. Plants (Basel) 5:9. https://doi.org/10.3390/plants5010009

Glick BR (2012) Plant growth-promoting bacteria: mechanisms and applications. Scientifica 2012:963401. https://doi.org/10.6064/2012/963401

Glick BR (2014) Bacteria with ACC deaminase can promote plant growth and help to feed the world. Microbiol Res 169:30–39. https://doi.org/10.1016/j.micres.2013.09.009

González-López MDC, Jijón-Moreno S, Dautt-Castro M, Ovando-Vázquez C, Ziv T, Horwitz BA, Casas-Flores S (2021) Secretome analysis of *Arabidopsis-Trichoderma atroviride* interaction unveils new roles for the plant glutamate:glyoxylate amino transferase GGAT1 in plant growth induced by the fungus and resistance against *Botrytis cinerea*. Int J Mol Sci 22:6804. https://doi.org/10.3390/ijms22136804

Grover M, Bodhankar S, Sharma A, Sharma P, Singh J, Nain L (2021) PGPR mediated alterations in root traits: way toward sustainable crop production. Front Sustain Food Syst 4:618230. https://doi.org/10.3389/fsufs.2020.618230

Guo Q, Li Y, Lou Y, Shi M, Jiang Y, Zhou J, Lai H (2019) *Bacillus amyloliquefaciens* Ba13 induces plant systemic resistance and improves rhizosphere micro ecology against tomato yellow leaf curl virus disease. Appl Soil Ecol 137:154–166. https://doi.org/10.1016/j.apsoil.2019.01.015

Haggag WM, Abouziena HF, Abd-El-Kreem F, El Habbasha S (2015) Agriculture biotechnology for management of multiple biotic and abiotic environmental stress in crops. J Chem Pharm Res 7:882–889

Han X, He X, He Qui W, Lu Z, Zhang Y, Chen S (2017) Pathogenesis related protein PR10 from *Salix matsudana Koidz* exhibits resistance to salt stress in transgenic *Arabidopsis thaliana*. Environ Exp Bot 141:74–82. https://doi.org/10.1016/j.envexpbot.2017.07.008

Hao Z, Xie W, Jiang X, Wu Z, Zhang X, Chen B (2019) Arbuscular mycorrhizal fungus improves rhizobium–glycyrrhiza seedling symbiosis under drought stress. Agronomy 9:572. https://doi.org/10.3390/agronomy9100572

Hassan R, Shaaban MI, Abdel Bar FM, El-Mahdy AM, Shokralla S (2016) Quorum sensing inhibiting activity of *Streptomyces coelicoflavus* isolated from soil. Front Microbiol 7:659. https://doi.org/10.3389/fmicb.2016.00659

Hayes S, Schachtschabel J, Mishkind M, Munnik T, Arisz SA (2021) Hot topic: thermosensing in plants. Plant Cell Environ 44:2018–2033. https://doi.org/10.1111/pce.13979

Helman Y, Chernin L (2015) Silencing the mob: disrupting quorum sensing as a means to fight plant disease. Mol Plant Pathol 16:316–329. https://doi.org/10.1111/mpp.12180

Hou S, Liu Z, Shen H, Wu D (2019) Damage-associated molecular pattern-triggered immunity in plants. Front Plant Sci 10:646. https://doi.org/10.3389/fpls.2019.00646

Jalmi SK (2020) Rhizosphere signaling nurturing phyto-microbiome niche. Trop Plant Res 7:522–528. https://doi.org/10.22271/tpr.2020.v7.i2.064

Jamil F, Mukhtar H, Fouillaud M, Dufossé L (2022) Rhizosphere signaling: insights into plant-rhizomicrobiome interactions for sustainable agronomy. Microorganisms 10(5):899. https://doi.org/10.3390/microorganisms10050899

Jiang L, Wu J, Fan S, Li W, Dong L, Cheng Q (2015) Isolation and characterization of a novel pathogenesis-related protein gene (GmPRP) with induced expression in soybean (*Glycine max*) during infection with *Phytophthora sojae* defence response by interacting with VpVDAC3. Plant Biotechnol J 16:1488–1501

Jochum MD, McWilliams KL, Borrego EJ, Kolomiets MV, Niu G, Pierson EA, Jo YK (2019) Bioprospecting plant growth-promoting rhizobacteria that mitigate drought stress in grasses. Front Microbiol 10:2106. https://doi.org/10.3389/fmicb.2019.02106

Johnson JM, Ludwig A, Furch ACU, Mithöfer A, Scholz S, Reichelt M, Oelmüller R (2019) The beneficial root-colonizing fungus *Mortierella hyaline* promotes the aerial growth of Arabidopsis and activates calcium-dependent responses that restrict *Alternaria brassicae*-induced disease development in roots. Mol Plant-Microbe Interact 32:351–363. https://doi.org/10.1094/MPMI-05-18-0115R

Joshi AU, Andharia KN, Patel PA, Kotadiya RJ, Kothari RK (2019) Plant growth promoting rhizobacteria: mechanism, application, advantages and disadvantages. Green biotechnology. Day Publishing House a Division of Astral International Pvt. Ltd., New Delhi, pp 13–40

Joshi B, Chaudhary A, Singh H, Kumar PA (2020) Prospective evaluation of individual and consortia plant growth promoting rhizobacteria for drought stress amelioration in rice (*Oryza sativa* L.). Plant Soil 457:225–240. https://doi.org/10.1007/s11104-020-04730-x

Kakkar A, Nizampatnam NR, Kondreddy A, Pradhan BB, Chatterjee S (2015) *Xanthomonas campestris* cell–cell signalling molecule DSF (diffusible signal factor) elicits innate immunity in plants and is suppressed by the exopolysaccharide xanthan. J Exp Bot 66:6697–6714. https://doi.org/10.1093/jxb/erv377

Kanchiswamy CN, Malnoy M, Maffei ME (2015) Bioprospecting bacterial and fungal volatiles for sustainable agriculture. Trends Plant Science 20:206–211. https://doi.org/10.1016/j.tplants.2015.01.004

Kapadia C, Kachhdia R, Singh S, Gandhi K, Poczai P, Alfarraj S (2022) *Pseudomonas aeruginosa* inhibits quorum sensing mechanisms of soft rot pathogen *Lelliottia amnigena* RCE to regulate its virulence factors and biofilm formation. Front Microbiol 13:977669. https://doi.org/10.3389/fmicb.2022.977669

Kaviya N, Upadhayay VK, Singh J, Khan A, Panwar M, Singh AV (2019) Role of microorganisms in soil genesis and functions. In: Mycorrhizosphere and pedogenesis. Springer Singapore, Singapore, pp 25–52. https://doi.org/10.1007/978-981-13-6480-8_2

Keswani C, Singh HB, García-Estrada C, Caradus J, He YW, Mezaache-Aichour S, Sansinenea E (2020) Antimicrobial secondary metabolites from agriculturally important bacteria as next-generation pesticides. Appl Microbiol Biotechnol 104:1013–1034. https://doi.org/10.1007/s00253-019-10300-8

Khan A, Singh AV (2021) Multifarious effect of ACC deaminase and EPS producing Pseudomonas sp. and Serratia marcescens to augment drought stress tolerance and nutrient status of wheat. World J Microbiol Biotechnol 37:198. https://doi.org/10.1007/s11274-021-03166-4

Khan A, Singh J, Upadhayay VK, Singh AV, Shah S (2019) Microbial biofortification: a green technology through plant growth promoting microorganisms. In: Sustainable green technologies for environmental management. Springer, Singapore, pp 255–269. https://doi.org/10.1007/978-981-13-2772-8_13

Khan N, Bano A, Ali S, Babar MA (2020) Cross talk amongst phytohormones from planta and PGPR under biotic and abiotic stresses. Plant Growth Regul 90:189–203. https://doi.org/10.1007/s10725-020-00571-x

Khan A, Panthari D, Sharma RS, Punetha A, Singh AV, Upadhayay VK (2023a) Biofertilizers: a microbial-assisted strategy to improve plant growth and soil health. In: Advanced microbial techniques in agriculture, environment, and health management. Elsevier, Amsterdam, pp 97–118. https://doi.org/10.1016/B978-0-323-91643-1.00007-7

Khan A, Singh AV, Pareek N, Arya P, Upadhayay VK, Kumar Jugran A, Mishra PK, Goel R (2023b) Credibility assessment of cold adaptive Pseudomonas jesenni MP1 and P. palleroniana N26 on growth, rhizosphere dynamics, nutrient status, and yield of the kidney bean cultivated in Indian Central Himalaya. Front Plant Sci 14:1042053. https://doi.org/10.3389/fpls.2023.1042053

Khati P, Parul, Bhatt P, Nisha Kumar R, Sharma A (2018) Effect of nanozeolite and plant growth promoting rhizobacteria on maize. 3 Biotech 8:141. https://doi.org/10.1007/s13205-018-1142-1

Khati P, Chaudhary P, Gangola S, Sharma P (2019) Influence of nanozeolite on plant growth promotory bacterial isolates recovered from nanocompound infested agriculture field. Environ Ecol 37:521–527

Kim ST, Yoo SJ, Weon HY, Song J, Sang MK (2022) *Bacillus butanolivorans* KJ40 contributes alleviation of drought stress in pepper plants by modulating antioxidant and polyphenolic compounds. Sci Hortic 301:111111. https://doi.org/10.1016/j.scienta.2022.111111

Kukreti B, Sharma A, Chaudhary P, Agri U, Maithani D (2020) Influence of nanosilicon dioxide along with bioinoculants on *Zea mays* and its rhizospheric soil. 3 Biotech 10:345. https://doi.org/10.1007/s13205-020-02329-8

Kumar A, Sharma A, Chaudhary P, Gangola S (2021a) Chlorpyrifos degradation using binary fungal strains isolated from industrial waste soil. Biologia 10:3071–3081. https://doi.org/10.1007/s11756-021-00816-8

Kumar G, Lal S, Maurya SK, Bhattacherjee AK, Chaudhary P, Gangola S, Rajan S (2021b) Exploration of *Klebsiella pneumoniae* M6 for paclobutrazol degradation, plant growth attributes, and biocontrol action under subtropical ecosystem. PLoS One 16:e0261338. https://doi.org/10.1371/journal.pone.0261338

Kumar G, Suman A, Lal S, Ram RA, Bhatt P, Pandey G, Chaudhary P, Rajan S (2021c) Bacterial structure and dynamics in mango (*Mangifera indica*) orchards after long term organic and conventional treatments under subtropical ecosystem. Sci Rep 11:20554. https://doi.org/10.1038/s41598-021-00112-0

Kumar K, Mandal SN, Neelam K, de Los Reyes BG (2022) MicroRNA-mediated host defense mechanisms against pathogens and herbivores in rice: balancing gains from genetic resistance with trade-offs to productivity potential. BMC Plant Biol 22:351. https://doi.org/10.1186/s12870-022-03723-5

Kumari S, Sharma A, Chaudhary P, Khati P (2020) Management of plant vigor and soil health using two agriusable nanocompounds and plant growth promotory rhizobacteria in Fenugreek. 3 Biotech 10:1–11. https://doi.org/10.1007/s13205-020-02448-2

Kumari H, Khati P, Gangola S, Chaudhary P, Sharma A (2021) Performance of plant growth promotory rhizobacteria on maize and soil characteristics under the influence of TiO2 nanoparticles. Pantnagar J Res 19:28–39

Kusale SP, Attar YC, Sayyed RZ, Malek RA, Ilyas N, Suriani NL, El-Enshasy HA (2021) Production of plant beneficial and antioxidants metabolites by *Klebsiella variicola* under salinity stress. Molecules 26:1894. https://doi.org/10.3390/molecules26071894

Kusari P, Kusari S, Spiteller M, Kayser O (2015) Implications of endophyte-plant crosstalk in light of quorum responses for plant biotechnology. Appl Microbiol Biotechnol 99:5383–5390. https://doi.org/10.1007/s00253-015-6660-8

Lakkis S, Trotel-Aziz P, Rabenoelina F, Schwarzenberg A, Nguema-Ona E, Clément C, Aziz A (2019) Strengthening grapevine resistance by *Pseudomonas fluorescens* PTA-CT2 relies on distinct defense pathways in susceptible and partially resistant genotypes to downy mildew and gray mold diseases. Front Plant Sci 10:1112. https://doi.org/10.3389/fpls.2019.01112

Lengai GMW, Muthomi JW, Mbega ER (2020) Phytochemical activity and role of botanical pesticides in pest management for sustainable agricultural crop production. Sci Afr 7:e00239. https://doi.org/10.1016/j.sciaf.2019.e00239

Leontidou K, Genitsaris S, Papadopoulou A, Kamou N, Bosmali I, Matsi T, Mellidou I (2020) Plant growth promoting rhizobacteria isolated from halophytes and drought-tolerant plants: genomic characterization and exploration of phytobeneficial traits. Sci Rep 10:14857. https://doi.org/10.1038/s41598-020-71652-0

Li B, Meng X, Shan L, He P (2016) Transcriptional regulation of pattern-triggered immunity in plants. Cell Host Microbe 19:641–650. https://doi.org/10.1016/j.chom.2016.04.011

Li Y, Guo Q, Li Y, Sun Y, Xue Q, Lai H (2019) *Streptomyces pactum* Act12 controls tomato yellow leaf curl virus disease and alters rhizosphere microbial communities. Biol Fertil Soils 55:149–169. https://doi.org/10.1007/s00374-019-01339-w

Li H, Guo Q, Jing Y, Liu Z, Zheng Z, Sun Y, La H (2020a) Application of *Streptomyces pactum* Act12 enhances drought resistance in wheat. J Plant Growth Regul 39:122–132. https://doi.org/10.1007/s00344-019-09968-z

Li S, Wang Z, Tang B, Zheng L, Chen H, Cui X, Liu D (2020b) A pathogenesis-related protein-like genes involved in the *Panax notoginseng* defense response to the root rot pathogen. Front Plant Sci 11:610176. https://doi.org/10.3389/fpls.2020.610176

Limpens E, van Zeijl A, Geurts R (2015) Lipochitooligosaccharides modulate plant host immunity to enable endosymbioses. Annu Rev Phytopathol 53:311–334. https://doi.org/10.1146/annurev-phyto-080614-120149

Liu DQ, Han Q, Shah T, Chen CY, Wang Q, Tang BF, Ge F (2018) A hybrid proline-rich cell-wall protein gene *JsPRP1* from *Juglans sigillata* Dode confers both biotic and abiotic stresses in transgenic tobacco plants. Trees 32:1199–1209. https://doi.org/10.1007/s00468-018-1702-3

Lyu D, Backer R, Subramanian S, Smith DL (2020) Phytomicrobiome coordination signals hold potential for climate change-resilient agriculture. Front Plant Sci 11:634. https://doi.org/10.3389/fpls.2020.00634

Ma H, Xiang G, Li Z, Wang Y, Dou M, Su L, Xu Y (2018) Grapevine VpPR10.1 functions in resistance to *Plasmopara viticola* through triggering a cell death-like defense pathways in susceptible and partially resistant genotypes to downy mildew and gray mold diseases. Front Plant Sci 10:1112

Ma Y, Rajkumar M, Oliveira RS, Zhang C, Freitas H (2019) Potential of plant beneficial bacteria and arbuscular mycorrhizal fungi in phytoremediation of metal-contaminated saline soils. J Hazard Mater 379:120813. https://doi.org/10.1016/j.jhazmat.2019.120813

Maillet F, Poinsot V, André O, Puech-Pagès V, Haouy A, Gueunier M, Denarie J (2011) Fungal lipochitooligosaccharide symbiotic signals in arbuscular mycorrhiza. Nature 469:58–63. https://doi.org/10.1038/nature09622

Malhi GS, Kaur M, Kaushik P (2021) Impact of climate change on agriculture and its mitigation strategies: a review. Sustainability 13:1318. https://doi.org/10.3390/su13031318

Manullang W, Chuang HW (2020) *Streptomyces* sp. mitigates abiotic stress response and promotes plant growth. J Plant Prot Res 60:263–274

Manzar N, Jayaswal D, Kishan G, Chauhan S, Joshi D, Singh J, Chaudhary P, Jamali H, Kushwaha P, Parveen H, Kashyap AS, Sahu PK, Singh UB (2022) Application of molecular ecology approaches in sustainable agriculture for a better understanding of plant–microbiome interactions. In: Singh UB, Sahu PK, Singh HV, Sharma PK, Sharma SK (eds) Rhizosphere microbes, Microorganisms for sustainability, vol 40. Springer, Singapore. https://doi.org/10.1007/978-981-19-5872-4_3

Meena H, Ahmed MA, Prakash P (2015) Amelioration of heat stress in wheat, *Triticum aestivum* by PGPR (*Pseudomonas aeruginosa* strain 2CpS1). Biosci Biotechnol Res Asia 8:171–174

Mesa-Marín J, Del-Saz NF, Rodríguez-Llorente ID, Redondo-Gómez S, Pajuelo E, Ribas-Carbó M, Mateos-Naranjo E (2018) PGPR reduce root respiration and oxidative stress enhancing *Spartina maritima* root growth and heavy metal rhizo accumulation. Front Plant Sci 9:1500. https://doi.org/10.3389/fpls.2018.01500

Mohanty P, Singh PK, Chakraborty D, Mishra S, Pattnaik R (2021) Insight into the role of PGPR in sustainable agriculture and environment. Front Sustain Food Syst 5:1–12. https://doi.org/10.3389/fsufs.2021.667150

Morcillo RJ, Singh SK, He D, An GI, Vílchez JI, Tang K, Zhang H (2020) *Rhizobacterium*-derived diacetyl modulates plant immunity in a phosphate-dependent manner. EMBO J 39:e102602. https://doi.org/10.15252/embj.2019102602

Moumbock AFA, Gao M, Qaseem A, Li J, Kirchner PA, Ndingkokhar B, Günther S (2021) Streptome DB 3.0: an updated compendium of *Streptomycetes* natural products. Nucleic Acids Res 49:600–604. https://doi.org/10.1093/nar/gkaa868

Mukhtar T, Rehman S, Smith D, Sultan T, Seleiman MF, Alsadon AA, Saad MA (2020) Mitigation of heat stress in *Solanum Lycopersicum* L. by ACC deaminase and exopolysaccharide producing *Bacillus cereus*: effects on biochemical profiling. Sustainability 12:2159. https://doi.org/10.3390/su12062159

Mukhtar T, Ali F, Rafique M, Ali J, Afridi MS, Chaudhary HJ (2022) Biochemical characterization and potential of *Bacillus safensis* strain SCAL1 to mitigate heat stress in *Solanum lycopersicum* L. J Plant Growth Regul 42:523. https://doi.org/10.1007/s00344-021-10571-4

Nie P, Chen C, Yin Q, Jiang C, Guo J, Zhao H, Niu D (2019) Function of miR825 and miR825 as negative regulators in *Bacillus cereus*. Int J Mol Sci 20:5032. https://doi.org/10.3390/ijms20205032

Omar SA, Fetyan NAH, Eldenary ME, Abdelfattah MH, Abd-Elhalim HM, Wrobel J, Kalaji HM (2021) Alteration in expression level of some growth and stress related genes after rhizobacteria inoculation to alleviate drought tolerance in sensitive rice genotype. Chem Biol Technol Agric 41:1. https://doi.org/10.1186/s40538-021-00237-4

Pandey SS, Singh S, Babu CS, Shanker K, Srivastava NK, Shukla AK, Kalra A (2016) Fungal endophytes of *Catharanthus roseus* enhance vindoline content by modulating structural and regulatory genes related to terpenoid indole alkaloid biosynthesis. Sci Rep 6:26583. https://doi.org/10.1038/srep26583

Pandey S, Fartyal D, Agarwal A, Shukla T, James D, Kaul T, Negi YK, Arora S, Reddy MK (2017) Abiotic stress tolerance in plants: myriad roles of ascorbate peroxidase. Front Plant Sci 8:581. https://doi.org/10.3389/fpls.2017.00581

Park YG, Mun BG, Kang SM, Hussain A, Shahzad R, Seo CW, Yun BW (2017) *Bacillus aryabhattai* SRB02 tolerates oxidative and nitrosative stress and promotes the growth of soybean by modulating the production of phytohormones. PLoS One 12:e0173203. https://doi.org/10.1371/journal.pone.0173203

Parveen H, Chaudhary A, Chaudhary P, Sultana R, Khati P, Kumar G, Khati P, Rana M, Bhatt P (2022) Omics approaches for the production of the microbial enzymes and applications. In: Bhatt P (ed) Industrial applications of microbial enzymes, 1st edn. CRC Press, Boca Raton. https://doi.org/10.1201/9781003202998-16

Pel MJ, Pieterse CM (2013) Microbial recognition and evasion of host immunity. J Exp Bot 64:1237–1248. https://doi.org/10.1093/jxb/ers262

Pérez-Montaño F, Alías-Villegas C, Bellogín RA, del Cerro P, Espuny MR, Jiménez-Guerrero I, Cubo T (2014) Plant growth promotion in cereal and leguminous agricultural important plants: from microorganism capacities to crop production. Microbiol Res 169:325–336. https://doi.org/10.1016/j.micres.2013.09.011

Pieterse CM, Van der Does D, Zamioudis C, Leon-Reyes A, Van Wees SC (2012) Hormonal modulation of plant immunity. Annu Rev Cell Dev Biol 28:489–521. https://doi.org/10.1146/annurev-cellbio-092910-154055

Pieterse CM, Zamioudis C, Berendsen RL, Weller DM, Van Wees SC, Bakker PA (2014) Induced systemic resistance by beneficial microbes. Annu Rev Phytopathol 52:347–375. https://doi.org/10.1146/annurev-phyto-082712-102340

Plett JM, Daguerre Y, Wittulsky S, Vayssières A, Deveau A, Melton SJ, Martin F (2014) Effector MiSSP7 of the mutualistic fungus Laccaria bicolor stabilizes the Populus JAZ6 protein and represses jasmonic acid (JA) responsive genes. Proc Natl Acad Sci U S A 111:8299–8304. https://doi.org/10.1073/pnas.1322671111

Poole P, Ramachandran V, Terpolilli J (2018) Rhizobia: from saprophytes to endosymbionts. Nat Rev Microbiol 16:291–303. https://doi.org/10.1038/nrmicro.2017.171

Poria V, Singh S, Nain L, Singh B, Saini JK (2021) Rhizospheric microbial communities: occurrence, distribution, and functions. In: Microbial metatranscriptomics belowground. Springer, Singapore, pp 239–271

Prakash R, Subramani R, Berde CV, Chandrasekhar T, Prathyusha A, Kariali E, Bramhachari PV (2022) Rhizobacteriome: plant growth-promoting traits and its functional mechanism in plant growth, development, and defenses. In: Understanding the microbiome interactions in agriculture and the environment. Springer, Singapore, pp 315–344

Pramanik K, Mitra S, Sarkar A, Maiti TK (2018) Alleviation of phytotoxic effects of cadmium on rice seedlings by cadmium resistant PGPR strain *Enterobacter aerogenes* MCC 3092. J Hazard Mater 351:317–329. https://doi.org/10.1016/j.jhazmat.2018.03.009

Puthiyottil P, Akkara Y (2021) Pretreatment with *Bacillus subtilis* mitigates drought induced photooxidative damages in okra by modulating antioxidant system and photochemical activity. Physiol Mol Biol Plants 27:945–957. https://doi.org/10.1007/s12298-021-00982-8

Rady MM, Boriek SHK, Abd El-Mageed TA, Seif El-Yazal MA, Ali EF, Hassan FAS, Abdelkhalik A (2021) Exogenous gibberellic acid or dilute bee honey boosts drought stress tolerance in *Vicia faba* by rebalancing osmoprotectants, antioxidants, nutrients, and phytohormones. Plants (Basel) 10:748. https://doi.org/10.3390/plants10040748

Rashid U, Yasmin H, Hassan MN, Naz R, Nosheen A, Sajjad M, Ahmad P (2022) Drought-tolerant *Bacillus megaterium* isolated from semi-arid conditions induces systemic tolerance of wheat under drought conditions. Plant Cell Rep 41:549–569. https://doi.org/10.1007/s00299-020-02640-x

Rawat N, Sharma M, Suyal DC, Singh DK, Joshi D, Singh P, Goel R (2019) Psychrotolerant bio-inoculants and their co-inoculation to improve *Cicer arietinum* growth and soil nutrient status for sustainable mountain agriculture. J Soil Sci Plant Nutr 19:639–647. https://doi.org/10.1007/s42729-019-00064-5

Rehman FU, Kalsoom M, Adnan M, Toor M, Zulfiqar A (2020) Plant growth promoting rhizobacteria and their mechanisms involved in agricultural crop production: a review. Suntext Rev Biotechnol 1:1–6

Riaz U, Murtaza G, Anum W, Samreen T, Sarfraz M, Nazir MZ (2021) Plant growth-promoting rhizobacteria (PGPR) as biofertilizers and biopesticides. In: Microbiota and biofertilizers. Springer, Singapore, pp 181–196

Riyazuddin R, Nisha N, Ejaz B, Khan MIR, Kumar M, Ramteke PW, Gupta R (2021) A comprehensive review on the heavy metal toxicity and sequestration in plants. Biomol Ther 12:43. https://doi.org/10.3390/biom12010043

Roshani KA, Singh AV, Upadhayay VU, Prasad B (2020) Development of potential microbial consortia and their assessment on wheat (Triticum aestivum) seed germination. Environ Ecol 38:6–16

Rosier A, Bishnoi U, Lakshmanan V, Sherrier DJ, Bais HP (2016) A perspective on inter-kingdom signaling in plant–beneficial microbe interactions. Plant Mol Biol 90:537–548. https://doi.org/10.1007/s11103-016-0433-3

Saleem AR, Brunetti C, Khalid A, Della Rocca G, Raio A, Emiliani G, Centritto M (2018) Drought response of Mucuna pruriens (L). PLoS One 13:e0191218. https://doi.org/10.1371/journal.pone.0191218

Santoyo G, Urtis-Flores CA, Loeza-Lara PD, Orozco-Mosqueda MDC, Glick BR (2021) Rhizosphere colonization determinants by plant growth-promoting rhizobacteria (PGPR). Biology 10:475. https://doi.org/10.3390/biology10060475

Sarkar J, Chakraborty U, Chakraborty B (2021) High temperature resilience in *Bacillus safensis* primed wheat plants: a study of dynamic response associated with modulation of antioxidant machinery, differential expression of HSPs and osmolyte biosynthesis. Environ Exp Bot 182:104315. https://doi.org/10.1016/j.envexpbot.2020.104315

Sasse J, Martinoia E, Northen T (2018) Feed your friends: do plant exudates shape the root microbiome? Trends Plant Sci 23:25–41. https://doi.org/10.1016/j.tplants.2017.09.003

Schikora A, Schenk ST, Hartmann A (2016) Beneficial effects of bacteria-plant communication based on quorum sensing molecules of the N-acyl homoserine lactone group. Plant Mol Biol 90:605–612. https://doi.org/10.1007/s11103-016-0457-8

Seleiman MF, Al-Suhaibani N, Ali N, Akmal M, Alotaibi M, Refay Y, Battaglia ML (2021) Drought stress impacts on plants and different approaches to alleviate its adverse effects. Plants (Basel) 10:259. https://doi.org/10.3390/plants10020259

Selvaraj A, Thangavel K, Uthandi S (2020) Arbuscular mycorrhizal fungi (*Glomus intraradices*) and diazotrophic bacterium (*Rhizobium* BMBS) primed defense in blackgram against herbivorous insect (*Spodoptera litura*) infestation. Microbiol Res 231:126355. https://doi.org/10.1016/j.micres.2019.126355

Shah J, Zeier J (2013) Long-distance communication and signal amplification in systemic acquired resistance. Front Plant Sci 4:30. https://doi.org/10.3389/fpls.2013.00030

Shah D, Khan MS, Aziz S, Ali H, Pecoraro L (2021) Molecular and biochemical characterization, antimicrobial activity, stress tolerance, and plant growth-promoting effect of endophytic bacteria isolated from wheat varieties. Microorganisms 10:21. https://doi.org/10.3390/microorganisms10010021

Sham A, Al-Ashram H, Whitley K, Iratni R, El-Tarabily KA, AbuQamar SF (2019) Metatranscriptomic analysis of multiple environmental stresses identifies RAP2.4 gene associated with Arabidopsis immunity to *Botrytis cinerea*. Sci Rep 9:17010. https://doi.org/10.1038/s41598-019-53694-1

Sharma P, Chouhan R, Bakshi P, Gandhi SG, Kaur R, Sharma A (2017) Amelioration of chromium-induced oxidative stress by combined treatment of selected plant-growth-promoting rhizobacteria and earthworms via modulating the expression of genes related to reactive oxygen species metabolism in *Brassica juncea*. Front Microbiol 13:802512. https://doi.org/10.3389/fmicb.2022.802512

Sharma M, Sudheer S, Usmani Z, Rani R, Gupta P (2020) Deciphering the omics of plant-microbe interaction: perspectives and new insights. Curr Genomics 21(5):343–362. https://doi.org/10.2174/1389202921999200515140420

Singh RP, Jha PN (2017) The PGPR Stenotrophomonas maltophilia SBP-9 augments resistance against biotic and abiotic stress in wheat plants. Front Microbiol 8:1945. https://doi.org/10.3389/fmicb.2017.01945

Singh HB, Keswani C, Reddy MS, Royano ES, García-Estrada C (2019a) Secondary metabolites of plant growth promoting rhizo microorganisms: discovery and applications. Springer-Nature, Singapore, p 392

Singh UB, Malviya D, Singh S, Kumar M, Sahu PK, Singh HV, Saxena AK (2019b) *Trichoderma harzianum* and methyl jasmonate-induced resistance to *Bipolaris sorokiniana* through enhanced phenylpropanoid activities in bread wheat (*Triticum aestivum* L.). Front Microbiol 10:1697. https://doi.org/10.3389/fmicb.2019.01697

Singh V, Singh D, Gautam JK, Nandi AK (2019c) RSI1/FLD is a positive regulator for defense against necrotrophic pathogens. Physiol Mol Plant Pathol 107:40–45. https://doi.org/10.1016/j.pmpp.2019.04.005

Singh M, Bhasin S, Madan N, Suyal DC, Soni R, Singh D (2021) Bioinoculants for agricultural sustainability. In: Microbiological activity for soil and plant health management. Springer, Singapore, pp 629–641

Singh J, Singh AV, Upadhayay VK, Khan A, Chandra R (2022) Prolific contribution of Pseudomonas protegens in Zn biofortification of wheat by modulating multifaceted physiological response under saline and non-saline conditions. World J Microbiol Biotechnol 38:227. https://doi.org/10.1007/s11274-022-03411-4

Sood G, Kaushal R, Sharma M (2020) Significance of inoculation with *Bacillus subtilis* to alleviate drought stress in wheat (*Triticum aestivum* L.). Vegetos 33:782–792. https://doi.org/10.1007/s42535-020-00149-y

Sözen C, Schenk ST, Boudsocq M, Chardin C, Almeida-Trapp M, Krapp A, Colcombet J (2020) Wounding and insect feeding trigger two independent MAPK pathways with distinct regulation and kinetics. Plant Cell 32:1988–2003. https://doi.org/10.1105/tpc.19.00917

Sultana S, Paul SC, Parveen S, Alam S, Rahman N, Jannat B, Karim MM (2020) Isolation and identification of salt-tolerant plant-growth-promoting rhizobacteria and their application for rice cultivation under salt stress. Can J Microbiol 66:144–160. https://doi.org/10.1139/cjm-2019-0323

Tang A, Haruna AO, Majid NMA, Jalloh MB (2020a) Potential PGPR properties of cellulolytic, nitrogen-fixing, phosphate-solubilizing bacteria in rehabilitated tropical forest soil. Microorganisms 8:442. https://doi.org/10.3390/microorganisms8030442

Tang Y, Kang H, Qin Z, Zhang K, Zhong Y, Li H, Mo L (2020b) Significance of manganese resistant *Bacillus cereus* strain WSE01 as a bioinoculant for promotion of plant growth and manganese accumulation in *Myriophyllum verticillatum*. Sci Total Environ 707:135867. https://doi.org/10.1016/j.scitotenv.2019.135867

Tena G, Boudsocq M, Sheen J (2011) Protein kinase signaling networks in plant innate immunity. Curr Opin Plant Biol 14:519–529. https://doi.org/10.1016/j.pbi.2011.05.006

Tiwari S, Prasad V, Chauhan PS, Lata C (2017) *Bacillus amyloliquefaciens* confers tolerance to various abiotic stresses and modulates plant response to phytohormones through osmoprotection and gene expression regulation in rice. Front Plant Sci 8:1510. https://doi.org/10.3389/fpls.2017.01510

Tyagi J, Chaudhary P, Jyotsana BU, Bhandari G, Chaudhary A (2022a) Chapter 17 Impact of endophytic fungi in biotic stress management. In: Soni R, Suyal D, Goel R (eds) Plant protection: from chemicals to biologicals. De Gruyter, Berlin, pp 447–462. https://doi.org/10.1515/9783110771558-017

Tyagi J, Chaudhary P, Mishra A, Khatwani M, Dey S, Varma A (2022b) Role of endophytes in abiotic stress tolerance: with special emphasis on *Serendipita indica*. Int J Environ Res 16:62. https://doi.org/10.1007/s41742-022-00439-0

Ullah A, Bano A, Khan N (2021) Climate change and salinity effects on crops and chemical communication between plants and plant growth promoting microorganisms under stress. Front Sustain Food Syst 5:1–16. https://doi.org/10.3389/fsufs.2021.618092

Upadhayay VK, Singh AV, Pareek N (2018) An insight in decoding the multifarious and splendid role of microorganisms in crop biofortification. Int J Curr Microbiol Appl Sci 7:2407–2418. https://doi.org/10.20546/ijcmas.2018.706.286

Upadhayay VK, Singh AV, Khan A, Pareek N (2021) Influence of zinc solubilizing bacterial co-inoculation with zinc oxide supplement on rice plant growth and Zn uptake. Pharma Innov J 10:113–116

Upadhayay VK, Khan A, Singh J, Singh AV (2022a) Bacterial assisted improved Zn consignment in root and shoot of rice plant by zinc solubilizing Serratia marcescens bearing plant probiotic traits. Adv Biol Res 13:01–08

Upadhayay VK, Singh AV, Khan A (2022b) Cross talk between zinc-solubilizing bacteria and plants: a short tale of bacterial-assisted zinc biofortification. Front Soil Sci 1:788170. https://doi.org/10.3389/fsoil.2021.788170

Upadhayay VK, Singh AV, Khan A, Singh J, Pareek N, Raghav A (2022c) FE-SEM/EDX based zinc mobilization analysis of Burkholderia cepacia and Pantoea rodasii and their functional annotation in crop productivity, soil quality, and zinc biofortification of paddy. Front Microbiol 13:852192. https://doi.org/10.3389/fmicb.2022.852192

Upadhayay VK, Singh AV, Khan A, Sharma A (2022d) Contemplating the role of zinc-solubilizing bacteria in crop biofortification: an approach for sustainable bioeconomy. Front Agron 4:903321. https://doi.org/10.3389/fagro.2022.903321

Upadhayay VK, Maithani D, Dasila H, Taj G, Singh AV (2023a) Microbial services for mitigation of biotic and abiotic stresses in plants. In: Advanced microbial techniques in agriculture, environment, and health management. Elsevier, Amsterdam, pp 67–81. https://doi.org/10.1016/B978-0-323-91643-1.00003-X

Upadhayay VK, Chitara MK, Mishra D, Jha MN, Jaiswal A, Kumari G (2023b) Synergistic impact of nanomaterials and plant probiotics in agriculture: a tale of two-way strategy for long-term sustainability. Front Microbiol 14:1133968. https://doi.org/10.3389/fmicb.2023.1133968

Venturi V, Keel C (2016) Signaling in the rhizosphere. Trends Plant Sci 21:87–98. https://doi.org/10.1016/j.tplants.2016.01.005

Weller DM, Mavrodi DV, van Pelt JA, Pieterse CM, van Loon LC, Bakker PA (2012) Induced systemic resistance in Arabidopsis thaliana against *Pseudomonas syringae* pv. tomato by 2,4-diacetylphloroglucinol-producing *Pseudomonas fluorescens*. Phytopathology 102:403–412. https://doi.org/10.1094/PHYTO-08-11-0222

Wu Y, Zhang D, Chu JY, Boyle P, Wang Y, Brindle ID, Despres C (2012) The *Arabidopsis* NPR1 protein is a receptor for the plant defense hormone salicylic acid. Cell Rep 1:639–647. https://doi.org/10.1016/j.celrep.2012.05.008

Wu J, Kim SG, Kang KY, Kim JG, Park SR, Gupta R, Kim ST (2016) Over expression of a pathogenesis-related Protein 10 enhances biotic and abiotic stress tolerance in rice. Plant Pathol J 32:552–562. https://doi.org/10.5423/PPJ.OA.06.2016.0141

Wu Y, Zhou J, Li C, Ma Y (2019) Antifungal and plant growth promotion activity of volatile organic compounds produced by Bacillus amyloliquefaciens. Microbiology 8(8):e00813. https://doi.org/10.1002/mbo3.813

Xu Z, Shimizu H, Ito S, Yagasaki Y, Zou C, Zhou G, Zheng Y (2014) Effects of elevated CO_2, warming and precipitation change on plant growth, photosynthesis and peroxidation in dominant species from North China grassland. Planta 239:421–435. https://doi.org/10.1007/s00425-013-1987-9

Xu Z, Jiang Y, Jia B, Zhou G (2016) Elevated-CO_2 response of stomata and its dependence on environmental factors. Front Plant Sci 7:657. https://doi.org/10.3389/fpls.2016.00657

Yasmin H, Bano A, Wilson NL, Nosheen A, Naz R, Hassan MN, Kennedy I (2022) Drought-tolerant *Pseudomonas* sp. showed differential expression of stress-responsive genes and induced drought tolerance in *Arabidopsis thaliana*. Physiol Plant 174:e13497. https://doi.org/10.1111/ppl.13497

Younas R, Gul S, Ahmad R, Khan AR, Khan M, Anwar T, Qureshi H (2021) Bioinoculants in technological alleviation of climatic stress in plants. In: Technology in agriculture, vol 271. IntechOpen, London. https://doi.org/10.5772/intechopen.97340

Zhang J, Wang F, Liang F, Zhang Y, Ma L, Wang H, Liu D (2018) Functional analysis of a pathogenesis related thaumatin-like protein gene TaLr35PR5 from wheat induced by leaf rust fungus. BMC Plant Biol 18:76. https://doi.org/10.1186/s12870-018-1297-2

Zhang W, Luo Q, Zhang Y, Fan X, Ye T, Mishra S, Bhatt P, Zhang L, Chen S (2020) Quorum quenching in a novel *Acinetobacter* sp. XN-10 bacterial strain against *Pectobacterium carotovorum* subsp. *carotovorum*. Microorganisms 8(8):1100. https://doi.org/10.3390/microorganisms8081100

Zipfel C, Oldroyd GED (2017) Plant signalling in symbiosis and immunity. Nature 543:328–336. https://doi.org/10.1038/nature22009

Benefits and Drawbacks of Microbial Inoculant in Terms of Human Health and the Environment

Modupe S. Ayilara, Saheed A. Akinola, and Mosimininuoluwa T. Adebajo

Abstract The application of soil amendment has become very important in recent years for increasing crop production. The need for this practice is owed to the exacerbation of pests and diseases which have become paramount due to climate change and other factors. Chemical amendments, such as pesticides, fertilizers, and stimulants, have played a large role in improving agricultural productivity over the years. However, this role also has a detrimental dimension which includes increased toxicity to the plants, animals, humans, and other aquatic and terrestrial inhabitants. It is therefore expedient to promote the use of biological amendments which would serve the same purpose as their chemical counterparts to ensure a sustainable environment. Despite the existence of different biological amendments made from plants, microbes, and biological nanoparticles, the use of microbes has been preferred because of their ease of production, multiplication, and manipulation. Microbes such as bacteria, viruses, fungi, and algae have been useful in the production of biofertilizers and biopesticides to control the pests and diseases on crops. This chapter discusses the pros and cons of microbial inoculants and further recommends how microbial inoculants can be improved to enhance agricultural production.

Keywords Agriculture · Microbes · Human health · Environment

M. S. Ayilara (✉)
Faculty of Science, Department of Biological Sciences, Kings University, Ode-Omu, Nigeria

Food Security and Safety Focus Area, Faculty of Natural and Agricultural Sciences, North-West University, Mmabatho, South Africa

S. A. Akinola
Food Security and Safety Focus Area, Faculty of Natural and Agricultural Sciences, North-West University, Mmabatho, South Africa

Department of Microbiology and Parasitology, School of Medicine and Pharmacy, College of Medicine and Health Sciences, University of Rwanda, Butare, Rwanda

M. T. Adebajo
Department of Criminology and Security Studies, Kings University, Ode-Omu, Nigeria

School of Political Science, University of Haifa, Haifa, Israel

P. Chaudhary, A. Chaudhary (eds.), *Microbial Inoculants*, Interdisciplinary Biotechnological Advances, https://doi.org/10.1007/978-981-97-0633-4_17

1 Introduction

Agricultural commerce plays a significant role in meeting the world's food demands and consequently impacts a nation's economic development at the same time. According to a study carried out by the Food and Agriculture Organization (FAO) of the United Nations, 17% of people in developing nations were undernourished in 1997–1999, down from 20% in 1990–1992. This suggests that switching from traditional to modern farming techniques ensures increased output and boosts the economy. Previously, humans employed the addition of nutrients such as in manures and minerals like in wood ash to soil to increase crop productivity (Romdhane et al. 2021). Crop rotation was subsequently used in the seventeenth century to observe its impact on plant development and increased output. At the same period, Humboldt suggested guano, or seabird excrement, be used in agricultural soil as a soil amendment since it includes the three nutrients such as potassium, phosphate, and nitrogen that are essential for plant development. In this way, a time when agricultural soil was amended with additives began (Patil and Solanki 2016).

In the past, fertilizer, a term that refers to a set of compounds that improve plant growth and yield when added to agricultural land, was used to describe soil additions. It often contains secondary nutrients together with potassium, phosphate, and nitrogen molecules, which greatly enhance the amount and nutritional value of agricultural produce (Chun et al. 2017). Nitrogenous fertilizers, such as urea, ammonium nitrate, nitric acid, and synthetic ammonia, and phosphatic fertilizers, such as triple superphosphate, regular superphosphate, ammonium phosphate, and phosphoric acid, are the most popular and commonly used types of fertilizers (Yahaya et al. 2023). However, to protect the environment and humanity, conservationists oppose the excessive and prolonged application of these chemical-based fertilizers, primarily because they lead to the eutrophication of water bodies, which pollutes the air and groundwater (Ren et al. 2022; Chaudhary et al. 2022b). As a result, agriculturalists have shifted their attention toward bio-fertilizers and biosafety while concentrating on "high-quality nutrient-rich food" using sustainable agricultural methods. Generally, microorganisms such as potassium solubilizers (K-solubilizer), phosphorus solubilizers (P-solubilizers), and nitrogen fixers (N-fixers) are used as bio-fertilizers (Agri et al. 2021, 2022; Chaudhary et al. 2021b, c, d, 2022a). These are unique microorganisms that have the capacity to dissolve insoluble phosphate located in soils by secreting organic acids and making phosphate readily available for plant uptake (Chaudhary et al. 2021a). These helpful microorganisms are occasionally used in conjunction with fungi or bacteria linked to plant roots, such as *Rhizobium*, which interacts symbiotically with legume roots, or rhizobacteria, which live on root surfaces or in rhizosphere soil (Adeleke et al. 2022a, b; Akinola and Babalola 2020; Ayilara et al. 2023).

In the same vein, pesticides are essential for defending plants against harmful intrusions from weeds, insects, and microbial pathogens. These are either biological or chemical agents with the capacity to control or eradicate undesirable organisms, pests, or insects that cause harm to plants or animals (Lushchak et al. 2018).

Pesticides were initially made from natural substances or extracts like red pepper, tobacco extracts, sulfurous rock, and salt. Up to the 1940s, arsenic, heavy metals, and petroleum oils were often used to manage pests and weeds. Afterward, it was replaced with synthetic and organic pesticides, the most popular of which was dichlorodiphenyltrichloroethane (DDT), a chlorinated hydrocarbon (Jayaraj et al. 2017).

Considering the health implication and environmental damage imposed by some of these chemicals, some classes of pesticides with a chemical base that are used on agricultural land have been prohibited by the US Environmental Protection Agency (EPA) and similar agencies around the world (Pathak et al. 2022). This simultaneously increased consumer interest in high-quality organic products and above-discussed sustainable agriculture techniques that eventually increased the demand for biological products, particularly microbial inoculants. The biopesticides and bio-fertilizers became more acceptable because they are easier to apply and biodegradable, more target specific, and less specific compared to their traditional equivalents (Kumar et al. 2021b, c, d, 2022; Singh et al. 2023a, b).

2 Bioinoculation

Since premodern times, productive soil is usually transferred from one area to another with the belief that the practice could positively improve the agricultural productivity of the recipient soil and/or area (Brodt et al. 2011). This prompted the initiation of bioinoculation in agricultural culture, which in further years was revolutionized as mixing of "naturally inoculated" soil and became a suggested activity of legume inoculation in the USA (Chaudhary et al. 2022c, d).

Bio-inoculants are living microorganisms that, when added to agricultural soil, encourage plant development by nourishing and protecting plants, boosting hormone synthesis, increasing mineral absorption, and weathering of soil minerals (Fig. 1). It often consists of either a single microbial strain or a consortium of many helpful microorganisms that have a good effect on plant development (Chaudhary et al. 2023a, b, c). For better results, these formulations are tailored based on the needs and the types of soil, cropping methods, and microorganisms. These bio-inoculants can first be divided into the categories of bacteria and fungus, with the intracellular and extracellular subcategories for root-associated fungi (RAF) and bacteria and arbuscular mycorrhizal (AM) and ectomycorrhizal (EcM) for fungi (Adeleke et al. 2022a, b).

In response to soil amendment, native microflora, particularly rhizobia and/or symbiotic and nonsymbiotic plant growth-promoting bacteria (PGPB), immediately respond to the bio-inoculant (Backer et al. 2018). These reactions differ based on the bacteria used as inoculant, soil type, plant species, density, and environmental factors. On agricultural land, besides barren soil, it is difficult for microbial inoculants to find an empty niche, which can cause a quick decline in the bacterial population of inoculants (Akinola et al. 2021a, 2022).

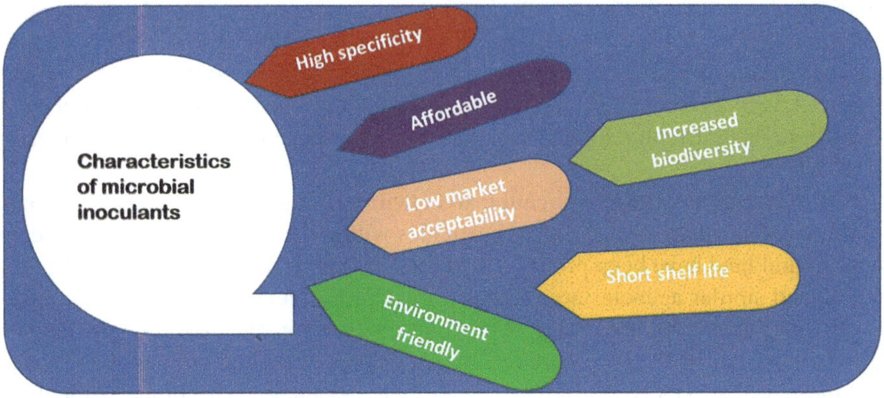

Fig. 1 Characteristics of bio-inoculants

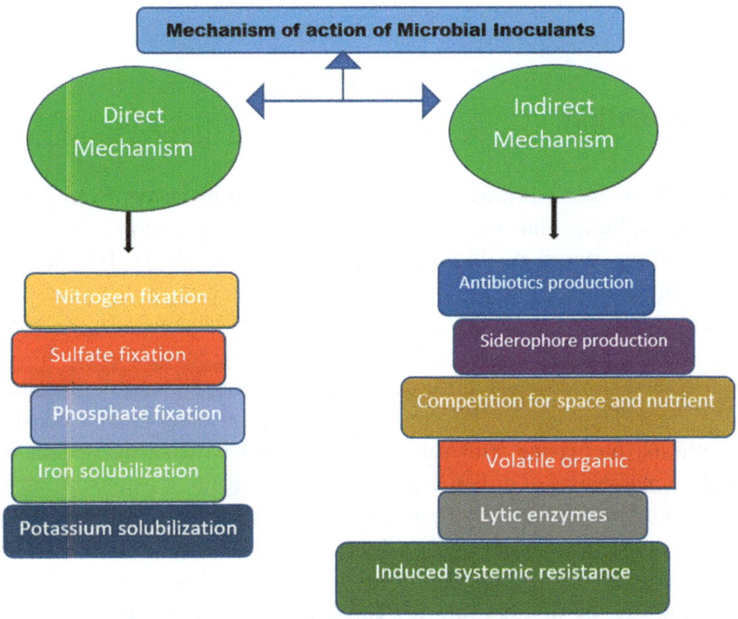

Fig. 2 Mechanism of action for microbial inoculants

The mechanism of action of microbial inoculum can be categorized into the direct or indirect mechanism (Fig. 2). In the direct mechanism, nitrogen, sulfur, and phosphorus are fixed, while iron is solubilized to a form where it can be absorbed by the plants. In the indirect mechanism, microbes compete for space and nutrient with the pathogen (Tyagi et al. 2022a, b).

Also, microbial inoculum involves the lytic enzyme production and the induction of systemic resistance. Lytic enzymes destroy the cell wall and structure of the pathogens which eventually leads to the pathogen's death (Alori and Babalola 2018). Microbial inoculum also produces phytoalexins, reactive oxygen species, etc. which help to induce systemic resistance in microbes (Chaudhary et al. 2022e).

2.1 *Bio-fertilization*

The capacity of the soil to permit seedling sprouting and root penetration by supplying nutrients, coupled with an appropriate soil structure to support the plant growth, is referred to as soil fertility (Yang et al. 2020). The foundation of effective farming techniques is the quality of the topsoil. Organic matter, minerals, and helpful microbes make up fertile soil, which also contains important amounts of potassium, phosphorus, nitrogen, and minerals including zinc, sulfur, molybdenum, magnesium, manganese, iron, copper, cobalt, chlorine, and boron. These elements significantly affect the chemical, physical, and soil processes and biological characteristics (Akinola and Babalola 2021).

Bio-fertilizers are substances that aid in the two crucial processes of mineralization and immobilization, which raise the soil's mineral and organic content and ultimately its fertility (Mahmud et al. 2021). Although immobilization refers to converting atmospheric nitrogen to ammonium for simple absorption by plants, mineralization results in the availability of soluble plant nutrients through the breakdown of organic materials in the soil (Grzyb et al. 2021). Bio-fertilizer is essentially a substance that may be applied to the soil, seeds, or surfaces of plants and is made up of living or dormant cells of advantageous microorganisms in a monoculture or mixed culture (Khati et al. 2018; Kumari et al. 2020). These microorganisms can encourage plant development by colonizing the rhizosphere or plant interior, which in turn improves the supply of essential minerals to the host plant in an assimilable form (Akinola et al. 2021b; Manzar et al. 2022).

On the other hand, there are certain drawbacks that reduce the competitiveness of bio-fertilizers, such as difficult transportation and storage, sensitivity to high temperatures, appropriate carrier material, and short shelf-life. It is also costlier than chemical fertilizers due to poor technology, low-skilled production, and improper utilization of copious waste (Fasusi et al. 2021). Due to the fact that biological formulations depend on the environment around the application region, their effectiveness is environment dependent. As a result, they exert a gradual impact and require special storage and carrier material mixing techniques in order to minimize side effects for prolonged usage (Patil and Solanki 2016). Small-scale compost production served as the foundation for this practice, which further demonstrated the effectiveness of bio-fertilizer.

3 Biopesticides

Advanced agricultural techniques have been employed in the contemporary agricultural period with the goal of improving sustainability by achieving better outcomes and using eco-friendly methods (Javaid et al. 2022). One of these methods is organic farming, which emphasizes maximum yields of exceptional quality and depends on tried-and-true techniques like biological pest management, compost, green manure, and crop rotation (Meemken and Qaim 2018). These environmentally friendly manufacturing techniques aid in promoting and/or enhancing soil biological activity and biodiversity. The healthy biological cycles are also positively impacted by these practices. The most well-liked and environmentally acceptable agricultural technique nowadays is the usage of biopesticide (Damalas and Koutroubas 2018; Chakraborty et al. 2023). According to "the US Environmental Protection Agency (US-EPA)," biopesticides are particular forms of insecticides produced from such natural materials as certain minerals, microbes, plants, and animals (Kumar et al. 2021a). Biochemical pesticides, plant-incorporated protectants, and microbial pesticides are the three subcategories of biopesticides based on formulation or active compositions (US-EPA) (Patil and Solanki 2016).

In the seventeenth century, the earliest biopesticide such as nicotine derived from plant extracts was employed to control plum beetles. Nonetheless, because of the rapid effects and nonspecific host efficacy, the usage of chemical-based pesticides substantially rose over the twentieth century (Kumar et al. 2021a). The drawback of chemical-based pesticides became apparent after a few decades as the development of insect resistance due to the use of synthetic chemicals to control pests eventually culminated in environmental and human health issues. It responds to the requirement for an efficient substitute with biodegradable and environmentally acceptable characteristics and causes a surge in biopesticides (Ayilara et al. 2023).

Microbial pesticides, or biopesticides, generated from beneficial microbes with pest-controlling abilities, are one of the more advanced and effective alternatives for managing plant diseases (Šunjka and Mechora 2022). Meanwhile, *Bacillus thuringiensis* (Bt) bacteria is the first and most often employed microbial pesticide till date. Shigetane Ishiwata of Japan, a biologist, recovered it from a sick silkworm (Kumar et al. 2021a). Ten years later, in Thüringen, Germany, biologist Ernst Berliner found it in a sick flour moth larva (BPIS-2015). Recently, microbes from genera such as *Trichoderma*, *Beauveria*, *Paecilomyces*, *Coniothyrium*, *Agrobacterium*, *Streptomyces*, *Pseudomonas*, and *Bacillus* were adopted as biopesticides worldwide (El-Saadony et al. 2022).

One of the key factors in biopesticides having an upper hand compared to their chemical counterparts is their capacity to kill target pests with great specificity, which indirectly promotes the survival of desirable species in treated crops (Patil and Solanki 2016). The awareness of biopesticides, particularly their mode of action, formulation, and ideal time for application, may make them more acceptable in this present climate. The following are a few biopesticides' positive and negative aspects.

Advantages of microbial pesticides:

1. Most of the time, they are environmentally friendly but poisonous to target pests.
2. They have high specificity and no effect on the pest's predators or parasites.
3. It works in extremely small doses and frequently degrades fast, causing relatively lesser exposure and pollution.
4. Application of nonhazardous residues is permitted even before harvesting.
5. Biopesticides can significantly reduce the usage of conventional pesticides in integrated pest management (IPM) programs, maintaining good crop yields.
6. For safe and successful use of biopesticides, the user must be knowledgeable about pest management and should carefully follow all label instructions.
7. Beneficial microorganisms (used in biopesticide) can occasionally establish themselves in the treated environment and continue to function over succeeding insect generations or seasons.
8. In addition, pest management could promote advantageous soil microflora, which boosts crop output.

Drawbacks of microbial pesticides:

1. Due to a narrow target range, a variety of pests that are present in the field and harm crops cannot be controlled.
2. As these materials are sensitive to UV exposure, desiccation, and heat, proper timing and methods are crucial for optimal application.
3. In some circumstances, special storage and formulation techniques are required, which may make the manufacture and delivery of some products more difficult.
4. Because of product specificity, some items may have a smaller potential market, resulting in lower availability or more expensive prices around the world.

4 Are Microbes the Most Reliable Option?

One of humanity's first profitable industries is agriculture, which heavily depends on healthy soil and stable environmental conditions. It significantly influences ecosystem, soil quality, water, and biological varieties. As previously stated, microbes are crucial to agriculture and significantly increase agribusiness. Rhizo-biomes are a group of microbes proximal to the root of plant. These microorganisms are thought to be essential for plant growth since they have a variety of potentials including ammonia volatilization, HCN, siderophore production, auxins, plant growth regulators, secondary metabolite production, antibiotic synthesis, phosphate solubilization, and atmospheric nitrogen fixation (Chaudhary and Sharma 2019; Khati et al. 2019; Kumari et al. 2021). Due to their capacity to digest contaminants including paints, dyes, hydrocarbons, pesticides, and other pollutants as well as dead plants and animal debris, microorganisms serve as natural scavengers. When the required type of microorganism is introduced in the right quantity and in its active state, these bacteria may function more effectively. When many strains of microorganisms are introduced, the performance is also impacted by the compatibility of each strain. It has been demonstrated in a study that microorganisms can cause plants to develop systemic acquired resistance (SAR) and induced systemic resistance (ISR), particularly when plant diseases are present. These microorganisms support plant development and assist in controlling plant pathogens (Chaudhary et al. 2021e, f, g; Tyagi et al. 2022a).

The rhizosphere interface offers a platform for the interactive connection of plant roots and soil microorganisms (Manzar et al. 2022; Chaudhary et al. 2021g). This heterotrophic microbial colony uses decaying plant materials and root exudates as a carbon source. Because of larger concentrations of alluring substances such as vitamins, amino acids, organic acids, and sugars, released by plant roots, the rhizoplane and rhizosphere are thought to have a higher microbial population than the soil with no vegetation. These compounds encourage competition and draw different kinds of bacteria, which results in a diversified microbial population at various rhizospheres (Hassan et al. 2019).

4.1 *Plant Responses in the Form of SAR and ISR*

Plants are frequently subjected to a variety of environmental stressors, and they have developed unique defense mechanisms to deal with these pressures. Every year, pests and phytopathogens destroy crops worth billions of dollars in yield and quality (Tyagi et al. 2022b). Pesticides were used in the modern agricultural system to combat infections, but their effectiveness is sometimes insufficient and, as was already said, there may be environmental health problems. Pathogen strains quickly acquire resistance to several new systemic pesticides, and the challenge of creating "environmentally benign" biopesticides has raised their price and decreased the number of options (Mishra et al. 2020). The market is being cleared of many of the advised pesticides, and those that remain can only be applied to specific crops for predetermined times.

Pesticide residues on food crops are subject to rigorous limitations imposed by agricultural importers, and for some pesticides, there is zero tolerance. Concerns over pesticide usage and residues on food items are growing among consumers and consumer advocacy groups (Wahab et al. 2022). To better manage both wild and agricultural plant populations, it has recently become of tremendous societal and biological interest to better understand the fundamental processes by which plants protect themselves. Plants have developed several structural and physical defenses to protect themselves from predators and live in harsh environments since they cannot physically escape these situations due to predation, a lack of nutrition, or poor environmental conditions (War et al. 2018). The emergence of transgenic plants, ISR (plant response), biocontrol, and growing adoption of disease-resistant plants leverages new technologies created by plant breeders. Advancements in biotechnology hold the promise of offering alternative ways of disease control that are efficient and affordable and minimize the need on pesticides (Šunjka and Mechora 2022).

The impact is systemic and frequently lasts for the duration of a year in an annual plant because the same mechanism for resistance is active in resistant and immunized plants. A plant's ability to defend itself against a wide range of pathogens and pests is enhanced by induced resistance, which develops following an appropriate stimulation. SAR and/or ISR refers to the enhanced resistance that develops as a result of an inducing substance following pathogen infection (Yu et al. 2022) (Fig. 3).

Rhizo-bacteria induce systemic resistance (ISR), whereas other agents are known for inducing systemic acquired resistance (SAR) (Zhu et al. 2022). It is the naturally occurring generalized (systemic) resistance that is enhanced or induced by plant-associated nonpathogenic plant growth-promoting microorganisms (PGPMs). There are no pathogenesis-related (PR) proteins produced since it is salicylic acid (SA)-independent, but genotype-dependent and plant-specific (Backer et al. 2019). Ethylene (ET), jasmonic acid (JA), and plant hormones all play essential roles in ISR. Plant response to biotic and abiotic stressors, such as pathogen attack, is regulated by JA. It can occasionally take the form of a volatile chemical that can reach surrounding plants and plant sections to warn off pathogen attacks and initiate plant defensive responses (Wang et al. 2020). The synthesis of ethylene is also stimulated by infection or injury, which starts the plant's defensive mechanisms. Rapid

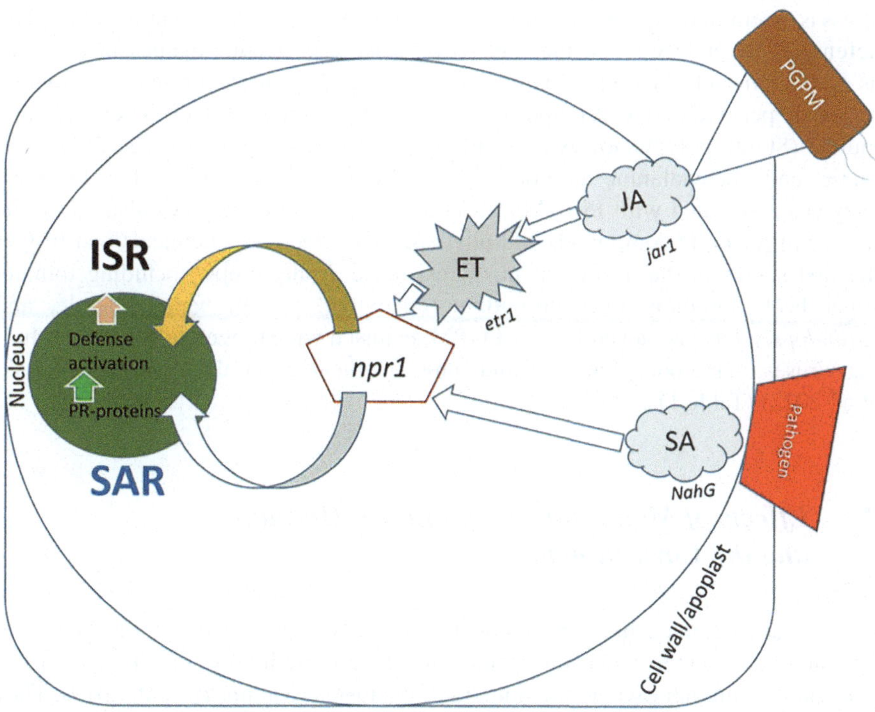

Fig. 3 Resistance triggered by plant growth-promoting microorganism

senescence, ripening, and abscission of infected tissue, such as the abrupt fall of leaves or fruits, are typical ethylene-induced visual defensive responses. SAR-secondary infection resistance is a direct defensive activation mechanism started by necrotizing pathogens (Ding et al. 2022). SAR is activated in response to a wide range of pathogens, including oomycetes, viruses, fungi, and bacteria. Salicylic acid builds up in the phloem tissue after necrosis, which causes the first hypersensitive reaction and induces SAR. In addition to controlling ethylene, salicylic acid is a plant hormone needed for the synthesis of PR proteins that prevent the expression of virulence factors. These proteins are created by PR genes that are triggered by pathogens. They are required to trigger plant defense, and some of them serve as antimicrobials by weakening pathogens' cell walls (Poór et al. 2019). Peroxidase, lysozyme, and chitinase enzyme are a few of these proteins that have lytic potential and are useful against a variety of infections. Moreover, they serve as messengers to alert the host to the pathogen attack, which further promotes the formation and deposition of lignin, forming a strong defense against infectious agents. SAR is hampered by SA buildup or biosynthesis inhibition. ISR is similar to SAR; however, unlike SAR, which is brought on by pathogens that cause necrosis, ISR is brought on by certain strains of rhizo-bacteria that promote plant development (Beneduzi et al. 2012). ISR, in contrast to SAR, is dependent on JA and ET, independent of SA, and unrelated to the expression of the PR gene. SAR and ISR are entangled at the molecular level in *Arabidopsis* by the NPR1 gene (Fig. 3). Early detection of the stress is essential for quick and effective reaction, which is a key component of plant defense. Once induced, resistance provides general protection against insect pests as well as harmful viruses, fungi, nematodes, and bacteria. Protease inhibitors, ascorbate peroxidase (APX), lipoxygenase (LOX), catalase (CAT), superoxide dismutase (SOD), polyphenol oxidase (PPO), β-1,3-glucanase, peroxidase (PO), chitinase, and phenylalanine ammonia lyase (PAL) are just a few of the defensive enzymes associated with ISR (Murali et al. 2021). These enzymes also cause the release of phenolic compounds and phytoalexins, which are chemicals that trigger the first stages in the development of resistance. Many crops, including tomato, sugar beet, sorghum, rice, groundnut, cucumber, cotton, banana, apple, and *Arabidopsis*, have achieved ISR by PGPM against a wide range of diseases, including viruses, nematodes, bacteria, and fungi (El-Saadony et al. 2022; Antoszewski et al. 2022) (Table 1).

4.2 Effects of Microbial Inoculants on Humans and the Environment

Organic amendments have both positive and negative effects on the health of humans and the environments. Organic amendments are highly specific in their target, and they are nontoxic to the nontarget organisms (Kaminsky et al. 2019). They help to promote organic agriculture and they increase biodiversity in the

Table 1 Studies explaining the bio-protective processes of different bio-inoculum

Microbial strain	Mechanism of action	Plant involved	Pathogens/diseases	References
Hypocrea/Trichoderma species	Accumulation of phenylalanine ammonia lyase, polyphenol oxidase, peroxidase, and total phenols	Tomato	*Rhizoctonia solani*/root rot	Solanki et al. (2011)
Trichoderma species	Cell wall-degrading enzymes	Tomato	*Alternaria solani*, *Xanthomonas euvesicatoria* early blight and bacterial spot	Fontenelle et al. (2011)
T. harzianum	Systemic resistance and nutrient absorption	Sunflower	*Plasmopara halstedii*/downy mildew	Nagaraju et al. (2012)
T. asperellum SKT-1	Increased expression levels of JA/ET	Rice	Seed-borne disease	Yoshioka et al. (2012)
Penicillium chrysogenum (PenC_JSB41) and *Trichoderma harzianum* (TriH_JSB27)	β-1,3-Glucanase and peroxidase	Tomato	*Ralstonia solanacearum*/bacterial wilt	Jogaiah et al. (2013)
Streptomyces vinaceusdrappus S5MW2	Chitinase	Tomato	*Rhizoctonia solani*/root rot	Yandigeri et al. (2015)
Streptomyces toxytricini vh6	Accumulation of total phenol and phenylalanine ammonia lyase	Tomato	*Rhizoctonia solani*	Patil et al. (2011)
Penicillium simplicissimum GP17-2	JA, SA, and ET genes	Tobacco and Arabidopsis	Cucumber mosaic virus	Elsharkawy et al. (2012)
B. antiquum, *E. ludwigii*, *A. tandoii*, *P. monteilii*, *P. plecoglossicida*	Siderophore	Sorghum	Charcoal rot	Gopalakrishnan et al. (2011)
P. fluorescens MPF47	Siderophore	Tomato	*Rhizoctonia solani*/root rot	Solanki et al. (2015)
P. fluorescens	Lipoxygenase phenylalanine ammonia lyase, guaiacol peroxidases, and polyphenol oxidase	Tomato	*Ralstonia solanacearum*/bacterial wilt	Vanitha and Umesha (2011)
P. fluorescens	2,4-Diacetylphloroglucinol	Groundnut	*Sclerotium rolfsii*/stem rot	Asadhi et al. (2013)

(continued)

Table 1 (continued)

Microbial strain	Mechanism of action	Plant involved	Pathogens/diseases	References
Pseudomonas fluorescens	2,4-Diacetylphloroglucinol	*Arabidopsis thaliana*	*Pseudomonas syringae pv.* tomato	Weller et al. (2012)
P. chlororaphis O6	Pyrrolnitrin	Tomato	*Phytophthora infestans*/leaf blight	Park et al. (2011)
Bacillus spp. SS 12.6 and SS-13.1	Turin A, bacillomycin, and surfactin	Apple fruits	Decay after harvesting	Dimkić et al. (2013)
B. subtilis SQR9	Iturin A, surfactin, fengycin, and bacillomycin	Cucumber	Fusarium wilt	Cao et al. (2011)
B. amyloliquefaciens W19	Bacillomycin D and iturin A	Banana	Fusarium wilt	Wang et al. (2013)
B. amyloliquefaciens MB101	Fungal lytic enzymes	Tomato	*Rhizoctonia solani*/root rot	Solanki et al. (2012)

environment. Unfortunately, due to changes in the environment, some microbes which have been tested in the laboratory to be beneficial could become pathogenic when exposed to an external environment. Hence, it is necessary to test microbes with potentials to be beneficial as plant amendments in different environments. When microbial amendments are used in excess, especially when they are used together with carriers such as peat and compost, they are washed into water bodies, thereby leading to eutrophication (Tsai and Chang 2020).

4.3 Soil Health

One naturally occurring nonrenewable resource that makes up the Earth's surface is soil. The soil is composed of a mixture of weathered rock and organic material. It provides a home for soil microbes and insects in addition to providing a foundation for all plants. The capacity of the soil to produce in terms of plant growth and yield is referred to as soil health (Kumar et al. 2023). Typically, it is based on the interplay between the soil's biological, chemical, and physical properties, which directly affect plant health and preserve the environment's quality. The consistency and balance of these health indicators maintain soil quality. Yet, it is occasionally believed that biological activity in the soil is what determines the health of the soil (Dror et al. 2022). Only a minority of people believe that soil directly affects plant growth and yield, while the majority believe that soil merely serves as a physical support for plants and that other factors, viz., pesticides and fertilizer, are responsible for growth and yield. By a variety of processes, including enrichment of soil organic matter, the decomposition of leftovers from dead animals and plants, and humus production, soil organisms significantly contribute to the health of the soil (Frouz 2018). These soil microorganisms transform organic nutrients into minerals by different enzymes that are simpler for plant roots to absorb (Jacoby et al. 2017; Parveen et al. 2022). They produce cellular debris and extracellular polysaccharides to hold soil aggregates together in order to increase soil's ability to retain water. They naturally compete with soilborne pathogens, which lowers the likelihood of plant diseases. It is crucial to preserve soil health in order to retain the stability of the terrestrial environment and to restore it following unpredictable disruptions like human anthropogenic activities, pollution, climate change, and drought (Tahat et al. 2020).

4.4 Plant Health

Plants contribute immensely to the world's food supply and economy in the form of lumber, vegetables, fruits, and grains. It also constitutes a vital environmental asset, share human heritage, and serve a special role in preserving the planet's biodiversity and ecological equilibrium (Shelef et al. 2017). Human health has received special attention for a long time, as seen by developments in medical diagnosis and ailment treatments. It is difficult to define plant health due to the lack of defined health markers; as a result, the absence of illness may be taken to mean that a plant is healthy. A plant is regarded healthy if its physiological function, which is governed by its environmental variations and genetic potential, is maintained (Huang et al. 2019). The rhizosphere contains helpful microbes that release exudates that encourage plant development and provide protection against plant diseases, which leads to healthy plant growth and vigor (Hakim et al. 2021).

Healthy plants are less vulnerable to pathogens and insects, which lowers the risk of acquiring diseases. A few suggested practices should be performed to maintain healthy plants (Rizzo et al. 2021). These include (a) sufficient watering, (b) appropriate mulching, (c) a good planting location, (d) prunes, and (e) fertilizers as necessary. There are several ways to compromise plant health, viz., the land usage, pressures associated with climate change, and emergence of new pests. Many factors, including the emergence of uncommon pests, stressors brought on by climate change, and changes in land usage, might compromise plant health (Singh et al. 2023a, b). Pests may enter the country through imported commodities and visitors, a natural method, or both. Plants may be able to modify the rhizosphere microbiome to their benefit by encouraging particular microbes with characteristics that are conducive to plant development and health (Manzar et al. 2022).

4.5 Human Health

Microbial inoculants increase the nutritional qualities of food consumed by humans and reduce the exposure of humans to chemicals arising from crops (Li et al. 2020) (Table 2). Since human's nutrient and health status highly depends on the safety and nutrient from foods they consumed, the contribution of microbes in enhancing food crops cannot be overemphasized. Improved nutritional value such as an increase in carbohydrate, calcium, potassium, phosphorus, zinc, iron, manganese, and magnesium has been reported in olive plant when microbial inoculants were used (Bizos et al. 2020). Similarly, an increased in the crude fiber, protein, ash, moisture, ether extract, dry matter, and moisture content was observed as a result of the presence of *Pseudomonas aeruginosa* and *Enterobacter xiangfangensis* on fennel plant (Dhiman et al. 2020). An increase in the protein, ether extract, fiber, and dry matter content of chickpea was observed when inoculated with *Bacillus cereus* and *Ensifer adhaerens* (Baliyan et al. 2022).

Microbes have been inoculated into harvested crops, especially during fermentation to reduce antinutrient properties. Since the presence of antinutrients in crops reduces the uptake of nutrient by the human body, its reduction will enhance nutritional status. *Lactobacillus acidophilus* and *Lactobacillus cereus* have been reported to reduce the antinutrient properties of soybean, especially the trypsin inhibitors (Madibana et al. 2022). Equally, the application of *Lactobacillus* sp. helped to reduce trypsin inhibitors in red lentil (Wagh et al. 2021).

Table 2 Effect of microbial inoculants on crops consumed by humans

Plant	Microbes	Effect	References
Maize	*Aspergillus flavus*, *Listeria monocytogenes*	Increased dry matter	Sharif et al. (2023)
Sweet and chili pepper	*Bacillus megaterium*	Improved protein, ash, and fiber contents	Bhatt and Maheshwari (2020)
Soybean	*Bradyrhizobium*	Enhanced carbohydrate, ash, protein, fiber, fat, and moisture contents	Ogbuehi (2020)
Barley	*Pseudomonas fluorescens*, *Bacillus megaterium*, and *Azotobacter chroococcum*	Improved protein and fat contents	Zayed et al. (2020)
Date	*Enterococcus faecium* and *Lactobacillus plantarum*	Improved crude protein, ether extract, nonfiber carbohydrate, and nitrogen-free extract contents	Farrag and El-Nahrawy (2021)
Sweet corn	*Lactobacillus plantarum*	Increased starch content	Ma et al. (2022)
Maize	*Arthrobacter* sp., *Sinomonas* sp., *Paenibacillus* sp., and *Lysinibacillus* sp.	Improved protein value	Ngone et al. (2023)
African star apple	*Glomus etunicatum*, *Funneliformis mosseae*, and *Rhizophagus irregularis*	Increased fat and crude fiber contents	Salawu et al. (2021)
Sour oranges	*G. mosseae*, *G. hoi*, and *G. intraradices*	Improved proline and carbohydrate concentrations	Hadian-Deljou (2020)
Apple	*Rhizophagus irregularis*	Increased sugar and proline contents	Huang et al. (2020)
Rice	*Serratia marcescens*	Increased zinc content	Upadhayay et al. (2022)

The ability of microbial inoculants to enhance antioxidant properties in plants has been reported, and this consequently help to improve the human health by helping to reduce and prevent cell damage. An elevation in the value of flavonoids, carotenoids, and polyphenols was observed in rice plant when inoculated with *Trichoderma asperellum* and *Pseudomonas fluorescens* (Singh et al. 2020). It is therefore very necessary to intensify research on how inoculants can be further used to enhance plant nutritional value to promote human health.

On the other hand, microbial inoculants could pose some health threats on humans. For instance, in a situation when composts are used too early, they contain pathogens that affect the health of humans who consumed food products on which the compost is applied (Ayilara et al. 2020). Although the high temperature experienced during composting inactivates many pathogenic microbes, organisms like *Salmonella enterica* and *E. coli* have been reported to be capable of surviving the heat (Alegbeleye and Sant'Ana 2020). *Salmonella enterica* causes gastroenteritis in humans (Alegbeleye and Sant'Ana 2020), while *E. coli* causes intestinal diseases (Luna-Guevara et al. 2019). *Mycobacterium* sp. and *Pseudomonas aeruginosa* are also compost-associated pathogens that were isolated from backyard and commercial composts (Mao et al. 2021). Therefore, it is necessary to ensure that composts are ripe and all the pathogens are properly destroyed before they are applied on plants.

5 Challenges with the Adoption of Microbial Inoculants

The use of microbial inoculants is impeded by several factors which include government policies, acceptability by farmers, scientific research, etc. Policies adopted and enacted by a government play a cogent role on the level of acceptability of microbial inoculants (Raimi et al. 2021). Onerous government regulatory requirements, unavailability of subsidies, and other support initiatives could reduce the production and adoption of microbial inoculants (Hodson et al. 2021). Another great challenge with the adoption of microbial inoculant is the sensitization and acceptability by farmers. Many farmers, especially in the developing countries, do not trust and understand the idea, process, and mechanisms of functioning of microbial inoculants; therefore, they do not use them (Sammauria et al. 2020; Seenivasagan and Babalola 2021).

Oftentimes, microbial inoculants lose their potency before getting to their site of action (Alori and Babalola 2018). This is majorly due to the alteration in environmental factors (such as temperature, nutrient, and pH) after they have been produced. The alteration occurs during storage, transportation, and marketing, which renders the microbes inactive during use.

Funds for research to support microbial inoculants are not readily available especially in the developing countries; hence, very little research are carried out on microbial inoculants in different environments (O'Callaghan et al. 2022). Different factors affect the viability of microbial inoculants; they include the environment, planting treatment, compatibility with different herbicides and pesticides, plant types, and soil types.

Microbial inoculants are highly specific in their targets, unlike the synthetic counterparts which have a broader spectrum (Alori and Babalola 2018). This should be an advantage because it would reduce toxicity to the nontarget species; unfortunately, it can also be seen as a disadvantage because it will reduce the cost of production for farmers. For instance, in the case of a pesticide, a single type of pesticide has to be purchased for each pest when a plant is being attacked by different pests.

6 Future Prospects and Conclusion

The use of microbial inoculant is a more affordable way of ensuring an affordable and sustainable agriculture. Microbial inoculant production serves as a source of income and innovation for the producers. Equally, it helps to improve the economy by creating job opportunities for the producers, marketers, and retailers. Due to the low cost of microbial inoculants, it helps to reduce the cost of production for farmers, consequently leading to a reduction in the cost of food and ensuring food security. Since the sources of microbial inoculants are organic, they help to enhance good health in humans by reducing the consumption of chemicals. They also prevent toxicity in the environment and ensure sustainable agriculture and environment.

Even with the positive effects of microbial inoculants which are discussed above, they still encounter some challenges, which if improved, would improve the utilization and acceptability of microbial inoculants. Developing countries most especially should intensify efforts to ease the regulations which are in place that makes the registration of microbial inoculants difficult. This is because, approvals from regulatory authorities will increase microbial inoculant's widespread and societal use, acceptability, and environmental sustainability while reducing health risks.

To improve the widespread use of microbial inoculants, it is both vital and logical that they are accepted by farmers. There is still a long way to go in sensitizing farmers to adopt the utilization of microbial inoculants, particularly among developing countries. Being a key stakeholder, that is, a primary consumer, farmers should be encouraged to accept microbial inoculants through initiatives, schemes, and drives like those under the umbrella of the United Nations' (UN) Sustainable Development Goals (SDGs). The prices of microbial inoculants should be further subsidized by the government; this will make them more affordable and reduce the cost of production for farmers. Furthermore, different regions should hold seminars, conferences, talks, etc. where the application of microbial inoculants will be demonstrated and advantages and benefits for the plants, animals, humans, and environment would be emphasized. In cases where farmers apply microbial inoculants inappropriately, there might be issues of product failure which will discourage the farmers from using the particular product and also accepting similar interventions.

More research should be carried out to increase existing knowledge on the use of microbial inoculants. Recently, nanoparticles have been used as an agent of delivery to coat and ensure the inoculants are intact from the period of production to the period of use. However, further studies are required to unravel new technologies which can be used as agents of delivery because nanoparticles if derived from synthetic sources could be dangerous to the environment.

Research should be intensified to vary the effect and compatibility of different environments, planting treatments, and plant and soil types with different environments. This will ensure that the right microbial species are used in the right environment and on the right plant species to improve crop yield and ensure environmental sustainability. The studies should be carried out on the use of a consortium of microbes or specific microbes with a wide spectrum effect which will help to increase the effectiveness of microbial inoculums.

References

Adeleke BS, Akinola SA, Adedayo AA, Glick BR, Babalola OO (2022a) Synergistic relationship of endophyte-nanomaterials to alleviate abiotic stress in plants. Front Environ Sci 10:2398. https://doi.org/10.3389/fenvs.2022.1015897

Adeleke BS, Ayilara MS, Akinola SA, Babalola OO (2022b) Biocontrol mechanisms of endophytic fungi. Egypt J Biol Pest Control 32(1):1–17. https://doi.org/10.1186/s41938-022-00547-1

Agri U, Chaudhary P, Sharma A (2021) In vitro compatibility evaluation of agriusable nanochitosan on beneficial plant growth-promoting rhizobacteria and maize plant. Natl Acad Sci Lett 44:555–559. https://doi.org/10.1007/s40009-021-01047-w

Agri U, Chaudhary P, Sharma A, Kukreti B (2022) Physiological response of maize plants and its rhizospheric microbiome under the influence of potential bioinoculants and nanochitosan. Plant Soil 474:451–468. https://doi.org/10.1007/s11104-022-05351-2

Akinola SA, Babalola OO (2020) The importance of adverse soil microbiomes in the light of omics: implications for food safety. Plant Soil Environ 66(9):421–430. https://doi.org/10.17221/118/2020-PSE

Akinola SA, Babalola OO (2021) The fungal and archaeal community within plant rhizosphere: a review on their contribution to crop safety. J Plant Nutr 44(4):600–618. https://doi.org/10.1080/01904167.2020.1845376

Akinola SA, Ayangbenro AS, Babalola OO (2021a) The immense functional attributes of maize rhizosphere microbiome: a shotgun sequencing approach. Agriculture 11(2):118. https://doi.org/10.3390/agriculture11020118

Akinola SA, Ayangbenro AS, Babalola OO (2021b) Metagenomic insight into the community structure of maize-rhizosphere bacteria as predicted by different environmental factors and their functioning within plant proximity. Microorganisms 9(7):1419. https://doi.org/10.3390/microorganisms9071419

Akinola SA, Adeyemo RO, Adebayo IA (2022) Overview and emergence of nanobiotechnology in plants. In: Diverse applications of nanotechnology in the biological sciences. Apple Academic Press, Palm Bay, pp 173–197

Alegbeleye OO, Sant'Ana AS (2020) Manure-borne pathogens as an important source of water contamination: an update on the dynamics of pathogen survival/transport as well as practical risk mitigation strategies. Int J Hyg Environ Health 227:113524. https://doi.org/10.1016/j.ijheh.2020.113524

Alori ET, Babalola OO (2018) Microbial inoculants for improving crop quality and human health in Africa. Front Microbiol 9:2213. https://doi.org/10.3389/fmicb.2018.02213

Antoszewski M, Mierek-Adamska A, Dąbrowska GB (2022) The importance of microorganisms for sustainable agriculture—a review. Metabolites 12(11):1100. https://doi.org/10.3390/metabo12111100

Asadhi S, Bhaskara Reddy B, Sivaprasad Y, Prathyusha M, Murali Krishna T, Vijay Krishna Kumar K, Raja Reddy K (2013) Characterisation, genetic diversity and antagonistic potential of 2,4-diacetylphloroglucinol producing *Pseudomonas fluorescens* isolates in groundnut-based cropping systems of Andhra Pradesh. India Arch Phytopathol Plant Prot 46(16):1966–1977. https://doi.org/10.1080/03235408.2013.782223

Ayilara MS, Olanrewaju OS, Babalola OO, Odeyemi O (2020) Waste management through composting: challenges and potentials. Sustainability 12(11):4456. https://doi.org/10.3390/su12114456

Ayilara MS, Adeleke BS, Akinola SA, Fayose CA, Adeyemi UT, Gbadegesin LA, Babalola OO (2023) Biopesticides as a promising alternative to synthetic pesticides: a case for microbial pesticides, phytopesticides, and nanobiopesticides. Front Microbiol 14:1040901. https://doi.org/10.3389/fmicb.2023.1040901

Backer R, Rokem JS, Ilangumaran G, Lamont J, Praslickova D, Ricci E, Subramanian S, Smith DL (2018) Plant growth-promoting rhizobacteria: context, mechanisms of action, and roadmap to commercialization of biostimulants for sustainable agriculture. Front Plant Sci 9:1473. https://doi.org/10.3389/fpls.2018.01473

Backer R, Naidoo S, van den Berg N (2019) The nonexpressor of pathogenesis-related genes 1 (NPR1) and related family: mechanistic insights in plant disease resistance. Front Plant Sci 10:102. https://doi.org/10.3389/fpls.2019.00102

Baliyan N, Qureshi KA, Jaremko M, Rajput M, Singh M, Dhiman S, Kumar A (2022) Bioformulation containing cohorts of *Ensifer adhaerens* MSN12 and *Bacillus cereus* MEN8 for the nutrient enhancement of *Cicer arietinum* L. Plants 11(22):3123. https://doi.org/10.3390/plants11223123

Beneduzi A, Ambrosini A, Passaglia LM (2012) Plant growth-promoting rhizobacteria (PGPR): their potential as antagonists and biocontrol agents. Genet Mol Biol 35:1044–1051. https://doi.org/10.1590/s1415-47572012000600020

Bhatt K, Maheshwari DK (2020) *Bacillus megaterium* strain CDK25, a novel plant growth promoting bacterium enhances proximate chemical and nutritional composition of Capsicum annuum L. Front Plant Sci 11:1147. https://doi.org/10.3389/fpls.2020.01147

Bizos G, Papatheodorou EM, Chatzistathis T, Ntalli N, Aschonitis VG, Monokrousos N (2020) The role of microbial inoculants on plant protection, growth stimulation, and crop productivity of the olive tree (*Olea europaea* L.). Plants 9(6):743. https://doi.org/10.3390/plants9060743

Brodt S, Six J, Feenstra G, Ingels C, Campbell D (2011) Sustainable agriculture. Nat Educ Knowl 3(10):1

Cao Y, Zhang Z, Ling N, Yuan Y, Zheng X, Shen B, Shen Q (2011) *Bacillus subtilis* SQR 9 can control Fusarium wilt in cucumber by colonizing plant roots. Biol Fertil Soils 47:495–506. https://doi.org/10.1007/s00374-011-0556-2

Chakraborty N, Mitra R, Pal S, Ganguly R, Acharya K, Minkina T, Sarkar A, Keswani C (2023) Biopesticide consumption in India: insights into the current trends. Agriculture 13(3):557. https://doi.org/10.3390/agriculture13030557

Chaudhary P, Sharma A (2019) Response of nanogypsum on the performance of plant growth promotory bacteria recovered from nanocompound infested agriculture field. Environ Ecol 37:363–372

Chaudhary A, Parveen H, Chaudhary P, Bhatt P, Khatoon H (2021a) Rhizospheric microbes and their mechanism. In: Bhatt P, Gangola S, Udayanga D, Kumar G (eds) Microbial technology for sustainable environment. Springer, Singapore. https://doi.org/10.1007/978-981-16-3840-4_6

Chaudhary P, Chaudhary A, Parveen H, Rani A, Kumar G, Kumar A, Sharma A (2021b) Impact of nanophos in agriculture to improve functional bacterial community and crop productivity. BMC Plant Biol 21:519. https://doi.org/10.1186/s12870-021-03298-7

Chaudhary P, Khati P, Chaudhary A, Maithani D, Kumar G, Sharma A (2021c) Cultivable and metagenomic approach to study the combined impact of nanogypsum and *Pseudomonas taiwanensis* on maize plant health and its rhizospheric microbiome. PLoS One 16:e0250574. https://doi.org/10.1371/journal.pone.0250574

Chaudhary P, Parveen H, Gangola S, Kumar G, Bhatt P, Chaudhary A (2021d) Plant growth promoting rhizobacteria and their application in sustainable crop production. In: Bhatt P, Gangola S, Udayanga D, Kumar G (eds) Microbial technology for sustainable environment. Springer, Singapore. https://doi.org/10.1007/978-981-16-3840-4_13

Chaudhary P, Khati P, Chaudhary A, Gangola S, Kumar R, Sharma A (2021e) Bioinoculation using indigenous *Bacillus* spp improves growth and yield of *Zea mays* under the influence of nanozeolite. 3 Biotech 11:11. https://doi.org/10.1007/s13205-020-02561-2

Chaudhary P, Khati P, Gangola S, Kumar A, Kumar R, Sharma A (2021f) Impact of nanochitosan and *Bacillus* spp on health, productivity and defence response in *Zea mays* under field condition. 3 Biotech 11:237. https://doi.org/10.1007/s13205-021-02790-z

Chaudhary P, Sharma A, Chaudhary A, Khati P, Gangola S, Maithani D (2021g) Illumina based high throughput analysis of microbial diversity of maize rhizosphere treated with nanocompounds and *Bacillus* sp. Appl Soil Ecol 159:103836. https://doi.org/10.1016/j.apsoil.2020.103836

Chaudhary A, Chaudhary P, Upadhyay A, Kumar A, Singh A (2022a) Effect of gypsum on plant growth promoting rhizobacteria. Environ Ecol 39(4A):1248–1256

Chaudhary P, Singh S, Chaudhary A, Agri U, Bhandari G (2022b) Chapter 3 Agrochemicals and their effects on soil microbial population. In: Soni R, Suyal D, Goel R (eds) Plant protection: from chemicals to biologicals. De Gruyter, Berlin, pp 45–58. https://doi.org/10.1515/9783110771558-003

Chaudhary P, Agri U, Chaudhary A, Kumar A, Kumar G (2022c) Endophytes and their potential in biotic stress management and crop production. Front Microbiol 13:933017. https://doi.org/10.3389/fmicb.2022.933017

Chaudhary P, Chaudhary A, Agri U, Khatoon H, Singh A (2022d) Recent trends and advancements for agro-environmental sustainability at higher altitudes. In: Goel R, Soni R, Suyal DC, Khan M (eds) Survival strategies in cold-adapted microorganisms. Springer, Singapore. https://doi.org/10.1007/978-981-16-2625-8_19

Chaudhary P, Singh S, Chaudhary A, Sharma A, Kumar G (2022e) Overview of biofertilizers in crop production and stress management for sustainable agriculture. Front Plant Sci 13:930340. https://doi.org/10.3389/fpls.2022.930340

Chaudhary P, Xu M, Ahamad L, Chaudhary A, Kumar G, Adeleke BS, Verma KK, Hu D-M, Širić I, Kumar P, Popescu SM, Abou Fayssal S (2023a) Application of synthetic consortia for improvement of soil fertility, pollution remediation, and agricultural productivity: a review. Agronomy 13:643. https://doi.org/10.3390/agronomy13030643

Chaudhary P, Chen S, Rajput VD, Jaiswal DK (2023b) Editorial: bioinoculants with nano-compounds to improve soil health: a step toward sustainable agriculture. Front Environ Sci 11:1270002. https://doi.org/10.3389/fenvs.2023.1270002

Chaudhary P, Chen S, Rajput VD, Jaiswal DK (eds) (2023c) Bioinoculants with nano-compounds to improve soil health: a step toward sustainable agriculture. Frontiers Media SA, Lausanne. https://doi.org/10.3389/978-2-8325-3287-4

Chun J-H, Kim S, Arasu MV, Al-Dhabi NA, Chung DY, Kim S-J (2017) Combined effect of nitrogen, phosphorus and potassium fertilizers on the contents of glucosinolates in rocket salad (*Eruca sativa* Mill.). Saudi J Biol Sci 24:436–443. https://doi.org/10.1016/j.sjbs.2015.08.012

Damalas CA, Koutroubas SD (2018) Current status and recent developments in biopesticide use. Agriculture 8(1):13. https://doi.org/10.3390/agriculture8010013

Dhiman S, Baliyan N, Maheshwari DK (2020) Buffalo dung-inhabiting bacteria enhance the nutrient enrichment of soil and proximate contents of *Foeniculum vulgare* Mill. Arch Microbiol 202:2461–2470. https://doi.org/10.1007/s00203-020-01969-x

Dimkić I, Živković S, Berić T, Ivanović Ž, Gavrilović V, Stanković S, Fira D (2013) Characterization and evaluation of two *Bacillus* strains, SS-12.6 and SS-13.1, as potential agents for the control of phytopathogenic bacteria and fungi. Biol Control 65(3):312–321. https://doi.org/10.1016/j.biocontrol.2013.03.012

Ding LN, Li YT, Wu YZ, Li T, Geng R, Cao J, Zhang W, Tan XL (2022) Plant disease resistance-related signaling pathways: recent progress and future prospects. Int J Mol Sci 23(24):16200. https://doi.org/10.3390/ijms232416200

Dror I, Yaron B, Berowitz B (2022) The human impact on all soil-forming factors during the Anthropocene. ACS Environ Au 2:11–19. https://doi.org/10.1021/acsenvironau.1c00010

El-Saadony MT, Saad AM, Soliman SM, Salem HM, Ahmed AI, Mahmood M, El-Tahan AM, Ebrahim AAM, Abd El-Mageed TA, Negm SH, Selim S, Babalghith AO, Elrys AS, El-Tarabily KA, AbuQamar SF (2022) Plant growth-promoting microorganisms as biocontrol agents of plant diseases: mechanisms, challenges and future perspectives. Front Plant Sci 13:923880. https://doi.org/10.3389/fpls.2022.923880

Elsharkawy M, Shimizu M, Takahashi H, Hyakumachi M (2012) Induction of systemic resistance against Cucumber mosaic virus by *Penicillium simplicissimum* GP17-2 in *Arabidopsis* and tobacco. Plant Pathol 61(5):964–976. https://doi.org/10.1111/j.1365-3059.2011.02573.x

Farrag HM, El-Nahrawy M (2021) Effect of bacterial inoculants on composition, quality characteristics, bacterial count and nutritional values of date palm frond silage Middle East. J Appl Sci 11:313–322

Fasusi OA, Cruz C, Babalola OO (2021) Agricultural sustainability: microbial biofertilizers in rhizosphere management. Agriculture 11(2):163. https://doi.org/10.3390/agriculture11020163

Fontenelle A, Guzzo S, Lucon C, Harakava R (2011) Growth promotion and induction of resistance in tomato plant against *Xanthomonas euvesicatoria* and *Alternaria solani* by *Trichoderma* spp. Crop Prot 30(11):1492–1500. https://doi.org/10.1016/j.cropro.2011.07.019

Frouz J (2018) Effects of soil macro- and mesofauna on litter decomposition and soil organic matter stabilization. Geoderma 332:161–172. https://doi.org/10.1016/j.geoderma.2017.08.039

Gopalakrishnan S, Humayun P, Kiran BK, Kannan IGK, Vidya MS, Deepthi K, Rupela O (2011) Evaluation of bacteria isolated from rice rhizosphere for biological control of charcoal rot of sorghum caused by *Macrophomina phaseolina* (Tassi) Goid. World J Microbiol Biotechnol 27:1313–1321. https://doi.org/10.1007/s11274-010-0579-0

Grzyb A, Wolna-Maruwka A, Niewiadomska A (2021) The significance of microbial transformation of nitrogen compounds in the light of integrated crop management. Agronomy 11(7):1415. https://doi.org/10.3390/agronomy11071415

Hadian-Deljou M, Esna-Ashari M, Mirzaie-asl A (2020) Alleviation of salt stress and expression of stress-responsive gene through the symbiosis of arbuscular mycorrhizal fungi with sour orange seedlings. Sci Hortic 268:109373. https://doi.org/10.1016/j.scienta.2020.109373

Hakim S, Naqqash T, Nawaz MS, Laraib I, Siddique MJ, Zia R, Mirza MS, Imran A (2021) Rhizosphere engineering with plant growth-promoting microorganisms for agriculture and ecological sustainability. Front Sustain Food Syst 5:617157. https://doi.org/10.3389/fsufs.2021.617157

Hassan MK, McInroy JA, Kloepper JW (2019) The interactions of rhizodeposits with plant growth-promoting rhizobacteria in the rhizosphere: a review. Agriculture 9(7):142. https://doi.org/10.3390/agriculture9070142

Hodson E, Niggli U, Kitajima K, Lal R, Sadoff C (2021) Boost nature positive production—a paper on action track 3. Sci Innov 3:221. https://sc-fss2021.org/

Huang H, Ullah F, Zhou D-X, Yi M, Zhao Y (2019) Mechanisms of ROS regulation of plant development and stress responses. Front Plant Sci 10:800. https://doi.org/10.3389/fpls.2019.00800

Huang D, Ma M, Wang Q, Zhang M, Jing G, Li C, Ma F (2020) Arbuscular mycorrhizal fungi enhanced drought resistance in apple by regulating genes in the MAPK pathway. Plant Physiol Biochem 149:245–255. https://doi.org/10.1016/j.plaphy.2020.02.020

Jacoby R, Peukert M, Succurro A, Koprivova A, Kopriva S (2017) The role of soil microorganisms in plant mineral nutrition—current knowledge and future directions. Front Plant Sci 8:1617. https://doi.org/10.3389/fpls.2017.01617

Javaid M, Haleem A, Singh RP, Suman R (2022) Enhancing smart farming through the applications of agriculture 4.0 technologies. Int J Intell Netw 3:150–164. https://doi.org/10.1016/j.ijin.2022.09.004

Jayaraj R, Megha P, Sreedev P (2017) Organochlorine pesticides, their toxic effects on living organisms and their fate in the environment. Interdiscip Toxicol 3–4:90–100. https://doi.org/10.1515/intox-2016-0012

Jogaiah S, Abdelrahman M, Tran LSP, Shin-Ichi I (2013) Characterization of rhizosphere fungi that mediate resistance in tomato against bacterial wilt disease. J Exp Bot 64(12):3829–3842. https://doi.org/10.1093/jxb/ert212

Kaminsky LM, Trexler RV, Malik RJ, Hockett KL, Bell TH (2019) The inherent conflicts in developing soil microbial inoculants. Trends Biotechnol 37(2):140–151. https://doi.org/10.1016/j.tibtech.2018.11.011

Khati P, Parul BP, Nisha Kumar R, Sharma A (2018) Effect of nanozeolite and plant growth promoting rhizobacteria on maize. 3 Biotech 8:141. https://doi.org/10.1007/s13205-018-1142-1

Khati P, Chaudhary P, Gangola S, Sharma P (2019) Influence of nanozeolite on plant growth promotory bacterial isolates recovered from nanocompound infested agriculture field. Environ Ecol 37:521–527

Kumar A, Sharma A, Chaudhary P, Gangola S (2021a) Chlorpyrifos degradation using binary fungal strains isolated from industrial waste soil. Biologia 10:3071–3081. https://doi.org/10.1007/s11756-021-00816-8

Kumar P, Kamle M, Borah R, Mahato DK, Sharma B (2021b) *Bacillus thuringiensis* as microbial biopesticide: uses and application for sustainable agriculture. Egypt J Biol Pest Control 31:95. https://doi.org/10.1186/s41938-021-00440-3

Kumar J, Ramlal A, Mallick D, Mishra V (2021c) An overview of some biopesticides and their importance in plant protection for commercial acceptance. Plants 10(6):1185. https://doi.org/10.3390/plants10061185

Kumar G, Lal S, Maurya SK, Bhattacherjee AK, Chaudhary P, Gangola S, Rajan S (2021d) Exploration of *Klebsiella pneumoniae* M6 for paclobutrazol degradation, plant growth attributes, and biocontrol action under subtropical ecosystem. PLoS One 16:e0261338. https://doi.org/10.1371/journal.pone.0261338

Kumar G, Lal S, Soni SK, Maurya SK, Shukla PK, Chaudhary P, Bhattacherjee AK, Garg N (2022) Mechanism and kinetics of chlorpyrifos co-metabolism by using environment restoring microbes isolated from rhizosphere of horticultural crops under subtropics. Front Microbiol 13:891870. https://doi.org/10.3389/fmicb.2022.891870

Kumar U, Sheleke RM, Singh R (2023) Editorial: soil-plant-microbe interactions: an innovative approach towards improving soil health and plant growth. Front Agron 5:1165328. https://doi.org/10.3389/fagro.2023.1165328

Kumari S, Sharma A, Chaudhary P, Khati P (2020) Management of plant vigor and soil health using two agriusable nanocompounds and plant growth promotory rhizobacteria in Fenugreek. 3 Biotech 10:1–11. https://doi.org/10.1007/s13205-020-02448-2

Kumari H, Khati P, Gangola S, Chaudhary P, Sharma A (2021) Performance of plant growth promotory rhizobacteria on maize and soil characteristics under the influence of TiO2 nanoparticles. Pantnagar J Res 19:28–39

Li H, Qiu Y, Yao T, Ma Y, Zhang H, Yang X (2020) Effects of PGPR microbial inoculants on the growth and soil properties of *Avena sativa, Medicago sativa*, and *Cucumis sativus* seedlings. Soil Tillage Res 199:104577. https://doi.org/10.1016/j.still.2020.104577

Luna-Guevara JJ, Arenas-Hernandez MM, Martínez de la Peña C, Silva JL, Luna-Guevara ML (2019) The role of pathogenic *E. coli* in fresh vegetables: behavior, contamination factors, and preventive measures. Int J Microbiol 2019:2894328. https://doi.org/10.1155/2019/2894328

Lushchak VI, Matviishyn TM, Husak VV, Storey JM, Storey KB (2018) Pesticide toxicity: a mechanistic approach. EXCLI J 17:1101–1136. https://doi.org/10.17179/excli2018-1710

Ma ZY, Ungerfeld E, Ouyang Z, Zhou X-L, Han X-F, Zeng Y-Q, Tan Z-L (2022) Effect of *Lactobacillus plantarum* inoculation on chemical composition, fermentation, and bacterial community composition of ensiled sweet corn whole plant or Stover. Fermentation 8(1):24. https://doi.org/10.3390/fermentation8010024

Madibana M, Fouché C, Davids J, Mthembu S, Motubatse M, Poni Z, Nemakhavhani R (2022) *Lactobacillus acidophilus* and *Lactobacillus cereus*-containing products do reduce soybean antinutritional factors. Agric Sci Technol 14:32–39. https://doi.org/10.15547/ast.2022.03.034

Mahmud AA, Upadhyay SK, Srivastava AK, Bhojiya AA (2021) Biofertilizers: a nexus between soil fertility and crop productivity under abiotic stress. Curr Res Environ Sustain 3:100063. https://doi.org/10.1016/j.crsust.2021.100063

Manzar N, Jayaswal D, Kishan G, Chauhan S, Joshi D, Singh J, Chaudhary P, Jamali H, Kushwaha P, Parveen H, Kashyap AS, Sahu PK, Singh UB (2022) Application of molecular ecology approaches in sustainable agriculture for a better understanding of plant–microbiome interactions. In: Singh UB, Sahu PK, Singh HV, Sharma PK, Sharma SK (eds) Rhizosphere microbes. Microorganisms for sustainability, vol 40. Springer, Singapore. https://doi.org/10.1007/978-981-19-5872-4_3

Mao Y, Akdeniz N, Nguyen TH (2021) Quantification of pathogens and antibiotic resistance genes in backyard and commercial composts. Sci Total Environ 797:149197. https://doi.org/10.1016/j.scitotenv.2021.149197

Meemken E-M, Qaim M (2018) Organic agriculture, food security, and the environment. Annu Rev Resour Econ 10:39–63

Mishra J, Dutta V, Arora NK (2020) Biopesticides in India: technology and sustainability linkages. 3 Biotech 10(5):210. https://doi.org/10.1007/s13205-020-02192-7

Murali M, Naziya B, Ansari MA, Alomary MN, AlYahya S, Almatroudi A, Thriveni MC, Gowtham HG, Singh SB, Aiyaz M, Kalegowda N, Lakshmidevi N, Amruthesh KN (2021) Bioprospecting of rhizosphere-resident fungi: their role and importance in sustainable agriculture. J Fungi (Basel) 7(4):314. https://doi.org/10.3390/jof7040314

Nagaraju A, Sudisha J, Murthy SM, Ito S (2012) Seed priming with *Trichoderma harzianum* isolates enhances plant growth and induces resistance against *Plasmopara halstedii,* an incitant of sunflower downy mildew disease. Australas Plant Pathol 41:609–620. https://doi.org/10.1007/s13313-012-0165-z

Ngone MA, Ajoacha DMB, Achiri DT, Tchakounté GVT, Ruppel S, Tening AS, Ngosong C (2023) Potential of bio-organic amendment of palm oil mill effluent manure and plant growth-promoting bacteria to enhance the yield and quality of maize grains in Cameroon. Soil Secur 11:100090. https://doi.org/10.1016/j.soisec.2023.100090

O'Callaghan M, Ballard RA, Wright D (2022) Soil microbial inoculants for sustainable agriculture: limitations and opportunities. Soil Use Manag 38(3):1340–1369. https://doi.org/10.1111/sum.12811

Ogbuehi HC (2020) Effect of nodumax inoculant on morpho-physiological parameters, nutrient content and yield of soybean (*Glycine max.* L). J Agric Food Sci 18(2):54–72. https://doi.org/10.4314/jafs.v18i2.4

Park J, Oh S, Anderson A, Neiswender J, Kim JC, Kim Y (2011) Production of the antifungal compounds phenazine and pyrrolnitrin from *Pseudomonas chlororaphis* O6 is differentially regulated by glucose. Lett Appl Microbiol 52(5):532–537. https://doi.org/10.1111/j.1472-765X.2011.03036.x

Parveen H, Chaudhary A, Chaudhary P, Sultana R, Khati P, Kumar G, Khati P, Rana M, Bhatt P (2022) Omics approaches for the production of the microbial enzymes and applications. In: Bhatt P (ed) Industrial applications of microbial enzymes, 1st edn. CRC Press, Boca Raton. https://doi.org/10.1201/9781003202998-16

Pathak VM, Verma VK, Rawat BS, Kaur B, Babu N, Sharma A, Dewali S, Yadav M, Kumari R, Singh S, Mohapatra A, Pandey V, Rana N, Cunill JM (2022) Current status of pesticide effects on environment, human health and it's eco-friendly management as bioremediation: a comprehensive review. Front Microbiol 13:962619. https://doi.org/10.3389/fmicb.2022.962619

Patil HJ, Solanki MK (2016) Microbial inoculant: modern era of fertilizers and pesticides. In: Microbial inoculants in sustainable agricultural productivity: research perspectives, vol 1. Springer, Berlin, pp 319–343. https://doi.org/10.1007/978-81-322-2647-5_19

Patil HJ, Srivastava AK, Singh DP, Chaudhari BL, Arora DK (2011) Actinomycetes mediated biochemical responses in tomato (*Solanum lycopersicum*) enhances bioprotection against *Rhizoctonia solani*. Crop Prot 30(10):1269–1273. https://doi.org/10.1016/j.cropro.2011.04.008

Poór P, Czékus Z, Tari I, Ördög A (2019) The multifaceted roles of plant hormone salicylic acid in endoplasmic reticulum stress and unfolded protein response. Int J Mol Sci 20(23):5842. https://doi.org/10.3390/ijms20235842

Raimi A, Roopnarain A, Adeleke R (2021) Biofertilizer production in Africa: current status, factors impeding adoption and strategies for success. Sci Afr 11:e00694. https://doi.org/10.1016/j.sciaf.2021.e00694

Ren C, Zhang X, Reis S, Gu B (2022) Socioeconomic barriers of nitrogen management for agricultural and environmental sustainability. Agric Ecosyst Environ 333:107950. https://doi.org/10.1016/j.agee.2022.107950

Rizzo DM, Lichtveld M, Mazet JAK, Togami E, Miller SA (2021) Plant health and its effects on food safety and security in a One Health framework: four case studies. One Health Outlook 3:6. https://doi.org/10.1186/s42522-021-00038-7

Romdhane L, Ebinezer LB, Panozzo A, Barion G, Dal Cortivo C, Radhouane L, Vamerali T (2021) Effects of soil amendment with wood ash on transpiration, growth, and metal uptake in two contrasting maize (*Zea mays* L.) hybrids to drought tolerance. Front Plant Sci 12:661909. https://doi.org/10.3389/fpls.2021.661909

Salawu O, Banjo E, Olubode A, Odeyemi O, Olubode O, Chikaleke V (2021) Seedling growth and nutrient uptake of African Star Apple (*Chrysophyllum albidum*) in response to soil type and AMF inoculum. Niger J Hortic Sci 26(1):77–88

Sammauria R, Kumawat S, Kumawat P, Singh J, Jatwa TK (2020) Microbial inoculants: potential tool for sustainability of agricultural production systems. Arch Microbiol 202(4):677–693. https://doi.org/10.1007/s00203-019-01795-w

Seenivasagan R, Babalola OO (2021) Utilization of microbial consortia as biofertilizers and biopesticides for the production of feasible agricultural product. Biology 10(11):1111. https://doi.org/10.3390/biology10111111

Sharif S, Hanif NQ, Ghazanfar S, Imran M, Naiel MA, Alagawany M (2023) Dominance of *Bacillus* sp. alter microbiological and nutritional quality and improve aerobic stability of the corn silage. Rendiconti Lincei Scienze Fisiche Naturali 34:1–11. https://doi.org/10.1007/s12210-022-01130-4

Shelef O, Weisberg PJ, Provenza FD (2017) The value of native plants and local production in an era of global agriculture. Front Plant Sci 8:2069. https://doi.org/10.3389/fpls.2017.02069

Singh DP, Singh V, Shukla R, Sahu P, Prabha R, Gupta A, Gupta VK (2020) Stage-dependent concomitant microbial fortification improves soil nutrient status, plant growth, antioxidative defense system and gene expression in rice. Microbiol Res 239:126538. https://doi.org/10.1016/j.micres.2020.126538

Singh S, Chaudhary P, Sharma A (2023a) Efficient decolourization of toluidine blue and methylene blue using *Bacillus cereus*. Vegetos. https://doi.org/10.1007/s42535-023-00747-6

Singh BK, Delgado-Baquerizo M, Egidi E, Guirado E, Leach JE, Liu H, Trivedi P (2023b) Climate change impacts on plant pathogens, food security and paths forward. Nat Rev Microbiol 21:640. https://doi.org/10.1038/s41579-023-00900-7

Solanki MK, Singh N, Singh RK, Singh P, Srivastava AK, Kumar S, Arora DK (2011) Plant defense activation and management of tomato root rot by a chitin-fortified *Trichoderma/Hypocrea* formulation. Phytoparasitica 39:471–481. https://doi.org/10.1007/s12600-011-0188-y

Solanki MK, Robert AS, Singh RK, Kumar S, Pandey AK, Srivastava AK, Arora DK (2012) Characterization of mycolytic enzymes of *Bacillus* strains and their bio-protection role against *Rhizoctonia solani* in tomato. Curr Microbiol 65:330–336. https://doi.org/10.1007/s00284-012-0160-1

Solanki MK, Singh RK, Srivastava S, Kumar S, Kashyap PL, Srivastava AK (2015) Characterization of antagonistic-potential of two *Bacillus* strains and their biocontrol activity against *Rhizoctonia solani* in tomato. J Basic Microbiol 55(1):82–90. https://doi.org/10.1002/jobm.201300528

Šunjka D, Mechora Š (2022) An alternative source of biopesticides and improvement in their formulation—recent advances. Plants 11(22):3172. https://doi.org/10.3390/plants11223172

Tahat MM, Alananbeh KA, Othman YI, Leskovar D (2020) Soil health and sustainable agriculture. Sustainability 12(12):4859. https://doi.org/10.3390/su12124859

Tsai C-C, Chang Y-F (2020) Nitrogen availability in biochar-amended soils with excessive compost application. Agronomy 10(3):444. https://doi.org/10.3390/agronomy10030444

Tyagi J, Chaudhary P, Jyotsana BU, Bhandari G, Chaudhary A (2022a) Chapter 17 Impact of endophytic fungi in biotic stress management. In: Soni R, Suyal D, Goel R (eds) Plant protection: from chemicals to biologicals. De Gruyter, Berlin, pp 447–462. https://doi.org/10.1515/9783110771558-017

Tyagi J, Chaudhary P, Mishra A, Khatwani M, Dey S, Varma A (2022b) Role of endophytes in abiotic stress tolerance: with special emphasis on *Serendipita indica*. Int J Environ Res 16:62. https://doi.org/10.1007/s41742-022-00439-0

Upadhayay VK, Khan A, Singh J, Singh AV (2022) Bacterial assisted improved Zn consignment in root and shoot of rice plant by zinc solubilizing *Serratia marcescens* bearing plant probiotic traits. Adv Biores 13:1–8. https://doi.org/10.15515/abr.0976-4585.13.1.108

Vanitha S, Umesha S (2011) *Pseudomonas fluorescens* mediated systemic resistance in tomato is driven through an elevated synthesis of defense enzymes. Biol Plant 55:317–322. https://doi.org/10.1007/s10535-011-0045-3

Wagh SK, Kurallu MS, Padul MV (2021) Analysis of enzymatic reduction of red lentil antinutrient trypsin inhibitors by *Lactobacilli* isolated from Idli batter. Int J Innov Res Technol 8:2349–6002

Wahab S, Muzammil K, Nasir N, Khan MS, Ahmad MF, Khalid M, Ahmad W, Dawria A, Reddy LKV, Busayli AM (2022) Advancement and new trends in analysis of pesticide residues in food: a comprehensive review. Plants (Basel) 11(9):1106. https://doi.org/10.3390/plants11091106

Wang B, Yuan J, Zhang J, Shen Z, Zhang M, Li R, Shen Q (2013) Effects of novel bioorganic fertilizer produced by *Bacillus amyloliquefaciens* W19 on antagonism of Fusarium wilt of banana. Biol Fertil Soils 49:435–446. https://doi.org/10.1007/s00374-012-0739-5

Wang J, Song L, Gong X, Xu J, Li M (2020) Functions of Jasmonic acid in plant regulation and response to abiotic stress. Int J Mol Sci 21(4):1446. https://doi.org/10.3390/ijms21041446

War AR, Taggar GK, Hussain B, Taggar AS, Nair RM, Sharma HC (2018) Plant defence against herbivory and insect adaptations. AoB Plants 10:ply037. https://doi.org/10.1093/aobpla/ply037

Weller DM, Mavrodi DV, van Pelt JA, Pieterse CM, van Loon LC, Bakker PA (2012) Induced systemic resistance in Arabidopsis thaliana against *Pseudomonas syringae* pv. tomato by 2,4-diacetylphloroglucinol-producing *Pseudomonas fluorescens*. Phytopathology 102(4):403–412. https://doi.org/10.1094/PHYTO-08-11-0222

Yahaya SM, Mahmud AA, Abdullahi M, Haruna A (2023) Recent advances in the chemistry of nitrogen, phosphorus and potassium as fertilizers in soil: a review. Pedosphere 33:385–406. https://doi.org/10.1016/j.pedsph.2022.07.012

Yandigeri MS, Malviya N, Solanki MK, Shrivastava P, Sivakumar G (2015) Chitinolytic *Streptomyces vinaceusdrappus* S5MW2 isolated from Chilika lake, India enhances plant growth and biocontrol efficacy through chitin supplementation against *Rhizoctonia solani*. World J Microbiol Biotechnol 31:1217–1225. https://doi.org/10.1007/s11274-015-1870-x

Yang T, Siddique KHM, Liu K (2020) Cropping systems in agriculture and their impact on soil health—a review. Glob Ecol Conserv 23:e01118. https://doi.org/10.1016/j.gecco.2020.e01118

Yoshioka Y, Ichikawa H, Naznin HA, Kogure A, Hyakumachi M (2012) Systemic resistance induced in *Arabidopsis thaliana* by *Trichoderma asperellum* SKT-1, a microbial pesticide of seedborne diseases of rice. Pest Manag Sci 68(1):60–66. https://doi.org/10.1002/ps.2220

Yu Y, Gui Y, Li Z, Jiang C, Guo J, Niu D (2022) Induced systemic resistance for improving plant immunity by beneficial microbes. Plants 11(3):386. https://doi.org/10.3390/plants11030386

Zayed M, Eissa N, Aly T, Hassanein M, Abdallah M (2020) Production of green barley sprouts on rice straw in the presence of microbial inoculants and elevated CO2 as an unconventional dual bio-food and fodders. Egypt J Microbiol 55:1–14

Zhu L, Huang J, Lu X, Zhou C (2022) Development of plant systemic resistance by beneficial rhizobacteria: recognition, initiation, elicitation and regulation. Front Plant Sci 13:952397. https://doi.org/10.3389/fpls.2022.952397